Excel

Get the Results You Want!

YEAR 12 PHYSICS

Mark Butler

CONTENTS

MODULE 5: ADVANCED MECHANICS

MODULE 6: ELECTROMAGNETISM

MODULE 7: THE NATURE OF LIGHT

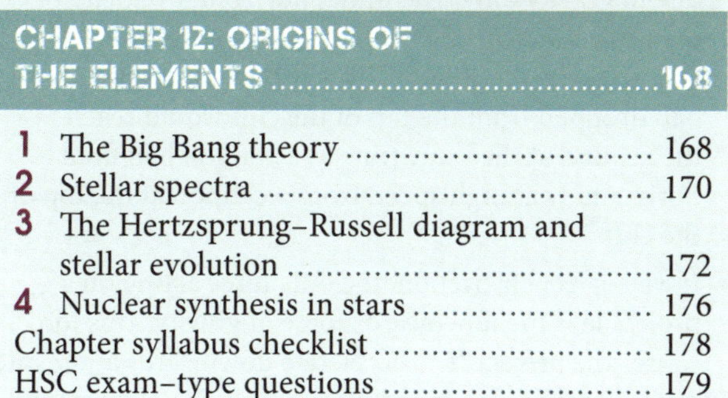

MODULE 8: FROM THE UNIVERSE TO THE ATOM

CHAPTER 1 PROJECTILE MOTION

MODULE 5 ADVANCED MECHANICS

INQUIRY QUESTION:

How can models that are used to explain projectile motion be used to analyse and make predictions?

The motion of a projectile moving under the influence of gravity can be predicted and analysed by using a model that treats the vertical and horizontal components of the motion independently. The horizontal motion is modelled as uniform rectilinear motion, and the vertical component is modelled as uniformly accelerated rectilinear motion.

1 Understanding projectile motion

» Students analyse the motion of projectiles by resolving the motion into horizontal and vertical components, making the following assumptions:
- a constant vertical acceleration due to gravity
- zero air resistance.

→ A projectile is an object that is moving freely under the influence of gravity (see Figure 1.1).

Figure 1.1 Relative motion of a projectile

→ Figure 1.1 shows a ball falling from the top of the mast of a boat travelling at constant velocity. To the person on the ship the ball appears to fall vertically downwards, but to an observer at rest on the shore, the ball appears to take a **parabolic path**. The parabolic path of the projectile seen by the stationary observer results from the combination of the downwards acceleration of the ball and the constant horizontal velocity of the ball.

parabolic path: a trajectory that has the shape of a parabola (i.e. a quadratic function such as $y = x^2$)

→ Galileo showed that projectile motion could be analysed by treating the horizontal and vertical components of the motion independently. Figure 1.2 shows how the displacement of a projectile is related to the vertical and horizontal displacement.

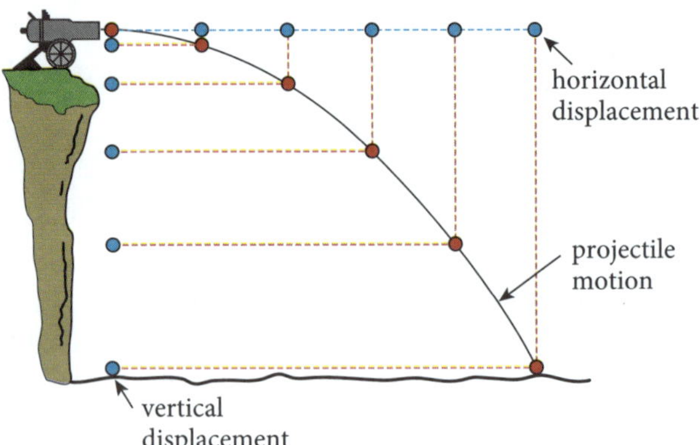

Figure 1.2 Illustration of how the path of a projectile is related to its vertical and horizontal displacement

Because the horizontal and vertical components of projectile motion are independent, the time of flight of the cannon ball shown in Figure 1.2 is determined by the height of the cliff and the local acceleration due to gravity only. The initial horizontal velocity of the projectile has no effect on the time the ball will take to reach the ground. This means that a cannon ball dropped from the top of the cliff would reach the ground at the same time as a cannon ball fired horizontally at high speed from a cannon at the top of the cliff.

→ If we neglect air friction, the only force acting on a projectile is the downwards force of gravity. This force causes the projectile to accelerate downwards at the rate of acceleration due to gravity. As there is no horizontal force acting on the projectile, the horizontal velocity must remain constant throughout the flight. Figure 1.3 shows how adding a constant horizontal velocity to a changing vertical velocity gives the actual velocity of the projectile at each instant. Note that the *horizontal* velocity vector remains constant throughout the flight, while the *vertical* velocity increases due to the force exerted on the projectile by gravity.

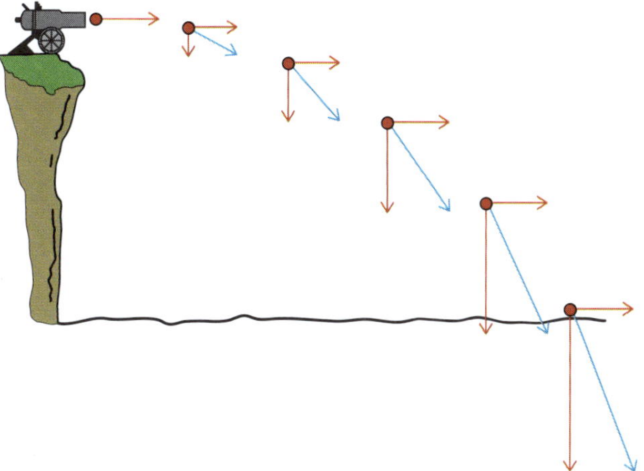

Figure 1.3 Velocity of a projectile is the sum of the vertical and horizontal velocities

- To find the position of a projectile at a specific time we add the horizontal displacement and vertical displacement at that time. Similarly, to find the velocity at a specific time, we add the horizontal and vertical velocity at that time.
- If we call the horizontal direction x and the vertical direction y, we can break up the **initial velocity** (u) into its **vertical and horizontal components**, as shown in Figure 1.4. The horizontal velocity will remain constant during the flight, while the vertical velocity will change because of the downwards acceleration due to gravity.

initial velocity: the initial velocity of a projectile is the velocity with which it is launched
initial horizontal velocity: the component of the initial velocity in the horizontal direction
initial vertical velocity: the component of the initial velocity in the vertical direction

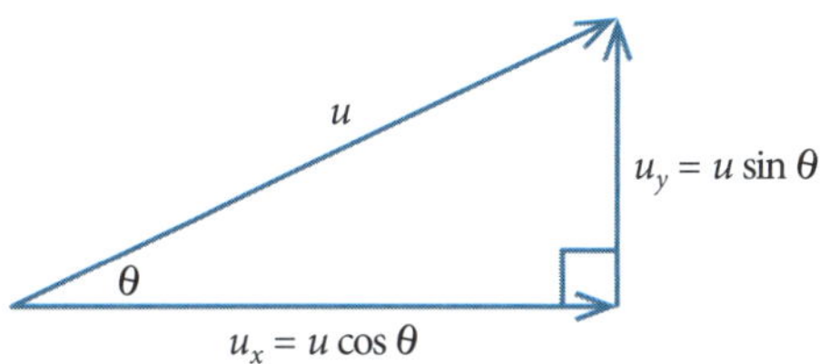

Figure 1.4 Breaking up the initial velocity (u) of a projectile into its vertical (u_y) and horizontal (u_x) components

- To take the vector nature of motion into account we call upwards vector quantities positive, and downwards vector quantities negative. For example, as the acceleration due to gravity is always directed downwards, we will always write it as a negative quantity (i.e. $g = -9.8$ ms^{-2}).
- When solving projectile motion problems, it is useful to remember that the vertical component of the velocity will be zero at the top of the flight, and that the magnitude of the initial vertical velocity and final vertical velocity will be equal if the projectile is launched and lands at the same height.

KEY QUESTIONS

1. **Why is the rate at which an object falls not related to its horizontal velocity?**
2. **How did Galileo analyse projectile motion?**
3. **How is the velocity of a projectile related to the vertical and horizontal components of the velocity at any instant during the flight?**

Answers ➲ p. 15

EXAMPLE 1

A ball is dropped by a boy inside a car travelling horizontally at 8 ms^{-1}. Find the velocity of the ball after 0.5 s with respect to:

a the boy in the car

b a stationary observer who is outside the car.

The path of a projectile will appear vertical to an observer moving with the same horizontal velocity as the projectile

Answer:

a The boy in the car will see the ball fall vertically downwards, from rest, with an acceleration of 9.8 ms^{-2}. To take the vector nature of the motion into account we call vector quantities directed upwards *positive*, and vector quantities directed downwards *negative*. We can then use the equations of motion to find the final velocity (v):

$$v = u + at = -9.8(0.5) + 0$$
$$= -4.9 \text{ ms}^{-1} \text{ or } 4.9 \text{ ms}^{-1} \text{ downwards}$$

b The observer at rest outside the car sees the car and ball travelling horizontally at 8 ms^{-1} when the ball is released. Because there is no horizontal acceleration, the horizontal velocity will not change throughout the flight. To determine the final velocity we add the final vertical velocity and the final horizontal velocity. That is, we add 4.9 ms^{-1} downwards to the 8 ms^{-1} horizontally, as shown in Figure 1.5.

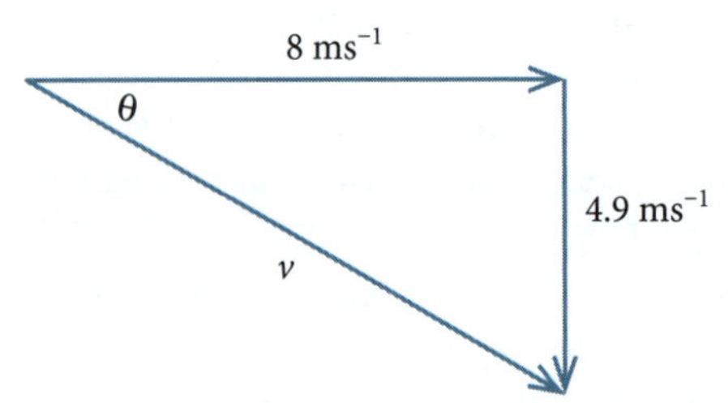

Figure 1.5 Vector diagram used to determine the final velocity

From the vector diagram shown in Figure 1.5 we see that:

$\theta = \tan^{-1}\left(\frac{4.9}{8}\right) = 31.5°$ and

$v = \sqrt{4.9^2 + 8^2} = 9.4\ \text{ms}^{-1}$ at 31.5° below the horizontal

2 Analysing and predicting the motion of a projectile

» Students apply the modelling of projectile motion to quantitatively derive the relationships between the following variables: initial velocity, launch angle, maximum height, time of flight, final velocity, launch height and horizontal range of the projectile.

➔ We saw in the section above that breaking the initial velocity vector into its vertical (u_y) and horizontal (u_x) components yields:

$$u_x = u\cos\theta$$
$$u_y = u\sin\theta$$

where u is the initial velocity
θ is the launch angle and
u_x and u_y are the respective horizontal and vertical components of the initial velocity

launch angle: the angle above or below the horizontal at which a projectile is launched

➔ Assuming there is no air friction, the only force operating on the projectile will be gravity. Applying the laws of motion to the vertical (y) and horizontal (x) components of the motion we can write:

For the horizontal component

$x = u_x t$

And for the vertical component

$y = u_y t + \frac{1}{2}gt^2$

$v_y = u_y + gt$

$v_y^2 = u_y^2 + 2gy$

➔ We can calculate the **maximum height** (y_{max}) that a projectile will reach by considering the vertical component of the motion.

maximum height: the maximum height above the launch point reached by the projectile

Using the relationship:

$v_y^2 = u_y^2 + 2gy$

At the top of the flight the vertical velocity will be zero ($v_y = 0$) and we can write:

$0 = u_y^2 + 2gy_{max}$

The maximum height reached by the projectile will therefore be given by:

$y_{max} = -u_y^2 / 2g = -u^2\sin^2\theta / 2g$

Note that the negative sign will disappear when we substitute $g = -9.8\ \text{ms}^{-2}$ for the acceleration due to gravity.

➔ To calculate the time of flight we also use the vertical component of the motion. For the vertical component we have:

$v_y = u_y + gt$

At the top of the motion the vertical velocity will be zero ($v_y = 0$) and hence:

$0 = u_y + gt$

And, the time to reach the top of the flight will be given by:

$t = -u_y / g = -u\sin\theta / g$

If there is no air friction and the projectile is launched and lands on a horizontal plane, the motion will be symmetric and the projectile will take the same time to reach the top of the flight as it will to fall down. Hence the total time of flight will be twice the time to go up:

$t = -2u_y / g = -2u\sin\theta / g$

Note that again the minus sign will disappear when we substitute $g = -9.8\ \text{ms}^{-2}$ for the acceleration due to gravity.

➔ To calculate the magnitude and direction of the **final velocity** we must add the final vertical velocity (v_y) to the final horizontal velocity (v_x). Note that the horizontal velocity does not change throughout the flight and hence $v_x = u_x$.

final velocity: the vector sum of the final horizontal and final vertical velocity of a projectile

In the special case of a projectile that is launched and lands at the same height, the magnitude of the final velocity will be the same as the magnitude of the initial velocity. However, the direction will be different. Because the direction of the vertical velocity is reversed, the final velocity will be at the same angle below the horizontal as the initial velocity was above the horizontal. This relationship is illustrated in Figure 1.6.

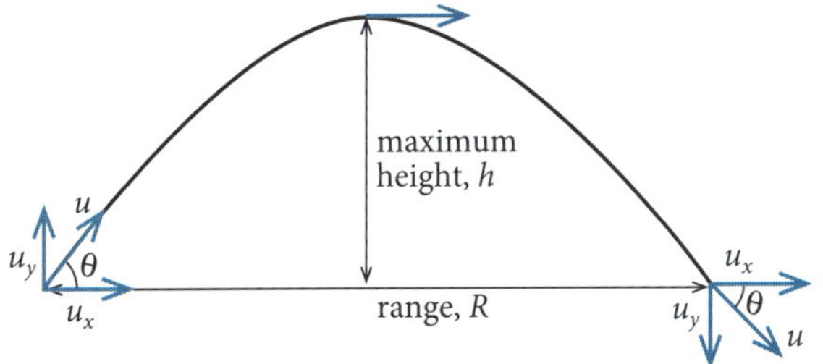

Figure 1.6 Relationship between initial and final velocity of a projectile

➔ The horizontal **range** of the projectile can be calculated using the horizontal component of the motion.

range: the horizontal distance travelled by a projectile before it reaches the ground

The horizontal displacement (x) is given by:

$x = u_x t = (u\cos\theta)t$

Substituting for the time of flight ($t = -2u\sin\theta/g$) we obtain:

$x = (u\cos\theta)t = -u^2(2\sin\theta\cos\theta)/g$

Using the trigonometric identity $2\sin\theta\cos\theta = \sin 2\theta$, we can simplify this to:

Range $= x = -u^2\sin 2\theta/g$

Note that the minus sign will again disappear when we substitute $g = -9.8$ ms^{-2} for the acceleration due to gravity.

➔ If the launch height and landing height of a projectile are different we can still calculate the path of the projectile by treating the vertical and horizontal components of the motion independently. Consider a projectile that lands at a height (h) above or below its launch height.

The maximum height above the launch point reached by the projectile would still be given by:

$y_{max} = -u^2\sin^2\theta/2g$

The time of flight can be calculated using:

$y = u_y t + \frac{1}{2}gt^2$

by substituting the final vertical displacement for y and solving the resulting quadratic equation for the time t. Once the time of flight has been calculated, the horizontal range can be found:

$x = u_x t = (u\cos\theta)t$

The time of flight can also be used to determine the final vertical velocity:

$v_y = u_y + gt = gt + u\sin\theta$

The final velocity can then be calculated by adding the initial horizontal velocity (which remains constant) and the final vertical velocity.

➔ KEY QUESTIONS

4 **How does the velocity change throughout the flight of a projectile?**

5 **What determines the range of a projectile?**

6 **What launch angle produces the maximum projectile range?**

Answers ➲ p. 15

EXAMPLE 2

A tennis player serves a ball horizontally from a height of 3m that just hits the top of the net, which is 0.914 m above the ground and 12 m from the tennis player. Find the initial velocity of the ball.

A ball thrown horizontally will reach the ground at the same time as a ball that is dropped from rest

Answer:

As there is no initial vertical velocity, the time of flight will be the same as the time it would take a ball at rest to fall to the height of the net. The vertical distance to the net is given by:

Vertical distance to net $= y = 3 - 0.914 = 2.086$ m

Hence using the vertical component of the motion:

$y = u_y t + \frac{1}{2}gt^2$ and, as the initial vertical velocity, $u_y = 0$

Thus $y = \frac{1}{2}gt^2$ or $t = \sqrt{(2y/g)} = \sqrt{\frac{2\times 2.086}{9.8}} = 0.65$ s

Using the horizontal component of the motion:

$x = u_x t$ and hence $u_x = x/t = \frac{12}{0.65} = 18.46$ ms^{-1}

EXAMPLE 3

A golf ball is hit with a velocity of 20 ms^{-1} at 30° above the horizontal. Find:

a **the initial horizontal velocity**

b **the initial vertical velocity**

c **the maximum height reached by the ball**

d **the time of flight**

e **the horizontal distance travelled by the ball before it hits the ground.**

The initial velocity can be broken into vertical and horizontal components and the components of the motion treated independently

Answer:

To solve this problem consider the horizontal and vertical components of the motion. First break the initial velocity into its horizontal and vertical components, as shown in Figure 1.7.

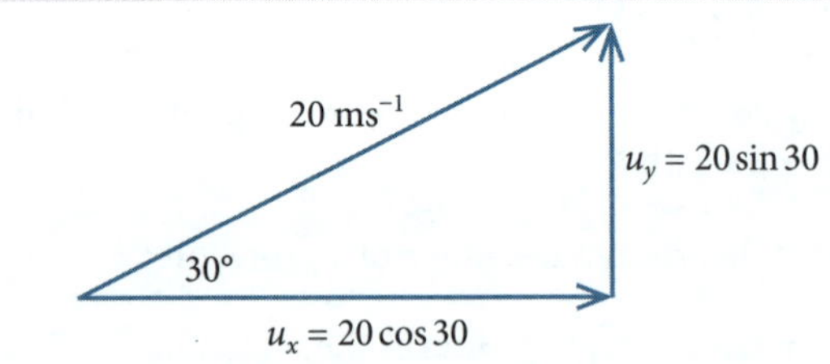

Figure 1.7 Vector diagram to break the initial velocity into its vertical and horizontal components

a From Figure 1.7:
$u_x = 20 \cos 30 = 17.32 \text{ ms}^{-1}$

b From Figure 1.7:
$u_y = 20 \sin 30 = 10 \text{ ms}^{-1}$

c We can apply the equations of motion to the vertical component of the motion to find the maximum height reached.
At the top of the motion the vertical velocity (v_y) will be zero.
$v_y^2 = u_y^2 + 2as$
But $v_y = 0$ and hence:
$s = -u_y^2/2a = \dfrac{-(10^2)}{(2 \times -9.8)} = 5.1 \text{ m}$

d As the flight of a projectile is symmetric, the magnitude of the initial and final vertical velocity will be equal and, as the velocities will be in opposite directions, we can write:
$v_y = -u_y$
Substituting this into the equation of motion
$v_y = u_y + at$
$-u_y = u_y + at$
$t = -2u_y/a = \dfrac{-2(10)}{(-9.8)} = 2.04 \text{ s}$

e Here we can use the horizontal component of the motion. As the horizontal velocity remains constant, the horizontal distance travelled after a time (t) will be given by:
$x = u_x t = (17.32)(2.04) = 35.3 \text{ m}$

EXAMPLE 4

A ball is kicked at a velocity of 25 ms^{-1} at an angle θ above the horizontal on a level field. If the ball lands 40 m from where the ball was kicked, what angle (θ) above the horizontal was the ball initially kicked at?

If we do not have enough information to analyse a problem by breaking the initial velocity into vertical and horizontal components, the range equation might be required

Answer:

Here we can use the expression for the range of a projectile on a plane:
Range $= x = -u^2 \sin 2\theta / g$
Hence $\theta = (\sin^{-1}(gx/u^2))/2 = \dfrac{(\sin^{-1}(9.8 \times 40/25^2))}{2} = 19.4°$

EXAMPLE 5

A ball is thrown with an initial velocity of 30 ms^{-1} at 60° above the horizontal from a 20 m cliff, as shown in Figure 1.8.

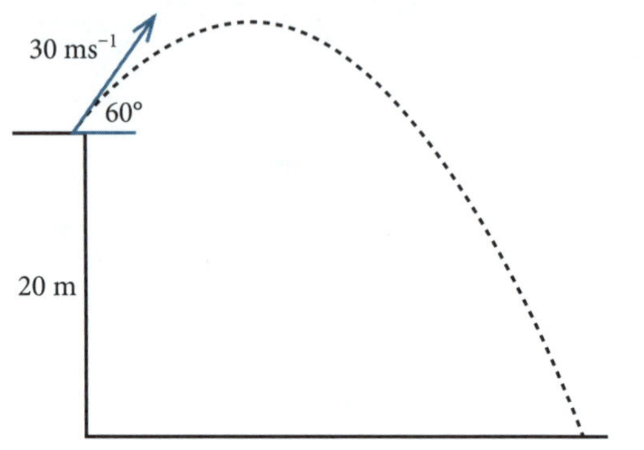

Figure 1.8 Ball projected from a cliff

Find:

a **the maximum height reached by the ball**

b **the time of flight**

c **the range of the ball**

d **the velocity of the ball after 3 s.**

Treat the displacement below the launch point as negative

Answer:

a Using the vertical component of the motion:
$v_y^2 = u_y^2 + 2gy$ and, as $v_y = 0$ at the top of the flight:
$0 = u_y^2 + 2gy$
Hence:
$y_{max} = -u_y^2/2g$
$= -u^2 \sin^2 \theta / 2g$
$= -(30^2 \sin^2 60°)/(-9.8 \times 2)$
$= 34.4$ m above the launch point

b Using the vertical component of the motion and noting that the final position is $y = -20$ m:
$y = u_y t + \frac{1}{2}gt^2$
Substituting in the values gives:
$-20 = (30 \sin 60°)t + \frac{1}{2}(-9.8)t^2$
This reduces to the quadratic equation:
$0 = -4.9t^2 + 25.98t + 20$
Applying the quadratic formula:
$$x = \frac{-b \pm \sqrt{(b^2 - 4ac)}}{2a}$$
$$t = \frac{-25.98 \pm \sqrt{(25.98^2 - (4 \times -4.9 \times 20))}}{2 \times -4.9}$$
We can neglect the negative answer to obtain the time of flight:
$t = 5.98$ s
Note we can also find the time of flight by adding the time to reach the maximum height to the time to fall from this height to the bottom of the cliff.

c Use the horizontal component of the motion to find the range:

$x = u_x t = (u\cos\theta)t$

Hence the range $= (30\cos 60°) \times 5.98 = 89.7$ m

d The horizontal velocity remains constant throughout the flight:

$v_x = u_x = u\cos\theta = 30\cos 60° = 15\ \text{ms}^{-1}$

The vertical velocity after 3 s will be given by:

$v_y = u_y + gt = 30\sin 60° + -9.8 \times 3$
$= -3.4\ \text{ms}^{-1} = 3.4\ \text{ms}^{-1}$ downwards

The velocity after 3 s is given by the vector sum of the horizontal and vertical velocities, as shown in Figure 1.9.

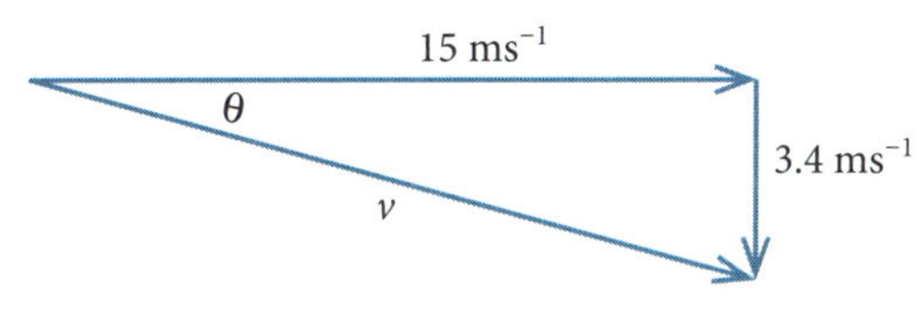

Figure 1.9 Vector diagram to find final velocity

$\theta = \tan^{-1}\left(\frac{3.4}{15}\right) = 12.8°$

$v = \sqrt{(15^2 + 3.4^2)} = 15.38\ \text{ms}^{-1}$ at 12.8° below the horizontal

3 Investigations

» Students conduct a practical investigation to collect primary data in order to validate the relationships derived above.

FIRSTHAND INVESTIGATION 1

Verifying independence of vertical and horizontal components of motion

This experiment proves that the horizontal and vertical components of projectile motion are independent by showing that the time of flight is independent of the initial horizontal velocity.

If the vertical and horizontal components of projectile motion are independent, the time for a projectile launched horizontally from a table to reach the ground should depend only on the height of the table and be independent of the launch velocity. Consider the experiment shown in Figure 1.10.

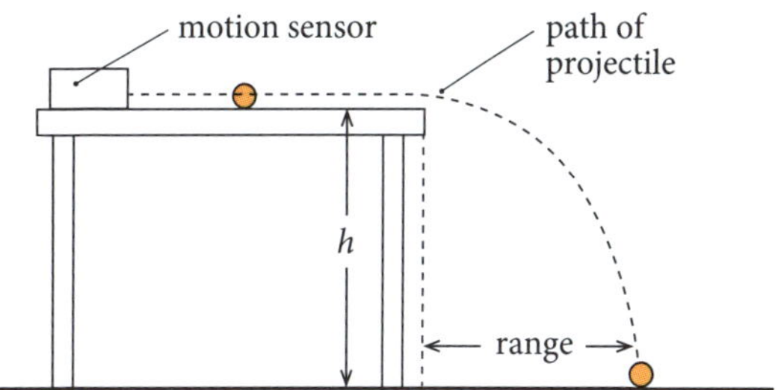

Figure 1.10 Experimental set up for projectile motion practical experiment

By measuring the range for various initial velocities we can verify that the time of flight is a constant and hence is independent of the initial horizontal velocity.

Typical data for this experiment, in which projectiles were launched horizontally from a table 90 cm high, are shown in Table 1.1.

Table 1.1 Initial velocity and range for a projectile launched horizontally

Initial horizontal velocity (ms^{-1})	Range (m)
0.5	0.22
1.0	0.42
1.5	0.63
2.0	0.86
2.5	1.10
3.0	1.28

Recall that the range is given by:

Range $= u_x t$, which is a linear equation of the form $y = mx$.

If the time is indeed a constant that is independent of the initial velocity, graphing the range against the initial velocity should give a straight line with the gradient equal to the time of flight (t). A graph of the data in Table 1.1 is shown in Figure 1.11, which has a gradient of 0.42 s.

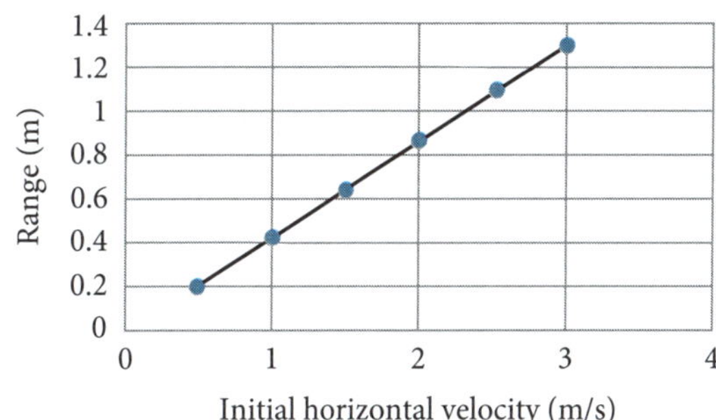

Figure 1.11 Range as a function of initial horizontal velocity

The time of flight should be equal to the time it takes for an object to fall from rest from the table to the floor. Using the vertical component of the motion:

$y = \frac{1}{2}gt^2 + u_y t$ and, as $u_y = 0$

$t = \sqrt{(2y/g)}$

In this case the table height was $y = 0.9$ m and hence the time of flight would be:

$$t = \sqrt{(2y/g)} = \sqrt{\frac{2 \times 0.9}{9.8}} = 0.43 \text{ s}$$

Note that the calculated time is very close to the value determined from the gradient of the graph.

If the gradient of the graph is equal to the time calculated from the vertical component, the results confirm Galileo's hypothesis that the vertical and horizontal components of the motion can be treated independently.

FIRSTHAND INVESTIGATION 2

Confirming the laws of projectile motion

The previous section derived the following equation for the range of projectile:

Range $= x = -u^2 \sin 2\theta / g$

To verify this equation experimentally, we need to show that when the initial velocity is kept constant a plot of range against $\sin 2\theta$ would be a straight line with gradient u^2/g. The data needed can be collected using a projectile launcher that fires a projectile with a specific initial velocity at various angles to the horizontal, as shown in Figure 1.12.

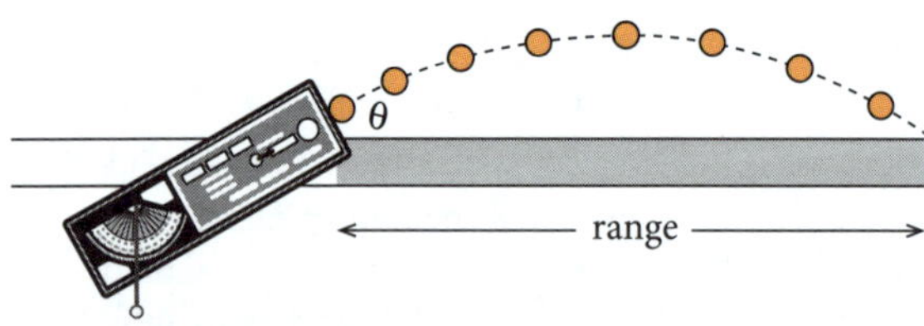

Figure 1.12 Measuring the range of a projectile

First find the initial velocity of the projectile by using two closely spaced timing gates and a data-logger placed just in front of the launcher, or by using the maximum height equation after launching the projectile vertically and measuring the height it reaches.

For example, if the projectile reached a height of $h = 60$ cm when it was launched vertically, we can write:

$v^2 = u^2 + 2as$ and, as $v = 0$ at the top of the flight:

$$\text{Initial velocity} = u = \sqrt{(-2as)}$$

$$= \sqrt{(-2 \times -9.8 \times 0.6)} = 3.4 \text{ ms}^{-1}$$

After measuring the initial velocity we keep it constant and measure the range as a function of the launch angle. Table 1.2 shows typical results. These data have been plotted in Figure 1.13.

Table 1.2 Range for different launch angle

Launch angle (degrees)	$\sin 2\theta$	Range (m)
10	0.34	0.40
15	0.50	0.60
20	0.64	0.74
25	0.77	0.93
35	0.94	1.03
45	1.0	1.17

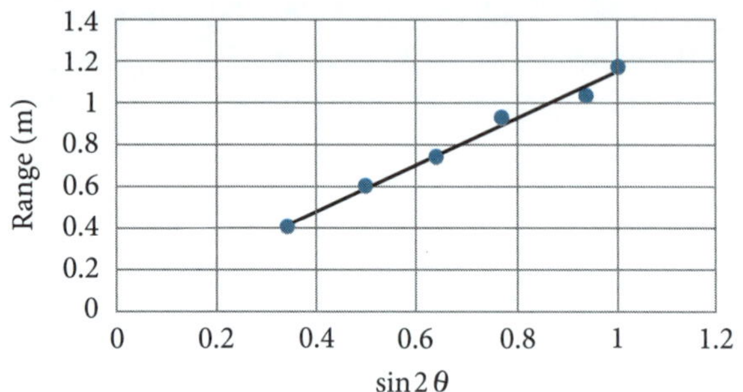

Figure 1.13 Range of the projectile plotted against $\sin 2\theta$

If the range is given by $x = -u^2 \sin 2\theta / g$, we would expect the graph to be a straight line, with gradient $u^2/g = \frac{3.4^2}{9.8} = 1.18$. The graph is a straight line as expected and the gradient of the line of best fit is 1.15 m, which is within 3% of the expected value. Experimental error would account for this small difference between the calculated and measured values.

FIRSTHAND INVESTIGATION 3

Investigating projectiles that land at different heights

This investigation uses projectile motion theory to calculate the range of a projectile that lands at a different height to its launch height, and then uses a projectile launcher to test the prediction. The apparatus used is shown in Figure 1.14.

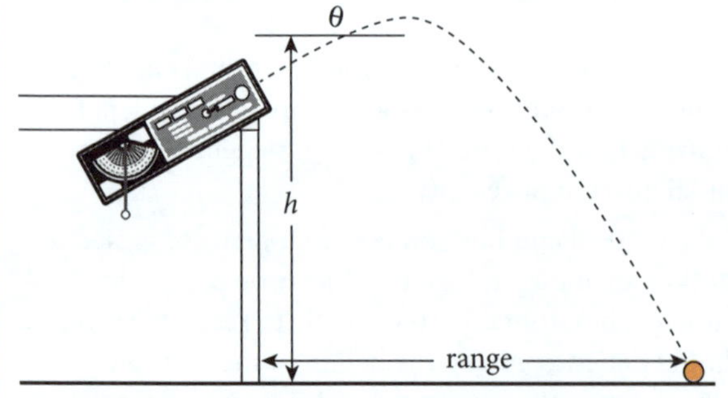

Figure 1.14 Experiment to test projectile motion calculations

First launch the projectile vertically and measure the maximum height (y_{max}), to calculate the initial velocity (u) using:

$u = \sqrt{(2gy_{max})}$

Use the equations of projectile motion and the height of the table (h) to find the range of the projectile when it is launched at 25°, 45° and 65° above the horizontal.
To test results, place a plastic cup at the measured distance and test to see if the projectile lands in the cup when launched at each angle.

4 Modelling the motion of a projectile

» Students solve problems, create models and make quantitative predictions by applying the equations of motion relationships for uniformly accelerated and constant rectilinear motion.

➔ We have seen that projectile motion can be described by analysing the vertical component and horizontal components of the velocity independently. It is also possible to combine the components to obtain a general equation for the path of the projectile. Consider the following relationships for the horizontal and vertical position of the particle:

$x = u_x t$

$y = u_y t + \frac{1}{2}gt^2$

By rearranging the first equation for time ($t = x/u_x$) and substituting it into the second equation to remove time we obtain:

$y = \frac{1}{2}g(x/u_x)^2 + u_y(x/u_x)$, or

$y = Ax^2 + Bx$

where $A = g/2u_x^{\ 2}$ and $B = u_y/u_x$

Recall from mathematics that this is an equation of a concave-down parabola that passes through the origin, as shown in Figure 1.15.

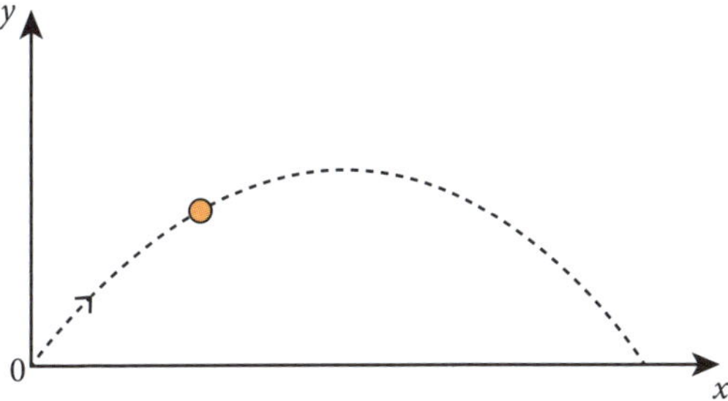

Figure 1.15 The path of a projectile is a concave-down parabola that passes through the origin

➔ We can use this general equation to derive an expression for range by finding the value of x when $y = 0$.

$y = Ax^2 + Bx$ and, setting $y = 0$ gives

$x = 0$ or $x = -B/A = -2u_x u_y/g = -(u^2 \sin 2\theta)/g$

Note that this is the same value for the range that we calculated by analysing the vertical and horizontal components independently.

➔ Because projectiles fall freely under the influence of gravity the total energy of a projectile will remain constant throughout the flight, provided air friction is neglected. Thus the sum of the kinetic and gravitational potential energies of the projectile will remain constant throughout the flight.

➔ KEY QUESTIONS

7 **What determines the maximum height of a projectile?**

8 **What determines the time of flight of a projectile?**

9 **How does our analysis of projectile motion change when the projectile lands at a different height to the launch height?**

Answers ➲ p. 15

CHAPTER SYLLABUS CHECKLIST

Are you able to answer these questions from the syllabus for this chapter? Tick each question as you go through the checklist if you are able to answer it. If you cannot answer a question, turn to the relevant page in the study guide to find the answer. For NESA key word meanings, go to www.educationstandards.nsw.edu.au and search 'key words'.

FOR A COMPLETE UNDERSTANDING OF THIS TOPIC:		PAGE NO.	✓
1	Can I explain how projectile motion can be analysed?	4	
2	Can I explain how and why the velocity of a projectile changes throughout its motion?	5	
3	Can I explain why the vertical component of a projectile's velocity changes but its horizontal component remains constant?	5	
4	Can I break the initial velocity of a projectile into its vertical and horizontal components?	5–6	
5	Can I determine the position and velocity of a projectile at any time during its flight?	5–6	
6	Can I apply the laws of motion to the vertical and horizontal components of projectile emotion?	6–7	
7	Can I derive relationships and solve problems to find the maximum height, time of flight and the range of a projectile that is launched and lands at the same height?	6–7	
8	Can I predict the maximum height, time of flight and range of a projectile that lands above or below its launch height?	7–8	
9	Can I design and carry out safe experiments to test the laws of projectile motion?	9–11	
10	Can I explain how the kinetic energy and potential energy of a projectile will change throughout its flight?	11	
11	Can I explain why the total energy of a projectile remains constant throughout the flight?	11	

All the **dark green panels** with information (like the one below) have the **syllabus inquiry questions** and all the **light green panels** (like the one below) have the **syllabus dot points** for each topic. Note the checklist questions in the table above are based on these.

INQUIRY QUESTION:

How can models that are used to explain projectile motion be used to analyse and make predictions?

The motion of a projectile moving under the influence of gravity can be predicted and analysed by using a model that treats the vertical and horizontal components of the motion independently. The horizontal motion is modelled as uniform rectilinear motion, and the vertical component is modelled as uniformly accelerated rectilinear motion.

1 Understanding projectile motion

» Students analyse the motion of projectiles by resolving the motion into horizontal and vertical components, making the following assumptions:
- a constant vertical acceleration due to gravity
- zero air resistance.

Physics Stage 6 Syllabus

Projectile Motion

Inquiry question: How can models that are used to explain projectile motion be used to analyse and make predictions?

Students:

- analyse the motion of projectiles by resolving the motion into horizontal and vertical components, making the following assumptions:
  - a constant vertical acceleration due to gravity
  - zero air resistance

HSC EXAM-TYPE QUESTIONS

Now for the real thing! The following questions are modelled on the types of questions you will face in the HSC Examination. Think about it: if you get extensive practice at answering these sorts of questions, you will be more confident in answering them in the actual HSC Examination. It makes sense, doesn't it?

Another advantage for your exam preparation is the format of the answers: they are deliberately structured to give you strategies on how to answer examination questions. This will help you aim for full marks!

- For each objective-response question you will have the correct answer and an explanation, and reasons why the other answers are incorrect.
- For each short-answer question you will have a detailed answer marked with ticks to indicate what part of the question gains which marks and also, when needed, an examiner's plan (Examiner Maximiser/ EM) to help you get full marks.

When you mark your work, highlight any questions you found difficult and earmark these areas for extra study.

Objective-response questions (1 mark each)

1 **For the projectile shown in Figure 1.16, which of the following statements for the positions marked *X* and *Y* is true?**

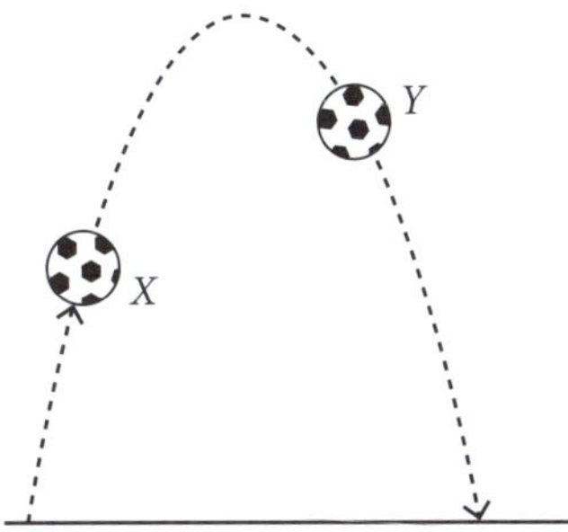

Figure 1.16 Two positions of a soccer ball during its flight

- A The acceleration at *X* is in the opposite direction as the acceleration at *Y*.
- B The magnitude of the vertical velocity is greater at *Y* than at *X*.
- C The total energy of the ball is greater at *Y* than at *X*.
- D The momentum of the ball is greater at *X* than it is at *Y*.

2 **A projectile is launched at 45° above the horizontal, as shown in Figure 1.17. If the projectile was launched with the same initial velocity at 30°, how would the maximum height, time of flight and range change?**

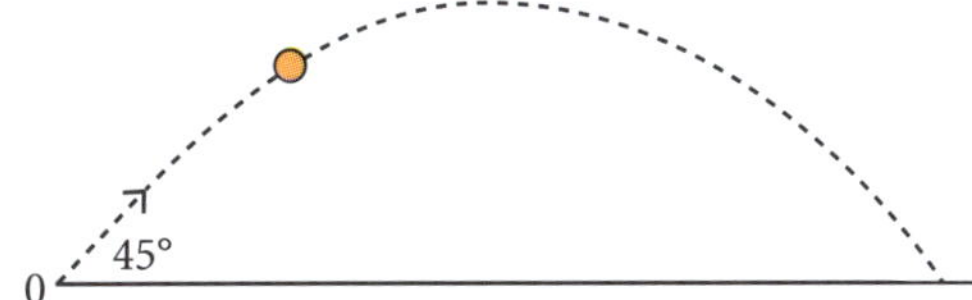

Figure 1.17 Projectile initially launched at 45°

	Maximum height	Time of flight	Range
A	Increased	Decreased	Decreased
B	Decreased	Decreased	Decreased
C	Decreased	Unchanged	Decreased
D	Unchanged	Decreased	Increased

3 **Which of the following statements must be true for two projectiles that are launched and land on a horizontal plane and have the same time of flight?**

- A The vertical component of the final velocity of both projectiles must be equal.
- B The projectiles must be launched at the same angle above the horizontal.
- C The projectiles must have the same range.
- D The projectiles must have the same initial velocity.

4 **What variables could be used to calculate the horizontal range of a projectile that was launched and landed at the same height?**

- A The initial kinetic energy of the projectile.
- B The launch angle and initial vertical component of the velocity.
- C The initial horizontal component of the velocity of the projectile.
- D The initial vertical component of the velocity.

5 **When the archer shown in Figure 1.18 fires an arrow at 45° above the horizontal with an initial velocity of *u*, it has a range of *R* metres.**

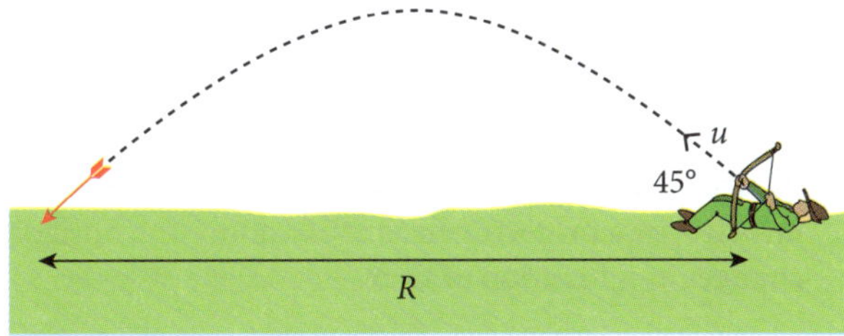

Figure 1.18 Archer firing an arrow from ground level

What would the range *R* change to if the angle was increased to 50°?

- A The range would increase.
- B The range would remain constant.
- C The range would decrease.
- D It is impossible to say from the information given.

Extended-response questions

6 A river flowing at 5 ms^{-1} horizontally produces a waterfall when it reaches a vertical drop of 20 m.

a How long will the water take to reach the bottom of the waterfall? (1 mark)

b How far from the base of the vertical cliff will the water land? (1 mark)

c If the speed of the water was halved and the height of the cliff doubled, how would your answers to parts a and b change? (2 marks)

7 A soccer ball is thrown horizontally by a passenger in a hot air balloon, as shown in Figure 1.19. The passenger notes that the ball is initially moving at 6 ms^{-1} with respect to the balloon, and an observer on the ground sees the balloon moving at 8 ms^{-1} with respect to the ground.

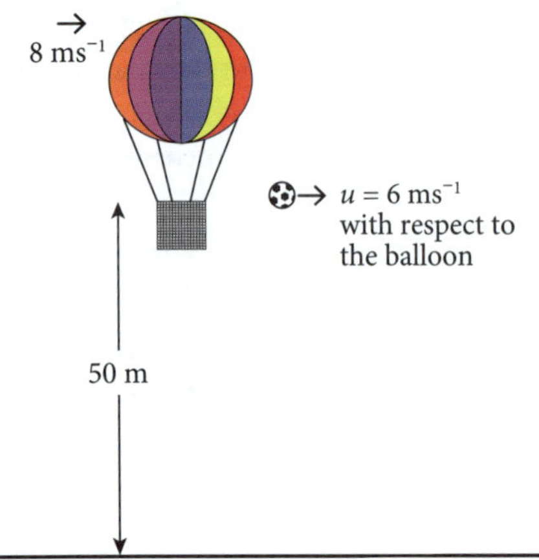

Figure 1.19 Soccer ball launched horizontally from a balloon moving at constant velocity

a Find the initial velocity of the ball with respect to a stationary observer on the ground. (1 mark)

b Calculate the time the ball would take to reach the ground. (1 mark)

c Determine the horizontal distance travelled by the ball with respect to an observer on the ground. (1 mark)

d Determine the horizontal distance travelled by the ball with respect to an observer in the balloon. (1 mark)

8 Figure 1.20 shows how the vertical component of the velocity for a projectile fired at 30° above the horizontal changes as a function of time.

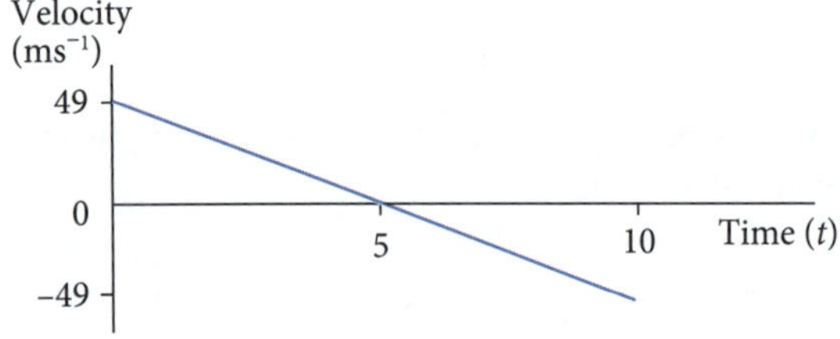

Figure 1.20 Vertical component of velocity as a function of time

Use the graph to find:

a the maximum height reached by the projectile. (1 mark)

b the initial horizontal velocity of the projectile. (2 marks)

c the range of the projectile. (1 mark)

9 A soccer ball is kicked (from ground level on a flat field) and, after being in flight for 2 s, is found to have a velocity of 15 ms^{-1} parallel with the ground. Find:

a the initial horizontal and vertical velocity. (2 marks)

b the maximum height reached by the ball. (1 mark)

c the magnitude and direction of the initial velocity of the ball. (2 marks)

d the horizontal range of the ball. (1 mark)

10 A basketball player throws a ball at 60° above the horizontal, as shown in Figure 1.21. The player is 5 m from the basketball hoop and the hoop is 1 m above the point where the ball left the player's hand.

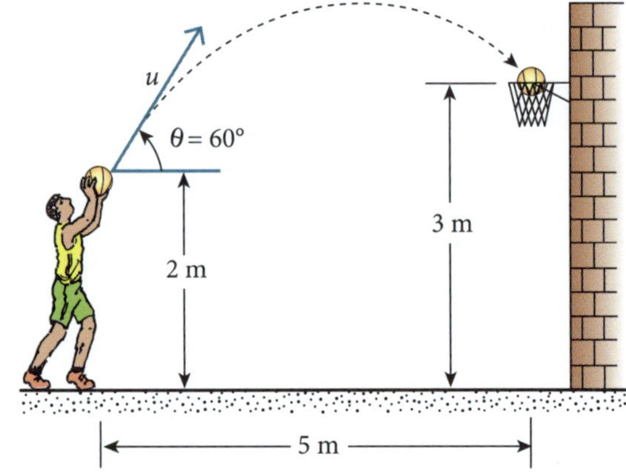

Figure 1.21 Basketball player throwing a ball into a hoop

Given that the ball took 0.8 s to reach the hoop, find:

a the horizontal component of the initial velocity. (1 mark)

b the vertical component of the initial velocity. (1 mark)

c the velocity of the ball at the instant it reached the hoop. (2 marks)

ANSWERS

KEY QUESTIONS

Key questions ➲ p. 5

1 The rate of fall depends only on gravity and, as the gravitational force has no component in the horizontal direction, gravity cannot affect the horizontal motion. Similarly the horizontal velocity has no component in the vertical direction and hence cannot affect the vertical motion.

2 Galileo treated the vertical and horizontal components of the motion independently.

3 The velocity of a projectile at any point is the vector sum of the vertical and horizontal velocities at that point.

Key questions ➲ p. 7

4 The horizontal component of the velocity remains constant but the vertical component changes at 9.8 ms^{-2} downwards. The velocity at any time is the vector sum of the two components.

5 The range of a projectile depends on the time of flight and the horizontal velocity.

6 As range $= -u^2 \sin 2\theta / g$, the maximum range occurs when $\sin 2\theta = 1$. That is, when the projectile is launched at 45° above the horizontal. Launching at smaller or greater angles will cause the projectile to have a shorter range.

Key questions ➲ p. 11

7 The maximum height of a projectile depends on the initial vertical velocity and the acceleration due to gravity.

8 The time of flight of a projectile depends on the initial vertical velocity and the acceleration due to gravity.

9 If a projectile lands at a different height to the launch height, the absolute value of the final velocity will not be equal to the absolute value of the initial velocity, and the time to reach the top of the flight will be different to the time to fall from the highest point to the landing site. These types of problems can be solved by using a quadratic equation to find the time of flight or by treating parts of the motion independently. For example, the time to reach the maximum height could be added to the time to fall from that height to find the total time of flight.

HSC EXAM-TYPE QUESTIONS

Objective-response questions

1 **D**. The ball will have the same horizontal velocity at both points but its vertical velocity and its total velocity and momentum will be greater at *X*. **A** is incorrect as the acceleration at both points is the same (i.e. the acceleration due to gravity). **B** is incorrect as the vertical velocity at *X* is greater than at *Y*, as point *Y* is above point *X*. **C** is incorrect as the total energy of the ball remains constant throughout the flight.

2 **B**. The vertical component of the initial velocity would be less, causing the projectile not to go as high, not to be in the air as long and not to travel as far horizontally. **A** is incorrect as the maximum height reached by the projectile would be less. **C** is incorrect because the time of flight would decrease. **D** is incorrect as the maximum height would be less and the range decreased.

3 **A**. The time of flight is determined by the initial vertical velocity. As the magnitude of the final vertical velocity is equal to the magnitude of the initial vertical velocity, both projectiles must have the same final vertical velocity. **B** is incorrect because the projectiles could be launched at different angles and have the same initial vertical component of velocity. **C** is incorrect because the range depends on the horizontal component of the velocity as well as the time of flight. **D** is incorrect as the projectiles could have the same vertical component of velocity if they had different initial velocities and launch angles.

4 **B**. The value of these variables would allow the horizontal component of the velocity and the time of flight to be determined. These variables determine the horizontal range of the projectile. **A** and **D** are incorrect as the projectile could be launched at different heights, which would change the range. **C** is incorrect as the projectile could have different initial vertical components of the velocity, which would change the time of flight and range.

5 **C**. Because the range $= -u^2 \sin 2\theta / g$, changing θ will change the range, provided the other variables remain constant. Hence increasing the range from 45° to 50° would reduce the value of $\sin 2\theta$ and therefore the range would decrease. **A** is incorrect as a launch angle of 45° produces the maximum range. **B** is incorrect as $\sin 2\theta$ and hence the range would change. **D** is incorrect as there is sufficient information given to answer the question.

Extended-response questions

6 **EM** This question tests projectile motion that involves only an initial horizontal velocity and tests student ability to apply the laws of projectile motion to a different context.

a As the initial vertical velocity is zero:

$y = u_y t + \frac{1}{2}gt^2$ and $u_y = 0$

Hence $t = \sqrt{(2y/g)} = \sqrt{\left(\frac{2 \times -20}{-9.8}\right)} = 2.02$ s ✓

Note the final vertical position is negative because it is below the launch site.

b Using the horizontal component of the motion:

$x = u_x t = 5 \times 2.02 = 10.1$ m ✓

c The time for the water to reach the ground will now be given by:

$t = \sqrt{(2y/g)} = \sqrt{\left(\frac{2 \times 2 \times -20}{-9.8}\right)} = 2.86$ s ✓

The distance travelled from the cliff is therefore:

$x = u_x t = (\frac{1}{2} \times 5)(2.86) = 7.15$ m ✓

7 **EM** This question tests students' ability to combine relative motion and projectile motion.

a The speed of the ball will be the sum of the speed of the ball and the speed of the balloon:

$8 + 6 = 14$ ms^{-1} ✓

b Using the vertical component of the motion and noting that the initial vertical velocity is zero:

$y = u_y t + \frac{1}{2}gt^2$ and $u_y = 0$

Hence $t = \sqrt{(2y/g)} = \sqrt{\left(\frac{2 \times -50}{-9.8}\right)} = 3.19$ s ✓

c An observer on the ground sees the ball travelling horizontally at 14 ms^{-1} and hence:

$x = u_x t = 14 \times 3.19 = 44.7$ m ✓

d An observer on the balloon sees the ball travelling horizontally at 6 ms^{-1} and hence:

$x = u_x t = 6 \times 3.19 = 19.14$ m ✓

8 EM This question tests students' ability to interpret graphical data and to combine the vertical and horizontal components of projectile motion.

a At the top of the flight the vertical velocity is zero. From the graph we see the vertical velocity is zero after 5 s and the initial vertical velocity (u_y) is 49 ms^{-1}. Using the vertical component of the motion we can write:

$y = u_y t + \frac{1}{2}gt^2 = 49 \times 5 + \frac{1}{2}(-9.8)5^2 = 122.5$ m ✓

Alternatively we could use:

$v_y^2 = u_y^2 + 2gy$ and, as $v_y = 0$ at the top of the flight:

$y = -u_y^2/2g = \frac{-49^2}{(-9.8 \times 2)} = 122.5$ m

b As the projectile was launched at 30° above the horizontal we can use the vector components of the initial velocity to find the initial horizontal velocity, as shown in Figure A1.1.

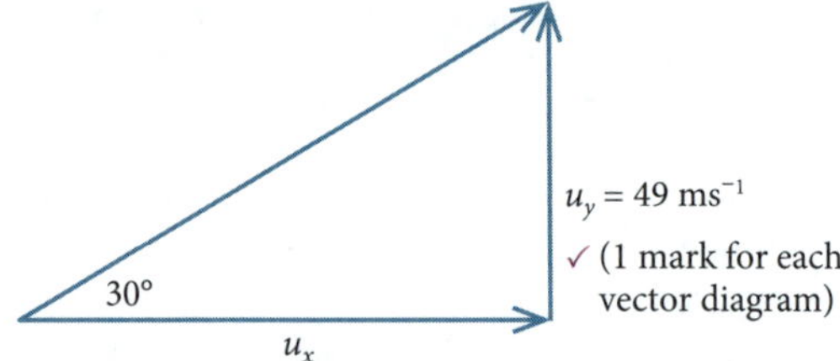

Figure A1.1 Vector diagram to find initial horizontal velocity

From Figure A1.1 we see that:

$\tan 30° = 49/u_x$

Hence $u_x = 49/\tan 30° = 84.9$ ms^{-1} ✓

c The total time of flight from the graph is 10 s. The range is therefore:

$x = u_x t = 84.9 \times 10 = 849$ m ✓

9 EM This question tests students' understanding of how the components of velocity and the total velocity of a projectile change throughout the flight.

a The horizontal velocity does not change and hence the initial horizontal velocity will be the same as the velocity at 2 s:

$u_x = 15$ ms^{-1} ✓

The question implies that the vertical velocity is zero after 2 s and hence the soccer ball must have taken 2 s to reach the top of its flight. Using the vertical component of the motion:

$v_y = u_y + gt$ and, as $v_y = 0$

$u_y = -gt = -(-9.8 \times 2) = 19.6$ ms^{-1} ✓

b Applying the appropriate equation of motion gives:

$v_y^2 = u_y^2 + 2gy$ and as $v_y = 0$ at the top of the flight:

$y = -u_y^2/2g = \frac{-19.6^2}{(-9.8 \times 2)} = 19.6$ m ✓

c We can now add the initial vertical and horizontal velocities to find the initial velocity, as shown in Figure A1.2.

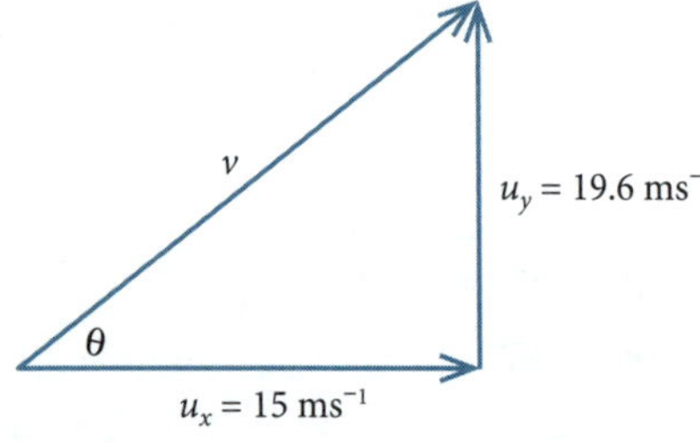

Figure A1.2 Vector diagram to add initial components of the velocity

Using the vector diagram shown in Figure A1.2:

$\tan\theta = \frac{19.6}{15}$ and hence $\theta = \tan^{-1}\left(\frac{19.6}{15}\right) = 52.6°$

$v = \sqrt{(19.6x^2 + 15^2)} = 24.7$ ms^{-1} ✓ at 52.6° above the horizontal ✓

d Using the horizontal component of the motion and noting that the total time of flight will be twice the time to reach the top of the flight (i.e. 2 × 2 = 4 s):

$x = u_x t = 15 \times 4 = 60$ m ✓

10 EM This question tests students' ability to analyse the motion of a projectile that has a different launch and landing height.

a Given that the ball takes 0.8 s to reach the hoop and that the hoop is 5 m horizontally from the player, we can calculate the horizontal velocity using:

$x = u_x t$ or $u_x = x/t = \frac{5}{0.8} = 6.25$ ms^{-1} ✓

b We can determine the initial vertical velocity using:

$y = u_y t + \frac{1}{2}gt^2$

And therefore:

$u_y = (y - \frac{1}{2}gt^2)/t$

$= \frac{(1 - \frac{1}{2} \times -9.8 \times 0.8^2)}{0.8} = 5.17$ ms^{-1} ✓

c At the hoop the horizontal velocity will still be $u_x = 6.25$ ms^{-1} but the vertical velocity will be:

$v_y = u_y + gt = 5.17 + (-9.8 \times 0.8)$

$= -2.67$ ms^{-1} $= 2.67$ ms^{-1} downwards

The vector addition of these two components is shown in Figure A1.3.

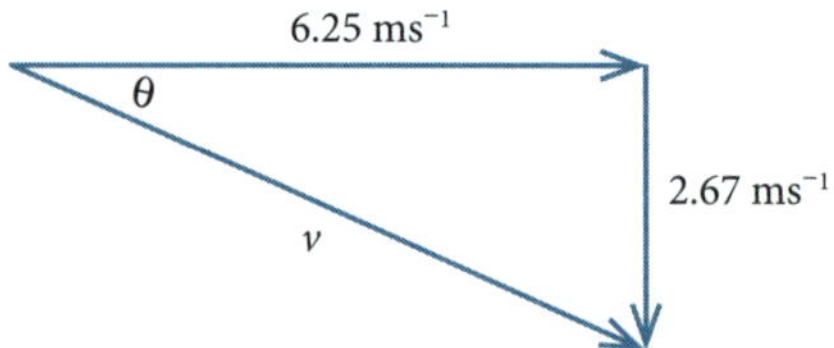

Figure A1.3 Vector diagram to find final velocity

$\tan\theta = \frac{2.67}{6.25}$ and hence $\theta = \tan^{-1}\left(\frac{2.67}{6.25}\right) = 30.6°$

$v = \sqrt{(2.67^2 + 6.25^2)} = 6.80$ ms^{-1} ✓ at 23.13° below the horizontal ✓

CHAPTER 2 CIRCULAR MOTION

MODULE 5 ADVANCED MECHANICS

INQUIRY QUESTION:

Why do objects move in circles?

Newton's first law tells us that an object will remain at rest or move at constant velocity in a straight line unless a net force acts on the object. An object must continually change the direction in which it is travelling to move in a circular path and hence circular motion requires a net force to be applied to the object. The force that causes an object to move in a circular path is called a *centripetal force* and is directed towards the centre of the circle.

1 Measuring centripetal force

» Students conduct investigations to explain and evaluate, for objects executing uniform circular motion, the relationships that exist between centripetal force, mass, speed and radius.

➔ **Uniform circular motion** refers to objects that move in a circle at constant speed. Because an object undergoing uniform circular motion is always moving at a tangent to the circle, we call the velocity of an object moving in a circle the **tangential velocity**. The tangential velocity of an object undergoing uniform circular motion is related to the radius of the circle and the time it takes the object to complete one cycle.

As velocity = distance/time
For an object undergoing uniform circular motion:

$$v = 2\pi r/T$$

where v is the tangential velocity
r is the radius and
T is the period of rotation (i.e. the time to go around the circle once)

uniform circular motion: the motion on an object moving in a circular path at constant speed
tangential velocity: the instantaneous velocity of an object undergoing circular motion

➔ Newton's first law of motion tells us that a moving object will not change its direction of motion unless a force is applied to the object. A force must therefore be acting on any object undergoing uniform circular motion as it is constantly changing its direction of motion. This force acts on the object towards the centre of the circle. We call a force that causes an object to move in a circle a **centripetal force** (F_c). The centripetal force and the tangential velocity (v) are always perpendicular to each other, as shown in Figure 2.1.

centripetal force: force directed towards the centre of the circle that enables an object to undergo circular motion; the centripetal force is always applied at 90° to the tangential velocity

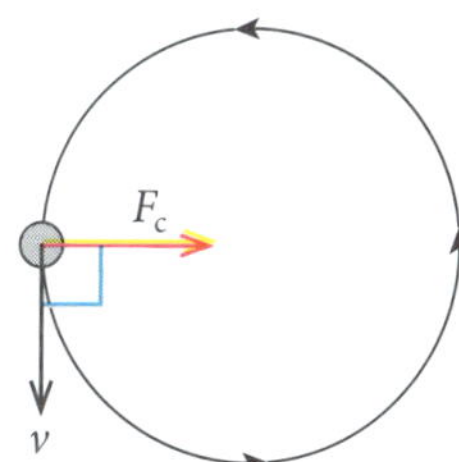

Figure 2.1 Direction of centripetal force (F_c) and tangential velocity (v)

➔ Note that because of Newton's third law there will be an equal and opposite **reaction force** that will oppose the centripetal force. The centripetal force acts on the body undergoing circular motion and the reaction force acts on the agent that is exerting the centripetal force. For example, if a person turns a mass on a string in circular motion the string exerts a centripetal force on the mass, and the mass exerts a reaction force, through the string, on the person. This results in a tension in the string as it exerts a force on the mass and on the person turning the string. Sometimes this force is called a centrifugal force but it is important to realise that this force is just a reaction force to the applied centripetal force.

reaction force: the force that operates in the opposite direction to the applied force, in accordance with Newton's third law of motion; the reaction force operates on the agent applying the force

→ KEY QUESTIONS

1 **Why does moving in a circle require a force to be applied?**

2 **In what direction must the centripetal force operate for an object to undergo circular motion?**

Answers ➲ p. 29

FIRSTHAND INVESTIGATION 1

Investigating centripetal force experimentally

This experiment can be used to find how the centripetal force acting on an object undergoing uniform circular motion is related to the mass and velocity of the object and the radius of its path. Because centripetal force is related to all three variables, we must be careful to examine the effect of one variable at a time, by keeping the other two variables constant. The experimental set up required is illustrated in Figure 2.2.

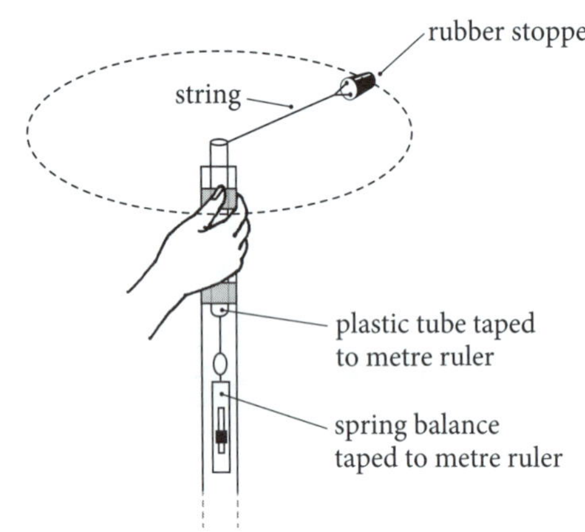

Figure 2.2 Simple experiment to directly measure centripetal force

Effect of velocity on centripetal force

To determine how velocity is related to centripetal force we spin the rubber stopper around above the apparatus and measure the centripetal force (i.e. the force on the spring balance) for different periods of oscillation. The velocity can be calculated from the period using the equation:

$v = 2\pi r/T$

In this part of the experiment we must keep the radius of the circular path and mass of the object constant. Typical results are shown in Table 2.1.

Table 2.1 Typical results for an experiment to determine relationship between centripetal force and velocity, for a radius of $r = 1$ m and a mass of $m = 0.05$ kg

Period (s)	Velocity = $2\pi r/T$ (ms^{-1})	Centripetal force (N)
1.2	5.2	1.5
1.0	6.3	2.0
0.8	7.9	3.0
0.5	12.7	8.0

Plotting the centripetal force against the velocity will produce a curve, while plotting the centripetal force against the square of the velocity will produce a straight line. This implies that the centripetal force is proportional to the square of the object's velocity. We can express this result as:

$F_c \alpha v^2$

where the symbol α (lowercase Greek letter *alpha*) means 'proportional to'

Effect of radius on centripetal force

In this part of the experiment students measure the centripetal force at different radii. Here it is important to keep the velocity and the mass of the object constant. The required period can be calculated from the velocity by rearranging:

$v = 2\pi r/T$, to give

$T = 2\pi r/v$

Typical results are shown in Table 2.2.

Table 2.2 Typical results for an experiment to determine relationship between centripetal force and radius, for a period of $v = 6$ ms^{-1} and a mass of $m = 0.05$ kg

Radius (m)	Period= $2\pi r/v$ (s)	Centripetal force (N)
0.3	0.4	6.0
0.6	0.6	3.0
0.8	0.8	2.2
1.0	1.0	1.8

Careful analysis of the results of this part of the experiment (e.g. by plotting F_c against $1/r$) will show that the centripetal force is inversely proportional to the radius, which we can write as:

$F_c \alpha 1/r$

Effect of mass on centripetal force

In this part of the experiment students measure the centripetal force produced with one, two and three rubber stoppers attached to the string. Again it is important to keep the velocity and the radius constant.

Students will find that doubling the mass will double the centripetal force, provided the radius and period of oscillation remain constant. Tripling the mass will require three times the centripetal force to be applied. These results imply that the centripetal force is proportional to the mass:

$F_c \alpha m$

Putting the three parts of the experiment together we see that:

$F_c \alpha (mv^2/r)$

By substituting values from the experiment we see the constant of proportionality is 1 and we can write:

$F_c = mv^2/r$

➔ KEY QUESTION

3 How is the magnitude of the centripetal force related to the velocity, mass and radius of curvature of an object undergoing uniform circular motion?

Answers ➲ p. 29

2 Analysing the motion of objects undergoing uniform circular motion

» Students analyse the forces acting on an object executing uniform circular motion in a variety of situations, for example:
- cars moving around horizontal circular bends
- a mass on a string
- objects on banked tracks.

» Students solve problems, model and make quantitative predictions about objects executing uniform circular motion in a variety of situations, using the following relationships:
- $a_c = \frac{v^2}{r}$
- $v = \frac{2\pi r}{T}$
- $F_c = \frac{mv^2}{r}$
- $\omega = \frac{\Delta\theta}{t}$

Uniform circular motion

➔ The **angular displacement** (θ) is the angle in radians, subtended by a line joining the object to the centre of the circle. Figure 2.3 shows the angular displacement and the rate of change of the angular displacement (ω), which we will examine shortly. As there are 2π radians in one complete cycle, the distance travelled by the object around the circle is related to the angular displacement and the radius by:

$\Delta s = (\Delta\theta)r$

angular displacement: angle through which an object undergoing circular motion moves between its initial position and final position; angular displacement is measured in radians (rad)

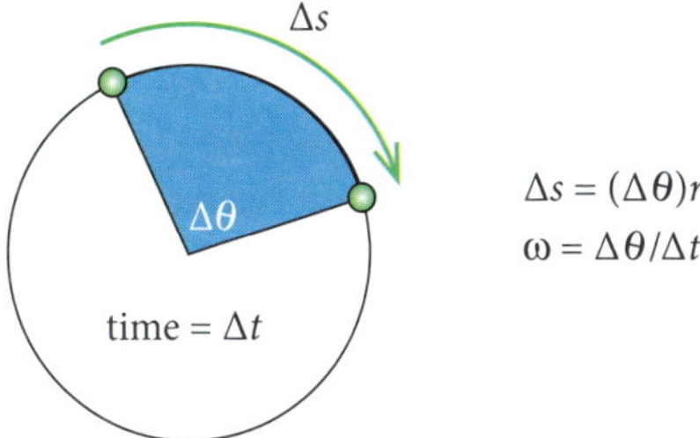

Figure 2.3 Change in angular displacement ($\Delta\theta$) after a time (Δt)

➔ In an analogous manner to linear velocity, which is defined as the time rate of change of displacement, we define **angular velocity** (ω) as the time rate of change of angular displacement (θ):

$$\omega = \Delta\theta/\Delta t$$

angular velocity: the rate of change of the angular displacement measured in radians per second

Note that the units for angular velocity would be radians per second. The angular velocity, like the period and frequency, is a measure of how fast the object is rotating. As there are 2π radians in every cycle, the number of cycles per second (f) multiplied by 2π will give the number of radians per second (ω). Hence the angular velocity is related to the period and frequency of the motion by:

$$\omega = 2\pi f = 2\pi/T$$

where ω is the angular velocity
f is the frequency and
T is the period

➔ The magnitude of the tangential velocity remains constant for an object undergoing uniform circular motion but the direction of the object's velocity continually changes. You will recall from *Excel Year 11 Physics* that acceleration is the time rate of change of the velocity (i.e. $a = \Delta v/\Delta t$) and that a change in the magnitude or direction of the velocity constitutes an acceleration. Because the direction of the velocity vector is continually changing when an object undergoes uniform circular motion, it must be continually accelerating.

- The instantaneous acceleration of an object undergoing uniform circular motion can be determined by finding the rate of change of the velocity over a small time-interval. Figure 2.4 shows how subtracting the final velocity from the initial velocity to determine the change in velocity results in a change of velocity ($\Delta v = v - u$) that is directed towards the centre of the circle. As the acceleration is in the direction of the change in velocity, the acceleration of an object undergoing uniform circular motion must be directed towards the centre of the circle. We call this acceleration the **centripetal acceleration**.

> **centripetal acceleration:** the acceleration experienced by an object undergoing circular motion; centripetal acceleration is measured in ms^{-2} and is always directed towards the centre of the circle

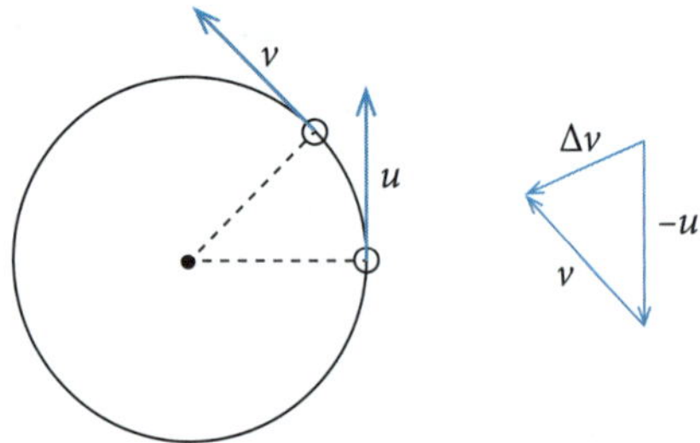

Figure 2.4 Change in velocity is directed towards the centre of the circle.

- Because the centripetal acceleration is directed towards the centre of the circle it is always perpendicular to the velocity of the object, as shown in Figure 2.5. This is a necessary condition for an object to undergo uniform circular motion: if the centripetal acceleration was not perpendicular to the tangential velocity the centripetal acceleration would have a component in the direction of the velocity, which would cause the object to speed up or slow down.

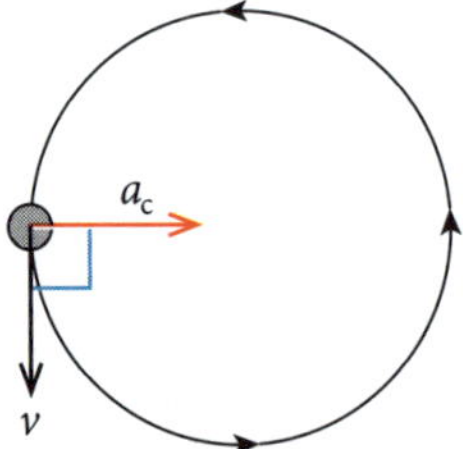

Figure 2.5 The centripetal acceleration is always perpendicular to the tangential velocity

- As we saw earlier, we can also understand the need for a centripetal acceleration by applying Newton's first law. Because the object is not moving in a straight line, it must have a net force applied to it and must therefore be accelerating.
- Figure 2.6 shows how the centripetal acceleration changes the direction of the tangential velocity for an object undergoing uniform circular motion. If there was no centripetal acceleration the object would continue to move in a straight line and after a time t, as there is no tangential acceleration, it would have travelled a distance, $s = vt$. But because of the centripetal acceleration (a_c) the object moved in a circular path rather than a straight line. As there is no initial velocity towards the centre of the circle, we can see that the centripetal acceleration must have caused the object to change its position by $s = \frac{1}{2}a_ct^2$.

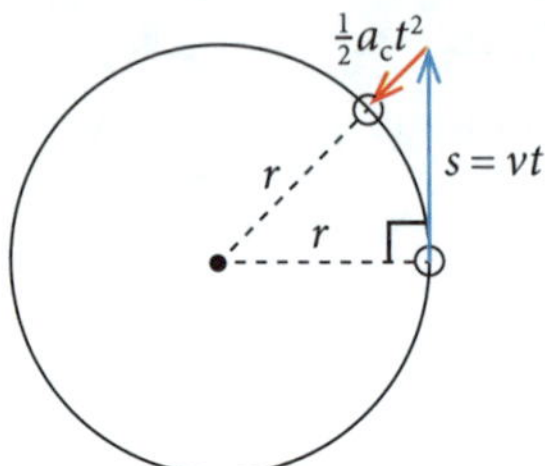

Figure 2.6 Relating the centripetal acceleration and tangential velocity

Applying Pythagoras's theorem to the right-angle triangle in the diagram yields:

$(r + \frac{1}{2}a_ct^2)^2 = (vt)^2 + r^2$

Removing the brackets and simplifying gives:

$\frac{1}{4}a_c^2t^2 + ra_c = v^2$

Now as we are after the instantaneous acceleration, we use a mathematical trick and consider what happens to the equation when we apply it over a very small time-interval (i.e. the limit when $t \to 0$). Clearly as time approaches zero the term with time in it disappears, leaving:

$ra_c = v^2$, or

$$a_c = v^2/r$$

where a_c is the centripetal acceleration
v is the tangential velocity and
r the radius of the motion

- Using Newton's second law we can derive an expression for the centripetal force that produces this acceleration. As $F = ma$:

$$F_c = ma_c = mv^2/r$$

where F_c is the centripetal force
r is the radius of the uniform circular motion and
v is the object's tangential velocity

Note that the centripetal force will be in the same direction as the centripetal acceleration; that is, towards the centre of the circle.

- We saw in the previous section that for an object moving in uniform circular motion, the period is related to the tangential velocity by:

$v = 2\pi r/T$

By substituting this expression for v into the relationships for centripetal force and acceleration we can write expressions for the centripetal force and acceleration in terms of the period (T):

$a_c = v^2/r = 4\pi^2 r/T^2$

$F_c = ma_c = m(4\pi^2 r/T^2)$

Now as the angular velocity $\omega = 2\pi/T$, we can replace $2\pi/T$ in the above equations to obtain:

$\mathbf{v = \omega r}$

$\mathbf{a_c = \omega^2 r}$**, and**

$\mathbf{F_c = m\omega^2 r}$

➔ KEY QUESTIONS

4 **How is the centripetal acceleration related to the tangential velocity?**

5 **What is the difference between angular velocity and tangential velocity?**

6 **How is the period of an object undergoing uniform circular motion related to its tangential velocity and angular velocity?**

Answers ➲ p. 29

EXAMPLE 1

A fisherman spins a fishing line in a horizontal circle of radius 1.2 m above his head. He uses a 0.2 kg lead sinker (mass) on the end of the line.

a If the string has a breaking strain of 50 N, what frequency of rotation would be required to break the line? You may neglect the effect of the weight of the sinker in your calculations.

b What was the speed of the sinker just before the line broke?

c What path would the sinker take after the line broke?

The centripetal force required by a mass to move in a circular path is related to the frequency at which the mass is rotated

Answer:

a If the centripetal force reaches 50 N, the string will break.
Now $F_c = m\omega^2 r$ and therefore $\omega = \sqrt{(F_c/mr)}$

$= \sqrt{\left(\frac{50}{0.2 \times 1.2}\right)}$

$= 14.4 \text{ rads}^{-1}$

The frequency $f = \omega/2\pi = 14.4/2\pi = 2.3$ Hz

b We can use $v = \omega r = 14.4 \times 1.2 = 17.28 \text{ ms}^{-1}$

c Viewed from above the sinker would move in a straight line at a tangent to the circle from the point at which the line broke. Viewed from the side the sinker would move in projectile motion with an initial horizontal velocity of 17.28 ms^{-1}.

EXAMPLE 2

It is possible to turn a bucket of water in a vertical circle, as shown in Figure 2.7, without the water spilling out of the bucket.

Figure 2.7 Turning a bucket of water in a horizontal circle

a If the bucket is turned slowly enough, the woman will not have to exert any force on the bucket when it is at the top of the circle and yet the water would stay in the bucket. Explain how this is possible.

b Write an expression for the force the water would exert on the bucket at the bottom of the motion.

c Find the minimum period a bucket containing water can be turned in vertical circle of radius 1 m without spilling the water at the top.

Gravity can provide a centripetal force when it is directed towards the centre of rotation

Answer:

a At the top of the motion the turning bucket and water will experience a centripetal acceleration downwards. If this acceleration was equal to the acceleration due to gravity the bucket and the water would be accelerating downwards at the same rate and hence there would be no force between the bucket and the water. Therefore, rather than the water falling out of the bucket, the bucket and the water will accelerate downwards together at the same rate. We can say that the centripetal force at the top is provided by gravity and hence:
$mg = mv^2/r$

b At the bottom of the motion, the force exerted by the water on the bucket would be the weight of the water added to the reaction force of the centripetal force applied to the water by the bucket:
$F_R = mg + mv^2/r$

c From part a we have:
$mg = mv^2/r$, or $v = \sqrt{(rg)} = \sqrt{(9.8)} = 3.13 \text{ ms}^{-1}$
Now as $v = 2\pi r/T$, $T = 2\pi r/v = 2\pi \times 1/3.1 = 2$ s

EXAMPLE 3

Figure 2.8 shows a device called a conical pendulum, which consists of a 0.5 kg steel ball on the end of a 2 m long string. The angle between the string and the vertical is 30° when the mass moves in a horizontal circle, with a period of 2.8 s.

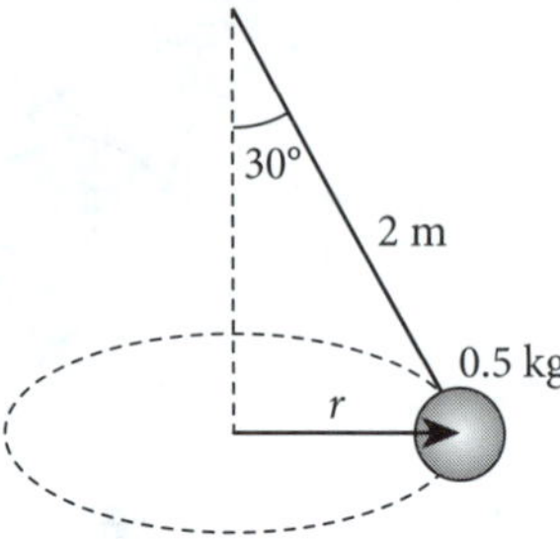

Figure 2.8 A conical pendulum

a Describe the forces acting on the ball and explain how they produce the centripetal force required to keep the ball moving in a circle?

b Calculate the centripetal force on the ball.

c Find the tension in the string when the conical pendulum is rotating.

When an object moves in a horizontal circle the net force on the object must be directed towards the centre of the circle

Answer:

a Two forces are operating on the ball: the weight force (mg), which operates downwards, and the tension in the string. Adding these two forces together produces a net force towards the centre of the circle. This net force is the centripetal force on the ball.

b The centripetal force is given by $F_c = mv^2/r$

The radius of curvature is $r = 2\sin\theta = 2\sin 30° = 1$ m

The velocity of the ball is $v = 2\pi r/T$

$$= \left(\frac{2\pi \times 1}{2.8}\right) = 2.24 \text{ ms}^{-1}$$

The centripetal force on the ball is:

$$F_c = mv^2/r = \left(\frac{0.5 \times (2.24)^2}{1}\right)$$

= 2.5 N towards the centre of the circle.

c The vector sum of the weight and the tension in the string must equal the net force (the centripetal force) on the ball. We can use the vector diagram shown in Figure 2.9 to find the tension in the string. Weight force = $mg = 0.5 \times 9.8 = 4.9$ N.

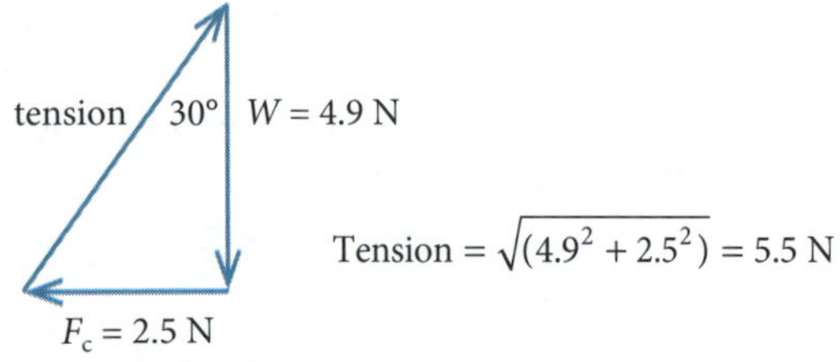

Figure 2.9 Adding the forces acting on the mass to find the centripetal force

Uniform circular motion and friction

➔ For a car to negotiate a corner safely on a horizontal road there must be enough friction between the tyres and the road to provide the required centripetal force. We can express this idea as follows:

$f_{\text{friction}} \geq mv^2/r$

Note that we refer to the maximum value of friction here and that in this case friction is a reaction force. Hence friction will provide the centripetal force for the car to negotiate the corner (not a greater force).

In the Year 11 Physics course we saw that the maximum force of friction is given by $f_{\text{friction}} = \mu mg$ and hence, for a car travelling at velocity v to safely negotiate a corner, the coefficient of friction (μ) must be:

$\mu \geq v^2/gr$

EXAMPLE 4

Consider a car of mass 1500 kg travelling at 20 ms^{-1} in a straight line. The car driver turns the steering wheel and the car turns to the left around a circular bend in the road. The bend has a radius of 30 m and the car maintains its speed as it moves through the bend.

a Find the magnitude of the centripetal force on the car when it is moving through the curve.

b What exerts the centripetal force on the car?

c Why do the passengers slide towards the right of the cabin when the car negotiates a left-hand bend?

d If the car encounters an oil spill on the road when it is halfway through the corner and the friction between the tyres and the road is reduced to zero, what path will the car take?

Recall that a force is required to change the direction of travel of a moving object

Answer:

a Applying the equation for centripetal force:

$$F_c = mv^2/r = \frac{1500(20)^2}{30} = 2.0 \times 10^4 \text{ N}$$

towards the centre of curvature of the bend.

b When the front wheels are turned to the left, the tyres exert an outwards force on the road and the reaction force due to friction produces an equal and opposite force on the tyres towards the centre of curvature of the bend, as shown in Figure 2.10.

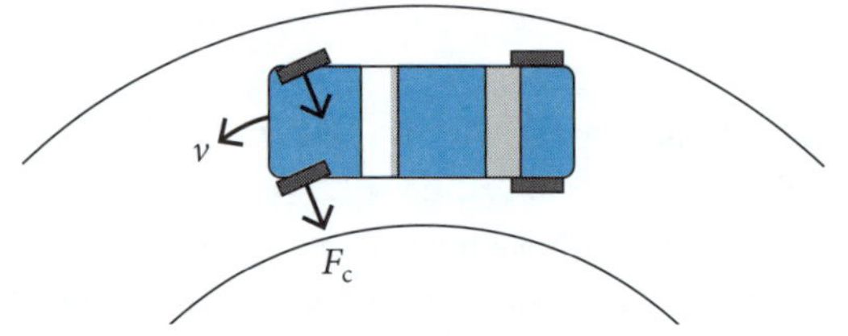

Figure 2.10 The reaction force provided by friction on the tyres produces a centripetal force on the car.

c The inertia of the passengers will cause them to continue to move in a straight line, while the car moves to the left underneath them. The passengers will exert a force outwards on the car as it moves to the left and the reaction force on the passengers provided by the car produces a centripetal force that causes the passengers to move around the corner with the car.

d Without friction the road will be unable to provide a reaction force to the front tyres and consequently there will be no centripetal force on the car. As there is no force on the car, Newton's first law tells us that the car's inertia will cause it to continue to move in a straight line. The line will be at a tangent to the circle from the point the car encounters the oil slick.

EXAMPLE 5

A car's tyres have a coefficient of resistance of $\mu = 0.8$ on a dry road and $\mu = 0.2$ on a wet road. Calculate the maximum speed the car could negotiate an unbanked corner of 50 m radius in:

a dry conditions.

b wet conditions.

Friction between the tyres and the road provides the centripetal force that enables a car to go round a corner

Answer:

a The frictional force ($f_{friction} = \mu F_N$) must be great enough to provide the centripetal force for the car to go round the corner if it is not to slip sideways, hence:

$\mu mg = mv^2/r$

And therefore:

$v = \sqrt{(\mu gr)} = \sqrt{(0.8 \times 9.8 \times 50)} = 19.8\ \text{ms}^{-1}$

b Changing the coefficient of friction to $\mu = 0.2$ and substituting into the equation above gives:

$v = \sqrt{(\mu gr)} = \sqrt{(0.2 \times 9.8 \times 50)} = 9.9\ \text{ms}^{-1}$

Banked corners

➔ Road engineers sometimes bank corners to reduce the friction required for the car to negotiate the corner. Some high-speed test tracks have highly banked corners to enable cars to negotiate the corners with very little sideways force on the car and occupants. Figure 2.11 illustrates how the horizontal component of the normal reaction force (F_N) on a car can provide the centripetal force required for the car to travel around the corner.

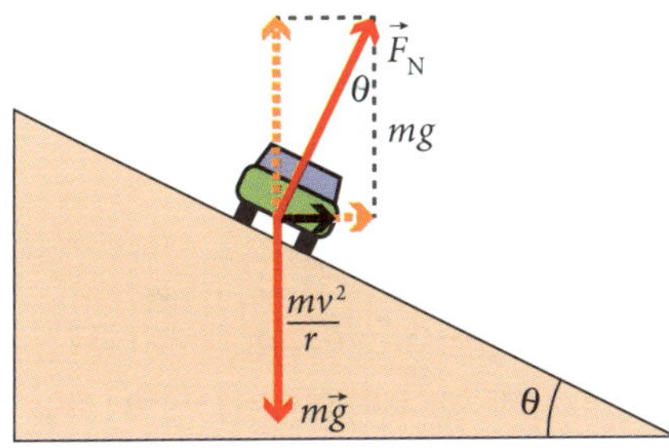

Figure 2.11 Forces on a car negotiating a banked corner

➔ The banking angle can be chosen to ensure the normal reaction force provides the entire centripetal force for a specific cornering speed. At this specific speed and banking angle no friction would be required between the road and the tyres for the car to go around the corner. With reference to Figure 2.11 the required angle of banking for a specific speed (v) will occur when:

$$\tan\theta = (mv^2/r)/mg = v^2/rg$$

If the car is going faster than v, the tires will have to provide a frictional force to keep the car moving around the corner and prevent it sliding up the incline. If the car is travelling slower that v, the tires will have to provide a frictional force to stop the car sliding down the banked track as it moves through the corner.

➔ KEY QUESTIONS

7 **Why is friction required for a car to go around a horizontal corner?**

8 **Why are banked corners on high-speed roads safer than unbanked corners?**

Answers ➲ p. 29

EXAMPLE 6

A corner on a high-speed railway line has a radius of 1000 m and is banked at an angle of 15°.

a Explain why the corner in the railway track is banked.

b Calculate the speed a train would have to travel at if it was to negotiate the corner without exerting any sideways force on the track.

c If the train travelled through the corner at the speed calculated in part b, what forces would the passengers experience as the train moved through the banked corner?

Banking a corner reduces the frictional force required for the vehicle to move around the curve

Answer:

a The corner is banked to ensure that a component of the normal reaction force on the train contributes to the centripetal force required to move around the corner. If the correct speed is chosen for the train it will not exert a sideways force on the track when the train negotiates the corner.

b There will be no sideways force when:

$\tan\theta = (mv^2/r)/mg = v^2/rg$

And hence:

$v = \sqrt{(rg\tan\theta)} = \sqrt{(1000 \times 9.8 \times \tan 15°)} = 51\ \text{ms}^{-1}$

c The passenger would be pushed down harder into their seat (i.e. increase their apparent weight) as they went around the corner but they would not experience any sideways force. This is because the normal reaction force at this speed counteracts the weight of the passenger and provides the centripetal force for the passenger to go around the corner.

3 Energy and uniform circular motion

» Students investigate the relationship between the total energy and work done on an object executing uniform circular motion.

➔ If the speed of a body remains constant its kinetic energy also remains constant. Hence a body undergoing uniform circular motion has a constant energy of:

$K = \frac{1}{2}mv^2$

As $v = \omega r$, the energy of a body undergoing uniform circular motion with a radius r can also be expressed as:

$K = \frac{1}{2}m\omega^2 r^2$

➔ Even though there is a net force on an object undergoing uniform circular motion, no work is done because the net force is directed towards the centre of the circle, perpendicular to the velocity, and hence the force has no component in the direction of motion. As the angle (θ) between the net force and the displacement is 90° for an object undergoing uniform circular motion:

$\text{Work} = Fs\cos\theta = Fs\cos(90°) = 0$

Thus a satellite in circular orbit does no work as it orbits the central body.

➔ Although it is beyond the scope of the HSC course, students might wish to investigate the energy stored in a rotating object like a spinning top. Rotating objects possess energy due to the kinetic energy of the moving mass in the object but, because the mass is distributed at different distances from the centre of rotation, each part of the object will be travelling at a different speed and this makes it more difficult to calculate the energy stored in the rotating object. To do this we must add the contribution of the mass at different radii and calculate quantity called the *moment of inertia* (I) of the rotating body. In an analogous way to mass being a measure of the difficulty to change the translational motion of an object, the moment of inertia is a measure of how difficult it is to change the rotational motion of an object.

➔ KEY QUESTION

9 **Why is no work done by the centripetal force operating on an object undergoing uniform circular motion?**

Answers ➲ p. 29

EXAMPLE 7

A car of mass 1800 kg travels at constant velocity around a horizontal circular track of radius 2 km (you may neglect rolling friction and air friction).

a If the car takes 10 minutes to complete each lap, calculate the energy of the car.

b If the car's energy was doubled, how long would it take to complete a lap of the track?

The energy of an object undergoing uniform circular motion is determined by the tangential velocity and mass of the object

Answer:

a The velocity of the car can be calculated using

$v = 2\pi r/T = \frac{2\pi(2000)}{(10 \times 60)} = 20.9\ \text{ms}^{-1}$

The energy of the car is therefore:

$E = \frac{1}{2}mv^2 = \frac{1}{2}(1800)(20.9^2) = 3.95 \times 10^5\ \text{J}$

b If the energy is doubled, the car would have an energy of $E = 2 \times 3.95 \times 10^5 = 7.9 \times 10^5$ J

Now using $E = \frac{1}{2}mv^2$, we have:

$v = \sqrt{(2E/m)} = \sqrt{\frac{(2 \times 7.9 \times 10^5)}{1800}} = 29.6\ \text{ms}^{-1}$

$T = \frac{2\pi r}{v} = \frac{2\pi \times 2000}{29.6} = 424.5\ \text{s} = 7\ \text{min}\ 4.5\ \text{s}$

4 Torque

» Students investigate the relationship between the rotation of mechanical systems and the applied torque: $\tau = r_\perp F = rF\sin\theta$

➔ A force that causes an object to rotate around an axis of rotation is called a **torque** (τ). We can think of torque as a turning force.

torque: the turning force exerted on an object; torque is measured in Nm as it is the product of applied force and the distance between the centre of rotation and the line of action of the applied force

➔ If we call the distance between the axis of rotation and the position where the force is applied the length of the lever arm (r), the torque is defined as the product of the length of the lever arm and the component of force perpendicular to the lever arm. The concept of torque is illustrated in Figure 2.12. We can express torque mathematically as:

$$\tau = r_\perp F = rF\sin\theta$$

where τ is the torque in newton metres
r is the distance to the axis of rotation and
$F\sin\theta$ is the component of the applied force perpendicular to a line joining the axis of rotation to the point of application of the force

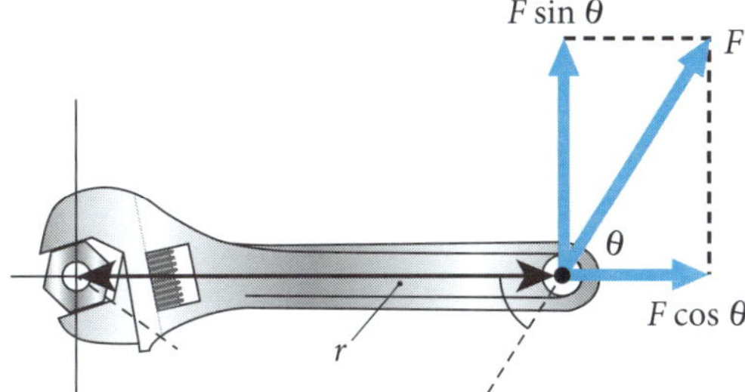

Figure 2.12 The concept of torque

- Engineers use the term **moment of force** rather than torque but the two terms mean the same thing and are essentially interchangeable.

moment of force: another term for torque; this term is often used by engineers who use it to describe the turning forces on static structures

- If the net torques operating on an object do not sum to zero around any point of rotation (axis), the object will rotate around that point. Thus for an object to remain at rest, the net forces acting on the object must sum to zero and the net torques operating on the body must sum to zero around any point of rotation.
- Note that no net torque is operating on a body that undergoes uniform circular motion because, if a net torque was acting, the angular velocity would change. That is, a net torque will produce an angular acceleration in an analogous way to the net force on an object producing a linear acceleration.
- The equations of uniform circular motion and for torque are summarised in Table 2.3.

Table 2.3 Circular motion quantities

Quantity	Symbol	Units	Formulas
Angular displacement	θ	rad	$\theta = s/r$
Angular velocity	Ω	rads^{-1}	$\omega = \Delta\theta/\Delta t$
Tangential velocity	v	ms^{-1}	$v = 2\pi r/T = 2\pi rf = \omega r$
Centripetal acceleration	a_c	ms^{-2}	$a_c = v^2/r = \omega^2 r$
Centripetal force	F_c	N	$F_c = mv^2/r = m\omega^2 r$
Energy	K	J	$K = \frac{1}{2}mv^2 = \frac{1}{2}m\omega^2 r^2$
Torque	τ	Nm	$\tau = rF\sin\theta$

KEY QUESTIONS

10 **Explain why the net torque operating on an object undergoing circular motion must be zero.**

11 **Outline the necessary conditions for an object to remain totally at rest, without moving or rotating.**

Answers ➲ p. 29

EXAMPLE 8

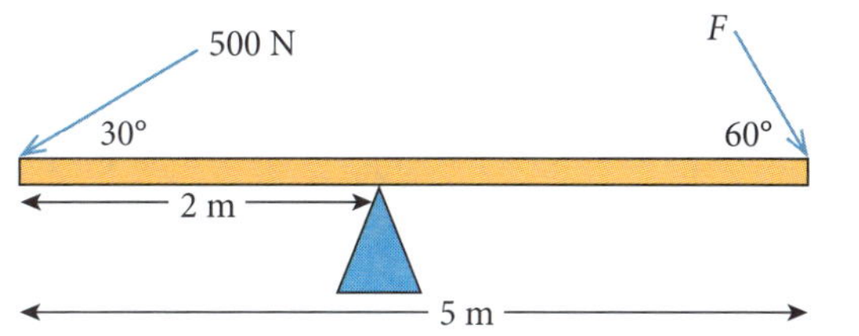

Figure 2.13 Forces acting on an aluminium plank

Consider the 5 m long aluminium plank shown in Figure 2.13. Find the force F that would be required to ensure the plank did not rotate. (You may neglect the weight of the plank.)

The sum of the torques acting on an object around any axis must be zero or the object would rotate around that axis

Answer:

If the plank is not to rotate, the torques around the axis of rotation must sum to zero. Hence:

Clockwise torque = anticlockwise torque

$(F\sin 60°) \times 3 = (500 \times \sin 30°) \times 2$

$$F = \left(\frac{(500 \times \sin 30°) \times 2}{(\sin 60°) \times 3}\right) = 192\text{ N}$$

EXAMPLE 9

A beam of uniform density is supended from the edge of a wall by a cable, as shown in Figure 2.14. The edge of the beam is pivoted at the wall. Find the tension in the cable.

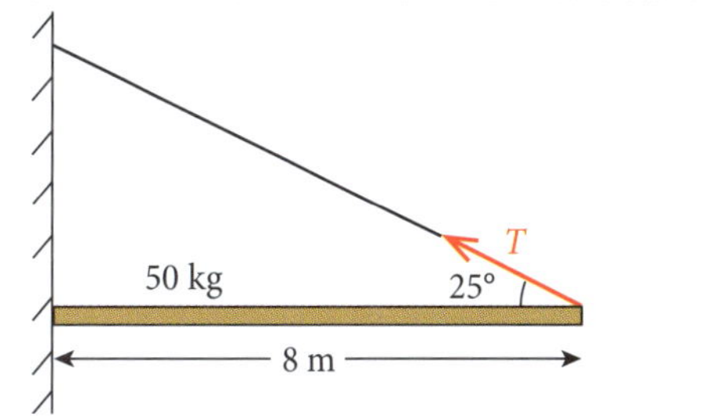

Figure 2.14 A beam suspended by a rope from a wall

The sum of the torques acting on a body around any axis of rotation must sum to zero or the body would rotate around that axis

Answer:

We may consider the weight of the beam to act through the centre of mass, which would be halfway along the beam (i.e. 4 m from the wall). If we take moments around the point where the beam is pivoted to the wall, the sum of the torques acting must be zero (otherwise the beam would rotate around that point):

Clockwise torque = $Fr = (mg)r = 50 \times 9.8 \times 4 = 1960$ Nm

Anticlockwise torque = $Fr = (T\sin 25°)(8) = 3.38T$

Equating the torques gives $1960 = 3.38T$ or

$$T = \frac{1960}{3.38} = 580\text{ N}$$

EXAMPLE 10

Consider the boys on the seesaw shown in Figure 2.15. The seesaw has a mass of 10 kg and is pivoted in the centre. The boy on the left has a mass of 60 kg and the two boys on the right have equal mass. When the boys sit in the positions shown, the seesaw is balanced and remains at rest.

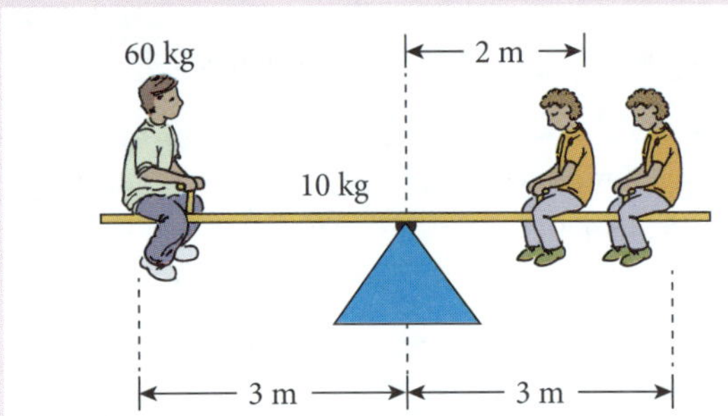

Figure 2.15 Three riders on a seesaw in equilibrium

a **Find the mass of the boy on the far right of the seesaw.**

b **Find the reaction force applied by the pivot.**

For an object to remain at rest the forces acting on the object must sum to zero, and for an object not to rotate, the torques acting on the body with respect to any axis of rotation must sum to zero

Answer:

a If the seesaw is in equilibrium the forces must sum to zero and the torques around any axis must sum to zero. Taking moments around the central pivot we have:

Clockwise torque = anticlockwise torque

$60 \times 3 = m \times 2 + m \times 3$ or $m = \dfrac{60 \times 3}{5} = 36$ kg

b The total downwards force = weight of the boys + the weight of the seesaw = total mass × g

Downwards force = (60 + 36 + 36 + 10)9.8 = 1391.6 N

Hence the reaction force would be 1391.6 N upwards.

CHAPTER SYLLABUS CHECKLIST

Are you able to answer these questions from the syllabus for this chapter? Tick each question as you go through the checklist if you are able to answer it. If you cannot answer a question, turn to the relevant page in the study guide to find the answer. For NESA key word meanings, go to www.educationstandards.nsw.edu.au and search 'key words'.

FOR A COMPLETE UNDERSTANDING OF THIS TOPIC:		PAGE NO.	✓
1	Can I explain why a centripetal force is required for an object to move in circular motion?	17	
2	Can I define centripetal displacement to centripetal velocity?	19	
3	Can I explain in terms of changing velocity why an object undergoing uniform circular motion is accelerating?	19–20	
4	Can I calculate centripetal force and acceleration for an object undergoing uniform circular motion?	20	
5	Can I explain how the tension in a string being used to hold a mass undergoing uniform circular is related to the magnitude of the mass, the radius of the motion and the velocity of the mass?	20	
6	Can I relate the velocity of an object undergoing uniform circular motion to the radius and period of the motion?		
7	Can I relate the angular velocity to the frequency and period of an object undergoing uniform circular motion?		
8	Can I explain how friction is used by a car to negotiate a horizontal circular bend?	22–23	
9	Can I explain why banked corners are used and calculate the angle required for a car to move around a circular corner without the tyres applying a sideways force to the road?	23	
10	Can I explain why a centripetal force does no work on an object undergoing uniform circular motion?	24	
11	Can I calculate the kinetic energy of an object undergoing uniform circular motion?	24	
12	Can I define torque?	24–25	
13	Can I calculate the torque produced by a force on an object?	25	
14	Can I specify the necessary conditions for the centre of mass of an object to remain at rest and the necessary condition for the object not to rotate?	25	
15	Can I use torque to solve problems involving mechanical systems at rest?	25–26	

HSC EXAM-TYPE QUESTIONS

Objective-response questions (1 mark each)

1 **Why is a centripetal acceleration required for an object to undergo uniform circular motion?**

- A to ensure the energy of the object continues to increase as it rotates
- B to increase the tangential velocity throughout each rotation
- C to produce a centripetal force
- D to continually change the direction of the tangential velocity

2 **Consider a mass m on the end of a string undergoing uniform circular motion in a horizontal circle of radius *r*. Which of the following answers best describes what would happen if the radius of the string was suddenly halved?**

- A The energy of the mass would double but the tension in the string would remain the same.
- B The energy would stay the same but the tension in the string would double.
- C The energy would stay the same but the tension in the string would halve.
- D The energy and the tension in the string would halve.

3 **Which of the following statements best describes what causes a car to change direction when the steering wheel is turned?**

- A The car tyres exert a friction force on the road towards the centre of the bend.
- B The road exerts a force towards the outside of the bend on the car tyres.
- C The wheels turn at a different speed, forcing the car to change direction.
- D The road exerts a frictional force on the tyres towards the centre of the bend.

4 **Consider the two masses suspended from the double pulley in Figure 2.16.**

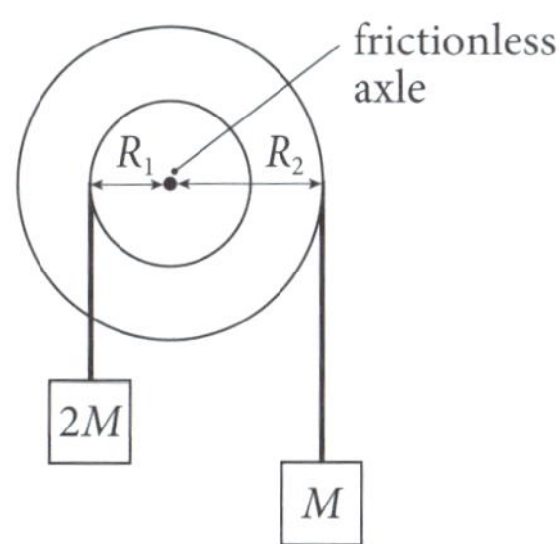

Figure 2.16 Two masses suspended from a double pulley

When released, the lighter mass (M) is observed to accelerate downwards. Assuming the strings are essentially weightless and the pulley is free to turn around the axle, what can we say about the radii of the pulleys?

- A $R_2 > 2R_1$
- B $R_2 < 2R_1$
- C $R_2 = 2R_1$
- D $R_2 < R_1$

5 **Why do railway engineers bank circular bends in railway lines?**

- A to reduce the sideways force exerted on the tracks by the train
- B to reduce the centripetal force on the train
- C to increase the sideways force exerted on the tracks by the train
- D to increase the centripetal force on the train.

Extended-response questions

6 **When a mechanic has trouble loosening a tight nut on a wheel (as shown in Figure 2.17) they sometimes change to a longer spanner.**

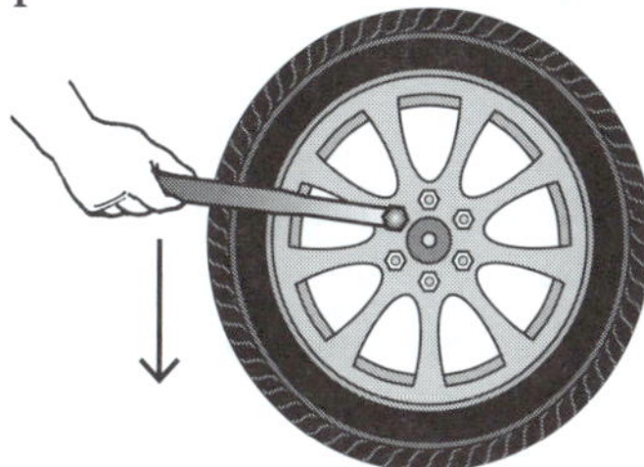

Figure 2.17 Using a spanner to loosen a wheel nut

Explain using physics principles why using a longer spanner is useful in this situation. (3 marks)

7 **Figure 2.18 shows an amusement park ride called The Rotor. The cylinder is spun rapidly and the floor is then lowered while the riders stay fixed to the wall of the rotor.**

Figure 2.18 The Rotor amusement park ride

The cylinder has a radius of 2.5 m and must be rotated at 0.5 Hz to ensure a 40 kg rider does not slide downwards when the floor is lowered.

a **What provides the centripetal force on the rider?** (2 marks)

b **Determine the tangential velocity of the rider.** (1 mark) c Calculate the centripetal force on the 40 kg rider when The Rotor turns with a frequency of 0.5 Hz. (1 mark)

d Determine the minimum coefficient of friction between the rider and the wall of The Rotor to ensure the 40 kg rider does not slide downwards as the floor is lowered when The Rotor is turning with a frequency of 0.5 Hz. (2 marks)

8 A 5 m long plank with a mass of 20 kg rests between two ladders labelled A and B, as shown in Figure 2.19.

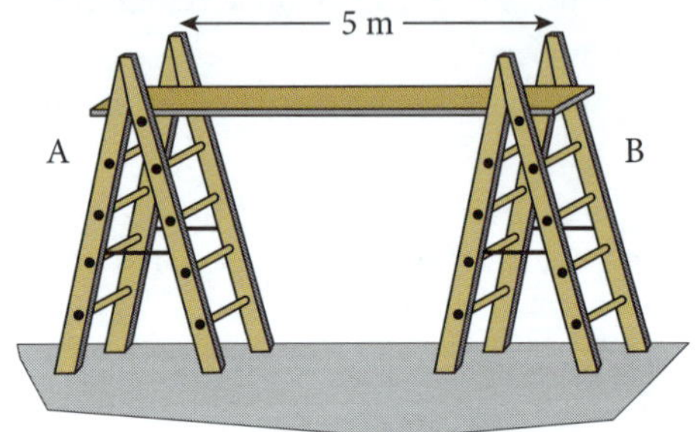

Figure 2.19 Plank of mass 20 kg supported by two ladders 5 m apart

a Determine the force applied by the plank to each ladder. You may assume the mass of the plank is distributed uniformly along the plank. (1 mark)

b If a 100 kg man stood on the plank in the middle, what would be the force exerted by the plank on each ladder? (1 mark)

c If the 100 kg man stood on the plank 2 m from ladder A, what force would the plank exert on each ladder? (2 marks)

9 Part of the fun of riding a rollercoaster is feeling your weight increase and decrease at various points throughout the ride. Consider the rollercoaster shown in Figure 2.20, which starts at rest from position A. Note that the radius of curvature of the rollercoaster track is the same at the points B, C and D.

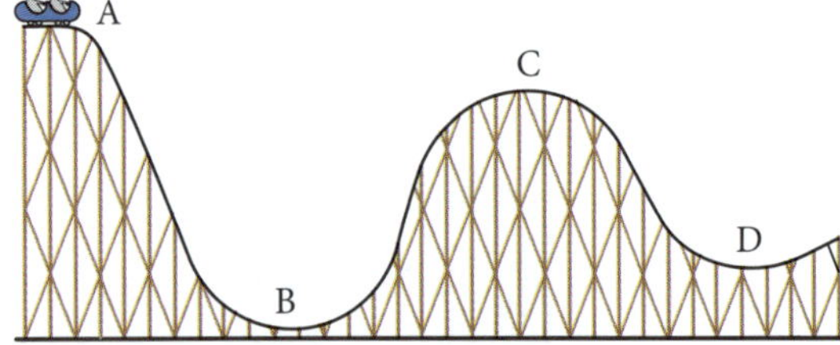

Figure 2.20 Rollercoaster

a Qualitatively compare the apparent weight of the rider at the points A, B, C and D. (1 mark)

b Justify your answer to part a by explaining why the apparent weight of the rider changes throughout the ride. (4 marks)

10 In another amusement park ride the riders sit in swings that rotate around, as shown in Figure 2.21.

Figure 2.21 Swing carousel

As the carousel turns faster, the riders swing out further and make a greater angle with the vertical. When the carousel takes 8 s to complete each rotation the riders make an angle of 26.7° with the vertical.

a Draw a vector diagram showing the forces acting on the riders when they make an angle of 26.7° with the vertical and indicate the direction of the net force operating on the riders in your diagram. (1 mark)

b Find the angular velocity of the riders when the carousel turns with a period of 8 s. (1 mark)

c If the radius of the horizontal path taken by the riders is 8 m, when the period is 8 s find the centripetal acceleration of the riders. (1 mark)

d If a rider had a mass of 60 kg, find the tension in the cable holding the rider when it made an angle of 26.7° with the vertical. (1 mark)

ANSWERS

KEY QUESTIONS

Key questions ➲ p. 18

1 A force is required because the object must continually change the direction in which it is travelling. A change in the magnitude or direction of the velocity constitutes acceleration and Newton's second law tells us that a force is required to produce acceleration.

2 The instantaneous change in velocity is directed towards the centre of the circle and hence the centripetal acceleration and force must also be directed towards the centre of the circle.

Key question ➲ p. 19

3 Centripetal force = $F_c = mv^2/r$. Hence centripetal force is proportional to the mass$_1$, inversely proportional to the radius and proportional to the square of the tangential velocity.

Key questions ➲ p. 21

4 Centripetal acceleration is the instantaneous change in tangential velocity (i.e. $a_c = \Delta v/\Delta t = v^2/r$).

5 Angular velocity is the rate of change of the angular displacement, measured in radians per second. Tangential velocity is the directed speed of an object undergoing circular motion, measured in metres per second.

6 Tangential velocity = v = distance/time = circumference/period = $2\pi r/T$. Now, as angular velocity = $\omega = 2\pi/T$, tangential velocity = ωr.

Key questions ➲ p. 23

7 To go around a corner a car exerts a force away from the centre of the circle on the road (by turning the front wheels into the corner), and friction produces a reaction force on the car, which provides the centripetal force required for the car to go round the corner.

8 Banked corners reduce the amount of friction required between the tyre and the road needed to provide the centripetal force required for the car to negotiate a corner.

Key question ➲ p. 24

9 Because the centripetal force is perpendicular to the tangential velocity it has no component in the direction of the movement and hence does no work on the object.

Key questions ➲ p. 25

10 If a torque is applied to an object it will rotate at an increasing rate. That is, a net torque causes the angular velocity to increase. As an object undergoing uniform circular motion maintains a constant angular velocity, the net torque operating on such an object must be zero.

11 For an object to remain stationary the net force on the object must be zero and the torques operating on the body must sum to zero around any axis. If there was a net force on the object it would accelerate and if there was a net torque around any axis the object would rotate at an increasing rate around that axis.

HSC EXAM-TYPE QUESTIONS

Objective-response questions

1 **D**. The object must continually change direction as it moves round the circle and the rate of change of direction is what we call the centripetal acceleration. **A** is incorrect because an object undergoing uniform circular motion has a constant energy. **B** is incorrect because the velocity remains constant when an object undergoes uniform circular motion. **C** is incorrect because the centripetal acceleration is produced by the centripetal force. The centripetal force is not produced by the centripetal acceleration.

2 **B**. Energy is conserved in a closed system and hence as no work is done on or by the system the energy must remain constant. For the energy to remain constant the velocity must remain constant (as $K = \frac{1}{2}mv^2$) but the tension will double because $T = mv^2/r$ and r has halved. **A** and **D** are incorrect because no work has been done on the system and the energy cannot change. **C** is incorrect as the tension must increase because the velocity remains constant but radius halves.

3 **D**. When the steering wheel is turned the tyres exert a sideways force on the road and the reaction force provided by friction on the tyres acts as a centripetal force that causes the car to move around the bend. **A** and **B** are incorrect because the centripetal force must be on the car (not on the road) for it to go around the bend. **C** is incorrect as the wheels do turn at a different rate but this is not what causes the car to go around the corner; it is a result of the car going around the corner.

4 **A**. For the pulley to turn clockwise the torque exerted by mass M must be greater than the torque exerted by the $2M$ mass (i.e. $MR_2 > 2MR_1$ and hence $R_2 > 2R_1$). **B** and **D** are incorrect because each would produce more torque on the left-hand side of the pulley, which would cause it to rotate anticlockwise rather than clockwise. **C** is incorrect because the net torque would be zero and the pulley would not rotate.

5 **A**. Without banking, all the centripetal force has to be provided by the track pushing the train wheels horizontally and the normal reaction force pushes the tracks sideways. Banked tracks enable some of the centripetal force to be provided by the track, pushing the train upwards as well as to the side. This increases the apparent weight of the train but reduces the sideways reaction force on the track. **B** and **D** are incorrect as the centripetal force required for the train to go around the corner is not influenced by banking as it only relates to the velocity of the train and the radius of the bend. **C** is incorrect because banking reduces the sideways force on the track.

Extended-response questions

6 EM This question tests students' understanding of torque.

If a nut does not rotate, a greater torque must be applied ✓ to overcome the frictional force preventing the nut from turning. Torque = $\tau = Fr\sin\theta$ ✓ and, provided the force is applied perpendicularly to the spanner, $\theta = 90°$. This equation reduces to $\tau = Fr$. Hence to increase the torque we must increase the applied force and/or apply the force further from the nut. A longer spanner will increase the point of application of the force from the nut (increase r) and hence it will apply a greater torque to the nut, ✓ which will hopefully free the nut and enable it to turn.

7 **EM** This question tests students' ability to apply the equations of circular motion and student understanding of centripetal force and the coefficient of friction.

a When the drum begins to rotate the rider will move in a straight line and hence apply a force to the wall of the drum. ✓ The reaction force from the wall of the drum on the rider ✓ is the centripetal force that causes the rider to move in a circle with the drum.

b Period $= T = 1/f = \frac{1}{0.5} = 2.0$ s

Now $v = 2\pi r/T = \frac{(2\pi \times 2.5)}{2} = 7.85\ \text{ms}^{-1}$ ✓

c $F_c = mv^2/r = \frac{40 \times (7.85)^2}{2} = 1.23 \times 10^3$ N ✓

d The frictional force between the rider and the wall must be great enough to support the weight of the rider. ✓ The normal reaction force experienced by the rider is the centripetal force calculated in part **c** above. Now the minimum frictional force will be:

$f_{\text{friction}} = \mu F_N = mg$

And hence:

$\mu = mg/F_N = \frac{40 \times 9.8}{1.23 \times 10^3} = 0.32$ ✓

8 **EM** This question examines students' understanding of the role of torque and force in systems in equilibrium.

a The weight of the plank will be distributed evenly between the ladders and hence the applied force on each ladder will be:

$F = mg/2 = \frac{20 \times 9.8}{2} = 98$ N ✓

b The weight would again be spread equally across the ladders and hence:

$F = mg/2 = \frac{(20 + 100) \times 9.8}{2} = 588$ N ✓

c If the plank and the man remain at rest, the net forces must sum to zero and hence:

Weight of man + weight of plank = reaction force at ladder A (F_A) + reaction force at ladder B (F_B).

Hence $100 \times 9.8 + 20 \times 9.8 = F_A + F_B$

Or $F_A + F_B = 1176$ N

We can think of the plank's weight acting on the centre of the plank, while the man's weight acts 2 m from ladder A. If we consider a pivot to be at the point the plank rests on ladder A, the torques around this point must sum to zero and hence:

Torque due to man's weight + torque due to weight of plank = torque due to the reaction force at ladder B

Therefore $(100 \times 9.8) \times 2 + (20 \times 9.8) \times 2.5 = (F_B) \times 5$ and $F_B = 490$ N ✓

Now, as we have established that $F_A + F_B = 1176$ N:

$F_A = 1176 - F_B = 1176 - 490 = 686$ N ✓

(Alternatively, we could have taken moments around the point of contact with ladder B to find F_A.)

9 **EM** This question examines students' understanding of apparent weight and centripetal force.

a The apparent weight of the rider at each point in the ride will be related by:

$W_B > W_D > W_A > W_C$ ✓

b The apparent weight will depend on the position of the trolley as the normal reaction force on the rider will depend on their weight and the centripetal force at that position. At B and D their weight will appear to increase as the rider experiences the reaction force due to gravity and the centripetal force but at C some of the rider's weight provides the centripetal force, which reduces the apparent weight.

The apparent weight of the rider at B and D will be the sum of the weight and the centripetal force (i.e. $mg + F_C$) ✓ on the rider and, as the rollercoaster is moving faster at B than D (because B is at a lower point), the centripetal force at B will be greater than at D. ✓ The apparent weight at A will simply be the weight (i.e. mg) as the trolley is not moving and therefore will be less than the apparent weight at B and D. ✓ The apparent weight at C will be the difference between the weight and the centripetal force (i.e. $mg - F_C$) and hence will be lower than the apparent weight at A. ✓

10 **EM** This question examines students' understanding of the vector nature of forces and students' ability to apply the equations of uniform circular motion.

a Riders move in a horizontal circle and the net force on each rider is towards the centre of the circle, as shown in Figure A2.1.

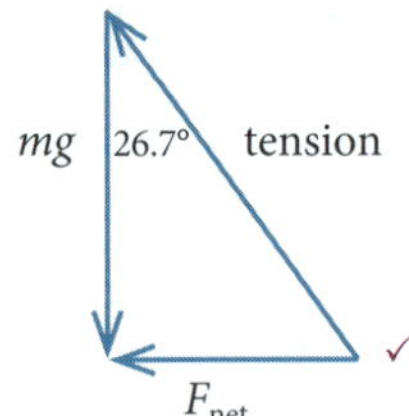

Figure A2.1 Forces on a rider on the carousel

b Angular velocity $= \omega = 2\pi/T = \frac{2\pi}{8} = 0.785\ \text{rads}^{-1}$ ✓

c $a_c = \omega^2 r = (0.785)^2 \times 8 = 4.93\ \text{ms}^{-2}$ ✓

d Centripetal force $= ma_c = 60 \times 4.93 = 296$ N

Weight force $= mg = 60 \times 9.8 = 588$ N

Now by applying Pythagoras's theorem to the vector diagram shown in part **a**:

Tension $= \sqrt{(9.76^2 + 588^2)} = 658$ N ✓

MODULE 5 ADVANCED MECHANICS

CHAPTER 3 MOTION IN GRAVITATIONAL FIELDS

INQUIRY QUESTION:

How does the force of gravity determine the motion of planets and satellites?

Satellites and planets orbit massive central bodies because gravity exerts a centripetal force on the planets and satellites. As the central body and orbiting body experience an equal force of gravitational attraction, they orbit one another around the centre of mass of the system. However, if the central body is much more massive than the orbiting body, we can assume the central body is at rest because the centre of mass of the system would be very close to the centre of the massive body.

1 Gravitational force and gravitational fields

» Students apply qualitatively and quantitatively Newton's law of universal gravitation to:

- determine the force of gravity between two objects $F = \frac{GMm}{r^2}$
- investigate the factors that affect the gravitational field strength $g = \frac{GM}{r^2}$
- predict the gravitational field strength at any point in a gravitational field, including at the surface of a planet.

➔ Newton showed that all objects with mass exert a force of gravitational attraction on another given by:

$$F = GMm/r^2$$

where M and m are the masses of the objects
r is the distance between the centre of mass of each object and
$G = 6.67 \times 10^{-11}$ Nm^{2}kg^{-2} is a constant called the *universal gravitational constant* or simply the *gravitational constant*

➔ Figure 3.1 illustrates the gravitational attraction between two bodies. Note that the force on each body is equal and opposite in accordance with Newton's third law. Thus the gravitational force that you exert on the Earth is equal to the gravitational force that the Earth exerts on you.

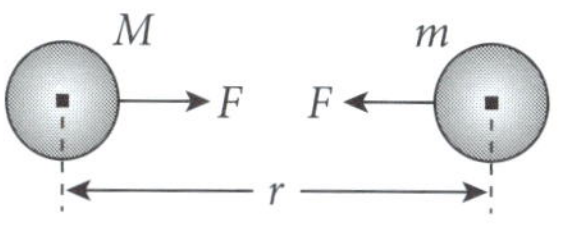

$F = GMm/r^2$

Figure 3.1 The gravitational force between two bodies of mass M and m separated by a distance r

➔ The force of gravity on an object is called the *weight of the object* and we saw in the Year 11 course that weight is related to mass as follows:

Weight = $w = mg$

where w is the weight

m is the mass and

g is the local acceleration due to gravity

➔ Equating the weight of an object near a planet to the force of gravity on the object enables us to derive an expression for the acceleration due to gravity at a distance r from the centre of planet of mass M:

Weight = force of gravity

$mg = GMm/r^2$

And hence $g = GM/r^2$

➔ In the same way that the electric field strength is defined as the force per unit charge, the **gravitational field strength** is defined as the force per unit mass, or:

Gravitational field strength = gravitational force/mass
= weight/mass = $mg/m = g$

gravitational field strength: the gravitational force that would be exerted on a 1 kg mass at the point (i.e. gravitational force per unit mass); the gravitational field strength is equal to the acceleration due to gravity at the point

Thus the gravitational field strength at a specific point in space is equal to the acceleration due to gravity at that point. We can understand this by recalling Newton's second law. As $F = ma$ the force on a unit mass (i.e. when $m = 1$ kg) is simply the acceleration. We also can confirm this by unit analysis because the units of gravitational field strength are:

Unit of force/unit of mass = kgms^{-2}/kg = ms^{-2} = unit of acceleration

As we saw above, the acceleration due to gravity near a body of mass M is given by:

$$g = GM/r^2$$

As the gravitational field strength is g, this expression also gives the gravitational field strength at a distance r from the centre of mass of an object of mass M. Note that like g, gravitational field strength is a vector directed towards the centre of the mass.

➔ The gravitational field strength (acceleration due to gravity) near a real planet may vary a little at different points around the planet. The acceleration due to gravity on the surface of the Earth varies by up to 0.7% at different places on the surface. These changes are due to variations in the thickness and density of the Earth's crust and to the Earth not being a perfect sphere. The rotation of the Earth causes the Earth to bulge at the equator, where the sea level is 21 km further from the centre of the Earth than at the poles.

➔ In the same way we draw electric lines of force around charges to represent electric fields, we can also draw gravitational lines of force around bodies to represent gravitation fields. The direction of the gravitational field at a specific point is the direction of the force of gravity that would be exerted on a mass if it was placed at the point. The density of the field lines reflects the intensity of the field and, like electric fields, gravitational field lines can never cross one another because gravitational forces add together to produce a single force on an object. Figure 3.2 shows the gravitational lines of force around the Earth, and Figure 3.3 shows gravitational field lines between the Earth and the Moon. Because the gravitational field is directed towards a point at the centre of the Earth, the field around the Earth is called a *radial gravitational field*.

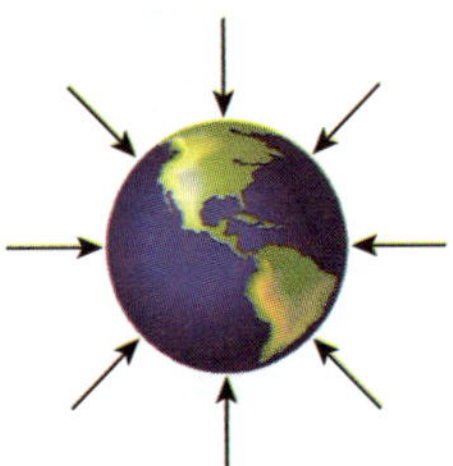

Figure 3.2 Gravitational field lines around the Earth

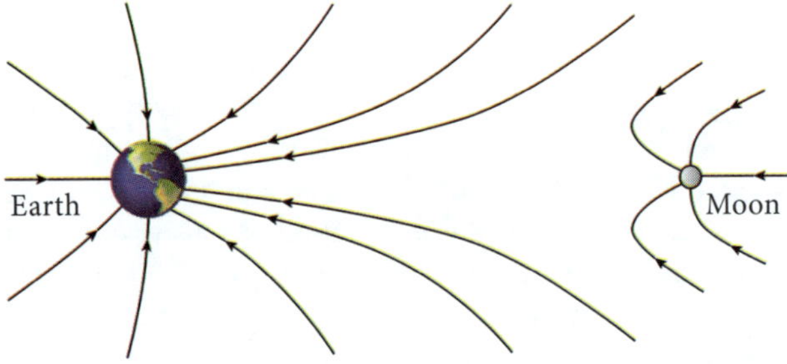

Figure 3.3 Gravitational field lines between the Earth and Moon

➔ Sometimes the points of equal gravitational potential in a field are joined together with lines called **equipotential lines**. These equipotential lines are always perpendicular to the field lines and hence would appear as concentric circles around the Earth, as shown in Figure 3.4. Note the strength of the field is reflected in the closeness of the equipotential lines. These concentric circles simply indicate that the gravitational field strength will be the same at every position that is a specific height above the Earth's surface.

equipotential lines: lines that join points that have the same potential energy in a field

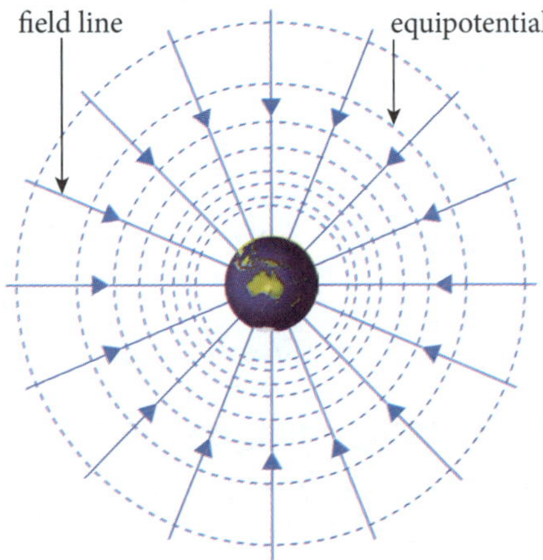

Figure 3.4 Equipotential gravitational lines around the Earth

➔ KEY QUESTIONS

1 **What factors affect the gravitational force of attraction between two bodies?**

2 **How is the gravitational field strength related to the acceleration due to gravity?**

3 **What does the spacing and direction of the gravitational field lines tell us about gravity in the region?**

Answers ➲ p. 48

EXAMPLE 1

The Earth has a mass of $M_E = 6.0 \times 10^{24}$ kg and a radius of $r_E = 6.371 \times 10^6$ m, and the Moon has a mass of $M_M = 7.35 \times 10^{22}$ kg and a radius of $r_M = 1.737 \times 10^6$ m. The average distance between the centre of the Moon and the centre of the Earth is 384 000 km. Use this information to answer the following questions.

a **Find the gravitational force exerted by the Earth on the Moon.**

b **Calculate the gravitational field strength on the surface of the Earth and on the surface of the Moon.**

c **Find the gravitational field strength midway between the Earth and the Moon.**

d **How much force would gravity exert on a 100 kg astronaut at the point midway between the Earth and the Moon?**

> Gravitational field strength is the force that a 1 kg object would experience if it was placed at the point; the gravitational field strength at a specific point in space is equal to the acceleration due to gravity at the point

Answer:

a Using Newton's law of gravitation:

$F = GM_E M_M / r^2$

$= \dfrac{(6.67 \times 10^{-11} \times 6.0 \times 10^{24} \times 7.35 \times 10^{22})}{(3.84 \times 10^8)^2}$

$F = 2 \times 10^{20}$ N

b The gravitational field strength is the gravitational force per kilogram. As gravity exerts a force of:

$F = GMm/r^2$ on a mass m

The force per kg will be $F/m = g = GM/r^2$

For the Earth:

$g = GM_E/(r_E)^2$

$= \dfrac{(6.67 \times 10^{-11} \times 6.0 \times 10^{24})}{(6.371 \times 10^6)^2}$

$= 9.86$ ms^{-2}

For the Moon:

$g = GM_M/(r_M)^2$

$= \dfrac{(6.67 \times 10^{-11} \times 7.35 \times 10^{22})}{(1.737 \times 10^6)^2}$

$= 1.62$ ms^{-2}

c We must find the vector sum of the force per unit mass towards the Earth and the force per unit mass towards the Moon, midway between the two bodies. That is, at a point, $r = \dfrac{3.84 \times 10^8}{2} = 1.92 \times 10^8$ m from the centre of each body.

The force of attraction per unit mass towards the Earth is:

$g = GM_E/r^2$

$= \dfrac{(6.67 \times 10^{-11} \times 6.0 \times 10^{24})}{(1.92 \times 10^8)^2}$

$= 1.086 \times 10^{-2}$ ms^{-2}

And towards the Moon is:

$g = GM_M/r^2$

$= \dfrac{(6.67 \times 10^{-11} \times 7.35 \times 10^{22})}{(1.92 \times 10^8)^2}$

$= 1.33 \times 10^{-4}$ ms^{-2}

Hence the field strength is:

$g = 1.086 \times 10^{-2} - 1.33 \times 10^{-4}$

$= 1.07 \times 10^{-2}$ ms^{-2} towards the Earth

d Force $= mg = 100 \times 1.07 \times 10^{-2} = 1.07$ N towards the Earth.

EXAMPLE 2

Travelling from the Earth to the Moon astronauts pass a point where the gravitational field strength is zero. Find the ratio of the distance from this point to the centre of the moon (r_1) and the distance from this point to the centre of the Earth (r_2).

> If the gravitational field strength at a point in space is zero, the net gravitational force on object placed at the point must also be zero

Answer:

As the net force acting on the astronauts at the point is zero, the force of attraction towards the Moon must be equal and opposite to the force of attraction towards the Earth at the point. Hence:

$GM_E/(r_2)^2 = GM_M/(r_1)^2$

Which simplifies to:

$r_1/r_2 = \sqrt{M_M/M_E}$

$= \sqrt{\dfrac{7.35 \times 10^{22}}{6.0 \times 10^{24}}}$

$= 0.11$

2 Orbital motion

» Students investigate the orbital motion of planets and artificial satellites when applying the relationships between the following quantities: gravitational force; centripetal force; centripetal acceleration; mass; orbital radius; orbital velocity; and orbital period.

- A satellite is on object that orbits a central body. We call manufactured satellites *artificial satellites*, and planets, asteroids, comets, etc. *natural satellites*.
- Many satellites and planets have circular or near circular orbits and undergo uniform circular motion. We can therefore describe and predict the motion of planets and satellites by applying the relationships for uniform circular motion, as derived in the previous chapter.
- This section deals only with satellites with circular orbits.
- The centripetal force holding satellites in their orbits is provided by the gravitational attraction of the satellite to the central body and hence:

> $F_c = GMm/r^2$
> And as $a_c = F_c/m = GM/r^2$

Note that the centripetal acceleration is independent of the mass of the satellite and is equal to the acceleration due to gravity at that distance from the central body.

- We can therefore think of a satellite as being in free fall towards the central body. The tangential velocity of the satellite is just fast enough for the satellite to remain at the same distance from the central body as it falls towards it. This concept is illustrated in Figure 3.5. If no force was applied the satellite would move in a straight line from its starting point to 1′, but because of the gravitational attraction it experiences, it instead ends up at point 1. Thus the satellite continually falls towards the central body but does not get closer to the body.

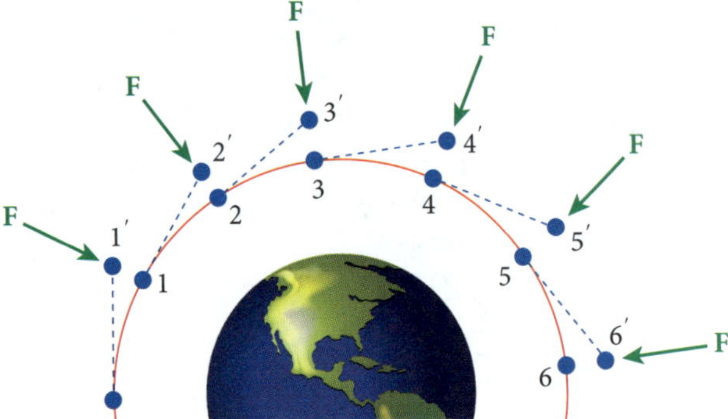

Figure 3.5 A satellite continually accelerates towards the central body but does not get closer to the central body

➔ You will recall from the previous chapter that for a body undergoing uniform circular motion the tangential velocity and period are related by:

$$v = 2\pi r/T$$

We also learnt that the centripetal force and acceleration were related to the radius and tangential velocity by:

$F_c = mv^2/r$ and $a_c = v^2/r$

These relationships can also be applied to satellite motion. We can derive other useful expressions by substituting the gravitational force for the centripetal force. For example:

$F_c = mv^2/r = GMm/r^2$

Simplifying and rearranging this expression gives the tangential velocity required to be in orbit at a specific orbital radius:

$$v = \sqrt{(GM/r)}$$

Note that the orbital velocity (v) is independent of the mass of the orbiting body and that the orbital velocity decreases with the orbital radius. That is, the further a satellite is from the central body, the slower it must be moving.

➔ We can derive an expression called the **satellite equation** by equating the orbital velocity in the equation above with an expression we derived for tangential velocity in Chapter 2:

$v = \sqrt{(GM/r)}$ and $v = 2\pi r/T$

Hence $2\pi r/T = \sqrt{(GM/r)}$

satellite equation: an equation that relates the period and orbital radius of an orbiting satellite

This can be written as:

$$r^3/T^2 = GM/4\pi^2$$

where r is the orbital radius
T is the orbital period and
M is the mass of the central body

➔ KEY QUESTION

4 **What force/forces operate on a satellite when it is in a stable orbit?**

Answers ➲ p. 48

EXAMPLE 3

The Earth is 150 million kilometres from the Sun and has a period of 365.25 days.

a **Find the orbital velocity of the Earth.**
b **Find the mass of the Sun.**
c **If the Earth was to split into two equally massive halves, how would its orbit change? Justify your answer.**

We can apply the relationships for uniform circular motion and the satellite equation to natural and artificial satellites in circular orbits

Answer:

a Changing to SI units and applying the circular motion equation that relates velocity and period:
$v = 2\pi r/T$
$= \dfrac{2\pi \times 150 \times 10^9}{(365.25 \times 24 \times 60 \times 60)}$
$= 2.99 \times 10^4\ \text{ms}^{-1} = 29.9\ \text{kms}^{-1}$

b Applying the satellite equation with the Sun as the central body and the Earth as its satellite:
$r^3/T^2 = GM/4\pi^2$
$M = (4\pi^2 r^3)/(GT^2)$
$= 4\pi^2 \times (150 \times 10^9)^3/(6.67 \times 10^{-11}) \times (365.25 \times 24 \times 60 \times 60)^2$
$= 2.0 \times 10^{30}$ kg
Note that we could also solve the problem by using $v = \sqrt{(GM/r)}$

c The orbital velocity is independent of the mass (as $v = \sqrt{(GM/r)}$) of the orbiting object (m). Hence the two halves would continue to orbit with the same period, tangential velocity and orbital radius as the Earth did before it split in half.

3 Planets and satellites

» Students predict quantitatively the orbital properties of planets and satellites in a variety of situations, including near the Earth and geostationary orbits, and relate these to their uses.

→ Artificial satellites have a wide range of applications. Different orbital paths (ground tracks) and orbital radii are employed for different applications. For example, satellites that need to pass over each point on the globe every day are placed in low Earth, polar orbits. Figure 3.6 shows how **low Earth orbit** satellites (LEOs) with polar orbits pass over a different section of the Earth each orbit as the Earth rotates below the satellite at a much slower rate than the satellite orbits.

low Earth orbits: Earth orbits with altitudes between 160 and 2000 km

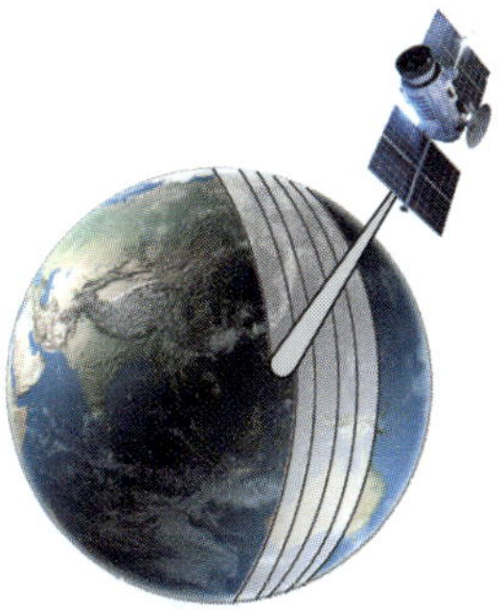

Figure 3.6 In a low Earth polar orbit a satellite passes over the poles and sweeps over a different section of the Earth's surface each orbit

→ Satellites are broadly classified in terms of their orbital radii, as shown in Table 1.1. These orbits are also shown in Figure 3.7. Note that Figure 3.7 is not drawn to scale as geostationary satellites must orbit at a height of 35 786 km, which is almost six Earth radii above the surface.

Table 1.1 Satellites classified in terms of orbital altitude

Type of satellite	Altitude (km)	Period (hours)	Tangential velocity (kms^{-1})
Low Earth orbit (LEO)	160 to 2000	1.5 to 2	7.8 to 6.9
Middle Earth orbit (MEO)	2000 < r < 357 86	2 to just below 24	6.9 to 3.0
Geosynchronous and geostationary orbit (GSO)	35 786	24	3.07

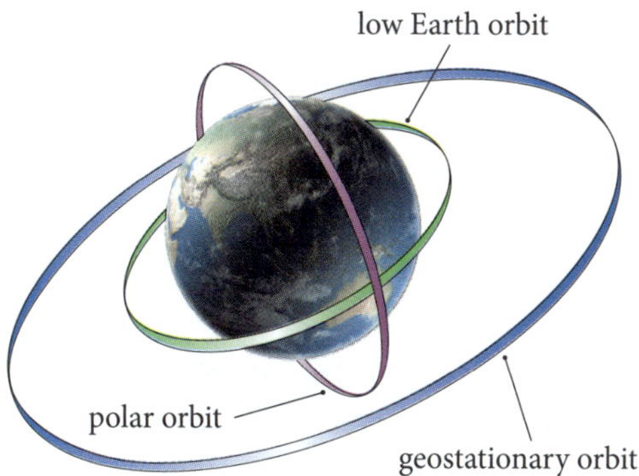

Figure 3.7 Examples of equatorial and polar satellite orbits

→ Low Earth orbit satellites are slowed slightly by drag from the thin upper atmosphere and consequently are called **degenerate orbits** because LEO satellites eventually spiral back into the atmosphere. Usually these short lifespan satellites burn up during re-entry. The lifetime of an LEO satellite depends on its altitude because the density of the atmosphere decreases with altitude. Without thrusters to counter drag, a satellite at an altitude of 160 km would undergo orbital decay in only a few days, but at an altitude of 900 km a satellite would remain in orbit for hundreds of years.

degenerate orbit: a low Earth orbit that decays due to atmospheric drag

→ **Geostationary** and geosynchronous orbits have a period of 24 hours, travel towards the east and hence move in synch with the Earth as it rotates. The difference between geostationary and geosynchronous orbits is that geostationary orbits are equatorial and hence geostationary satellites remain above one point on the equator as the Earth rotates. This makes geostationary satellites very useful for telecommunication applications, such as satellite TV. Geosynchronous satellites also have a period of 24 hours but unlike geostationary satellites they are placed in non-equatorial orbits. This causes them to move above and below the equator, tracing out a figure-8 path (ground track) above the Earth's surface as they orbit, as shown in Figure 3.8.

geostationary orbit: equatorial Earth orbit with a period of 24 hours

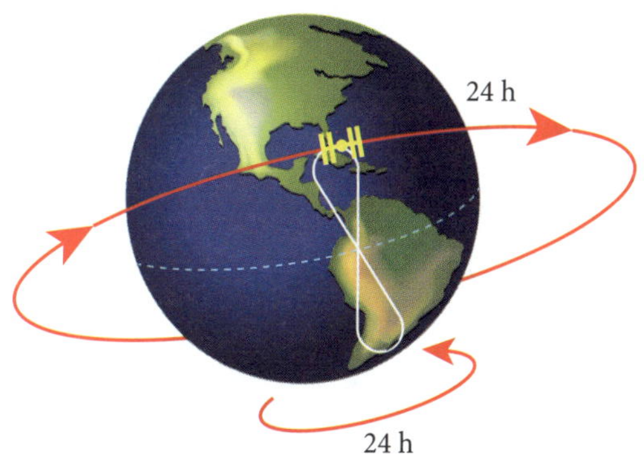

Figure 3.8 A geosynchronous satellite traces out a figure-8 path on the Earth's surface as it moves through its orbit

→ There are over 2000 artificial satellites orbiting the Earth. Satellites have important applications in a wide range of areas, including communications, navigation, scientific investigation and weather prediction. Table 1.2 provides more detail on the applications of satellites at different altitudes.

Table 1.2 Satellite applications

Type of orbit	Satellite applications
Low Earth	Remote sensing (measurement) of a wide range of parameters on the Earth, such as gravitational fields, ocean currents, temperature, icecap coverage, etc. (remote sensing), spying, space stations (ISS), space telescopes, weather prediction (polar orbits)
Medium	Navigation (global positioning system), communication, remote sensing
Geostationary	Communication (TV and satellite phones), weather prediction

➔ The orbital properties of satellites can be determined by applying the equations of uniform circular motion, the equation for orbital velocity and the satellite equation.

➔ KEY QUESTION

5 **What is the difference between a low Earth orbit satellite and a geostationary satellite?**

Answers ➲ p. 48

EXAMPLE 4

The International Space Station (ISS) orbits the Earth with a period of 92 minutes and 39 seconds. Given that the Earth has a mass of 6.0×10^{24} kg and a radius of 6371 km, find:

a **the orbital radius and altitude of the ISS.**

b **the tangential speed of the ISS.**

Remember to convert to SI units before substituting values into equations

Answer:

a First we convert the period to seconds:

$T = (92 \times 60) + 39 = 5559$ s

Using the satellite equation:

$r^3/T^2 = GM/4\pi^2$

$r = \sqrt[3]{GMT^2/4\pi^2}$

$= \sqrt[3]{\dfrac{6.67 \times 10^{-11} \times 6 \times 10^{24} \times 5559^2}{4\pi^2}}$

$= 6\,791\,571 \text{ m} = 6791.6 \text{ km}$

Altitude = orbital radius – Earth radius

$= 6791.6 - 6371 = 421$ km

b The tangential velocity can be determined using $v = 2\pi r/T$ or by using $v = \sqrt{(GM/r)}$

Using the first equation:

$v = \dfrac{2\pi(6\,791\,571)}{(5559)} = 7676 \text{ ms}^{-1} = 7.7 \text{ kms}^{-1}$

EXAMPLE 5

Two spherical satellites A and B have the same radius but satellite A has a mass of M while satellite B has a mass of $10M$. Both satellites are orbiting the Earth at an altitude of 200 km. Note that the Earth has a radius of 6371 km.

a **Find the initial period of the satellites when they are placed in orbit.**

b **Compare the atmospheric drag on each satellite. Justify your answer.**

c **Explain how the atmospheric drag will affect the lifetime of the satellites.**

Atmospheric drag, like air friction, depends on the size and velocity of the moving object

Answer:

a Orbital radius will be $r = 6371 + 200$

$= 6571 \text{ km} = 6.571 \times 10^6 \text{ m}$

Now applying the satellite equation:

$r^3/T^2 = GM/4\pi^2$

$T = \sqrt{4\pi^2 r^3/GM}$

$= \sqrt{\dfrac{4\pi^2(6.571 \times 10^6)^3}{6.67 \times 10^{-11} \times 6 \times 10^{24}}}$

$= 5290 \text{ s} = 88 \text{ min } 10 \text{ s}$

b The drag force depends on the size and shape of the satellite and its velocity. Because both satellites are the same size and shape and are travelling at the same speed, both satellites will experience the same drag force.

c The drag force will cause both satellites to slow down and eventually spiral back into the atmosphere. Satellite B will remain in orbit longer than satellite A because satellite A will be slowed down much more rapidly by the drag force than satellite B. The same drag force operates on both satellites but because satellite A is 1/10th as massive as satellite B, it will decelerate 10 times faster (as $a = F/m$).

4 Kepler's laws of planetary motion

» Students investigate the relationship of Kepler's laws of planetary motion to the forces acting on, and the total energy of, planets in circular and non-circular orbits using:

- $v = \dfrac{2\pi r}{T}$
- $\dfrac{r^3}{T^2} = \dfrac{GM}{4\pi^2}$

➔ After painstakingly analysing a huge number of astronomical observations early in the 17th century, Johannes Kepler proposed three laws of planetary motion. Kepler's laws were empirical laws that were based on observation rather than from some underlying theory. Fifty year later, Newton gave Kepler's laws a theoretical foundation by showing that Kepler's laws could be derived by applying the universal law of gravitation to planetary motion.

- Kepler's three laws of planetary motion are:

  First law: The orbit of a planet is an ellipse with the Sun at one of the two **foci**.

  Second law: An imaginary line joining a planet to the Sun will always sweep out an equal area during an equal time interval.

  Third law: The square of the orbital period of a planet (T^2) is proportional to the cube of the semi-major axis of the planet's orbit (r^3). (Note that to apply this law to circular orbits we replace the semi-major axis with the orbital radius.)

foci: one of two fixed points in an ellipse that are used to define the curve; an ellipse is the locus of the points formed when the sum of the distance from each focus to the point is a constant

- Let's look more closely at Kepler's laws. Kepler's first law states that the orbits of the planets are elliptical rather than circular and that the Sun is at one of the foci rather than at the centre of the ellipse. You will recall from mathematics that a circle is the round shape traced out by a moving point on a plane that is a specific distance (the radius) from a given point (the centre of the circle). An ellipse is the oval shape traced out by a point moving in a plane such that the sum of the distance from two given points (the two foci) is constant.
- To draw an ellipse, we could use two pins and a piece of string, as shown in Figure 3.9. By placing the two pins (foci) closer together the ellipse begins to look more like a circle in shape. Moving the foci further apart will make the ellipse less circular. We call this action *increasing the eccentricity of the ellipse*. Planetary orbits are almost circular but the orbital paths of comets around the Sun can be very eccentric. Note that the central body is located at one focus of the elliptical orbit and there is nothing at the other focus. The point in the orbit where a satellite is closest to the Earth is called the **perigee**, while the point where it is furthest from the Earth is called the **apogee**.

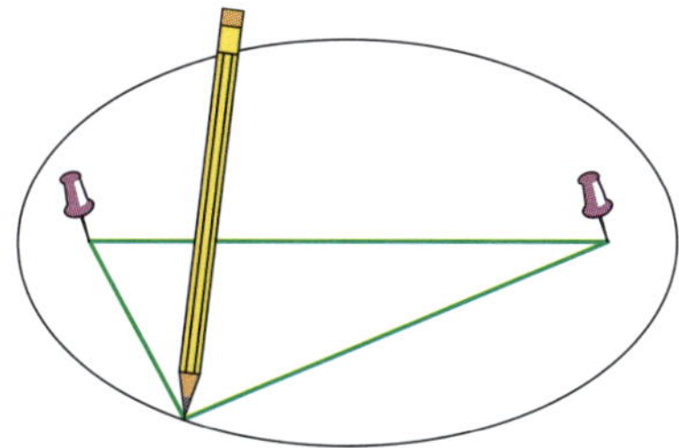

Figure 3.9 Using a pencil inside a loop of string placed around two pins (foci) to draw an ellipse; if this was an orbit, the central body would be at the position of one of the pins

perigee: the point of an elliptical Earth orbit that is closest to the Earth

apogee: the point of an elliptical Earth orbit that is furthest from the Earth

- Kepler's second law is illustrated in Figure 3.10. To sweep out equal areas in equal time intervals requires the planet to move faster when it is near the Sun and slower when it is further from the Sun. We have already seen that satellites in circular orbits must move faster when they are near the central body to remain in orbit (i.e. $v = \sqrt{(GM/r)}$).

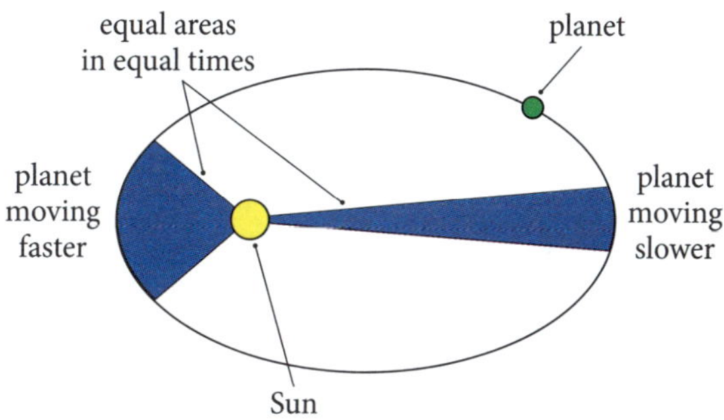

Figure 3.10 A line joining a planet to the Sun would sweep out equal areas in equal time intervals

- We can understand why the velocity changes by looking at the direction of the force of attraction on the orbiting body. In a circular orbit gravity always exerts a force on the orbiting body that is perpendicular to the velocity and hence the magnitude of the velocity remains constant. But in an elliptical orbit the force of attraction will not be perpendicular to the velocity and hence the force will have some component parallel to the velocity, which will change the magnitude of the velocity. When the orbiting object moves towards the Sun there will be a component of force of attraction on the body in the direction of the motion and this will increase its velocity. But when the orbiting body is moving away from the Sun this component of the force on the body will oppose its motion and reduce the object's velocity.

  Newton later showed that a force of attraction that was proportional to the inverse square of distance between the bodies would cause an orbiting body to change velocity at the rate Kepler's second law predicted.
- Kepler's third law states that the period is related to the length of the semi-major axis of the orbit. The line drawn through both foci of an ellipse at the greatest diameter of the ellipse is called the *major axis* and half this line is called the *semi-major axis*. We express the third law as:

  $r^3 \alpha T^2$ or as r^3/T^2 = a constant

If we use astronomical units (1 AU = distance between the Earth and the Sun) for the radius of the orbit and Earth years for the period, the constant = 1, which simplifies our calculations when dealing with heliocentric (Sun-centred) orbits.

Note also that for any two bodies (1 and 2) orbiting the same central body we can write:

$$\frac{r_1^3}{T_1^2} = \frac{r_2^3}{T_2^2}$$

To apply Kepler's law to circular orbits we simply use orbital radius (r) rather than the semi-major axis because for a circle the semi-major axis is the radius.

By applying his universal law of gravitation to planetary orbits, Newton could derive Kepler's third law and show that the constant of proportionality in Kepler's third law was:

r^3/T^2= constant = $GM/4\pi^2$

This is the satellite equation derived previously in the chapter.

➔ Provided a satellite's orbit is high enough above the atmosphere of the central body not to be affected by atmospheric drag, the total energy of the satellite will remain constant. We saw in the last chapter that the energy of an object undergoing uniform circular motion remains constant because the magnitude of its velocity remains constant. The total energy of a body in an elliptical orbit also remains constant but in this case energy moves between kinetic (K) and gravitational potential (U) as it is a free-falling body in a gravitational field. We can express this relationship as:

Total energy = E_{total} = a constant = $K + U$

As the orbiting body moves towards the central body its velocity and hence its kinetic energy increases but its gravitational potential energy decreases at the same rate to ensure that the total energy remains constant. When the orbiting body moves away from the central body its velocity and kinetic energy decrease and its gravitational energy increases again at a rate to ensure the total energy of the orbiting body remains constant.

➔ KEY QUESTION

6 **How does the energy of a satellite in an elliptical orbit change throughout the orbit?**

Answers ➲ p. 48

EXAMPLE 6

Consider the elliptical orbit shown in Figure 3.11.

a **Compare qualitatively the total energy, kinetic energy and potential energy at the points a to e. Justify your answer.**

b **Compare qualitatively the velocity of the satellite at positions a and d to the velocity of a satellite in a circular orbit at the same altitude above the planet. Justify your answer.**

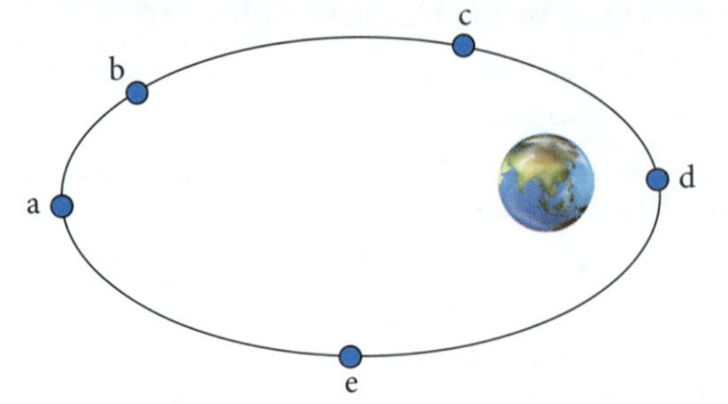

Figure 3.11 The orbit of a satellite around the Earth

The velocity of a satellite is greater when the satellite is closer to the central body

Answer:

a Because the satellite is a free-falling body its total energy remains constant and hence the total energy of the satellite at each point in the orbit will be the same. Because the velocity is greatest when the satellite is closest to the central body, kinetic energy (K) at the points will be related by $K_d > K_c > K_e > K_b > K_a$. The gravitational potential energy (U) is lower for a satellite closer to the planet as it would require less energy to raise it to that point and therefore the gravitational potential energy at the positions is related by $U_a > U_b > U_e > U_c > U_d$.

b At position d the tangential velocity is greater than the tangential velocity of a satellite in a circular orbit as the velocity causes the satellite to move to a greater altitude (rather than remain at the same altitude). At position a, the tangential velocity is lower than the velocity of a satellite in a circular orbit at the same altitude because the velocity is not great enough to prevent the satellite falling to a lower altitude after it passes position a.

EXAMPLE 7

Table 3.3 shows the orbital radius and period of four of Jupiter's moons. Calculate the missing values in the table.

Table 3.3 Orbital data for the moons of Jupiter

Moon	Orbital radius (km)	Period (Earth days)
Io	421 700	1.77
Europa		3.55
Ganymede	1 070 412	
Callisto		16.69

The ratio r^3/T^2 is a constant for all the satellites orbiting the same central body

Answer:

We can use combination of units with Kepler's law of periods as long as we use the same units on both sides. In this example we will use kilometres and Earth days.

Applying Kepler's law of periods to Io (2) and Europa (1):

$$\frac{r_1^3}{T_1^2} = \frac{r_2^3}{T_2^2}$$

$$r_1^3 = \frac{T_1^2 r_2^3}{T_2^2} = \frac{(3.55)^2 \times (4.217 \times 10^5)^3}{(1.77)^2} \text{ km}^3$$

$r_1^3 = 3.017 \times 10^{17}$ and hence $r = 6.71 \times 10^5$ km

Applying Kepler's law of periods to Io (2) and Callisto (1):

$$\frac{r_1^3}{T_1^2} = \frac{r_2^3}{T_2^2}$$

$$r_1^3 = \frac{T_1^2 r_2^3}{T_2^2} = \frac{(16.7)^2 \times (4.217 \times 10^5)^3}{1.77^2} = 1.06 \times 10^{28} \text{ km}^3$$

$r_1^3 = 6.676 \times 10^{28}$ and hence $r = 1.88 \times 10^6$ km

Finally, applying Kepler's law of periods to Io (2) and Ganymede (1):

$$\frac{r_1^3}{T_1^2} = \frac{r_2^3}{T_2^2}$$

$$T_1^2 = \frac{r_1^3 T_2^2}{r_2^3} = \frac{(1.07 \times 10^6)^3 (1.77)^2}{(4.217 \times 10^5)^3}$$

$T_1^2 = 51.178$ and hence $T_1 = 7.15$ days

FIRSTHAND INVESTIGATION

Drawing ellipses

Students can draw a series of ellipses by placing a piece of a piece of paper over a pinboard and inserting two pins into the board to act as foci, as shown previously in Figure 3.9. By placing a loop of string over the nails a pencil can be used to trace out an ellipse.

A family of ellipses with different eccentricities can be produced by progressively bringing one of the pins (foci) closer to the other.

Students with appropriate skills may also wish to simulate this process by coding a computer to draw ellipses.

5 Energy in radial gravitational fields

» Students derive quantitatively and apply the concepts of gravitational force and gravitational potential energy in radial gravitational fields to a variety of situations, including but not limited to:

- the concept of escape velocity $v_{esc} = \sqrt{\frac{2Gm}{r}}$
- total potential energy of a planet or satellite in its orbit $U = \frac{-GMm}{r}$
- total energy of a planet or satellite in its orbit $U + K = \frac{-GMm}{2r}$
- energy changes that occur when satellites move between orbits
- Kepler's laws of planetary motion.

Gravitational potential energy

- In the Year 11 course we saw that the change in potential energy of a mass that was raised a distance h near the Earth's surface was given by $\Delta U = mg\Delta h$. This equation could only be applied to the change in gravitational potential energy near the surface of the Earth because we assumed that the force on the mass (mg) was constant when we derived the equation. This is a good approximation near the surface of a planet but will overestimate the change in potential energy if the height through which the mass moves becomes large because the gravitational force on the mass (i.e. the weight) decreases with height.
- To obtain the correct change in gravitational potential we must accommodate this changing force in our calculations. To do this we first define a position an infinite distance from the planet as having zero gravitational potential. We then calculate how much work is done to take a mass from this point to a point a distance r from the planet. This requires a mathematical technique called *integration*, which is beyond the scope of this course but results in the following expression for the gravitational potential energy:

$$U = -GMm/r$$

where U is the gravitational potential energy
G is the universal constant of gravitation
M is the mass of the central body
m is the mass of the body under consideration and
r is the distance from the centre of the planet

Note that gravitational potential energy is inversely proportional to the distance from the centre of the planet and is negative. The negative sign comes from our definition as the work is done against gravity to bring the mass to the point r. To move from r to infinity we would have to do a positive amount of work but for the mass to move from infinity to r, gravity would do work on the mass. We can think of this as doing negative work as the mass has lost gravitational potential energy as it falls.

Another way to understand this is to think that a negative potential energy means the mass is trapped by the gravitational field of the planet and an amount of work equal to the potential energy must be done to free the mass. It is a bit like a mass placed in a hole in the ground. If we took the ground level as our zero of gravitational potential energy, the mass in the hole would have a negative potential energy because it would require work to be done against gravity to raise the mass to ground level.

→ KEY QUESTION

7 **Why is gravitational potential energy always negative?**

Answers ➲ p. 48

EXAMPLE 8

Figure 3.12 shows the gravitational potential energy as a function of distance (r) between the Earth and the Moon. The red line shows the potential energy of a spacecraft with respect to the Earth, the black line shows the potential energy with respect to the Moon, and the blue line shows the combined potential due to the Earth and the Moon.

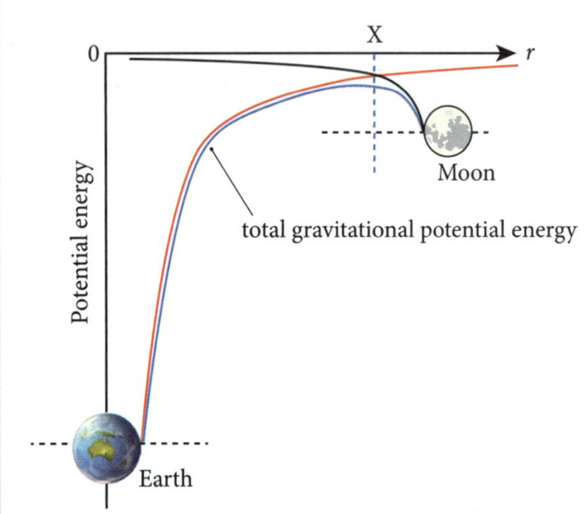

Figure 3.12 Gravitational potential energy between the Earth and the Moon

a **Would it require less energy to fly from the Earth to the Moon or from the Moon to the Earth? Justify your answer with reference to Figure 3.12.**

b **The three-stage *Saturn V* rockets used to launch the *Apollo* lunar missions shut down and were jettisoned shortly after the craft left the Earth's atmosphere. Explain qualitatively how the velocity of the *Apollo* spacecraft would change as it moved from the Earth to the Moon and explain the significance of the position marked X on Figure 3.12.**

c **On which sections of the journey would the astronauts experience weightlessness? Justify your answer.**

> Work must be done by a rocket against the gravitational field to move away from a planet but work is done by the gravitational field on the rocket when it moves closer to a planet

Answer:

a It would take more energy to fly to the Moon because, as shown in Figure 3.12, the Earth has a much lower gravitational potential energy than the Moon. Thus a spacecraft would have to do a great deal more work to fly from the Earth to the Moon than it would have to do in flying from the Moon to the Earth.

b The velocity would increase while the rocket engines were operating but once switched off, the velocity would decrease as the spacecraft moved further from the Earth because the spacecraft's kinetic energy is continually being converted into gravitational potential energy. At the point marked X the velocity would begin to increase because at this point the force of attraction towards the Moon becomes greater than the retarding force towards the Earth that was slowing the spacecraft.

c The astronauts will be weightless as soon as the rocket engines are switched off and will remain weightless for the entire journey because they and their spacecraft will be accelerated at the same rate as they are moving freely in a gravitational field.

Total energy of a satellite in orbit

- → Because a satellite is trapped by the Earth's gravitational field we would expect it to have a total energy that was negative. This is because work would have to be done on the satellite (energy added) to free the satellite from the Earth's gravitational field.
- → The total energy of a satellite in orbit is the sum of its kinetic energy and its gravitational potential energy: $E_{total} = U + K$
- → You will recall that:
  Kinetic energy = $K = \frac{1}{2}mv^2$
  And that a satellite in a circular orbit has a velocity of:
  $v = \sqrt{(GM/r)}$
  Combining these two equations and simplifying gives the kinetic energy of a satellite:

$$K = GMm/2r$$

- You will also recall that the gravitational potential energy of a satellite is:
  $U = -GMm/r$
  The total energy of a satellite is therefore:
  $E_{total} = U + K = -GMm/r + GMm/2r$
  Which simplifies to:

  $$E_{total} = U + K = -GMm/2r$$

  where G is the universal constant of gravity
  M the mass of the central body
  m the mass of the satellite and
  r is the orbital radius

  Note that the total energy is a negative quantity as expected and that the total energy of a satellite would increase (become a smaller negative number) if it was moved to a higher orbit.
- To place a satellite in orbit we must do work on the satellite to raise it to the altitude required and to give it enough tangential velocity to remain in orbit at that altitude.
- To reduce the energy required to place artificial satellites in Earth orbit, rocket engineers make use of the Earth's orbital velocity by launching towards the east from launch sites near the equator. As the Earth's rotation causes it to move at 460 ms^{-1} towards the east at the equator, satellites launched at the equator will have an initial velocity towards the east of 460 ms^{-1} and will therefore require less work to be done to accelerate the satellite to orbital velocity.
- The Earth's orbital velocity (30 ms^{-1}) is used when launching space probes into Sun-centred orbits to reach other planets. By launching with the Earth's orbital motion, the probe leaves the Earth with a velocity that is too large to remain in a circular Earth-orbit and the probe would move into an elliptical orbit with the Sun at the near focus. When the probe reached the other side of the Sun from its launch point it would be much further from the Sun than the Earth's orbital path and could intercept an outer planet (see Figure 3.13). By launching in the opposite direction to the Earth's orbital motion the probe would be going too slow to remain in an Earth orbit and consequently would move into an elliptical orbit with the Sun at the far focus. This would enable the probe to intercept an inner planet.

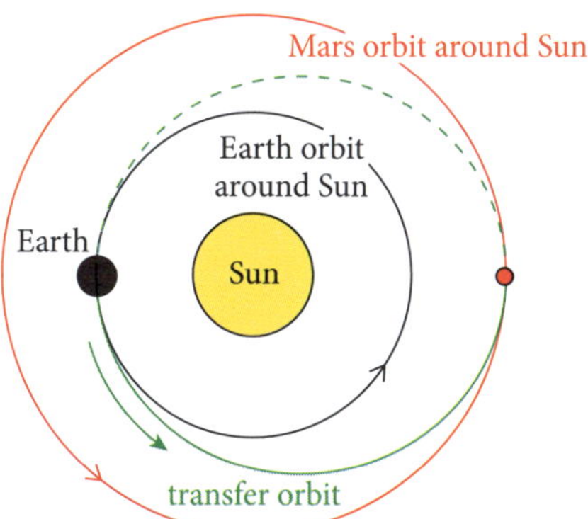

Figure 3.13 Launching with the Earth's orbital motion to reach Mars

→ KEY QUESTIONS

8 Why is the total energy of a satellite always negative?

9 Why are satellites generally launched near the equator?

10 How is the Earth's orbital velocity used to assist spacecraft to reach other planets?

Answers ⊃ p. 48

EXAMPLE 9

All geostationary satellites have equatorial orbits and have an orbital radius of 42 164 km. Note also that the Earth has a mass of 6.0×10^{24} kg and a radius of 6371 km.

a Calculate the kinetic energy, potential energy and total energy of a 500 kg geostationary satellite.

b How much work would have to be done to lift this satellite from the Earth's surface to this orbital radius?

c Calculate the total energy required to lift the satellite into orbit from the Earth's surface and accelerate it to orbital velocity. (You may assume the initial velocity is zero.)

The work done on a satellite is equal to the change in the energy of the satellite

Answer:

a The gravitational potential energy is given by:

$$U = -GMm/r = \frac{-(6.67 \times 10^{-11} \times 6.0 \times 10^{24} \times 500)}{4.2164 \times 10^{7}}$$
$$= -4.75 \times 10^{9}\text{ J}$$

The kinetic energy is given by:

$$K = GMm/2r = -U/2 = \frac{-(-4.7 \times 10^{9})}{2} = 2.37 \times 10^{9}\text{ J}$$

Total energy is given by:

$E_{total} = -GMm/2r = -K = -2.37 \times 10^{9}$ J

b The work done to lift the satellite to the correct altitude will be equal to the change in gravitational potential energy.

The potential energy on the Earth's surface is given by:

$$U_{Earth} = -GMm/r = \frac{-(6.67 \times 10^{-11} \times 6.0 \times 10^{24} \times 500)}{6.371 \times 10^{6}}$$
$$= -3.14 \times 10^{10}\text{ J}$$

The change in potential energy to raise the satellite will be:

Work done = $\Delta U = U_{final} - U_{initial} = U_{orbit} - U_{Earth}$

$= -4.75 \times 10^9 - (-3.14 \times 10^{10})$

$= 2.665 \times 10^{10}$ J

c The total work done will be the work to lift the satellite into orbit (calculated in part **b**) plus the work to give the satellite the required kinetic energy (calculated in part **a**):

Work = $\Delta U + \Delta K$

Work = $2.665 \times 10^{10} + 2.37 \times 10^9 = 2.902 \times 10^{10}$ J

Escape velocity

➔ We saw earlier that the magnitude of the gravitational potential energy of an object on the surface of a planet is equal to the energy required to free the object from the gravitational attraction of the planet. If we neglect air friction an object with enough kinetic energy would be able to escape from the planet totally. An object thrown upwards with this **escape velocity** or with a higher velocity would never fall back to the planet. The initial kinetic energy required to escape from a planet will be equal to the work done against gravity to reach the position where the gravitational potential energy is zero. That is:

$K = \frac{1}{2}mv^2 = |U| = GMm/r$

escape velocity: the minimum velocity with which an object must be projected upwards from the surface of a planet and not return to the planet; that is, the minimum velocity to escape the gravitational influence of the planet

➔ Rearranging this equation gives us the escape velocity from the planet:

$$v_{esc} = \sqrt{(2GM/r)}$$

where v_{esc} is the minimum velocity at the surface of a planet (neglecting air friction) required to escape from the gravitational attraction of a planet
G is the universal constant of gravity
M is the mass of the planet and
r is the radius of the planet

➔ Note that the escape velocity is independent of the mass of the body being ejected from the planet.

EXAMPLE 10

The Earth has a mass of 6.0×10^{24} kg and a radius of 6371 km, and the Moon has a mass of 7.35×10^{22} kg and a radius of 1737 km.

a Find the escape velocity for the Earth and for the Moon.

b If a planet had 18 times the mass and twice the radius of the Earth how would the escape velocity on the planet compare to the escape velocity on the Earth?

Recall that escape velocity is proportional to the square root of the mass of the planet and inversely proportional to the square root of the radius of the planet

Answer:

a The escape velocity is given by:

$v_{esc} = \sqrt{(2GM/r)}$

For the Earth:

$$v_{esc} = \sqrt{\frac{(2 \times 6.67 \times 10^{-11} \times 6.0 \times 10^{24})}{(6.371 \times 10^6)}}$$

$= 11.2$ kms^{-1}

For the Moon:

$$v_{esc} = \sqrt{\frac{(2 \times 6.67 \times 10^{-11} \times 7.35 \times 10^{22})}{(1.737 \times 10^6)}}$$

$= 2.38$ kms^{-1}

b Escape velocity is proportional to the square root of the mass divided by the radius of the planet and hence for a planet 18 times more massive and with twice the radius, the escape velocity will be related to the escape velocity from the Earth (v_E):

$$v_{esc} = v_E\sqrt{\frac{18}{2}} = 3v_E$$

Therefore the escape velocity would be three times greater from the planet than from the Earth.

Changing orbits

➔ If a satellite in a stable orbit uses its rocket engine to increase its velocity we say it has performed a **posigrade burn**. The increase in velocity caused by a posigrade burn moves the satellite into a new orbit because it will be moving too fast to remain in its original orbit. As shown in Figure 3.14, the satellite will still pass through the point each orbit where the initial posigrade burn occurred (the perigee in this case) but will be further from the Earth at all other points in the orbit.

posigrade burn: a rocket thrust that increases the velocity and kinetic energy of a spacecraft

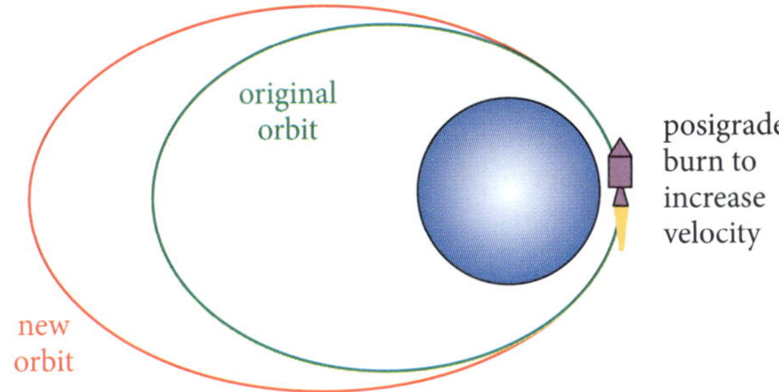

Figure 3.14 Effect of performing a posigrade burn on the orbit of a satellite

→ If a satellite applies its rocket engines in reverse to slow the satellite we say the satellite has performed a **retrograde burn**. After a retrograde burn the satellite would be travelling too slowly to remain in the original orbit and would enter a new orbit that would be lower than the original orbit at all points except the position at which the burn occurred. Again, the satellite will return each orbit to the position where the initial retrograde burn occurred.

retrograde burn: a rocket thrust that decreases the velocity and kinetic energy of a spacecraft

→ If an Earth satellite in a circular orbit performed a posigrade burn, the satellite would move into an elliptical orbit with the Earth at the near focus. In contrast a retrograde burn would move the satellite into an elliptical orbit with the Earth at the far focus.

→ To move from an orbit to a higher orbit, two short rocket burns are required. As shown in Figure 3.16 the first short burn moves the satellite into an elliptical orbit, called a **transfer orbit**, and the second short burn moves the satellite from the transfer orbit into the new circular orbit. The most energy-efficient way to do this is to perform the first burn at the perigee and the second burn at the apogee of the transfer orbit, as shown in Figure 3.15. This type of transfer orbit is called **Hohmann transfer orbit** and is the most energy efficient way to change from one orbit to another.

transfer orbit: the intermediate orbit taken by a spacecraft moving from one orbit to higher or lower orbit
Hohmann transfer orbit: most energy-efficient transfer orbit for moving from one circular orbit to another circular orbit; the apogee and perigee of a Hohmann orbit intersects the two circular orbits

→ With reference to Figure 3.15, a more energetic posigrade burn at the perigee could have been used to change the lower orbit to a higher circular orbit.

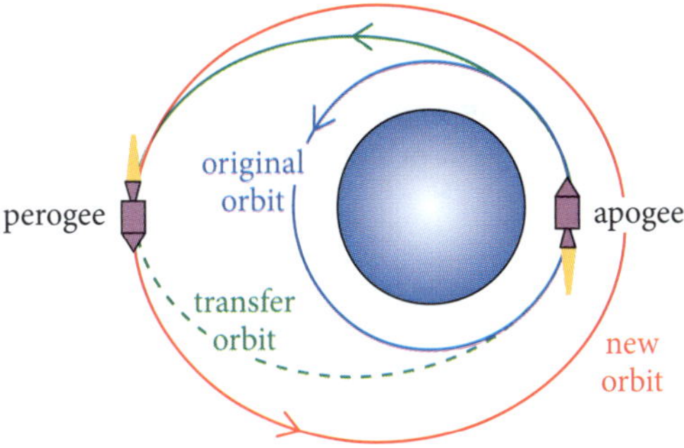

Figure 3.15 Transferring a satellite to a higher orbit

→ The total energy of the satellite shown in Figure 3.15 increases (becomes a smaller negative number) when it moves to the high orbit because it would require less work to remove the satellite at higher altitude from the Earth's gravitation pull. Because satellites move more slowly in higher altitude orbits, the kinetic energy must decrease so the increase in total energy is due to a relatively large increase in gravitational potential energy. At first it seems strange that two posigrade burns would result in a lower final kinetic energy but we must remember that as the satellite moves further from the Earth its gravitational potential increases at the expense of its kinetic energy and it slows down.

→ To move a satellite into a lower altitude orbit, two short retrograde burns are used. The first moves the satellite into an elliptical transfer orbit with an apogee at the altitude of the new orbit and the second burn moves the satellite from the transfer orbit to the new orbit at the lower altitude. In this case the net energy decreases (becomes a larger negative number) because the satellite is more tightly bound to the Earth and will require more work to be done to free it from the Earth. The potential energy decreases as the satellite is closer to the Earth but its kinetic energy increases as the satellite must have a higher orbital velocity in the low altitude orbit. Even though the retrograde burns slowed the satellite twice, its final velocity is greater than its initial velocity because its potential energy is converted to kinetic energy as it moves closer to the Earth, while changing from the higher to the lower orbit.

→ KEY QUESTIONS

11 **How is gravitational energy related to escape velocity?**

12 **How can a satellite move from one circular orbit to another circular orbit?**

Answers ➲ p. 48

EXAMPLE 12

The satellite shown in Figure 3.16 uses its thrusters to move from a low Earth orbit (LEO) to a geostationary orbit.

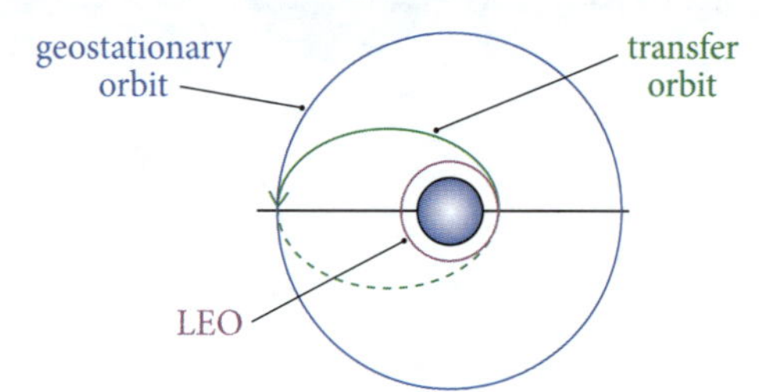

Figure 3.16 A satellite moving from a low Earth orbit to a geostationary orbit

a Explain how the rocket engines on the satellite could be used to move the satellite from the LEO to the geostationary orbit.

b Qualitatively compare the total energy, potential energy, kinetic energy and acceleration of the satellite when it is in the low Earth orbit with the total energy, potential energy, kinetic energy and acceleration of the satellite when it is in the geostationary orbit.

Changing the speed of a satellite in orbit will change the orbit of the satellite

Answer:

a The satellite would initiate a short posigrade burn to increase its velocity, which would change its circular LEO to an elliptical transfer orbit. When it reaches the required altitude, it would initiate another posigrade burn to increase its velocity to move it from the elliptical transfer orbit to the new higher-altitude circular orbit.

b The total energy and potential energy of the satellite will both be greater (smaller negative numbers) when the satellite is in the geostationary orbit, while the kinetic energy and centripetal acceleration of the satellite will be greater when the satellite is in the low Earth orbit. Note that the total energy increases because the rocket engines do work on the rocket during the two posigrade burns.

CHAPTER SYLLABUS CHECKLIST

Are you able to answer these questions from the syllabus for this chapter? Tick each question as you go through the checklist if you are able to answer it. If you cannot answer a question, turn to the relevant page in the study guide to find the answer. For NESA key word meanings, go to www.educationstandards.nsw.edu.au and search 'key words'.

	FOR A COMPLETE UNDERSTANDING OF THIS TOPIC:	PAGE NO.	✓
1	Can I calculate the force of gravity between two objects?	31	
2	Can I draw gravitational fields around a planet and between objects?	32	
3	Can I define *gravitational field strength*, relate it to acceleration due to gravity and describe the factors that affect it?	31–32	
4	Can calculate the gravitational field strength at the surface and at different heights above the surface of a planet?	32	
5	Can I calculate the centripetal force on a satellite and the centripetal acceleration of a satellite?	33	
6	Can I relate the velocity, period and orbital radius of a satellite?	34	
7	Can I use gravitational force and orbital velocity to account for the motion of a satellite?	34	
8	Can I determine the orbital velocity of a satellite with a specific orbital radius?	34	
9	Can I compare LEO satellites with geostationary satellites?	35–36	
10	Can I recall Kepler's three laws of planetary motion and use Newton's law of gravity to explain Kepler's laws?	37–38	
11	Can I derive and use the satellite equation to quantitatively relate the orbital radius of a satellite to its period?	38	
12	Can I explain, in terms of the work done, why the gravitational potential energy is a negative quantity but the change in gravitational potential may be positive or negative?	39–40	
13	Can I define and calculate the gravitational potential energy of an object on the surface and at various heights above the surface of a planet?	39–40	
14	Can I calculate the gravitational potential energy, the kinetic energy and the total energy of a satellite?	40–41	
15	Can I explain how the Earth's rotation is used to reduce the energy required to place satellites in orbit and how the Earth's orbital motion is used to reduce the energy needed for probes to reach other planets?	41	
16	Can I explain how and why the kinetic energy and gravitational potential energy change as a satellite moves in an elliptical orbit?	42–43	
17	Can I calculate the escape velocity from a planet and explain how escape velocity is related to an object's initial potential energy?	42	
18	Can I explain how a satellite can move between elliptical and circular orbits?	42–43	
19	Can I explain how a satellite can be efficiently moved from one circular orbit to another circular orbit with a different orbital radius?	43	

HSC EXAM-TYPE QUESTIONS

Objective-response questions (1 mark each)

1 **The gravitational field strength on the surface of a planet X is 10 Nkg^{-1}. What would the gravitational strength be on the surface of a planet that was four times more massive but had half the radius of planet X?**

A 160 N
B 80 ms^{-2}
C 160 ms^{-2}
D 20 Nkg^{-1}

2 **A planet exerts a force of attraction on a satellite. Why does this force not cause the satellite to fall to the surface of the planet?**

A The satellite continually falls towards the planet but it has a tangential velocity that prevents it getting closer to the planet.
B A centripetal force opposes the gravitational attraction and hence there is no net force on the satellite.
C Satellites are so far from the planet that the force of gravity on the satellite is negligible.
D The tangential velocity opposes the gravitational attraction, resulting in a net force of zero on the satellite.

3 **Figure 3.17 shows a satellite moving from a low altitude circular orbit *A* to a higher altitude circular orbit *B*.**

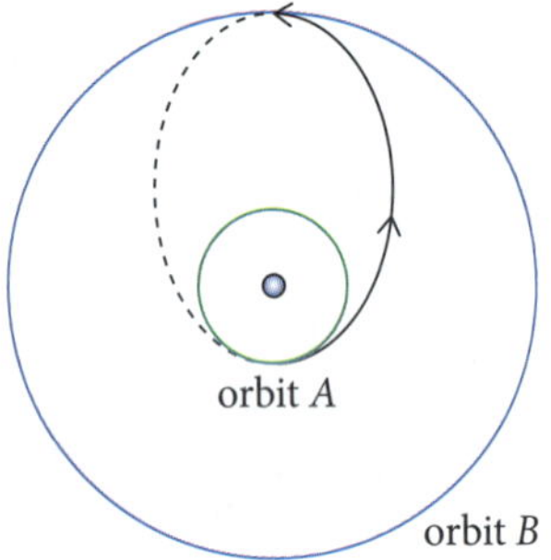

Figure 3.17 Satellite moving from one circular orbit to another circular orbit

Which of the following statements best describes how the energy of the satellite changes when it moves between orbits?

A The kinetic energy decreases and remains positive. The potential energy and total energy increase but both remain negative.
B The kinetic energy decreases and remains positive. The potential energy and total energy increase and both remain positive.
C The kinetic energy and the potential energy increase, producing a net increase in the total energy of the satellite.
D The potential energy increases, but this is offset by a large decrease in the kinetic energy resulting in a small decrease in the total energy.

4 **A satellite is in a circular geostationary orbit around the Earth when it initiates a short retrograde rocket burn and slows its velocity a little. How will the orbit be changed by this rocket burn?**

A The satellite will move into an elliptical orbit with the Earth at the focus closest to the burn but will still pass through the point where the burn occurred each orbit.
B The satellite will move into an elliptical orbit with the Earth at the focus furthest from the burn but will still pass through the point where the burn occurred each orbit.
C The satellite will spiral slowly towards the Earth.
D The satellite will move into a lower circular orbit.

5 **The rocket engine that propelled the *Apollo* astronauts to the Moon operated for only the first 20 minutes of the mission. How did the total energy and velocity of the craft change from the time the rocket engine was switched off until it approached Moon orbit?**

A The velocity and total energy remained constant.
B The velocity and energy decreased until the probe was near the Moon and then both began to increase.
C The total energy of the probe remained constant but the velocity decreased until the probe was near the Moon and then began to increase.
D The velocity remained constant throughout the flight but the energy increased until it reached a maximum and began to decrease as the craft neared Moon orbit.

Extended-response questions

6 **Newton used a thought experiment to explain satellite motion. He imagined a cannon firing a cannonball horizontally from the top of a mountain that was higher than the atmosphere. Some of the possible trajectories of cannonballs fired with different initial velocities are shown in Figure 3.18.**

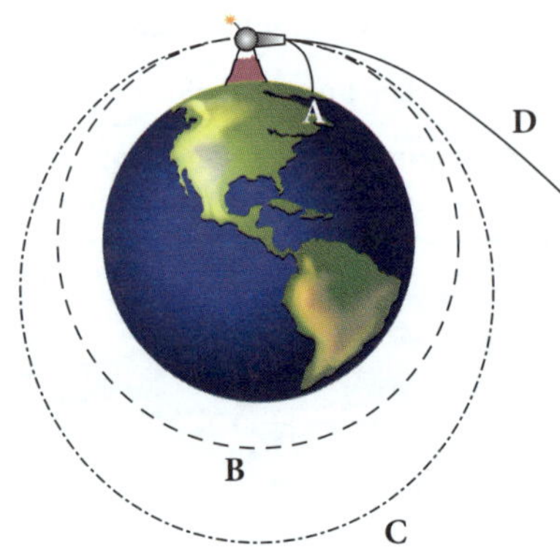

Figure 3.18 Newton's thought experiment on satellite orbits

a. Use Newton's law of gravity to derive an equation that would enable you to find the velocity of the cannonball if it followed a circular orbit (i.e. orbit B). (2 marks)

b. Compare the initial velocity of the orbits A and C to the initial velocity of the cannonball B that follows a circular orbit. (2 marks)

c. Given that trajectory D is not an orbital path, what can we conclude about the initial velocity of the cannonball in this case? (1 mark)

d. Which of the cannonballs has the greatest total energy? Justify your answer. (2 marks)

7 The Earth has a mass of 6.0×10^{24} kg and it is 384 400 km from the Moon, which has a mass of 7.35×10^{22} kg. Note also that the Earth has a radius of 6371 km and the Moon has a radius of 1737 km.

a. Calculate the orbital velocity and centripetal acceleration of the Moon. (2 marks)

b. Determine the period of the Moon's orbit. (1 mark)

c. Explain how the Moon's orbit would change if it was hit by a meteorite that decreased the Moon's orbital velocity to zero. Justify your answer. (2 marks)

8 Consider the two small moons orbiting the large planet shown in Figure 3.19. Note that Moon A has three times the mass and twice the orbital radius of Moon B. The moons are so far apart we may ignore the gravitational force between them.

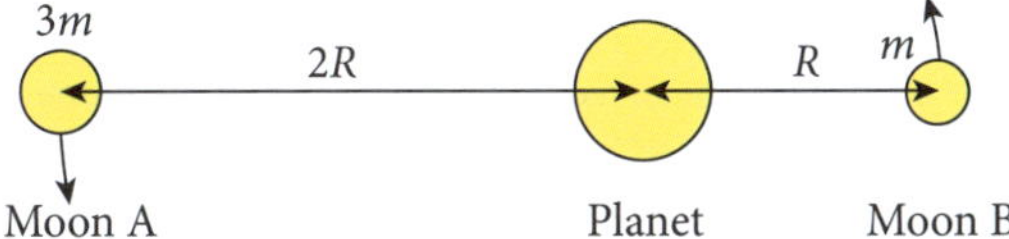

Figure 3.19 Two moons orbiting a planet

a. Determine the ratio of the gravitational potential energy of Moon A to the gravitational potential of Moon B. (1 mark)

b. Determine the ratio of the centripetal force on Moon A to the centripetal force on Moon B. (1 mark)

c. Sketch a diagram to show the most energy-efficient path for a spacecraft to travel from Moon B to Moon A. (2 marks)

d. In the time it takes Moon A to complete one orbit, how many orbits would Moon B have completed? (2 marks)

9 Relate the work done and energy changes involved in moving a spacecraft from the surface of the Earth through the atmosphere to a circular geostationary orbit around the Earth. You may use equations in your discussion but no calculations are required. (5 marks)

10 A satellite has a mass of 320 kg and orbits with a radius of 20 000 km around the Earth, which has a mass of 6.0×10^{24} kg.

a. Find the kinetic and potential energy of this satellite. (2 marks)

b. Determine the tangential velocity of the satellite. (1 mark)

c. What minimum change in velocity would be required for this satellite to leave its geostationary orbit and escape the Earth's gravitational pull entirely? (2 marks)

ANSWERS

KEY QUESTIONS

Key questions ⊃ p. 32

1 Bodies attract one another with a force of gravity that is proportional to the product of the bodies' masses and inversely proportional to the square of the distance between the bodies.

2 The gravitational field strength at any point in space is equal to the acceleration due to gravity at that point.

3 The direction of the gravitational field is the direction of the force that a mass would experience if placed in the field at that point. The proximity of the lines indicates the relative strength of the field.

Key question ⊃ p. 34

4 The only force operating on a satellite is gravity. This gravitational force acts towards the central body and provides the centripetal force that keeps the satellite in its orbit.

Key question ⊃ p. 36

5 A low Earth orbit has an altitude between 160 km and 2000 km, a short period and, for lower Earth orbits, a specific lifetime due to its interaction with the atmosphere. Low Earth orbits can also have any orientation from polar to equatorial. Geostationary satellites have equatorial orbits, are slower moving, have a period of 24 hours and all orbit at an altitude of approximately 36 000 km.

Key question ⊃ p. 38

6 The total energy remains constant and is the sum of the kinetic energy and the gravitational potential energy of the satellite. As the satellite moves closer to the central body some of its potential energy is converted into kinetic energy and, as it moves away from the central body, some of its kinetic energy is converted into gravitational potential energy.

Key question ⊃ p. 40

7 Because we define the zero of gravitational potential energy as a point an infinite distance away from the central body (where the gravitational force is zero), all points closer to the central body must have a negative potential. This is because it would require work to be done on the body to lift it to the point where the gravitational potential was zero. This is analogous to the gravitational potential energy being negative for objects below ground level if we were to call ground level the zero of potential.

Key questions ⊃ p. 41

8 A satellite has negative potential energy as it is trapped in an orbit around the planet. For the satellite to be free of the planet and reach a point where it had zero gravitational potential energy, work would have to be done on the satellite. The orbiting satellite therefore must have less than zero energy (i.e. a negative energy).

9 Because the surface of the Earth is moving fastest at the equator (460 ms^{-1}), satellites launched near the equator towards the east have to do less work to reach orbital velocity as they start with an initial velocity equal to the Earth's rotational velocity.

10 Launching an interplanetary probe in the direction of the orbital velocity (30 kms^{-1}) of the Earth will ensure the probe's initial velocity with respect to the Sun will be too great to remain in an Earth orbit. The probe will instead move into an elliptical orbit with the Sun at the near focus, enabling the probe to reach the outer planets. Launching in a direction opposite to the Earth's orbital motion will mean the probe will be going far too slowly to remain in an Earth orbit. In this case the probe will move into an elliptical orbit with the Sun at the far focus, enabling the probe to reach the inner planets.

Key questions ⊃ p. 43

11 The escape velocity is the minimum velocity required to ensure a probe leaving the surface of a planet has a kinetic energy equal to its gravitational potential energy on the surface. Any object leaving a planet at the escape velocity of the planet or at a greater velocity will have enough kinetic energy to free the probe from the gravitational attraction of the planet and the probe will therefore never fall back to the planet.

12 To move to a higher-altitude circular orbit the probe would have to increase its velocity twice by performing two posigrade rocket burns: the first to place the satellite into an elliptical transfer orbit that will take it further from the planet and the second to change to a circular orbit at the higher altitude. To move to a lower altitude two retrograde burns would be needed: one to slow the satellite to move it into an elliptical transfer orbit closer to the planet, and a second to change from an elliptical orbit to a circular orbit at the lower altitude.

HSC EXAM-TYPE QUESTIONS

Objective-response questions

1 **C**. Gravitation field strength is proportional to the mass of the planet and inversely proportional to the square of the radius; that is, $g \alpha m/r^2 \alpha 4/(1/2)^2 = 16$ times greater. Hence $g = 16 \times 10 = 160$ ms^{-2}. **B** and **D** are incorrect as they have the wrong numerical value and **A** is incorrect as it has the wrong units.

2 **A**. The satellite would fall towards or move away from the planet if the orbital velocity was less or greater than the correct value, respectively. **B** is incorrect as the gravitational force is the centripetal force. **C** is incorrect as satellites experience a significant force of gravity or they would move in a straight line rather than in an orbit. **D** is incorrect because the satellite must have a net force operating on it or it would move in a straight line in accordance with Newton's first law.

3 **A**. The kinetic energy of a satellite is inversely proportional to the orbital radius, and the total energy and gravitational potential energy of a satellite are always negative and inversely proportional to the orbital radius. This means that moving to a higher orbit makes the total energy and gravitational potential energy smaller negative numbers. **B** is incorrect as the gravitational potential energy and total energy of all satellites is negative. **C** is incorrect because the kinetic energy must decrease as satellites with larger orbital radii travel slower than satellites with lower orbital radii (as $v = \sqrt{(GM/r)}$). **D** is incorrect because the total energy increases with orbital radius as the satellite is closer to being free from the gravitational attraction of the planet (i.e. $E_{total} \alpha \frac{-1}{r}$).

4 **B**. The satellite will be moving too slowly at the burn point to continue in an Earth orbit. It will instead move into an elliptical orbit that will bring it closer to the Earth, which means the Earth will be at the focus of the ellipse furthest from the point where the burn occurred. **A** is incorrect because this orbit would take the satellite further from the Earth and this would require more energy.

The retrograde burn slowed the satellite, reducing its energy. **C** is incorrect because this would require a continual loss of satellite energy throughout the downwards spiral. **D** is incorrect as a single posigrade or retrograde satellite burn always results in an orbit that passes through the point at which the burn occurred.

5 C. The spacecraft was a free-falling body in a gravitational field. In this situation the total energy remains constant but as the craft moved away from the Earth it would be slowed by gravitational attraction to the Earth until it was close enough to the Moon for the force of attraction to the Moon to exceed the retarding force from the Earth. From this point the velocity would increase. **A** and **D** are incorrect as the velocity would change as the ship moved away from the Earth because it would experience a net force towards the Earth, slowing the craft. **B** Is incorrect as the total energy cannot change because the craft is in freefall in a gravitational field.

Extended-response questions

6 EM This question tests students' understanding of how the initial velocity of a satellite determines its orbit or trajectory, and how the relationship between centripetal force and the force of gravity on a satellite can be used to derive the orbital velocity.

a The centripetal force on a satellite is provided by gravity and hence:
$mv^2/r = GMm/r^2$; ✓ simplifying and rearranging
gives $v = \sqrt{(GM/r)}$ ✓

b The initial velocity of orbit *A* is lower than the initial velocity of orbit *B* ✓ but the initial velocity of orbit *C* is greater than the initial velocity of orbit *B*. ✓

c For the cannonball to follow path D the initial velocity must be equal to or greater than the escape velocity from the Earth at the height of the cannon. ✓

d Cannonball D has the greatest energy ✓ because when launched all the cannonballs have the same gravitational potential energy but D is moving fastest. As the total energy is the sum of the potential and kinetic energy, D must have the greatest total energy. ✓

7 EM This question tests students' ability to calculate the variables involved with the motion of satellites and their understanding of orbital velocity and its role in maintaining orbits.

a Orbital velocity is given by:

$$v = \sqrt{(GM/r)}$$

$$= \sqrt{\frac{6.67\times10^{-11}\times6.0\times10^{24}}{3.844\times10^{8}}} = 1020\ \text{ms}^{-1} = 1.02\ \text{kms}^{-1}\ ✓$$

Because the centripetal force is provided by gravity:
$F_c = mv^2/r = GMm/r^2$ and as $a_c = F_c/m$ we have:

$$a_c = GM/r^2 = \frac{(6.67\times10^{-11}\times6.0\times10^{24})}{(3.844\times10^{8})^2} = 2.7\times10^{-3}\text{ms}^{-2}\ ✓$$

b Employing the satellite equation:
$r^3/T^2 = GM/4\pi^2$ and hence:

$$T^2 = \text{constant} = r^3 4\pi^2/GM$$

$$T = \sqrt{\frac{(3.844\times10^{8})^3\times4\pi^2}{(6.67\times10^{-11}\times6.0\times10^{24})}}$$

$$= 2.367\times10^{6}\ \text{seconds or 27.4 days}\ ✓$$

c If the Moon stopped moving in its orbit it would continue to be attracted to the Earth with an initial acceleration of $2.7\times10^{-3}\ \text{ms}^{-2}$ towards the Earth. It would therefore begin moving in a straight line towards the Earth. ✓ As the Moon fell towards the Earth the gravitational force on the Moon would increase and its acceleration would approach $9.8\ \text{ms}^{-2}$ as it reached the Earth. ✓

8 EM This question tests students' understanding of ratios based on physical relationships, transfer orbits and Kepler's law of periods.

a The gravitational potential is given by:
$U = -GMm/r$
Hence $U_A/U_B = (-GMm_A/r_A)/(-GMm_B/r_B)$

$$= m_A r_B/m_B r_A = \frac{3\times1}{1\times2} = \frac{3}{2}\ ✓$$

b The centripetal force is the gravitational force and hence the ratio of the gravitational force on Moon A to the gravitational force on Moon B is given by:
$F_A/F_B = (GM3m/(2r)^2)/(GMm/r^2) = ¾$ ✓

c The most energy-efficient transfer orbit would have the moons at either extreme of the elliptical transfer orbit, as shown in Figure A3.1.
(1 mark for drawing an elliptical transfer orbit and 1 mark for placing Moon B at the opposite side of the ellipse to Moon A.)

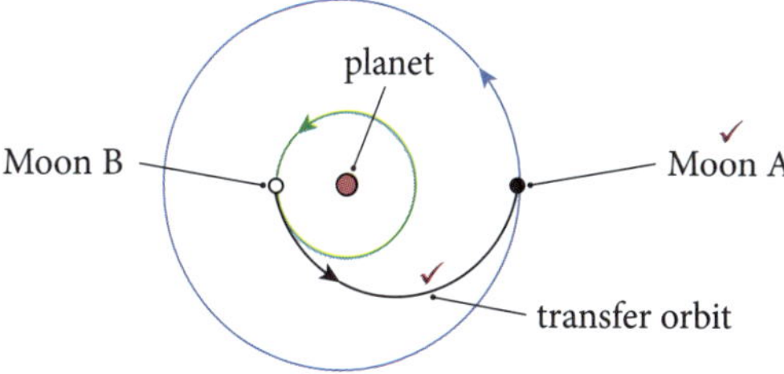

Figure A3.1 Transfer orbit for moving from Moon B to Moon A

d From Kepler's law of periods we have:

$$\frac{r_A^3}{T_A^2} = \frac{r_B^3}{T_B^2}\ ✓$$

Given that $r_A = 2r_B$, the ratio of the orbital periods would be:

$$\frac{T_B}{T_A} = \sqrt{\left(\frac{1}{2}\right)^3} = 0.354$$

Because the period is the time to complete one orbit, when Moon A completes one orbit Moon B completes:

$$\frac{1}{0.354} = 2.83\ \text{orbits}\ ✓$$

9 EM This question test students' understanding of the work done and energy changes involved in placing a satellite in orbit.
To place a satellite in orbit the satellite must be raised to the required altitude and accelerated to the correct orbital velocity for that orbital radius. To achieve a geostationary orbit the satellite must be moved to an altitude of approximately 36 000 km and be placed in an equatorial orbit. ✓ Therefore moving a satellite into orbit requires work to be done on the satellite to increase its kinetic energy ✓ and to increase its gravitational potential energy. ✓ Some work is also done against atmospheric drag moving the satellite through the atmosphere.

The work required to reach the appropriate altitude is equal to the change in gravitational potential energy between the ground and the orbital height: ✓

$\Delta U = U_{orbit} - U_{surface} = GMm(1/r_{surface} - 1/r_{orbit})$

The work required to accelerate the satellite to orbital velocity will be equal to the change in kinetic energy moving from the Earth's velocity to the orbital velocity: ✓

$\Delta K = K_{orbit} - K_{surface}$

By launching towards the east, near the equator, the work required to reach the required orbital kinetic energy is reduced by the kinetic energy the craft already has on the surface, due to the Earth's rotational velocity.

10 EM This question tests students' understanding of escape velocity and students' ability to apply energy equations to satellite motion.

a The kinetic energy is given by:

$K = GMm/2r = \frac{(6.67 \times 10^{-11} \times 6.0 \times 10^{24} \times 320)}{(2 \times 2.0 \times 10^{7})} = 3.2 \times 10^{9}$ J ✓

The gravitational potential energy is given by:

$U = -GMm/r = -2K = -(2 \times 3.2 \times 10^{9}) = -6.4 \times 10^{9}$ J ✓

b The orbital velocity can be calculated using $K = \frac{1}{2}mv^2$ or by using $v = \sqrt{(GM/r)}$

Using the second equation:

$$v = \sqrt{\frac{(6.67 \times 10^{-11} \times 6.0 \times 10^{24})}{2.0 \times 10^{7}}} = 4473 \text{ ms}^{-1} ✓$$

c To escape the Earth's gravitational attraction the satellite must have at least enough kinetic energy to reach the point in space where the potential energy is zero. That is, the initial kinetic energy must be equal to the magnitude of the satellite's potential energy. Thus the required kinetic energy (K_{esc}) is:

$K_{esc} = |U| = 6.4 \times 10^{9}$ J ✓

Now as $K = \frac{1}{2}mv^2$, $v_{esc} = \sqrt{(2K/m} = \sqrt{\frac{2 \times 6.4 \times 10^{9}}{320}} = 6325 \text{ ms}^{-1}$

Thus to escape from the Earth the satellite velocity would have to be increased by:

$\Delta v = 6325 - 4473 = 1852 \text{ ms}^{-1}$ ✓

CHAPTER 4 CHARGED PARTICLES, CONDUCTORS, AND ELECTRIC AND MAGNETIC FIELDS

INQUIRY QUESTION:

What happens to stationary and moving charged particles when they interact with an electric or magnetic field?

All charged particles experience a force if placed in an electric field. If the particle is positively charged the force is in the direction of the electric field; if the particle is negatively charged the force is in the opposite direction to the electric field. Magnetic fields only exert a force on moving charged particles that have a component of their velocity perpendicular to the magnetic field. The direction of this force is perpendicular to both the velocity of the charged particle and the direction of the magnetic field.

1 Charged particles in electric fields

» Students investigate and quantitatively derive and analyse the interaction between charged particles and uniform electric fields, including:

- electric field between parallel charged plates $E = \frac{V}{d}$
- acceleration of charged particles by the electric field $\vec{F}_{\text{net}} = m\vec{a}, \vec{F} = q\vec{E}$
- work done on the charge $W = qV, W = qEd, K = \frac{1}{2}mv^2$.

➔ You will recall from the Year 11 course that charge (q) is measured in coulombs (C) and that charged particles exert a force on one another that can be calculated using Coulomb's law:

$$F = \frac{1}{4\pi\varepsilon_0}\frac{q_1q_2}{r^2}$$

➔ You will also recall that charged objects are surrounded by electric fields (E fields) and that we can draw lines of force to picture these electric fields. The direction of the electric field at a point in space is the direction that a positively charged particle would experience a force, if placed at the point. Figure 4.1 shows the electric field around a positively charged particle, a negatively charged particle, a pair of particles with unlike charges (a dipole) and between oppositely charged parallel plates.

electric charge: physical property of matter that causes it to experience a force when placed in an electric field; electric charge has the SI unit of coulombs

electric field: region where a charged particle experiences a force; the electric field strength has the SI units NC^{-1} or Vm^{-1}

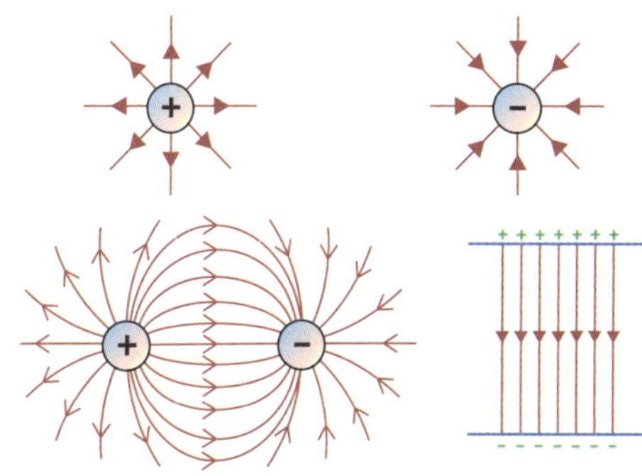

Figure 4.1 Electric field lines around individual positive and negative changes, around a pair of unlike charges (dipole) and between oppositely charged parallel plates

➔ The strength of the electric field at a point in space is defined as the force per unit positive charge at the point. We express this relationship as:

$E = F/q$

Now as the electric field strength is the force per unit charge, the force on a particle with charge q placed in an electric field of strength E will be given by:

$$F = qE$$

This equation just tells us that if E is the force on a particle with 1 coulomb of charge, the force on a particle with q coulombs of charge must be the number of coulombs (q) times the force per unit charge (E). Note that if q is negative the force will be negative, indicating the force is in the opposite direction to the electric field.

➔ KEY QUESTIONS

1 How can we determine the direction of an electric field?

2 How are charged particles affected by electric fields?

Answers ➲ p. 62

- If a charged particle is placed in an electric field the force on the charged particle will cause it to accelerate. We can use Newton's second law to find the rate at which the charged particle will accelerate:

  $F = ma$ and $F = qE$

  Hence $ma = qE$, or

  $a = qE/m$

  The charged particle will accelerate in the direction of the electric field if it is positively charged and in the opposite direction if it is negatively charged.

- You will also recall that we defined the **electrical potential difference** (V) between two points in an electric field as the work done on a unit positive charge when it is moved between the points:

  $V = \text{work}/q = \Delta U/q$

electrical potential difference: the work done moving a unit positive charge from one point in an electric field to another point in the field; it is therefore a measure of the electrical energy difference between two points in an electric field and has the SI units of JC^{-1} or volts

Note that rearranging this definition gives us an expression for how much work is done when a charged particle is moved between two points with an electrical potential difference of V between them:

$$\text{Work} = W = qV$$

- Consider for example the parallel plates separated by a distance d, shown in Figure 4.2.

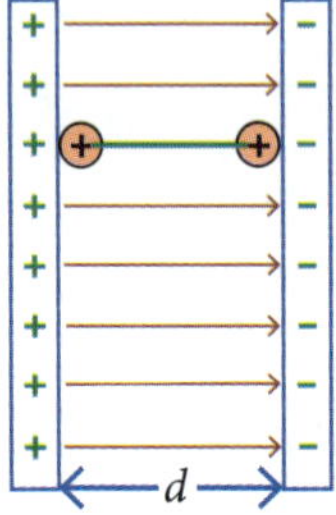

Figure 4.2 Moving a positively charged particle between parallel charged plates

If a positively charged particle was placed between the plates, the electric field would exert a force $F = qE$ towards the negative plate on the particle. A positively charged particle released at the positive plate would therefore be accelerated towards the negative plate. The work done on the particle would be:

$$\text{Work} = F_{\|}s$$
$$W = (qE)d$$

Substituting this expression for work into the equation for electrical potential difference gives:

$V = \text{work}/q = (qEd)/q = Ed$

This equation relates the electric field between the plates to the potential difference (voltage) across the plates.

- We say the positive plate is at a *higher electrical potential* (i.e. has more potential energy) than the negative plate because the field would do work on the positively charged particle, moving it from the positively charged plate to the negatively charged plate. Even though a negatively charged particle would have a lower potential energy at the positive plate, we still say the positive plate has the higher electrical potential energy because electrical potential, like the electric field and conventional current, is defined for positive charge.
- This expression ($V = E/d$) gives the potential difference between any two points in a uniform electric field when moving parallel to the field. By rearranging the equation we obtain a different expression for the electric field:

$$E = V/d$$

We can therefore think of the electric field as the force per unit charge with units of newtons per coulomb (NC^{-1}), or as the electrical potential difference per unit length in the direction of the field with units of volts per metre (Vm^{-1}).

- We saw when we studied mechanics that motion problems can often be solved by considering the forces and accelerations involved or by using work and energy. Similarly, the motion of charged particles in electric fields can be analysed using forces and/or energy. The method chosen will depend on the problem under consideration. Some problems will require force or energy to be used, some problems will require both and some will be able to be solved using either method.
- Consider the positively charged particle shown in Figure 4.2. The force on the charged particle will be $F = qE$ and the charged particle will accelerate towards the negative plate at:

  $a = F/m = qE/m$

  Now as $v^2 = u^2 + 2as$ and the particle started from rest ($u = 0$), the particle will reach the negative plate travelling with a velocity of:

  $v = \sqrt{(2as)} = \sqrt{(2qEd/m)}$

  Alternatively we could consider the work done on the particle as it moves to the negative plate:

  Work done = kinetic energy gained

  $\text{Work} = qV = \frac{1}{2}mv^2$

  $v = \sqrt{(2qV/m)}$

  Note that as $V = Ed$ this expression is identical to the one derived above using force and acceleration.

→ KEY QUESTIONS

3 **How is electrical potential difference related to the electric field?**

4 **How is the electric field between oppositely charged parallel plates related to the voltage across the plates?**

5 **How is work related to the electric field and the electrical potential difference?**

Answers ➲ p. 62

EXAMPLE 1

A cathode ray gun accelerates electrons ($m = 9.11 \times 10^{-31}$ kg and $q = -1.602 \times 10^{-19}$ C) across a potential of 220 V in a vacuum. Find the final velocity of the electron.

The work done on change by an electric field is equal to the kinetic energy gained by the charge

Answer:

The work done on the electron by moving across the potential difference will be equal to the kinetic energy gained by the electron:

Work done = kinetic energy gained

$qV = \frac{1}{2}mv^2$

$$v = \sqrt{(2qV/m)} = \sqrt{\frac{2 \times 1.602 \times 10^{-19} \times 220}{9.11 \times 10^{-31}}} = 8.8 \times 10^6 \text{ ms}^{-1}$$

EXAMPLE 2

A particle of mass 2×10^{-4} kg carrying a charge of $+2\ \mu$C is found to levitate when placed between a pair of oppositely charged parallel plates, separated by 2 cm, as shown in Figure 4.3.

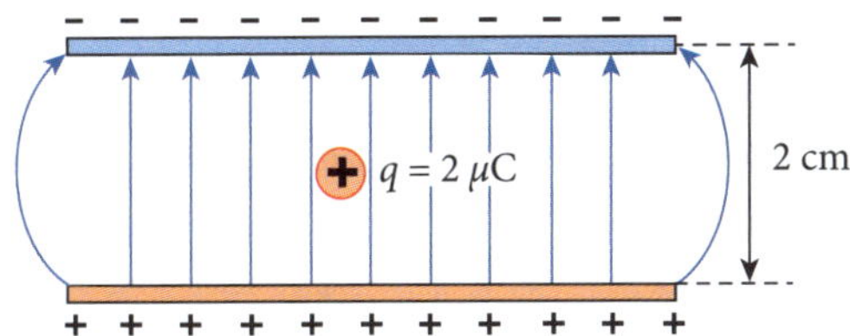

Figure 4.3 Charged particle of mass 2×10^{-4} kg levitated by a pair of charged plates

a **Determine the strength of the electric field between the plates.**

b **Determine the electrical potential difference between the plates.**

c **Explain what would happen to the particle if the direction of the electric field was reversed.**

Remember: the vector sum of the forces on a body at rest must be zero

Answer:

a If the charged particle is at rest, the net force on the particle must be zero and therefore the downwards force due to gravity must be equal to the upwards force due to the electric field:

$mg = qE$

$$E = mg/q = \frac{(2 \times 10^{-4} \times 9.8)}{2 \times 10^{-6}} = 980 \text{ NC}^{-1} \text{ upwards}$$

b As it is a uniform electric field:

$V = Ed = 980 \times 0.02 = 19.6$ V

c If the field direction was reversed, both forces on the particle would be directed downwards and the particle would accelerate downwards at:

$a = F_{net}/m = (mg + qE)/m$

$$= \frac{(2 \times 10^{-4} \times 9.8 + 2 \times 10^{-6} \times 980)}{(2 \times 10^{-4})}$$

$= 19.6 \text{ ms}^{-2}$ downwards

2 The motion of a charged particle in an electric field

» Students model qualitatively and quantitatively the trajectories of charged particles in electric fields and compare them with the trajectories of projectiles in a gravitational field.

→ In a uniform electric field, the force exerted on a charged particle ($F = qE$) remains constant in magnitude and direction, and hence the particle will accelerate at a constant rate parallel to the field. The motion of a charged particle in a uniform electric field is therefore analogous to the motion of a projectile in a uniform gravitational field and can be analysed using the same techniques; that is, by treating the components of the motion parallel to the field and perpendicular to the field independently.

→ Note that when applying the equations of motion to charged particles in electric fields, we need to replace the acceleration due to gravity with the acceleration in the field. We showed in the previous section this was given by:

$$a = qE/m$$

where a is the acceleration
q is the charge on the particle
E is the electric field strength and
m is the mass of the particle

→ Note also that the direction of the acceleration of a charged particle in an electric field, unlike a mass in a gravitational field, will depend on the sign of the charge on the particle.

→ KEY QUESTION

6 How is projectile motion related to the motion of a charged particle in a uniform electric field?

Answers ➲ p. 62

EXAMPLE 3

An electron ($m = 9.11 \times 10^{-31}$ kg and $q = -1.602 \times 10^{-19}$ C) is inserted vertically into the uniform electric field of 25 NC^{-1} between two parallel charged plates, as shown in Figure 4.4. The electron enters the field with an initial velocity of 6×10^5 ms^{-1}. When completing the answer, you may neglect gravity in your calculations.

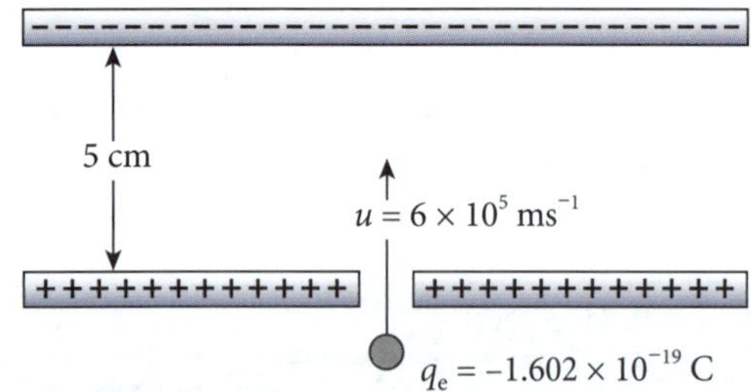

Figure 4.4 An electron fired vertically into a uniform electric field

a Qualitatively describe the motion of the electron in the electric field.

b Given the plates are separated by 5 cm, will the electron reach the top plate? Justify your answer with appropriate calculations.

A charged particle moving parallel to an electric field will experience a force parallel to the velocity, just like a mass moving vertically in a gravitational field

Answer:

a As the electron is a negatively charged particle the electric field will exert a downwards force on the electron when it is in the field. This force will cause the electron to accelerate at:

$a = qE/m$

$= \dfrac{-(1.602 \times 10^{-19} \times 25)}{9.11 \times 10^{-31}}$

$= -4.4 \times 10^{12}$ ms^{-2}

$= 4.4 \times 10^{12}$ ms^{-2} downwards

b The distance the electron would travel in a uniform electric field of this magnitude before the field would bring the electron to rest ($v = 0$) can be calculated using:

$v^2 = u^2 + 2as$, with $v = 0$ and, calling upwards positive

$s = -u^2/2a = \dfrac{-(6 \times 10^5)^2}{2 \times (-4.4 \times 10^{12})} = 0.041 \text{ m} = 4.1 \text{ cm}$

As the plates are 5 cm apart the electron will stop 0.9 mm from the top plate before moving downwards.

EXAMPLE 4

This question refers to Figure 4.5, which shows a proton ($q = +1.602 \times 10^{-19}$ C and $m = 1.67 \times 10^{-27}$ kg) travelling at 4×10^4 ms^{-1} horizontally entering midway between two horizontal plates separated from one another by 10 cm. A voltage of 0.5 V is applied to the plates and the top plate positively charged.

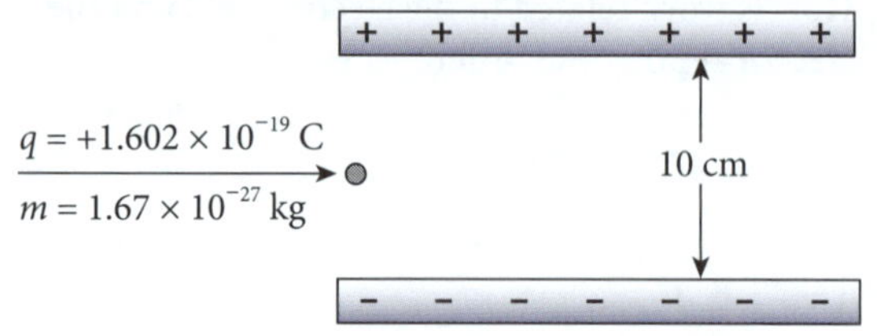

Figure 4.5 A proton moving into a region of uniform electric field between two charged plates

a Find the electric field between the plates.

b Compare the gravitational and electric forces on the proton when it is between the plates.

c Find the acceleration of the particle in the field.

d Describe the path the proton will take between the charged plates.

e How long would it take for an electric field with this strength to move the charged particle 5 cm downwards?

f Given that the plates are 50 cm long, will the proton escape from the parallel plates? Justify your answer.

A charged particle moving perpendicular to an electric field will experience a constant force that will cause it to follow a parabolic path, like a projectile in a gravitational field

Answer:

a The electric field would be given by:

$E = V/d = \dfrac{0.5}{0.1} = 5$ Vm^{-1} downwards

b The force due to gravity is the weight $= mg$

$= 1.67 \times 10^{-27} \times 9.8 = 1.6 \times 10^{-26}$ N downwards

The electric force is given by $F = qE$

$= 1.602 \times 10^{-19} \times 5 = 8 \times 10^{-19}$ N downwards

The force of gravity is negligible in comparison to the force due to the electric field, which is 5×10^7 greater than gravity.

c Applying Newton's second law:

$a = F/m = (qE)/m$

$= \dfrac{(1.602 \times 10^{-19} \times 5)}{(1.67 \times 10^{-27})}$

$= 4.8 \times 10^8$ ms^{-2}

Note that we have not included gravity in this calculation because it is negligible compared with the electric force.

d Because there is a constant downwards force on the proton, it will describe a parabolic path towards the lower plate (like a projectile).

e Using the vertical component of the motion, we can find the time for the particle to travel downwards 5 cm using:

$s = ut + \frac{1}{2}at^2$ and, as the initial vertical velocity is zero ($u = 0$),

$$t = \sqrt{(2s/a)} = \sqrt{\frac{2 \times 0.05}{4.8 \times 10^8}} = 1.44 \times 10^{-5}\ \text{s}$$

f Using the horizontal component of the motion to find the horizontal distance travelled in this time:

$s = ut = 4 \times 10^4 \times 1.44 \times 10^{-5} = 0.58\ \text{m} = 58\ \text{cm}$

The proton will therefore exit the right-hand side before colliding with the negatively charged plate.

EXAMPLE 5

In an early model of the hydrogen atom, Ernest Rutherford suggested that an electron orbited a proton and that the electrostatic attraction between the oppositely charged particles held the electron in its orbit. (You will find the mass and charge of the electron and proton in Examples 3 and 4, respectively.)

Assuming the electron orbits the proton with a radius of 5.0×10^{-11} m, find:

a the centripetal force on the electron.

b the orbital velocity of the electron.

c the orbital period of the electron and the frequency of its orbit.

A charge moving in a radial electric field will experience a centripetal force that could cause it to move in a circle

Answer:

a The centripetal force will be provided by the electrostatic attraction and hence:

$$F = \frac{1}{4\pi\varepsilon_0}\frac{q_1 q_2}{r^2}$$

$= 9 \times 10^9 \times (1.602 \times 10^{-19})^2/(5.0 \times 10^{-11})^2$

$= 9.2 \times 10^{-8}\ \text{N}$

b Using the equation for centripetal force:

$F_c = mv^2/r$ and hence

$$v = \sqrt{(F_c r/m)} = \sqrt{\frac{9.2 \times 10^{-8} \times 5 \times 10^{-11}}{9.11 \times 10^{-31}}}$$

$= 2.25 \times 10^6\ \text{ms}^{-1}$

c The period can by determined from the radius and velocity as:

$v = 2\pi r/T$ and hence

$$T = 2\pi r/v = \frac{2\pi(5 \times 10^{-11})}{2.25 \times 10^6} = 1.4 \times 10^{-16}\ \text{s}$$

Frequency can then be determined using:

$$f = 1/T = \frac{1}{1.4 \times 10^{-16}} = 7.2 \times 10^{15}\ \text{Hz}$$

3 Charged particles in magnetic fields

» Students analyse the interaction between charged particles and uniform magnetic fields, including:
- acceleration, perpendicular to the field, of charged particles
- the force on the charge $F = qv_\perp B = qvB\sin\theta$.

FIRSTHAND INVESTIGATION

Charges moving in electric and magnetic fields

A cathode ray tube is an evacuated tube that contains two electrodes. When a large voltage is placed across the electrodes a stream of electrons flows from the negative electrode to the positive electrode. Students should use a cathode ray tube to conduct a practical investigation on the forces on charged particles in electric and magnetic fields. Figure 4.6 shows how a pair of parallel, oppositely charged plates can be used to produce an electric field to deflect the electron beam. Students should investigate the effect of changing the voltage applied to the plates and reversing the polarity of the applied voltage.

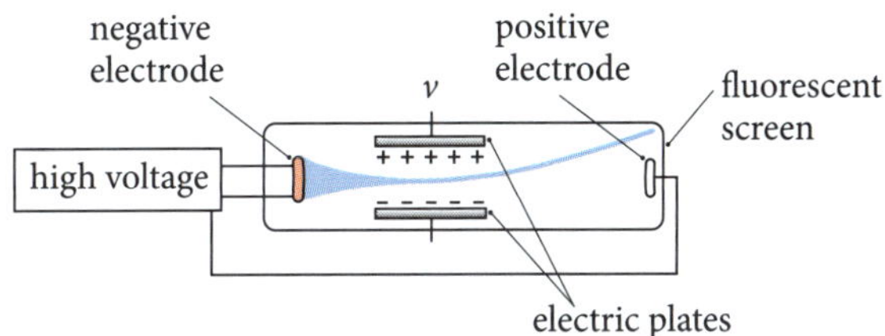

Figure 4.6 Using a cathode ray tube to examine the deflection of a beam of electrons in an electric field

Figure 4.7 shows how a fluorescent screen can be used to investigate the deflection of the electron beam by a magnetic field. Students should investigate the direction of the force on the moving electrons produced by the magnetic field, and the effect of the applied magnetic field strength on the amount of deflection of the beam.

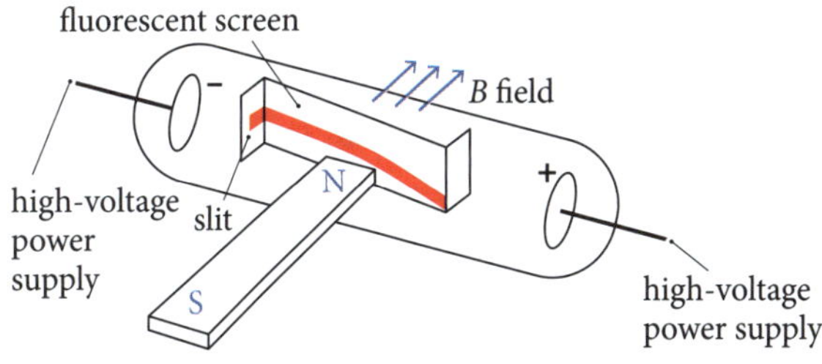

Figure 4.7 A bar magnet placed near a cathode ray tube with a fluorescent screen to deflect a beam of electrons

- You will recall from the Year 11 course that a magnetic field (B) is a vector field with a specific magnitude and direction at each point in space. The magnetic field strength is also called the **magnetic flux density** (B) and is measured in tesla (T). Note that a 1 tesla magnetic field is quite an intense field as the Earth's magnetic field is only 30 to 50 μT at the surface.

magnetic flux density: a measure of the strength and direction of the magnetic field at a given point, measured in tesla

- You will also recall that if a magnet is placed in a magnetic field it will experience a force. The magnet experiences a force because there are aligned spinning-charged particles in regions called *domains* within the magnet upon which the magnetic field exerts a force. Careful analysis of magnetic fields has shown that they are always produced by moving charged particles and that they only exert a force on charged particles that are moving with respect to the magnetic field.
- A magnetic field only exerts a force on a charged particle if the particle is moving with some component of its velocity perpendicular to the magnetic field. Thus a magnetic field does not exert a force on a stationary charged particle or a charged particle moving parallel to the magnetic field.
- The magnitude of force on a charged particle moving in a magnetic field is given by:

$$F = qv_{\perp}B = qvB \sin \theta$$

where q is the charge in coulombs
v is the velocity of the charged particle
B is the magnetic flux density in tesla and
θ is the angle between the velocity vector and the magnetic field vector

Figure 4.8 shows how the quantity $v \sin \theta$ in this equation represents the component of the velocity perpendicular to the field.

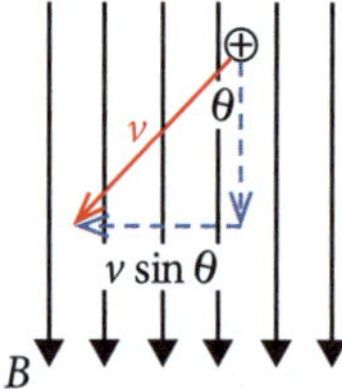

Figure 4.8 Component of the velocity perpendicular to the magnetic field

- The direction of the force exerted on the moving charged particle in a magnetic field is perpendicular to both the velocity vector and the magnetic field vector. We can determine the direction by applying the **right-hand palm rule** shown in Figure 4.9. The thumb points in the direction of the component of the charged particle's velocity that is perpendicular to the magnetic field, the fingers point in the direction of the magnetic field and the force on a positive particle would be out of the palm. The force on a negative particle would be in the opposite direction.

right-hand palm rule: a rule that gives the direction of the force on a moving positive charge in a magnetic field; if the thumb points in the direction of the positive charge velocity and the fingers point in the direction of the magnetic field, the force on a positive charge would be directed out of the palm (the force on a moving negative charge is in the opposite direction)

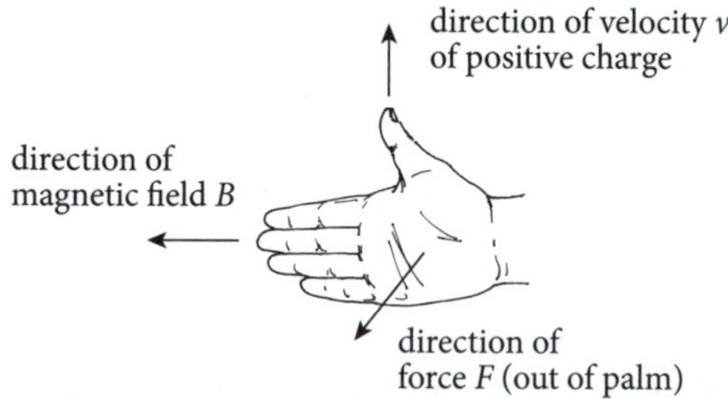

Figure 4.9 Right-hand palm rule gives the direction of the force on a positively charged particle moving in a magnetic field

- To draw fields in three dimensions we use the convention that crosses drawn on the page represent a field directed into the page, while dots drawn on the page represent a field that is directed out of the page. To help remember this, think of the dots as the points of arrows coming towards you and the crosses as the tail feathers of arrows moving away from you.
- Because the magnetic force is perpendicular to the charged particle's velocity it will act as a centripetal force on the particle and result in a centripetal acceleration that will cause the direction of the velocity to change constantly. Because it is a centripetal force, the force exerted on a moving charged particle by a magnetic field does not make the particle move faster or slower; it only changes the direction in which the particle moves. Figure 4.10 shows how a positively charged particle entering a uniform magnetic field perpendicular to the field will move in uniform circular motion in the field.

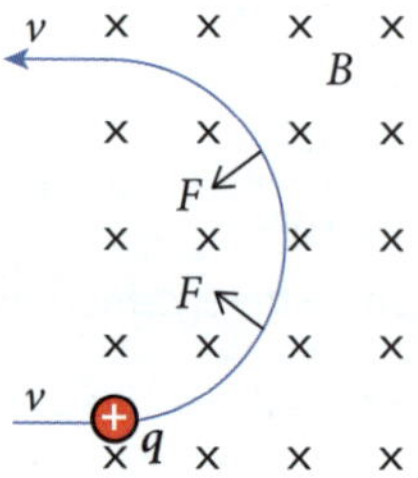

Figure 4.10 Circular path of a positively charged particle entering a uniform magnetic field at 90° to the field

- As the magnetic force exerts a centripetal force on the particle and the particle is moving at 90° to the field, $\sin\theta = 1$ and we can write:

  $qvB = mv^2/r$

  Hence $r = mv/qB$

  where r is the radius of curvature of the charged particle's path in the magnetic field
- If a charged particle is moving at an angle other than 0° or 90° to the magnetic field, the particle will follow a helical path in the field, as shown in Figure 4.11.

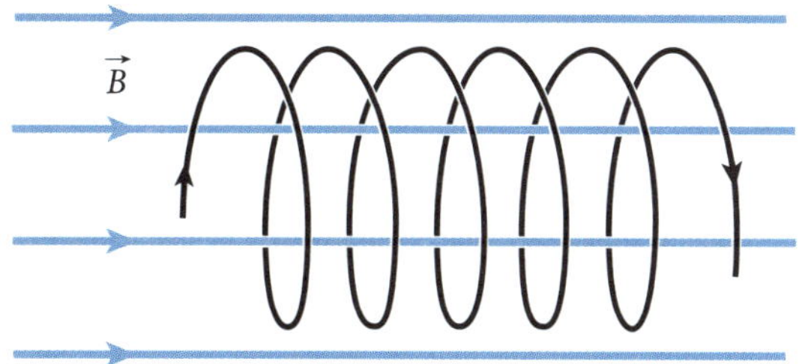

Figure 4.11 Helical motion of a charged particle moving in a magnetic field

- We can understand this by breaking the initial velocity into two components, one parallel with the field and the other perpendicular to the field, as shown in Figure 4.12.

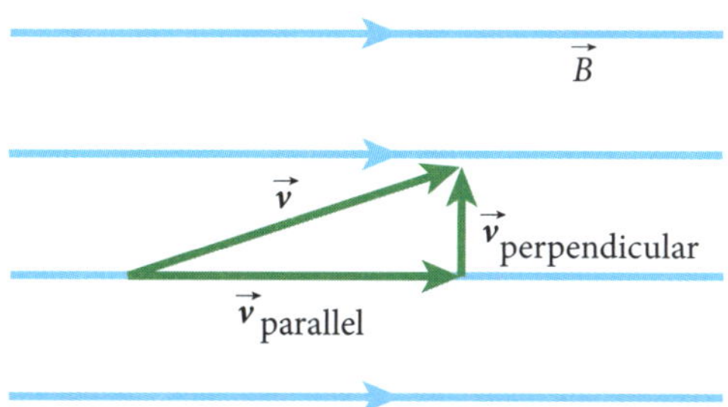

Figure 4.12 Breaking the velocity of the charged particle into components parallel and perpendicular to the field

The perpendicular component of the velocity will cause the charged particle to move in uniform circular motion, while the parallel component will cause the particle to continue to move at constant velocity parallel to the field. As the charged particle moves in a circle and drifts down the field it will trace out a helix, as shown in Figure 4.11.

→ KEY QUESTIONS

7 How can we determine the direction of the force on a charged particle moving in a magnetic field?

8 How can we determine the magnitude of the force on a charged particle in a magnetic field?

9 How is the magnetic force on a moving charged particle related to centripetal force?

Answers ➲ p. 62

EXAMPLE 6

A proton, a neutron and an electron are fired after one another with a velocity of 10^4 ms^{-1} at 90° to a uniform magnetic field of strength 5×10^{-2} T directed into the page, as shown in Figure 4.13. The mass and charge on each particle is shown in Table 4.1.

Particle	Mass (kg)	Charge (C)
Proton	1.67×10^{-27}	$+1.602 \times 10^{-19}$
Neutron	1.67×10^{-27}	0
Electron	9.11×10^{-31}	-1.602×10^{-19}

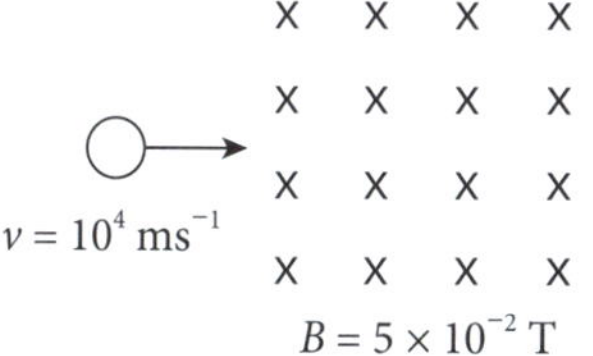

Figure 4.13 Particles moving with velocity 10^4 ms^{-1} entering a magnetic field of strength 5×10^{-2} T

a Find the magnitude and direction of the force that would be exerted on each particle when it entered the field.

b Compare the path of each particle in the field.

The oppositely charged particles moving perpendicular to a magnetic field will experience a centripetal force in opposite directions

Answer:

a As the angle between the velocity vector and the field vector is 90°, $\sin\theta = 0$ and hence:

  i for the proton, $F = qvB = 1.602 \times 10^{-19} \times 10^4 \times 5 \times 10^{-2} = 8.0 \times 10^{-17}$ N upwards (from right-hand palm rule).

  ii for the neutron, $q = 0$ and hence the field does not exert a force on the particle.

  iii for the electron, $F = qvB = 1.602 \times 10^{-19} \times 10^4 \times 5 \times 10^{-2} = 8.0 \times 10^{-17}$ N downwards.

b The neutron will move straight through the field without being deviated from its initial path but the electron and proton will move in circular paths in the field. By applying the right-hand rule we see that the positive proton will curve upwards and the negative electron will curve downwards. We can calculate the radius of each path using:

  $qvB = mv^2/r$

  And hence $r = mv/qB$

  $$r_{proton} = mv/qB = \frac{1.67 \times 10^{-27} \times 10^4}{1.602 \times 10^{-19} \times 5 \times 10^{-2}} = 2.08 \times 10^{-3}\text{ m} = 2.08\text{ mm}$$

  $$r_{electron} = mv/qB = \frac{9.11 \times 10^{-31} \times 10^4}{1.602 \times 10^{-19} \times 5 \times 10^{-2}} = 1.14 \times 10^{-6} = 1.14\ \mu\text{m}$$

4 Moving charged particles in magnetic and electric fields

» Students compare the interaction of charged particles moving in magnetic fields to:
- the interaction of charged particles with electric fields
- other examples of uniform circular motion.

➔ Charged particles are affected quite differently by electric and magnetic fields.

➔ In an electric field all positively charged particles experience a force in the direction of the field, and all negatively charged particles experience a force in the opposite direction to the electric field. As we saw earlier a charged particle moving at an angle to a uniform electric field moves in a parabolic motion, analogous to the motion of a projectile in a gravitational field. We also saw that a radial electric field can provide a centripetal force that would enable a charged particle to move with uniform circular motion, like the motion of a satellite in the radial gravitational field of the Earth.

➔ In contrast, a magnetic field does not exert a force on a charged particle unless the charged particle is moving and has some component of its velocity perpendicular to the magnetic field. The force that appears on a positively charged particle is perpendicular to both the velocity and the magnetic field in a direction given by the right-hand palm rule. The direction of the force on a negative charged particle would be in the opposite direction experienced by a positive particle. Because the force on the charged particle is always perpendicular to the velocity of the charged particle, the force acts as a centripetal force, causing the particle to move in uniform circular motion in the field.

➔ A charged particle at rest in a magnetic field does not experience a force and hence remains at rest. Similarly a charged particle moving parallel to the magnetic field continues to move in the same direction because it has no perpendicular component of velocity and hence is not affected by the magnetic field. If a charged particle is moving at an angle to the magnetic field it will follow a helical path because the component of the velocity parallel to the field will not be affected by the field, while the perpendicular component will cause the charged particle to follow a circular path. A helix is produced as the particle moves in a circle perpendicular to the magnetic field as it drifts at constant velocity in a direction parallel to the field.

➔ The uniform circular motion of charged particles moving in magnetic fields is like other forms of circular motion, but instead of the centripetal force being provided by the tension in a string or gravity, it is provided by the magnetic field. Another difference is that the faster a charged particle is moving perpendicularly to a magnetic field, the greater the centripetal force on the charged particle. You should also note that the radius of curvature of the path of the charged particle is proportional to the mass and the velocity of the particle but inversely proportional to the magnetic flux density and the charge on the particle (i.e. $r = mv/qB$).

➔ KEY QUESTION

10 **How does the force on a charged particle in an electric field differ from the force on a charged particle in a magnetic field?**

Answers ➲ p. 62

EXAMPLE 7

Consider the charged particle fired horizontally into the electric and magnetic fields, illustrated in Figure 4.14. The charged particle had an initial velocity of 4×10^2 ms^{-1} and a charge of +2 μC. Because the force of gravity is very small compared with the other forces we can neglect the force of gravity in this example.

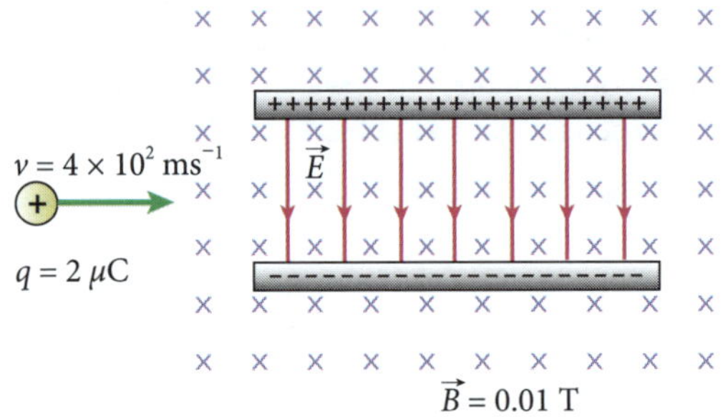

Figure 4.14 Charged particle fired horizontally into perpendicular magnetic and electric fields

a **Find the magnitude and direction of the magnetic force on the particle.**

b **What electric field would be required for the charged particle to pass straight through the field without being deflected?**

c **If the electric field was set at a strength to ensure the charged particle passed straight through the plates, what would happen to the particle if it entered the region between the plates with:**
   i **a velocity less than 4×10^2 ms^{-1}?**
   ii **a velocity greater than 4×10^2 ms^{-1}?**

The force on a charged particle moving perpendicular to a magnetic field is proportional to its velocity but the force on a charged particle in an electric field is not related to its velocity

Answer:

a Applying the right-hand palm rule shows the magnetic force is upwards. We can calculate the magnitude of the force using:

$F = qvB\sin\theta$ and, as $\theta = 90°$

$= 2 \times 10^{-6} \times 4 \times 10^2 \times 0.01 = 8 \times 10^{-6}$ N upwards

b For the particle not to be deflected the net force must be zero and hence the force due to the electric field must be equal and opposite to the force due to the magnetic field. We can therefore write:

$F_E = F_B$

$qE = F_B$

$E = F_B/q = \dfrac{8 \times 10^{-6}}{2 \times 10^{-6}} = 4\ \text{NC}^{-1}$ downwards

(Note that because the forces on the charged particle are equal when it passes straight through, we can write $qE = qvB$ or $v = E/B$. Thus the velocity required to pass straight through the fields without being deflected is equal to the ratio of the field strengths. This means that if the fields had equal strengths, a particle travelling at 1 ms^{-1} would pass through them and if the electric field was twice as strong as the magnetic field, the particle would have to travel at 2 ms^{-1} to pass straight through the fields without being deflected.)

c **i** The force due to the electric field would not change but the force due to the magnetic field would decrease (as the magnetic force is proportional to the velocity). This would result in a net downwards force on the charged particle, causing it to move towards the lower plate.

ii The magnetic force would increase in this case. Thus the upwards force due to the magnetic field would be greater than the downwards force due to the electric field and the particle would move towards the upper plate.

CHAPTER SYLLABUS CHECKLIST

Are you able to answer these questions from the syllabus for this chapter? Tick each question as you go through the checklist if you are able to answer it. If you cannot answer a question, turn to the relevant page in the study guide to find the answer. For NESA key word meanings, go to www.educationstandards.nsw.edu.au and search 'key words'.

FOR A COMPLETE UNDERSTANDING OF THIS TOPIC:		PAGE NO.	✓
1	Can I draw the electric field lines around charged objects and between oppositely charged parallel plates?	51	
2	Can I determine the magnitude and direction of the force that appears on a charged particle in an electric field and the acceleration that the force will produce?	51–52	
3	Can I define electrical potential difference?	52	
4	Can I relate work done when a charge moves between two points in an electric field to electrical potential difference between the points?	52	
5	Can I explain how energy is transformed between electrical potential energy and kinetic energy when a charged particle moves freely in an electric field?	52	
6	Can I determine the strength of the electric field between oppositely charged parallel plates given the plate separation and the potential difference between the plates?	52	
7	Can I use the equations of projectile motion to predict the trajectory of a moving charged particle in an electric field?	53–54	
8	Can I compare satellite motion to the circular motion of a charged particle in a radial electric field?	55 & 58	
9	Can I determine the magnitude and direction of the force exerted on a moving charged particle in a magnetic field?	56	
10	Can I relate the acceleration of a charged particle moving in a magnetic field to centripetal acceleration?	56	
11	Can I describe the possible paths of moving charged particles in a magnetic field?	56–57	
12	Can I relate the force on a charged particle in a magnetic field to centripetal force?	57	
13	Can I determine the radius of the circular path taken by charged particle moving in a magnetic field?	57	
14	Can I compare the force and resulting path of a charged particle in an electric field with the force and resulting path of a charged particle in a magnetic field?	58	
15	Can I calculate the net force on a stationary or moving charged particle in the presence of more than one field; for example, in a magnetic and electric field or in an electric and gravitational field?	58	

HSC EXAM-TYPE QUESTIONS

Objective-response questions (1 mark each)

1 **A negatively charged particle is placed in an electric field, as shown in Figure 4.15.**

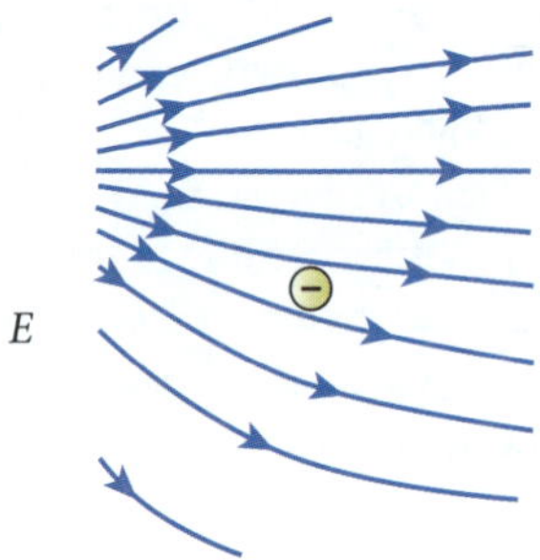

Figure 4.15 A negatively charged particle in an electric field

What will happen to the negatively charged particle in the field?

A It will move in the direction of the field with a decreasing acceleration.

B It will move in the opposite direction of the field with an increasing acceleration.

C It will move in the same direction of the field with an increasing acceleration.

D It will move in the opposite direction of the field with a decreasing acceleration.

2 **A charged particle q enters a magnetic field at 90° to the field and follows a circular path of radius R, as shown in Figure 4.16.**

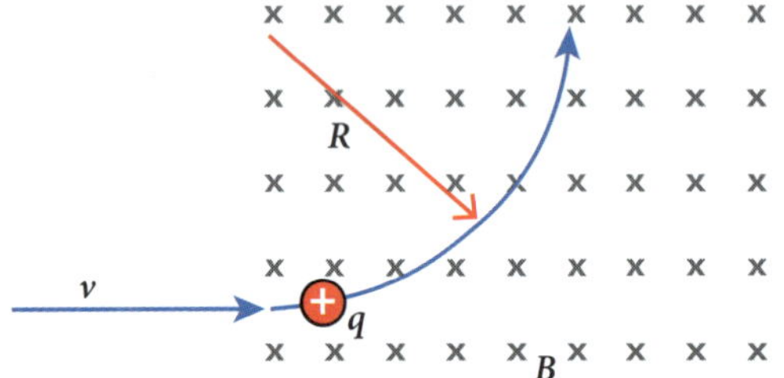

Figure 4.16 A charged particle q entering a magnetic field B with velocity v

If the initial velocity of the charged particle was doubled while the strength of the B field was halved, what would the new radius of curvature be?

A $2R$

B $4R$

C $R/2$

D $R/4$

3 **High-velocity ions and electrons are constantly emitted from the Sun towards the Earth. In what direction will a high-speed electron directed towards the equator from the Sun experience a force when it first encounters the Earth's magnetic field?**

A south

B north

C east

D west

4 **A cloud chamber is a device that can be used to view the path of charged particles. The particle track shown in Figure 4.17 illustrates energy being converted into a negative electron and a positive electron (called a *positron*). The cloud chamber used in this experiment employed a uniform magnetic field to produce the curved tracks shown in the figure. What is the most probable conclusion that can be drawn from these particle tracks?**

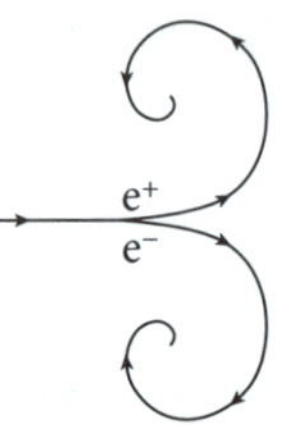

Figure 4.17 Electron (e^-) and positron (e^+) created from energy in a cloud chamber

A The magnetic field is directed into the page and the particles are slowing down.

B The magnetic field is directed into the page and the particles are speeding up.

C The magnetic field is directed out of the page and the particles are slowing down.

D The magnetic field is directed out of the page and the particles are speeding up.

5 **Figure 4.18 shows a proton ($q = +1.602 \times 10^{-19}$ C) travelling horizontally about to enter an electric field midway between two charged plates. When answering you may ignore the gravitational force on the particle because it is negligible compared with the electric force on the particle.**

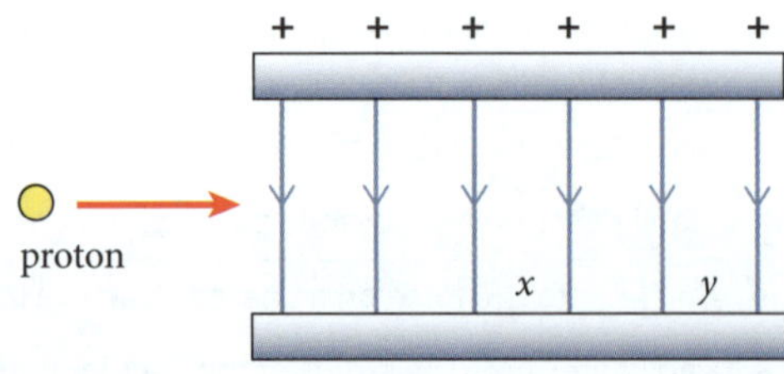

Figure 4.18 A proton moving at constant velocity moving into the region between two charged plates

The proton is observed to follow a parabolic path and collides with the lower plate at point x. Which of the

following changes could move the collision point from point x to point y?

A The separation between the plates could be decreased or the initial velocity of the proton could be increased.

B The initial velocity of the proton could be increased or the plates could be moved further apart from each other.

C The strength of the electric field between the plates could be decreased or the initial velocity of the particle could be decreased.

D The voltage across the plates could be increased or the velocity of the particle increased.

Extended-response questions

6 Two horizontal charged plates separated from one another by 2 cm produce an upwards electric field of 200 NC^{-1} between the plates. A proton ($m = 1.67 \times 10^{-27}$ kg and $q = +1.6 \times 10^{-19}$ C) is released at the lower plate. You may neglect gravity in your calculations.

a Determine the electrical potential difference between the plates and state which of the plates is at the higher potential. (2 marks)

b Calculate the force on the proton and the rate at which it will accelerate. (2 marks)

c Determine the velocity of the proton just before it hits the top plate. (1 mark)

D Determine the work done by the field on the proton as it moves between the charged plates. (1 mark)

7 Two parallel charged plates are separated by 10 cm and have a potential difference of 200 V across the plates.

a Sketch a graph of the electric field strength against the distance from the positive plate to the negative plate. (2 marks)

b Sketch a graph of the potential difference against distance from the positive plate to the negative plate. (2 marks)

c Sketch a graph of the velocity against time for a positively charged particle released from rest at the positive plate. (2 marks)

8 Figure 4.19 shows the path of two particles (X and Y) that entered a uniform magnetic field at 90° to the field. The magnetic flux density was 0.2 T, the charged particles were moving with an initial velocity of 120 ms^{-1}, both charged particles had a mass of 2×10^{-12} kg and both moved in circular path of diameter d = 0.2 mm in the field.

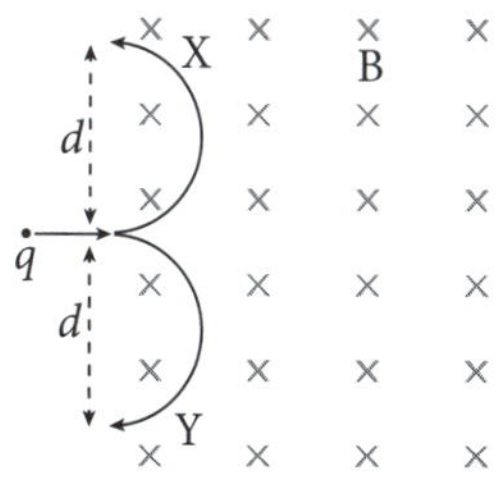

Figure 4.19 Two charged particles moving into a magnetic field at 90° to the field

a Find the charge on X and Y. (2 marks)

b Has the magnetic field done work on either charge particle? Justify your answer. (2 marks)

c The field changes the direction of the velocity of both particles in the example shown in Figure 4.19. Explain how a charged particle could be fired into the field and continue to move in a straight line. (2 marks)

9 Figure 4.20 shows an electron travelling horizontally at 200 ms^{-1} entering an electric field of strength 4 NC^{-1}. The electron was observed to move vertically 1 cm when it was between the plates.

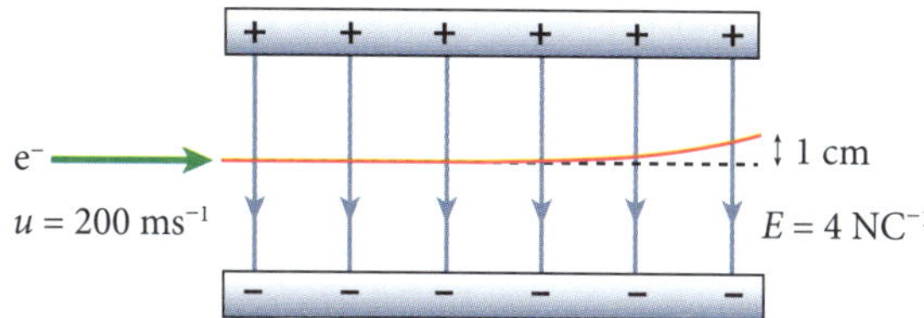

Figure 4.20 Deflection of an electron initially travelling at 200 ms^{-1} by an electric field

a If the separation between the plates was doubled and all other variables were kept constant, how far would the electron now move vertically while in the field? (2 marks)

b If the original plate separation was used but the electron's initial velocity was halved, how far would be it be displaced vertically by the field? (2 marks)

c A magnetic field is now added with the original plate separation that reduces the vertical deflection to zero. Determine the magnetic flux density required to enable the charge to move in a straight line between the plates. (2 marks)

10 Figure 4.21 shows an alpha particle ($m = 6.68 \times 10^{-27}$ kg and $q = +3.204 \times 10^{-19}$ C) moving anticlockwise around a horizontal circle of radius 2 mm with period of 2×10^{-4} s in a magnetic field.

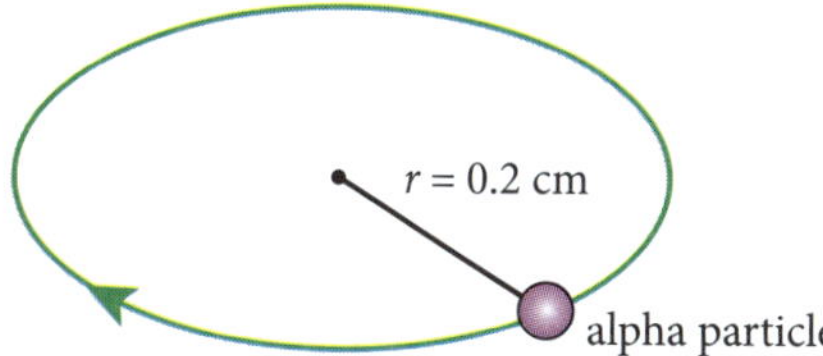

Figure 4.21 Alpha particle moving in a horizontal circle in a magnetic field (field not shown)

a Find the tangential velocity of the alpha particle. (1 mark)

b Determine the centripetal force on the alpha particle. (1 mark)

c Find the magnitude and direction of the magnetic field. (2 marks)

ANSWERS

KEY QUESTIONS

Key questions ➲ p. 51

1 The direction of the electric field at a point in space is the direction that a positively charged particle placed at the point would experience a force.

2 Positively charged particles experience a force in the direction of the electric field, while negatively charged particles experience a force in the opposite direction to the electric field. The force experienced by the charged particle is proportional to the charge and to the strength of the electric field (i.e. $F = qE$).

Key questions ➲ p. 53

3 In a uniform electric field, the potential difference between two points in the direction of the field is proportional to the distance between the points (i.e. $V = Ed$). We can therefore think of the electric field strength as being related to how much the potential difference changes per unit length in the field (i.e. $E = V/d$).

4 The electric field between parallel charged plates is constant and is related to the voltage across the plates by $E = V/d$ where d is the plate separation.

5 The work done in moving a unit positive charge from one point to another in an electric field is the electrical potential difference between the points (i.e. $W = Fs = qEd = qV$).

Key question ➲ p. 54

6 A charged particle moving in an electric field is analogous to a mass moving in a gravitational field. If the field is uniform it will exert a constant force on the charged particle, causing it to accelerate at a constant rate in one direction in the same way gravity causes masses to accelerate at a constant rate in one direction near the surface of the Earth. Note that, near the Earth's surface, gravity causes all masses to fall with an acceleration of 9.8 ms^{-2}, while charged particles in an electric field accelerate at a rate given by $a = qE/m$.

Key questions ➲ p. 57

7 If a moving charged particle has some component of its velocity perpendicular to the magnetic field, a force will act on the charged particle in a direction that is perpendicular to the velocity of the particle and perpendicular to the magnetic field. The right-hand palm rule is used to find the direction of the force on the particle.

8 The magnitude of the force on a charged particle moving in a magnetic field is proportional to the size of the charge, the component of the velocity perpendicular to the field and to the strength of the field (i.e. $F = qvB\sin\theta$).

9 Because the force on a charged particle moving in a magnetic field is always perpendicular to the particle's velocity, the force is a centripetal force ($F_B = F_C$ or $qvB\sin\theta = mv^2/r$).

Key question ➲ p. 58

10 A charged particle in an electric field experiences a constant force either in the direction of the field (for a positively charged particle) or in the opposite direction of the field (for a negatively charged particle), whether the charged particle is moving or not. In contrast, a charged particle in a magnetic field only experiences a force if it is moving with some component of is velocity perpendicular to the direction of the field. Unlike the electric field force, the force on a charged particle in a magnetic field is perpendicular rather than parallel to the field lines. Both fields cause the charged particle to accelerate but an electric field changes the magnitude of the velocity, while a magnetic field changes the direction of the particle's velocity. Thus an electric field increases the kinetic energy of the particle but the magnetic field simply changes the direction of travel of the particle.

HSC EXAM-TYPE QUESTIONS

Objective-response questions

1 **B**. A negatively charged particle experiences a force in the opposite direction of the electric field and, as the field lines are more closely spaced in this direction, the force and hence the acceleration of the particle will increase as it moves. **A** and **C** are incorrect as they give the direction of the force on a positively charged particle rather than on a negatively charged particle. **D** is incorrect as it implies the charged particle is moving into a region with a decreasing field strength and this would be true only if the field lines were further apart.

2 **B**. The radius of curvature is given by $qvB = mv^2/r$ and hence $r = mv/qB$. As r is proportional to v/B, doubling v and halving B will increase the radius by a factor of 4. **A**, **C** and **D** are incorrect as all give the wrong numerical answer.

3 **D**. The Earth's magnetic field is directed from the south to the north (unlike the field of a bar magnet). Applying the right-hand grip rule gives a force towards the east for a positive particle and hence the force must be towards the west on a negative particle. **A** and **B** are incorrect as the force must always be perpendicular to the field. **C** is incorrect as it is the direction that a positively charged particle would experience a force.

4 **A**. The force on the positively charged particle is upwards when it enters the field and this would only happen if the field was directed into the page (by the right-hand palm rule). The radius of curvature ($r = mv/qB$) is proportional to the velocity of the charged particle and, as the radius is decreasing, the charged particles must be slowing down. **B** is incorrect as the radius would continually increase if the velocity of the particles was increasing. **C** and **D** are incorrect because the right-hand palm rule shows that the charged particles would experience a force in the opposite direction to that shown in the diagram if the magnetic field was directed outwards.

5 **B**. This is because it is like projectile motion. To increase the range of the projectile we could increase the initial speed, decrease the rate at which it accelerates downwards or increase the vertical distance it must fall before hitting the ground. Moving the plates further apart would decrease the strength of the electric field between the plates (as $E = V/d$), which would decrease the downwards acceleration and increase the range. It would also increase the vertical distance to the lower plate, which would increase the range. **A** and **C** are incorrect as decreasing the initial velocity would decrease the range. **D** is incorrect because increasing the voltage would increase the strength of the electric field between the plates, which would decrease the range.

Extended-response questions

6 EM This questions tests students' understanding of how a pair of oppositely charged parallel plates produces a uniform electric field and their ability to calculate various aspects of the motion of a charged particle in an electric field.

a $E = V/d$ and hence $V = Ed = 200 \times 0.02 = 4$ V. ✓ As the electric field is directed upwards the lower plate will be at a higher potential ✓ (as the field would do work on a positively charged particle, moving it from the lower to the upper plate).

b $F = qE = 1.602 \times 10^{-19} \times 200 = 3.204 \times 10^{-17}$ N upwards. ✓

$a = F/m = \dfrac{3.204 \times 10^{-17}}{1.67 \times 10^{-27}} = 1.92 \times 10^{10}$ ms^{-2} upwards. ✓

c We could use the equations of motion ($v^2 = u^2 + 2as$) or conservation of energy to calculate the final velocity. Applying the law of conservation of energy:

Kinetic energy gained = potential energy lost

$\frac{1}{2}mv^2 = qV$

Or

$v = \sqrt{(qV/m)} = \sqrt{\dfrac{(2 \times 1.602 \times 10^{-19} \times 4)}{1.67 \times 10^{-27}}} = 2.77 \times 10^4$ ms^{-1} ✓

d The work done $= W = qV = 1.602 \times 10^{-19} \times 4 = 6.41 \times 10^{-19}$ J ✓

7 EM This question tests students' understanding of how the electric field, electrical potential and force on a charge varies at different points between two oppositely charged parallel plates.

a The electric field $E = V/d = \dfrac{200}{0.1} = 2000$ (Vm^{-1}) will be constant at all points between the charged plates, as shown in Figure A4.1. (1 mark for correctly labelled axes and 1 mark for horizontal line)

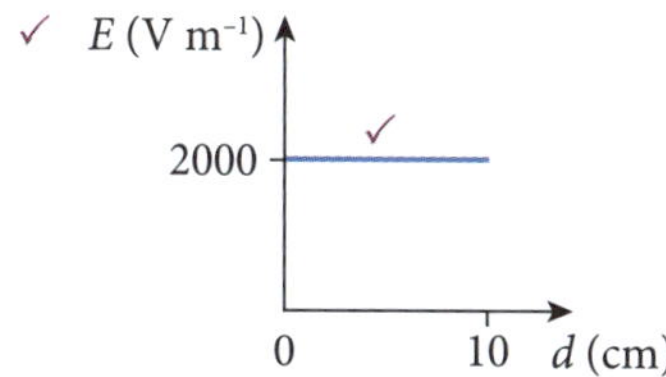

Figure A4.1 Electric field against distance

b The potential difference will vary at a constant rate from 200 V to zero at the negative plate, as shown in Figure A4.2. (1 mark for correctly labelled axes and 1 mark for straight diagonal line)

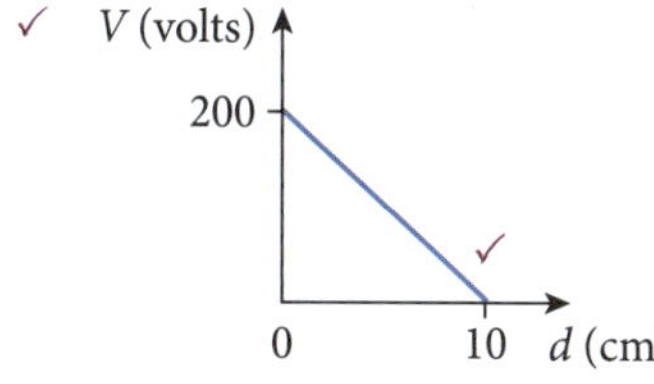

Figure A4.2 Potential difference as a function of position for the positively charged plate

c Figure A4.3 illustrates that the charged particle will accelerate at a constant rate from the positive plate until it collides with the negative plate at its maximum velocity (v_{max}).

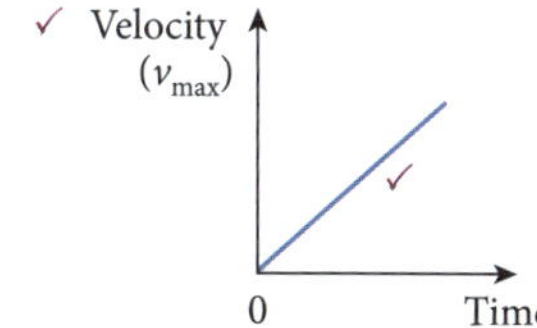

Figure A4.3 Velocity of a positively charged particle released from the positive plate against time

8 EM This question tests students' understanding of the force on a charged particle in a magnetic field, circular motion and the work.

a The magnetic force is a centripetal force:

$qvB = mv^2/r$ and hence

$q = mv/Br = \dfrac{2 \times 10^{-12} \times 120}{(0.2 \times 1 \times 10^{-4})} = 1.2 \times 10^{-5}$ C = 12 μC

$q_X = +12\ \mu$C ✓ and $q_Y = -12\ \mu$C ✓

b No work has been done by the magnetic field ✓ on the charged particle as the force is perpendicular to the displacement of the particle. ✓ Work $= F_{\parallel}s = Fs\cos\theta = 0$

c The charged particle would have to have to be moving parallel ✓ to the magnetic field as a magnetic field exerts a force only on charged particles that have some component perpendicular to the field. ✓

9 EM This question tests students' ability to relate the equations of projectile motion to a charge moving in an electric field, and the net force produced on a moving charged particle by crossed electric and magnetic fields.

a The acceleration of the charged particle in the field will be given by:

$a = F/m = qE/m$

Now as $E = V/d$, doubling the separation of the plates will halve the strength of the electric field between the plate and hence the acceleration will also halve. ✓

As the initial velocity is zero, the vertical displacement of the charged particle will be given by:

$s = \frac{1}{2}at^2$

Now as the acceleration is halved, the vertical displacement will also be halved and hence the charge will move 0.5 cm upwards. ✓

b If the initial velocity is halved, the time the charged particle is in the field will be doubled because the particle will take twice as long to reach the other end of the plates. ✓ Now as the vertical displacement is given by $s = \frac{1}{2}at^2$, if the time is doubled, the displacement will be increased by a factor of 4. Thus the vertical displacement would be $4 \times 1 = 4$ cm. ✓

c To pass straight through, the net force on the electron must be zero and hence the magnetic force must be equal and opposite to the electric force:

$qvB = qE$ and hence

$B = E/v = \dfrac{4}{200} = 0.02$ T ✓ into the page ✓ (by the right-hand palm rule)

10 EM This question test students' ability to apply their understanding of uniform circular motion to the motion of a moving charge in a magnetic field.

a Applying the equation for the velocity of an object undergoing uniform circular motion:

$v = 2\pi r/T = \dfrac{2\pi \times 0.002}{2 \times 10^{-4}} = 62.83$ ms^{-1} ✓

b Applying the equation for centripetal force:

$F_c = mv^2/r = \dfrac{(6.68 \times 10^{-27} \times (62.83)^2)}{0.002} = 1.32 \times 10^{-20}$ N ✓

c As the magnetic force is a centripetal force:

$F_c = F_B = qvB$, hence

$B = F_c/qv = \dfrac{1.32 \times 10^{-20}}{(3.204 \times 10^{-19} \times 62.83)}$

$= 6.56 \times 10^{-4}$ T ✓ upwards ✓ (by right-hand palm rule)

CHAPTER 5

THE MOTOR EFFECT

MODULE 6
ELECTROMAGNETISM

INQUIRY QUESTION:

Under what circumstances is a force produced on a current-carrying conductor in a magnetic field?

A force is exerted on a current-carrying wire in a magnetic field when the direction of the current in the wire has a component that is perpendicular to the direction of the magnetic field. When a magnetic field exerts a force on the current-carrying wire, the force is in a direction that is perpendicular to the magnetic field and perpendicular to the current direction (i.e. to the wire). The direction of the force can be determined using the right-hand palm rule.

1 Force on a current-carrying wire in a magnetic field

» Students investigate qualitatively and quantitatively the interaction between a current-carrying conductor and a uniform magnetic field $F = lI_{\perp}B = lIB \sin\theta$ to establish:

- conditions under which the maximum force is produced
- the relationship between the directions of the force, magnetic field strength and current
- conditions under which no force is produced on the conductor.

Investigating the force on a current-carrying wire in a magnetic field

➔ We saw in Chapter 4 that a force is exerted on a charge that has some component of its velocity perpendicular to a magnetic field. The charges moving in a current-carrying wire in a magnetic field will also experience a force and this may result in a net force appearing on the wire.

➔ Students should conduct a range of experiments to investigate the force on a current-carrying wire in a magnetic field. They should investigate:

- how the strength of the applied magnetic field and the magnitude of the current in the wire affect the force on the wire
- how the direction of the force on the wire is related to the direction of the applied magnetic field and the direction of the current
- how the magnitude of the force on the wire is related to the relative orientation of the wire and the applied magnetic field.

➔ Any of the following four experiments could be used by students to investigate all the questions specified above, but students should try to conduct as many as possible because each experiment will enable students to examine the force on the wire in a different context. The four investigations appear in order from the most straightforward to the most experimentally demanding.

FIRSTHAND INVESTIGATION 1

Determining the direction of the force on a current-carrying wire in a magnetic field

Students use a current-carrying wire or a coil made up of multiple turns of wire to investigate the force on a current-carrying wire when it is placed in a magnetic field. Figure 5.1 shows the direction of the force on the wire that will result.

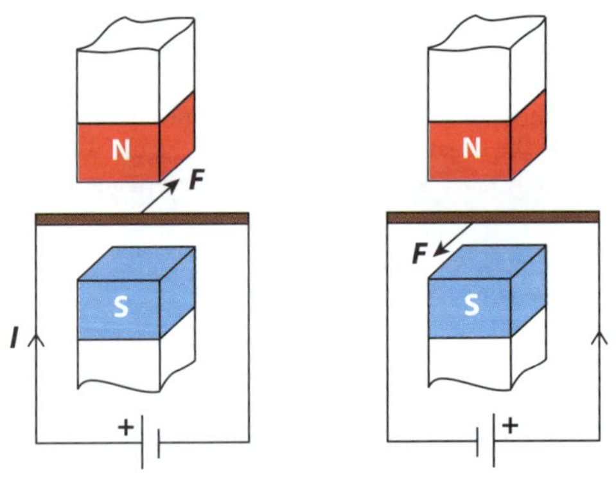

Figure 5.1 Simple experiment to investigate the force on a current-carrying wire in a magnetic field

Students should find that the greatest force is exerted when the wire is perpendicular to magnetic field, and that the direction of the force is given by the right-hand palm rule with the thumb pointing in the direction of the current in the wire (i.e. the direction of positive charge movement). Note that a rheostat might need to be placed in the circuit to reduce the current if the power source becomes overloaded.

FIRSTHAND INVESTIGATION 2

Estimating the magnitude of the force on a current-carrying wire in a magnetic field

Students can use the pendulum method illustrated in Figure 5.2 to investigate the force on a current-carrying conductor in a magnetic field. Provided the horseshoe magnet is moved to ensure the horizontal wire remains in the centre of the magnetic field, the angular displacement of the pendulum from the vertical can be used to estimate the strength of the force on the wire. Students may qualitatively (or quantitatively) investigate the importance of the strength of the magnetic field, the magnitude of the current in the wire and the length of the wire in the magnetic field (e.g. by using two horseshoe magnets adjacent to each other) on the force that appears on the suspended wire.

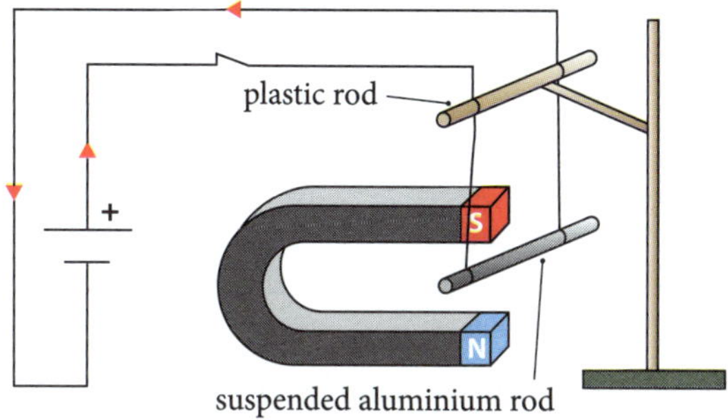

Figure 5.2 Apparatus for investigating the magnitude of the force on a current-carrying wire in a magnetic field

FIRSTHAND INVESTIGATION 3

Measuring the force on a current-carrying wire in a magnetic field

Figure 5.3 shows how an electronic balance can be used to measure the force on a current-carrying wire in a magnetic field. The wire in this case is a rigid copper rod supported by a pair of retort stands and clamps.

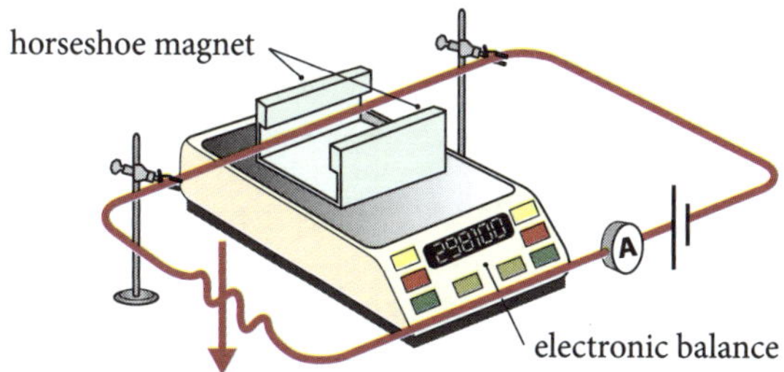

Figure 5.3 Using an electronic balance to measure the force on a current-carrying wire in a magnetic field

Because of Newton's third law, the force on the wire will be accompanied by an equal and opposite force on the magnet. The change in the measured mass of the magnet (Δm) that occurs when the current is switched on is therefore due to the reaction force to the force exerted on the wire by the magnetic field. This experiment can be used to investigate the variables that determine the size of the force on the wire. For example, by plotting the force on the wire ($F = \Delta mg$) against the current in the wire, students will obtain a straight-line graph indicating the force on the wire is proportional to the magnitude of the current in the wire.

FIRSTHAND INVESTIGATION 4

Alternate method of measuring the force on a current-carrying wire in a magnetic field

Students may investigate the force on a current-carrying wire in a magnetic field using a current balance such as the one shown in Figure 5.4. A current balance uses the weight of a small mass to balance the force exerted on the wire by the magnetic field. The weight (mg) of the mass added to balance the system will be equal to the force (F_{wire}) exerted on the wire by the magnetic field (i.e. $F_{wire} = mg$).

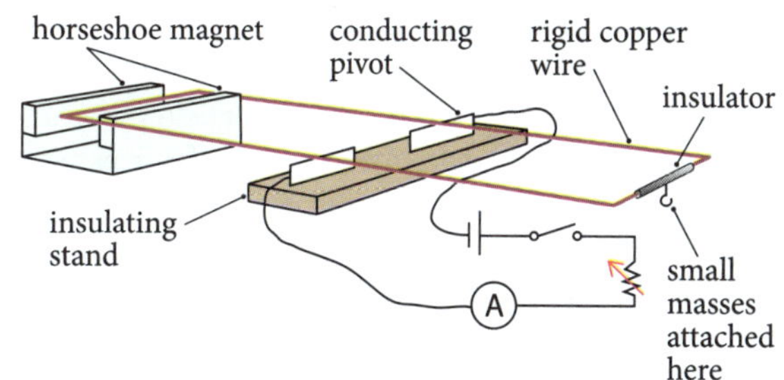

Figure 5.4 Typical current balance apparatus

KEY QUESTION

1 **Under what circumstances does a current-carrying wire experience a force when placed in a magnetic field?**

Answers p. 75

Analysing the force on a current-carrying wire in a magnetic field

- The **motor effect** is the term used to describe the force exerted on a current-carrying wire in a magnetic field.

motor effect: term used to describe the force that appears on a current-carrying wire in a magnetic field

- We can determine the force on a current-carrying wire by recalling that the force on a moving charge is given by:

  $F = qv_{\perp}B = qvB\sin\theta$

  If we call the length of wire in the magnetic field l and imagine it takes the charges a time t to travel this distance, the velocity of the charges will be $v = l/t$. Substituting this value into the above equation gives:

  $F = q(l/t)B\sin\theta = F = (q/t)lB\sin\theta$

  As $q/t = I$ we can write this expression as:

  $F = lIB\sin\theta$

- Thus the magnitude of the force on a straight current-carrying wire in a magnetic field is given by:

$$F = lI_{\perp}B = lIB\sin\theta$$

where l is the length of wire in the magnetic field
$I_{\perp} = I\sin\theta$ is the component of the current moving perpendicular to the magnetic field, and
B is the magnetic flux density

magnetic flux density: the intensity of the magnetic field, measured in the SI unit of tesla

- From this relationship we see that the maximum force on the wire occurs when the current-carrying wire is perpendicular to the external magnetic field (i.e.) and that no force will be exerted on the wire if it is parallel to the external field. We can understand this by recalling that the force on the individual charges will be maximum when they move perpendicular to the magnetic field and zero if they move parallel to the field.
- The direction of the force on the current-carrying wire, just like the force on a moving positive charge in a magnetic field, is perpendicular to the direction of charge movement and perpendicular to the applied magnetic field. To find the direction we apply the **right-hand palm rule**, as shown in Figure 5.5. We again point the thumb in the direction of the positive charge velocity (i.e. the current), the fingers in the direction of the applied magnetic field, and the force direction is out of the palm.

right-hand palm rule: a rule used to find the direction of the force on a moving charge or current-carrying wire in a magnetic field

- Another way to visualise the force direction is to consider the superposition of the applied magnetic field and magnetic field produced by the current in the wire. The force on the wire is in a direction from the region of high field intensity to the direction of low field intensity, as shown in Figure 5.6.

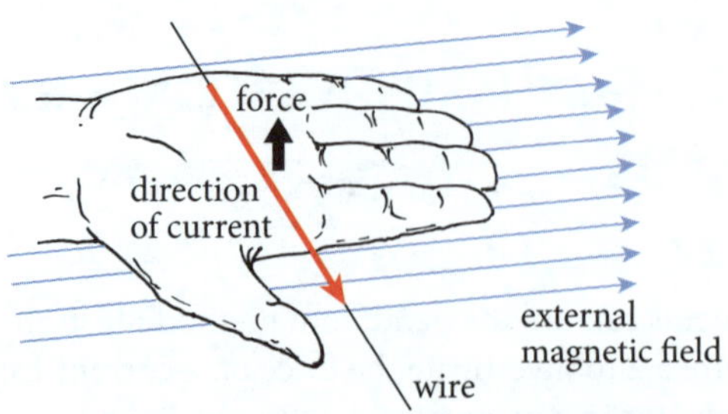

Figure 5.5 Applying the right-hand palm rule to find the direction of the force on a current-carrying wire in a magnetic field

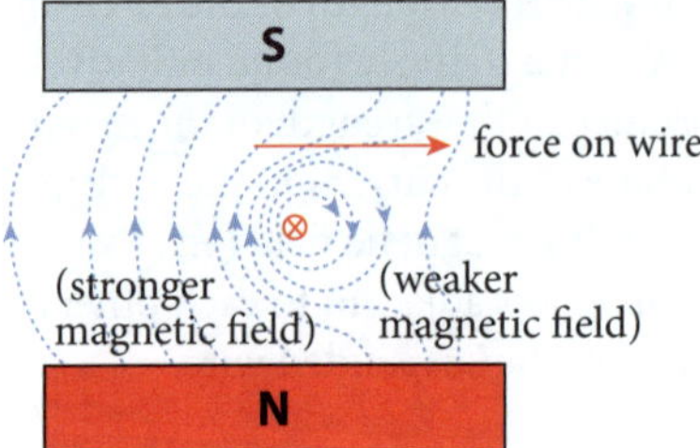

Figure 5.6 Relationship between magnetic field intensity and the force on a current-carrying wire in a magnetic field

→ KEY QUESTIONS

2 **What factors determine the size of the force that a magnetic field exerts on a current-carrying wire?**

3 **How can we determine the direction of the force on a current-carrying wire in a magnetic field?**

Answers ⮌ p. 75

- Because the motor force involves three dimensions and requires the application of the right-hand palm rule, students often have trouble solving problems involving the motor force. Carefully working through the following examples will help to give students the confidence needed to independently solve motor effect problems.

EXAMPLE 1

A 20 V potential difference is placed across a 50 cm long wire with a resistance of 2.5 Ω in a 0.05 T magnetic field, as shown in Figure 5.7. Find the magnitude and direction of the force on the wire.

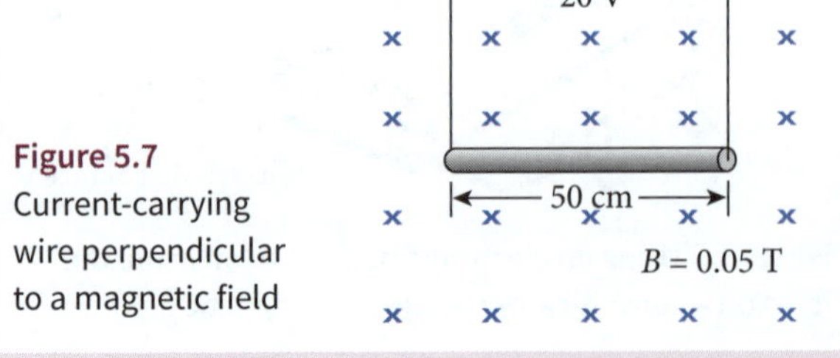

Figure 5.7 Current-carrying wire perpendicular to a magnetic field

Ohm's Law relates the current, resistance and potential difference

Answer:

We first use Ohm's law to find the current in the wire:

$I = V/R = \frac{20}{2.5} = 8\ \text{A}$

Now the force on the wire will be given by:

$F = lIB\sin\theta = 0.5 \times 8 \times 0.05 \times \sin(90°) = 0.2\ \text{N}$

Applying the right-hand palm rule, we see the force is directed up the page.

EXAMPLE 2

A simple electromagnetic accelerator is shown in Figure 5.8. The device consists of a 20 g rod carrying a current of 500 A resting on a pair of parallel, horizontal, frictionless rails. The rails are separated by 25 cm and a vertical magnetic field is applied to the rod. The rod accelerates for 1 m and leaves the device with a velocity of 50 ms^{-1}. Calculate the magnetic field applied to the rod.

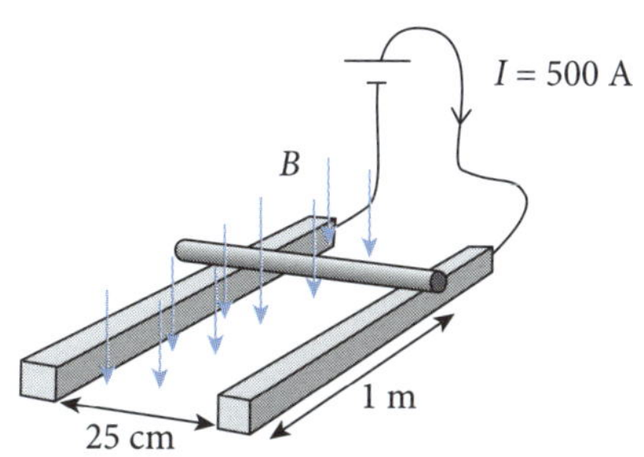

Figure 5.8 Current-carrying rod resting on two parallel, frictionless rails

If the mass and acceleration of an object are known, we can determine the force on the object using Newton's second law

Answer:

We first need to find the acceleration of the rod using:

$v^2 = u^2 + 2as$ and as $u = 0$, $a = v^2/2s = \frac{50^2}{2 \times 1} = 1250\ \text{ms}^{-2}$

Now the force required to produce this acceleration is:

$F = ma = 0.02 \times 1250 = 25\ \text{N}$

This force is produced by the magnetic field:

$F = lIB\sin\theta$ and, as $\theta = 90°$,

$B = F/\ lI = \frac{25}{0.25 \times 500} = 0.2\ \text{T}$ downwards

EXAMPLE 3

In the motor effect experiment shown in Figure 5.9, the electronic balance measured the mass of the magnets to be 2.600 kg when a current of 10 A was flowing in the circuit. Note that the magnetic field strength was 0.1 T between the magnets and that the rod is fixed at the points A and B.

What mass would the balance measure if the current in the wire was switched off?

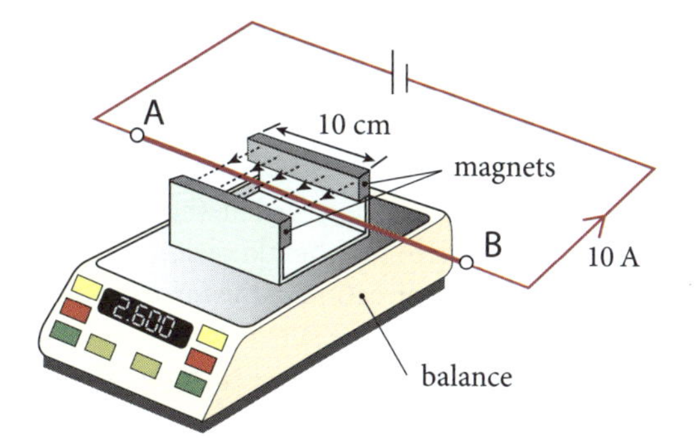

Figure 5.9 Motor effect experiment

The force on a current-carrying wire in a magnetic field will produce an equal and opposite force on the magnet that produced the field

Answer:

The force on the magnets will be equal and opposite to the force on the current-carrying wire in the magnetic field. Applying the right-hand palm rule to the wire shows that the force on the wire is downwards. Hence the reaction force on the balance will be upwards, reducing the weight of the magnets.

To find the size of the force:

$F = lIB\sin\theta = 0.1 \times 10 \times 0.1 = 0.1\ \text{N}$ upwards

The weight measured by the balance when the current was flowing was:

$w = mg = 2.6 \times 9.8 = 25.48\ \text{N}$

This is the weight of the magnets minus the upwards force on the magnets. If the current was switched off, this upwards force would disappear and the magnets would have a weight of:

$w = 24.50 + 0.1 = 25.58\ \text{N}$

The actual mass of the magnets (i.e. the mass measured by the balance with zero current flow) is therefore:

$m = w/g = \frac{25.58}{9.8} = 2.610\ \text{kg}$

2 Investigating the force between parallel current-carrying wires

» Students conduct a quantitative investigation to demonstrate the interaction between two parallel current-carrying wires.

➔ Because a current-carrying wire is surrounded by a magnetic field, if a second current-carrying wire is placed in the field of the first wire, it will experience a force. The force between the wires is maximum when the wires are parallel to each other. Students should conduct firsthand investigations of parallel current-carrying wires to examine:

- how the magnitude of the force is related to the current in the wires and the separation between the wires
- how the direction of the force on the wires is determined by the wires carrying current in the same direction or in opposite directions.

➔ Before conducting the suggested firsthand investigations, students should try to use their knowledge of magnetic fields to predict the direction of the force on each of the parallel wires. To do this, students must first think about the magnetic field surrounding a current-carrying wire (i.e. apply the right-hand grip rule) and the direction of the force that would appear on a second current-carrying wire if it was placed in the magnetic field (i.e. apply the right-hand palm rule).

FIRSTHAND INVESTIGATION 5

Demonstrating the force on parallel current-carrying conductors

A pair of parallel strips of aluminium foil, as shown in Figure 5.10, can be used to investigate the force on parallel current-carrying conductors.

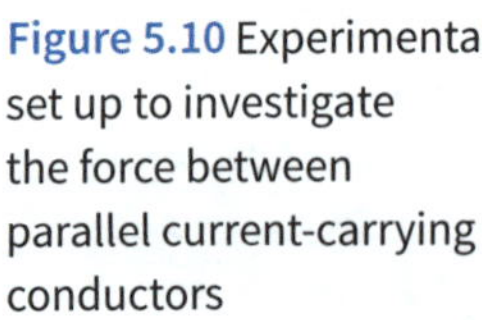

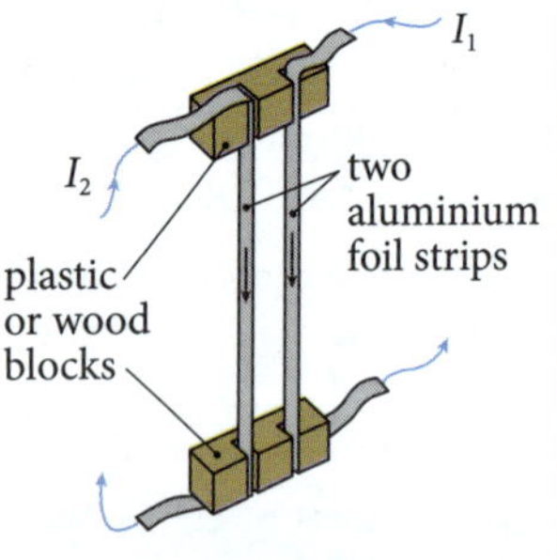

Figure 5.10 Experimental set up to investigate the force between parallel current-carrying conductors

The insulating blocks should be supported by a retort stand to hold the aluminium strips loosely (i.e. tensioned) and the current should then be passed through the aluminium strips. When current flows in the same direction in both strips, students will observe that the foil strips will bow inwards towards each other, indicating a force of attraction between the conducting strips (i.e. wires). Students will observe that the foils bow outwards when current flows in the opposite direction in each strip, indicating a force of repulsion between the strips.

Students could investigate the effect of increasing the current in the wires, changing the length of the foil strips and changing the separation between the strips on the force between the strips.

Students should also try to explain their observations by considering the magnetic field around the current-carrying foil strips.

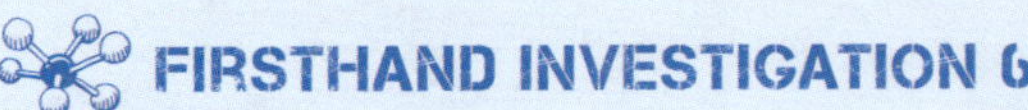

FIRSTHAND INVESTIGATION 6

Investing the force between parallel current-carrying conductors quantitatively

Quantitative analysis of the force between parallel current-carrying conductors can be undertaken with an experiment like the one shown in Figure 5.11. In this experiment a large square coil of wire, clamped at the points marked A and B on the diagram, is used to produce a larger magnetic field and hence intensify the force on the suspended parallel wire.

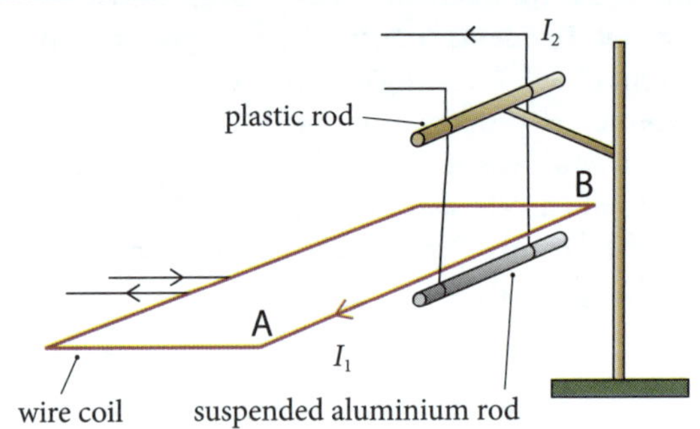

Figure 5.11 Set-up of an experiment to quantitatively investigate the force between parallel current-carrying conductors

If the sides of the square coil are large compared with the separation between the suspended rod and the coil, we can neglect the influence of the far side of the coil on the suspended rod. When current flows in opposite directions in the rod and in the side of the coil near the rod, the rod will swing away from the coil until it reaches an equilibrium. If the mass of the rod is known we can calculate the force on the rod by measuring the angle of deflection when the suspended rod reaches equilibrium. Figure 5.12 shows the relevant vector diagram for the forces on the rod when it is in equilibrium.

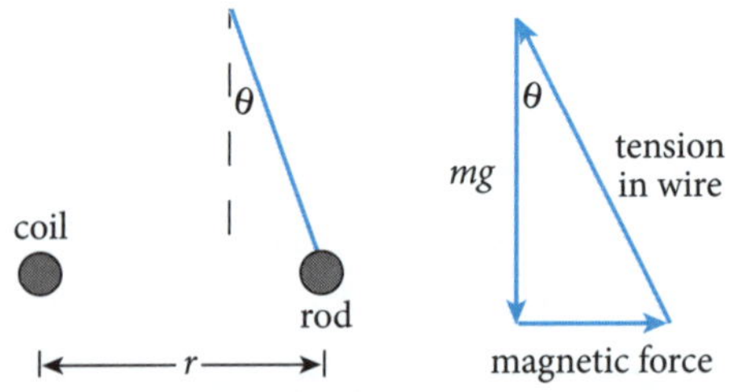

Figure 5.12 End view of suspended rod and force diagram for the rod

Calling the magnetic force F, we see from the vector diagram shown in Figure 5.12 that:

$\tan\theta = F/mg$ and hence $F = mg\tan\theta$

This experiment can be used to measure the force on the rod and to investigate the effect of the size of the current in the parallel wires (I_1 and I_2), the initial separation between the wires (r) and the common length of the wires (by using a longer aluminium rod). Note that the force on the rod measured with this technique is the force produced when the parallel wires are separated by the distance between the coil and the final position of the rod, not the original separation (r).

EXAMPLE 4

In the experiment illustrated in Figure 5.13, two parallel current-carrying wires that are carrying current in opposite directions repel each other and come to rest in a position where the angle between the suspending wires is 20°.

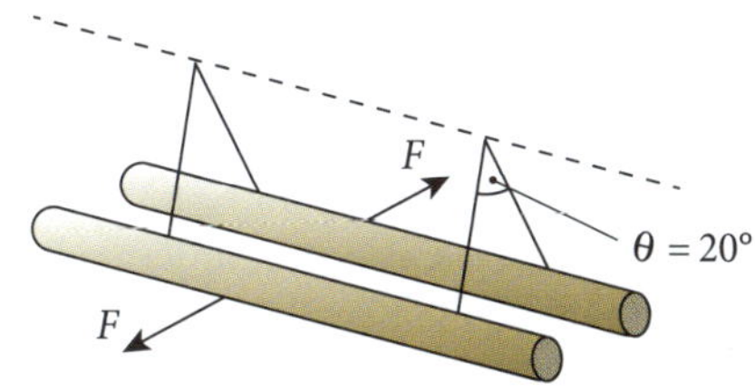

Figure 5.13 Two suspended, parallel current-carrying wires in equilibrium

Given that both rods have a mass of 20 g, find the magnetic force in each wire.

Recall that the vector sum of the forces acting on an object at rest must be zero

Answer:

By Newton's third law the force on both rods will equal and operate in the opposite direction. We can calculate the force by solving the vector diagram shown in Figure 5.14.

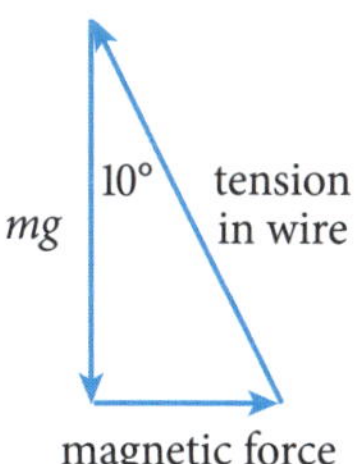

Figure 5.14 Vector diagram showing forces on one of the suspended rods

Calling the magnetic force F, we see from the vector diagram shown in Figure 5.14 that:

$F = mg\tan\theta = 0.02 \times 9.8 \times \tan 10° = 3.46 \times 10^{-2}$ N

3 Analysing the force between parallel current-carrying conductors

» Students analyse the interaction between two parallel current-carrying wires $\frac{F}{l} = \frac{\mu_0}{2\pi}\frac{I_1 I_2}{r}$ and determine the relationship between the International System of Units (SI) definition of an ampere and Newton's third law of motion.

→ You will recall from the Year 11 course that a long straight current-carrying wire is surrounded by a magnetic field that is directed in concentric circles around the wire. The direction of the field is given by applying the right-hand grip rule, as shown in Figure 5.15.

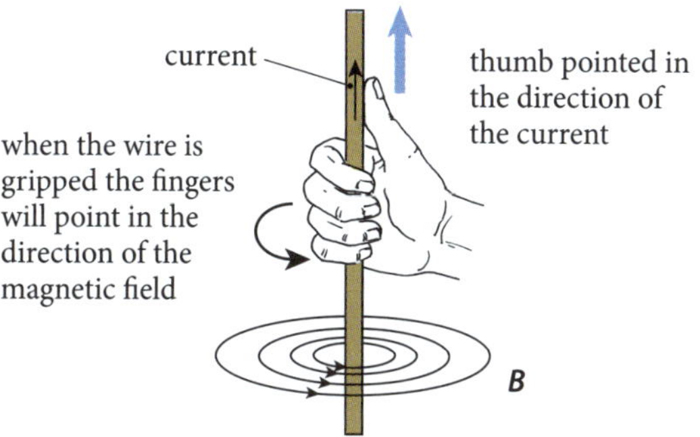

Figure 5.15 Right-hand grip rule and magnetic field around a current-carrying wire

→ You will also recall that the strength of the magnetic field at a distance r from the wire is given by:

$$B = \frac{\mu_0}{2\pi}\frac{I}{r}$$

where I is the current in the wire and $\mu_0 = 4\pi \times 10^{-7}$ NA^{-2} is a constant called the *magnetic constant or vacuum permeability*

→ Consider a long, straight, current-carrying wire carrying a current I_1. This wire will be surrounded with a magnetic field:

$$B = \frac{\mu_0}{2\pi}\frac{I_1}{r}$$

Now if we place a second current-carrying wire carrying a current I_2 parallel to and a distance r from the first wire, the second wire will experience a force given by:

$$F = lI_2 B\sin\theta = lI_2 B$$

Note that as the magnetic field from the first wire is perpendicular to the current in the second wire, $\sin\theta = \sin 90° = 1$. Substituting the value of B from the first equation into the second and rearranging gives an expression for the force per unit length on the second wire:

$$\frac{F}{l} = \frac{\mu_0}{2\pi}\frac{I_1 I_2}{r}$$

where F/l is the force per unit length on the wires
I_1 and I_2 are the currents in each wire
r is the separation between he wires and
$\mu_0 = 4\pi \times 10^{-7}$ NA^{-2} is the magnetic constant (called the *permeability of free space*)

- An equal and opposite force must appear on the first wire by Newton's third law. We can see how this occurs by considering the first current-carrying wire as being in the magnetic field produced by the second current-carrying wire. Rederiving the equation above with this assumption just changes I_1 to I_2 and I_2 to I_1, which gives the same value for the force. That is, the force on wire 1 is equal in magnitude to the force on wire 2.
- We can predict the direction of the forces on parallel current-carrying wires by considering the magnetic fields around the wires. We can consider one wire to be in the magnetic field produced by the other wire, as shown in Figure 5.16. Applying the right-hand grip rule enables us to find the magnetic field direction at the second wire, and applying the right-hand palm rule gives the direction of the force on the wire. By applying these techniques we find that wires carrying current in the same direction attract one another and wires carrying current in opposite directions repel one another.

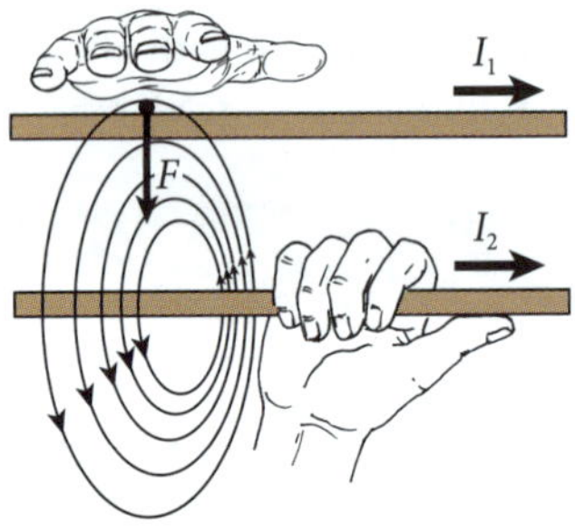

Figure 5.16 Appling both right-hand rules to find the direction of the force on parallel current-carrying wires

- We can also prove that parallel current-carrying conductors attract or repel one another by considering the superposition of the fields produced by each wire, as shown in Figure 5.17. When the current in the wires is flowing in the same direction, the combined magnetic field on the outside of the wires is more intense than the magnetic field between the wires. This would cause the wires to attract each other, as shown in diagram **a** in Figure 5.17. Current flowing in opposite directions will result in a more intense magnetic field between the wires, causing the wires to repel each other, as shown in diagram **b** in Figure 5.17.

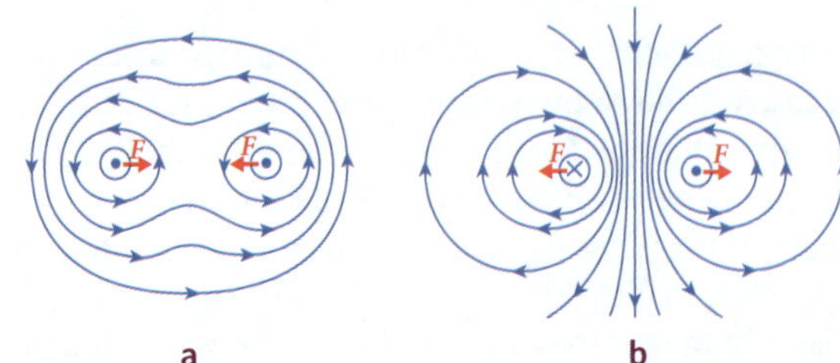

Figure 5.17 Magnetic field around two parallel current-carrying wires, carrying current in the same direction **a** and in opposite directions **b**

- The force between parallel current-carrying wires is used to define the SI unit of current, the **ampere**. One ampere is the current that must flow in two long, straight, parallel wires separated by 1 m to produce a force of attraction of 2×10^{-7} N per metre between the wires. This definition provides a standard value for the ampere that is reproducible anywhere in the world.

ampere: the SI unit of current and fundamental SI unit of electricity

- The ampere is the fundamental (or base) SI unit of electricity. Other derived electrical units are based on the ampere and the fundamental SI units of motion (m, kg and s). For example, the SI unit of charge is the coulomb = ampsecond = As, and the SI unit of electrical potential difference is the volt = joule per coulomb = JA^{-1}s^{-1} = kgm^{2}s^{-3}A^{-1}.

KEY QUESTIONS

4 **Under what circumstances do current-carrying wires exert a force on one another?**

5 **How can we determine the direction of the force on parallel current-carrying wires?**

6 **What determines the size of the force per unit length between parallel current-carrying wires?**

7 **How is the SI unit of current, the ampere, defined?**

8 **Why is the ampere chosen as the base SI unit for electricity?**

Answers ➲ p. 75

EXAMPLE 5

Figure 5.18 shows an experiment that was conducted to investigate the force between two parallel current-carrying wires. Each of the parallel rods is 1 m long and has a mass of 10 g. When the battery was connected, the top rod rose 2 cm and came to rest 4 cm above the lower rod.

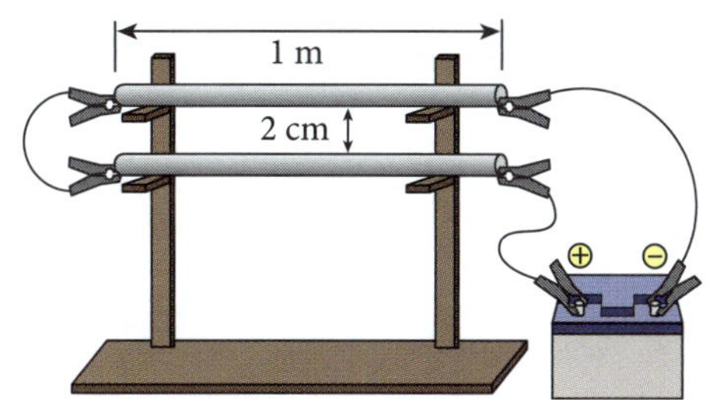

Figure 5.18 Investigating the force between parallel current-carrying wires

a **Explain why the top rod rose upwards when the current was switched on.**

b **Explain why the top rod came to rest 2 cm above its supports.**

c **Find the current in the rods that would cause the top rod to levitate 4 cm above the lower rod.**

The force between parallel current-carrying conductors decreases when the separation between the conductors is increased

Answer:

a Because the parallel rods are carrying current in opposite directions they will produce a magnetic field that will exert a force of repulsion between the rods. If this force is greater than the weight of the top rod (i.e. *mg*), there will be a net upwards force on the top rod that will cause it to accelerate upwards.

b Because the repulsive force between the rods is greater than the weight of the top rod, the top rod will initially accelerate upwards. But because the force between the rods decreases as the distance between the rods increases, the net upwards force on the top rod will decrease as it rises. When the distance between the rods is great enough for the upwards force to equal the weight of the rod, the net force on the rod will be zero and the rod will come to rest.

c The top rod will levitate when the net force on the rod is zero; that is, when the weight equals the repulsive force between the rods. Now as the rods have a 1 m length in parallel and the same current flows in each rod (i.e.):

$$mg = \frac{\mu_0}{2\pi}\frac{I^2}{r}$$

$$\text{or } I = \sqrt{\left(\frac{2\pi mgr}{\mu_0}\right)}$$

$$= \sqrt{\frac{2\pi \times 0.01 \times 9.8 \times 0.04}{4\pi \times 10^{-7}}}$$

$$= 140 \text{ A}$$

EXAMPLE 6

Figure 5.19 shows a long straight wire near a fixed rectangular loop of wire.

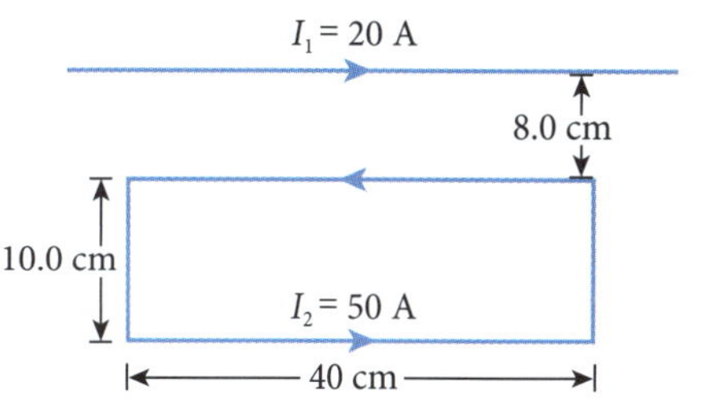

Figure 5.19 Current-carrying wire and rectangular loop

a **Without performing any calculations, explain how the direction of the net force on the long straight wire can be determined?**

b **Calculate the force on the long straight wire.**

The relative current direction on parallel current-carrying conductors determines the direction of the force between wires

Answer:

a The long straight wire will be repelled by the top horizontal wire in the rectangular loop because the wires are carrying currents in opposite directions. The wire will be attracted to the lower horizontal wire in the loop because it is carrying current in the same direction as the long straight wire. Because the top wire in the rectangular loop is closer to the long straight wire it will exert a greater force on the wire and hence the net force on the long wire will be upwards.

b We can find the net force by subtracting the downwards force on the wire from the upwards force. Calling the distance from the straight wire to the top parallel arm of the rectangular coil r_1 and the distance to the lower parallel wire r_2:

$$F_{net} = \frac{\mu_0}{2\pi}\frac{I_1 I_2 l}{r_1} - \frac{\mu_0}{2\pi}\frac{I_1 I_2 l}{r_2}$$

Note that the common parallel length (40 cm) and currents are the same for both pairs of wires and hence:

$$F_{net} = \frac{\mu_0 I_1 I_2 l}{2\pi}\left(\frac{1}{r_1} - \frac{1}{r_2}\right)$$

$$= (2 \times 10^{-7} \times 20 \times 50 \times 0.4)\left(\frac{1}{0.08} - \frac{1}{0.18}\right)$$

$$= 5.56 \times 10^{-4} \text{ N upwards}$$

CHAPTER SYLLABUS CHECKLIST

Are you able to answer these questions from the syllabus for this chapter? Tick each question as you go through the checklist if you are able to answer it. If you cannot answer a question, turn to the relevant page in the study guide to find the answer. For NESA key word meanings, go to www.educationstandards.nsw.edu.au and search 'key words'.

FOR A COMPLETE UNDERSTANDING OF THIS TOPIC:		PAGE NO.	✓
1	Can I describe an experiment that could be used to measure the force on a current-carrying wire in a magnetic field?	64–65	
2	Can I apply the right-hand palm rule to determine the direction of the force on a current-carrying wire in a magnetic field?	66	
3	Can I draw the magnetic field around a current-carrying wire in a magnetic field?	66	
4	Can I solve quantitative problems using the motor effect equation?	66–67	
5	Can I draw the magnetic field around a current-carrying wire and calculate the strength of the field at any distance from the wire?	69	
6	Can I use the right-hand grip rule and right-hand palm rule to predict the direction of the force on parallel current-carrying conductors?	70	
7	Can I draw the magnetic field lines around a pair of parallel current-carrying conductors?	70	
8	Can I outline how the force on parallel current-carrying conductors and Newton's third law are used to define the SI unit of current, the ampere?	70	
9	Can I solve quantitative problems involving the force on parallel current-carrying conductors?	70–71	

HSC EXAM-TYPE QUESTIONS

Objective-response questions (1 mark each)

1 **If a high-voltage DC transmission line near the equator carried current towards the north-east in the Earth's magnetic field, it would experience a force. What is the direction of this force?**

A towards the north-west
B vertically upwards
C vertically downwards
D towards the south-east

2 **A large current flows in a loop of wire, as shown in Figure 5.20. Which of the following statements best describes the force on the wire?**

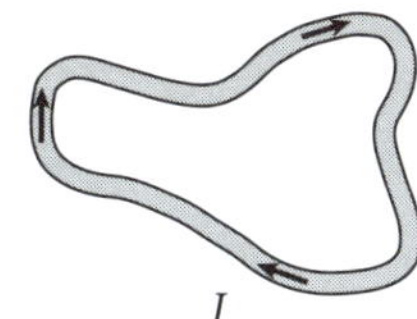

Figure 5.20 Current flowing in a loop of wire

A A force will appear that opposes the current and may cause the loop to rotate in the opposite direction to the current.
B No force is exerted on the wire.
C An inwards force is exerted on the wire that could cause the loop to compress.
D An outwards force is exerted on the wire, which could cause the loop to expand.

3 **Three wires are illustrated in Figure 5.21. If the top wire exerts a force X upwards on the bottom wire, what would be the net force on the centre wire?**

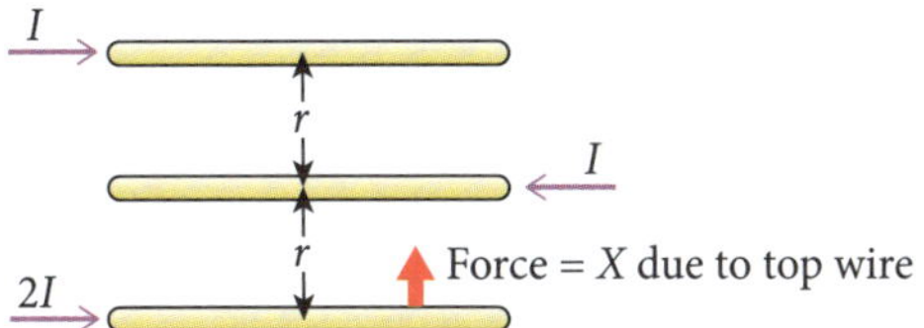

Figure 5.21 Three current-carrying parallel wires

A $3X$ upwards
B $2X$ upwards
C X upwards
D X downwards

4 **Consider the three wires shown in blue in Figure 5.22. The wires carry equal current but are not connected. If we only consider the forces on the wire due to the external magnetic field (i.e. we neglect the forces between the wires), which of the following statements about the forces on the wires is correct?**

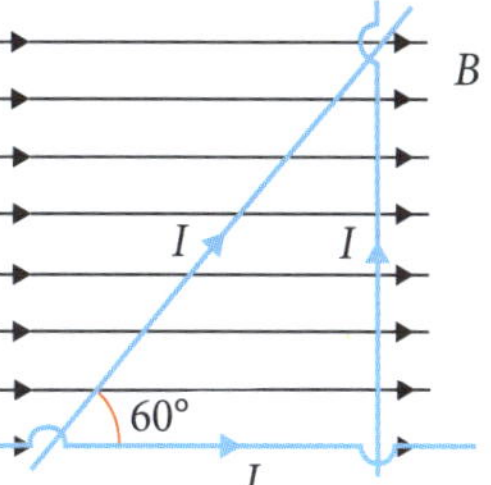

Figure 5.22 Three wires carrying equal currents in a magnetic field

A The magnetic field exerts an equal force on each wire.
B The magnetic field exerts the greatest force on the wire directed perpendicular to the field.
C The magnetic field exerts the greatest force on the wire directed parallel to the field.
D The magnetic field exerts an equal force on the wire perpendicular to the field and on the wire oriented at 60° to the field.

5 **Which of the following best defines the ampere?**

A the amount of current that must flow in two parallel wires 1 m apart to produce a force on the wires of 2×10^{-7} Nm^{-1}
B the amount of current that must flow in two long parallel wires 1 m apart to produce a force on the wires of 2×10^{-7} N
C the amount of current that must flow in two long parallel wires 1 m apart to produce a force on the wires of 1 Nm^{-1}
D the amount of charge that passes a point per second

Extended-response questions

6 **A student draws the magnetic field around a wire (carrying current out of the page) and the external magnetic field applied by a magnet, shown in Figure 5.23.**

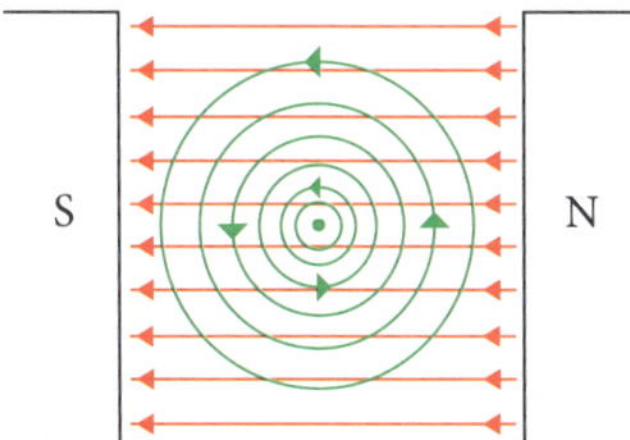

Figure 5.23 Magnetic fields drawn by a student

The diagram is incorrect because field lines should not cross one another.

a **In what direction would the field exert a force on the wire?** (1 mark)
b **Redraw the diagram to show the correct shape of the magnetic field.** (2 marks)

7 In an experiment to demonstrate the motor effect a teacher fixes a solid copper rod to the top of a laboratory bench and places a horseshoe magnet of mass 0.2 kg and width 2 cm over the rod, as shown in Figure 5.24. The magnetic field strength between the magnetic poles was measured to be 0.05 T.

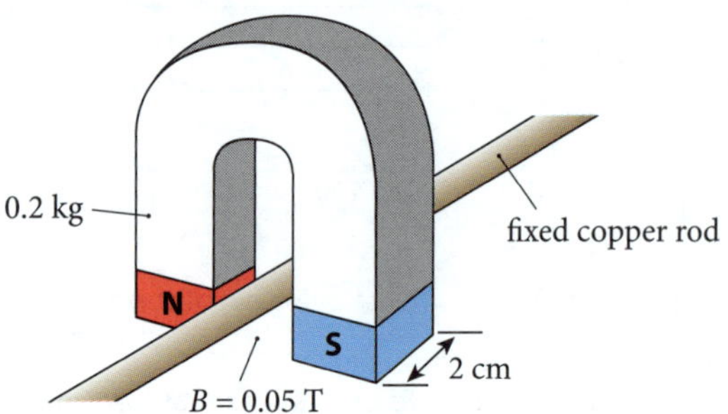

Figure 5.24 Motor effect experiment

The teacher hopes the motor effect will cause the magnet to lift upwards when current is passed through the copper rod.

a What direction would current need to flow in the rod to lift the magnet? (1 mark)

b What is the minimum current required to lift the magnet? (2 marks)

c Explain why this demonstration would be difficult to conduct in practice. (1 mark)

8 Consider a transmission line made up of one direct current (DC) line and one 50 Hz alternating current (AC) line, as shown in Figure 5.25.

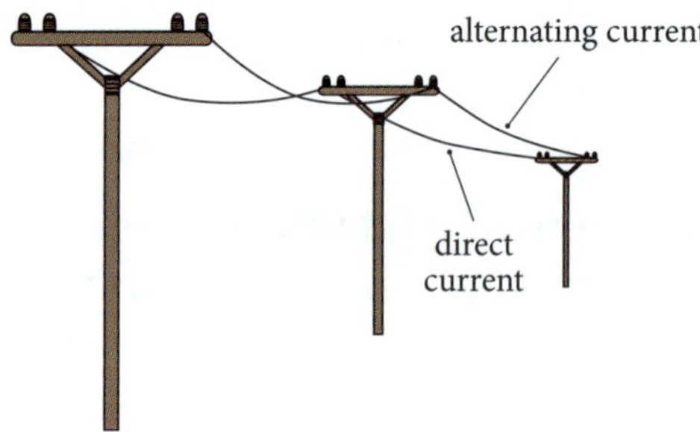

Figure 5.25 Transmission line

a Describe the force that the lines would produce on each other and explain why the force acts in this way. (2 marks)

b If the AC line was replaced with a DC line, how would the force on the lines change? (2 marks)

c If the separation between the DC lines was halved and the current in the lines doubled, how would the force on the lines change? (1 mark)

9 Three long straight parallel wires, each carrying a current of 2 A and separated from each other by 10 cm, are shown in Figure 5.26.

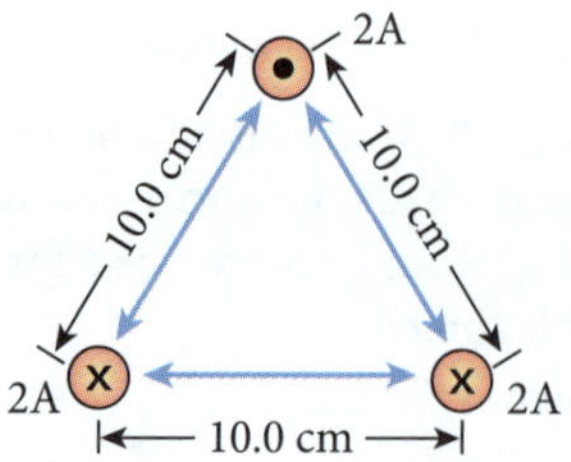

Figure 5.26 End view of three parallel current-carrying wires

With reference to Figure 5.26:

a If the current in the bottom left wire was switched off, what would the force per metre be on the top wire? (2 marks)

b If a current of 2 A flowed in each of the wires in the directions shown in Figure 5.26, what would the force per metre be on the top wire? (2 marks)

10 A square loop with sides 20 cm and carrying a current of 20 A is placed at 30° to a uniform magnetic field of strength 0.1 T, as shown in Figure 5.27. Note that the magnetic field and the loop of wire are in the plane of the page.

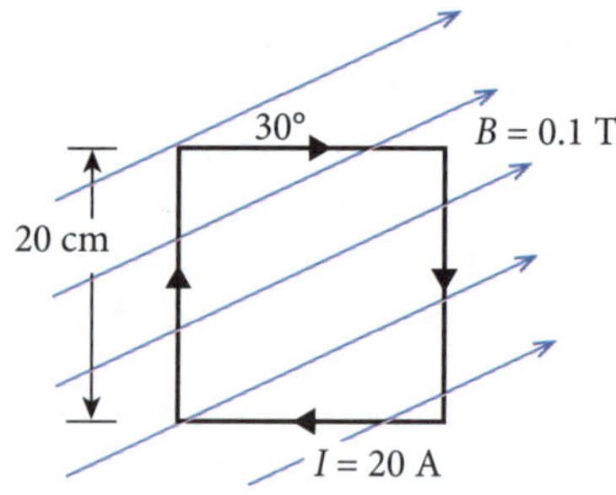

Figure 5.27 Square current-carrying coil in a magnetic field

a Find the magnitude and direction of the force on the top side of the coil. (2 marks)

b Find the magnitude and direction of the force on the left-hand side of the coil. (2 marks)

c Find the net force on the loop. (1 mark)

ANSWERS

KEY QUESTIONS

Key questions ➲ p. 65

1 A current-carrying wire will experience a force in a magnetic field if the current direction has a non-zero component perpendicular to the magnetic field.

Key questions ➲ p. 66

2 The force on a current-carrying wire in a magnetic field is proportional to the magnetic field intensity, the current, the length of wire in the magnetic field, and the sine of the angle between the magnetic field and the current direction.

3 The right-hand palm rule can be used to determine the direction of the force on the wire. When the thumb is pointed in the direction of the current and the fingers pointed in the direction of the magnetic field, the force on the wire will be directed out of the palm.

Key questions ➲ p. 70

4 Current-carrying wires will exert a force on each other if one wire carries current in a direction with a non-zero component parallel to the current in the other wire.

5 The force can be determined by applying the right-hand grip rule to one wire to determine the direction of the magnetic field on the other wire. The right-hand palm rule can then be applied to find the direction of the force on the second wire. This technique shows that parallel wires carrying current in the same direction attract each other, while parallel wires carrying current in opposite directions repel each other.

6 The force per unit length between parallel current-carrying wires is proportional to the product of the current in each wire and inversely proportional to the distance between the wires.

7 The ampere is defined as the current that must flow in two long straight parallel wires separated by 1 m to produce a force of attraction of 2×10^{-7} N per metre between the wires.

8 The ampere is the only SI unit in electricity that is defined using an experiment. This is important because it means that the standard ampere can be measured at any time, anywhere, by simply performing an experiment. All other SI electrical units can then be derived from the ampere and the other base units, such as the metre, kilogram or second. For example, the SI unit of charge: 1 coulomb = 1 ampsecond.

HSC EXAM-TYPE QUESTIONS

Objective-response questions

1 **B**. Applying the right-hand palm rule to the easterly component of the current and recalling that the Earth's magnetic field is directed from the south to the north on the surface gives the force direction as upwards. **A** and **D** are incorrect as the force must be perpendicular to the Earth's magnetic field. **C** is incorrect because the current would have to run towards the west to produce a downwards force at the equator.

2 **D**. Because the current on opposite sides of the loop is moving in opposite directions, opposite sides will repel each other. **B** is incorrect because the loop consists of sections of parallel wires carrying current in opposite directions; therefore there must be a force on the wire in the loop. **C** is incorrect as it implies the current is moving in the same directions of opposite sides of the loop. **A** is incorrect because a magnetic field exerts a force perpendicular to the current direction and hence has no component in the direction of current movement or in the opposite direction to the current movement.

3 **C**. Because the force between the wires is proportional to $\frac{I_1 I_2}{r}$, the top wire will exert a downwards force of X on the middle wire and the bottom wire will exert an upwards force of $2X$ on the middle wire, giving a net force of X upwards. **A**, **B** and **D** are incorrect as they each imply that the force between each pair of wires is not proportional to $\frac{I_1 I_2}{r}$.

4 **D**. Because the length of the component of the wire at 60° to the field that is perpendicular to the magnetic field will be the same length as the vertical wire (i.e. the wire that is perpendicular to the field) and, because the same current flows in both wires, the force on both wires will be equal. There is no force on the horizontal wire as it is parallel to the magnetic field.

A and **C** are incorrect as there would be no force on the horizontal wire (as the current moves parallel to the field) but a force on the other two wires. **B** is incorrect as the same force will be exerted on the wire that is perpendicular to the field and on the wire that is oriented 60° to the field because of the different lengths of the wires.

5 **A**. This is the correct definition of the ampere. **B** is incorrect as the length of the wires is not specified. **C** is incorrect as millions of amps would have the flow in the wires to produce a force large enough on the wires. **D** is incorrect as it is the definition of current not the ampere.

Extended-response questions

6 EM This question tests students' understanding of the superposition of magnetic fields and their ability to apply the right-hand palm rule.

a By applying the right-hand palm rule we see the force on the wire would be down the page. ✓

b As shown in Figure A5.1, when the two fields are in the same direction the resulting field will be stronger (field lines closer together) ✓ and when the two fields appose each other the field is less intense (field lines further apart). ✓

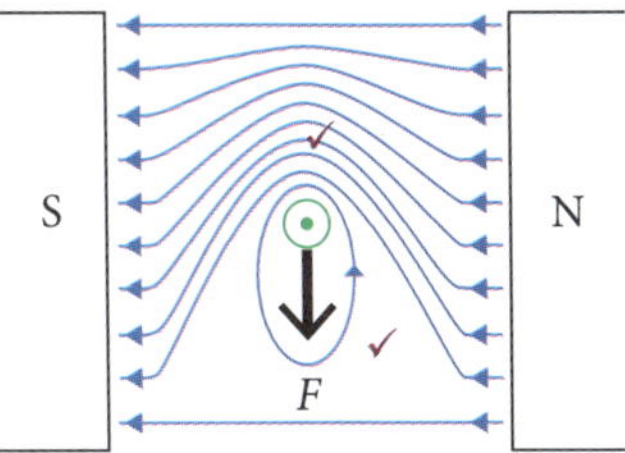

Figure A5.1 Superimposed fields of a magnet and current-carrying wire

7 EM This question tests students' understanding of the motor effect, reaction force and balancing forces.

a To place an upwards force on the magnet, a downwards force would need to be exerted on the rod. For this to occur, by the right-hand palm rule, current would have to flow along the rod into the page. ✓

b The force on the rod would be equal and opposite to the force on the magnet. ✓ Just to lift the magnet this force would have to be equal to the weight:

$lIB = mg$ or $I = mg/lB = \dfrac{0.2 \times 9.8}{0.02 \times 0.05} = 1960$ A ✓

c The current required is extremely large ✓ and would not be available in a school laboratory.

8 EM This question tests students' understanding of the factors that affect the magnitude and direction of the force between current-carrying conductors.

a Because the parallel wires would be carrying current in the same direction and opposite direction each half-cycle of the alternating current, the wires would alternately repel and attract one another ✓ 50 times per second. ✓

b The force between the wires would be a constant force of attraction if the DC currents in each wire were in the same direction, ✓ and a constant force of repulsion if the currents were in the opposite direction. ✓

c Because the force is proportional to the product of the currents and inversely proportional to their distance apart, the force between the wires would increase by $\dfrac{2 \times 2}{(1/2)} = 8$ times. ✓

9 EM This question tests students' understanding of the vector nature of the force between parallel current-carrying conductors.

a The force per metre would be given by:

$\dfrac{F}{l} = \dfrac{\mu_0}{2\pi}\dfrac{I_1 I_2}{r}$

$= \dfrac{2 \times 10^{-7}(2 \times 2)}{0.1}$

$= 8 \times 10^{-6}$ N ✓ away from the wire on the bottom right ✓ at a bearing of N330°

b The top wire will be repelled by both of the bottom wires. Figure A5.2 shows we can add these forces to find the net force on the wire.

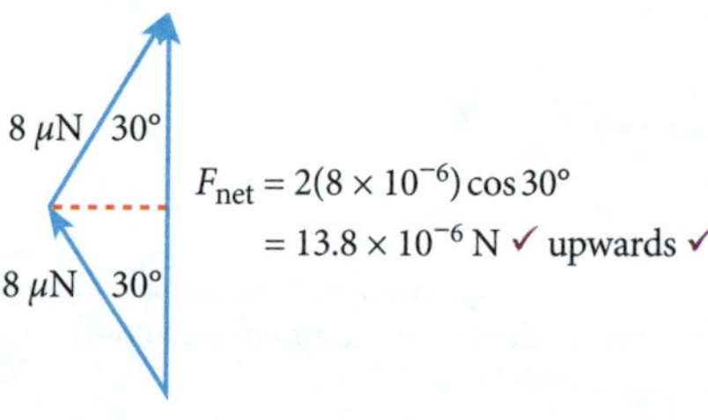

Figure A5.2 Vector diagram showing the addition of the forces acting on the top wire

10 EM This question tests students' ability to solve quantitative problems in which a current-carrying wire is at an angle to a magnetic field.

a The force on the top wire will be given by:

$F = lIB \sin\theta$

$= 0.2 \times 20 \times 0.1 \times \sin 30°$

$= 0.2$ N ✓ out of the page ✓ (by right-hand grip rule)

b The current is moving at 60° to the field on the left-hand side of the loop and hence:

$F = lIB \sin\theta$

$= 0.2 \times 20 \times 0.1 \times \sin 60°$

$= 0.35$ N ✓ into the page ✓ (by right-hand grip rule)

c Because the force on each of the opposite sides is equal and opposite, the net force on the loop is zero.

CHAPTER 6 ELECTROMAGNETIC INDUCTION

INQUIRY QUESTION:

How are electric and magnetic fields related?

Electric and magnetic fields are interrelated; they can exist independently or as a combined electromagnetic field. For example, a stationary charged particle is surrounded by an electric field and a permanent magnet is surrounded by a magnetic field. However, a moving charge particle is surrounded by a combined electromagnetic field. The close relationship between electric and magnetic fields is demonstrated when one of the fields is changed. When an electric field is changed it produces a magnetic field and when a magnetic field is changed it produces an electric field.

1 Magnetic flux

» Students describe how magnetic flux can change, with reference to the relationship $\Phi = B_{||}A = BA\cos\theta$.

- The magnetic field in space (as opposed to the field inside a ferromagnetic material) is called the *magnetic flux density* (B) and is measured in tesla (T).
- To understand how electricity is generated we must introduce a new quantity called **magnetic flux** (Φ). Magnetic flux is a measure of the magnetic flux density passing through a specific surface. Magnetic flux is defined as the product of the magnetic flux density perpendicular to a surface and the area of the surface. Hence for a surface of area A that is perpendicular to the magnetic field, as shown in Figure 6.1, the magnetic flux passing through the area would be given by $\Phi = BA$.

magnetic flux: a measure of the lines of magnetic field passing through a surface; it is defined as the product of the surface area and the component of the magnetic flux density perpendicular to the surface

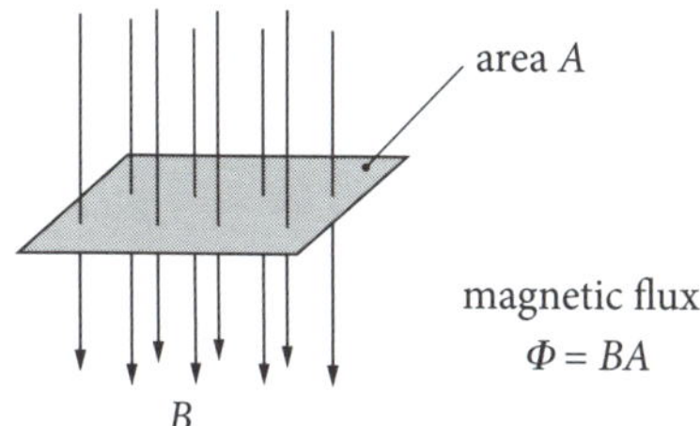

Figure 6.1 Magnetic flux linking an area perpendicular to the magnetic field

- If the magnetic field and area are not perpendicular we need to determine the component of the magnetic field that is perpendicular to the surface. To do this we consider the area to be a vector perpendicular to the surface and then find the component of the magnetic field that is parallel to the **area vector**, as shown in Figure 6.2.

area vector: the vector of a plane surface; the area vector has a magnitude equal to the area of the surface and a direction that is perpendicular to the plane of the surface

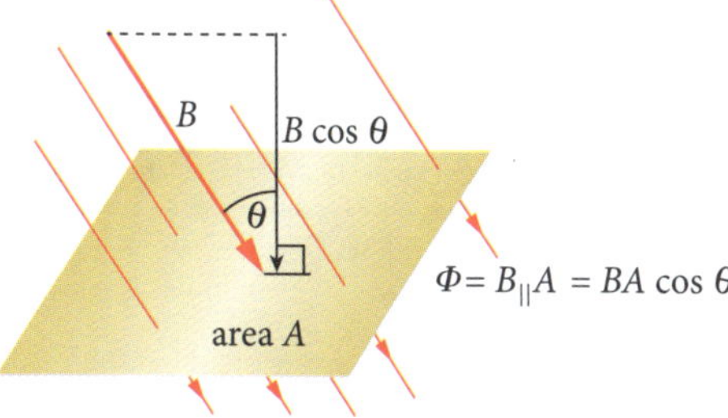

Figure 6.2 Magnetic flux passing through a surface that is not perpendicular to the magnetic field

- The magnetic flux is therefore given by the product of the component of the magnetic field parallel to the area vector and the area of the surface. That is:

$$\Phi = B_{||}A = BA\cos\theta$$

where Φ is the magnetic flux
$BA\cos\theta$ is the component of the magnetic field perpendicular to the surface
A is the area of the surface and
θ is the angle between the magnetic field and the area vector (a line drawn normal to the surface)

- Magnetic flux is a scalar quantity with the SI units of tesla metres squared (Tm^2) or webers (Wb).
- We will see in the following sections that changing the magnetic flux linking a loop or coil of wire will produce an electrical potential difference across the loop or coil.
- From the definition of magnetic flux, we see that we could produce a change in magnetic flux by changing the intensity of the applied magnetic field, the area of the loop, and/or the angle between the area vector and the magnetic field.

→ KEY QUESTIONS

1 **What is the difference between magnetic flux density and magnetic flux?**

2 **How can we calculate the magnetic flux through a surface if the magnetic field is not perpendicular to the plane of the surface?**

Answers ➲ p. 90

EXAMPLE 1

A square loop of wire with sides 30 cm is placed in a 0.02 T magnetic field. Initially the plane of the square loop of wire was perpendicular to the magnetic field. The square loop was then rotated by 30° from its initial position, changing the orientation of the magnetic field from 90° from the plane of the loop to 60° to the plane of the loop. Find the change in magnetic flux through the loop.

The area vector is perpendicular to the plane of the surface and it is the angle between this vector and the magnetic field that is used to calculate the magnetic flux through the surface (i.e. not the angle between the plane of the surface and the magnetic field)

Answer:

As the magnetic field is perpendicular to the surface, the area vector and the magnetic field will be parallel, $\theta = 0$, and the magnetic flux will be:

$\Phi = BA\cos\theta = 0.02(0.3 \times 0.3) = 1.8 \times 10^{-3}$ Wb

If the loop is turned through 30° the area vector will now be at 30° to the magnetic field and the magnetic flux will become:

$\Phi = BA\cos\theta = 0.02(0.3 \times 0.3)\cos 30° = 1.56 \times 10^{-3}$ Wb

The change in flux will be given by:

$$\Delta\Phi = \Phi_{final} - \Phi_{initial} = 1.56 \times 10^{-3} - 1.8 \times 10^{-3} = -2.4 \times 10^{-4} \text{ Wb}$$

The negative sign tells us that the magnetic flux through the coil decreased by 2.4×10^{-4} Wb when the loop was rotated.

2 Electromagnetic induction

» Students analyse qualitatively and quantitatively, with reference to energy transfers and transformations, examples of Faraday's law and Lenz's law $\varepsilon = -N\dfrac{\Delta\Phi}{\Delta t}$, including but not limited to:

- the generation of an electromotive force (emf) and evidence for Lenz's law produced by the relative movement between a magnet, straight conductors, metal plates and solenoids
- the generation of an emf produced by the relative movement or changes in current in one solenoid in the vicinity of another solenoid.

Faraday's law

- Students should conduct a series of investigations on electromagnetic induction to examine the induced current produced by moving a wire relative to a magnetic field, moving a solenoid relative to a magnetic field and changing the current in a solenoid near another solenoid.

FIRSTHAND INVESTIGATION 1

Moving a wire relative to a horseshoe magnet

In this experiment students move a wire connected to a **galvanometer** between the poles of a strong magnet or a magnet near a stationary wire, as shown in Figure 6.3.

galvanometer: a sensitive, centre-reading ammeter; galvanometers usually measure current in microamps (μA)

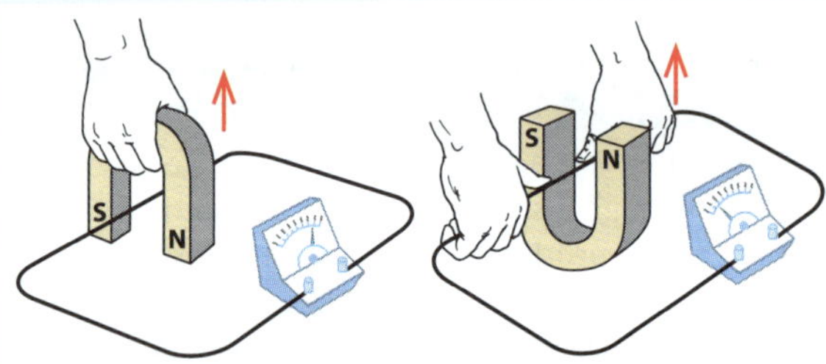

Figure 6.3 Relative motion between a wire and a magnet producing a current

Students should investigate the effect of the relative speed and the strength of the magnetic field on the current produced. They should also investigate how the relative motion affects the direction of the induced current. Students should try to apply the right-hand palm rule to the free charges in the wire to predict the direction that current will flow in the circuit.

Students should find that current is only generated when there is relative motion between the wire and the solenoid, and that changing the direction of the relative motion changes the direction of the induced current in the wire. They will also find that moving the magnet produces the same amount of current as moving the wire; that is, that the amount of current generated increases with the strength of the magnet used and with the relative speed of the motion between the wire and the magnet.

Moving a bar magnet relative to a solenoid

Students should investigate the production of current by moving a bar magnet into and out of a coil connected to a galvanometer, as shown in Figure 6.4.

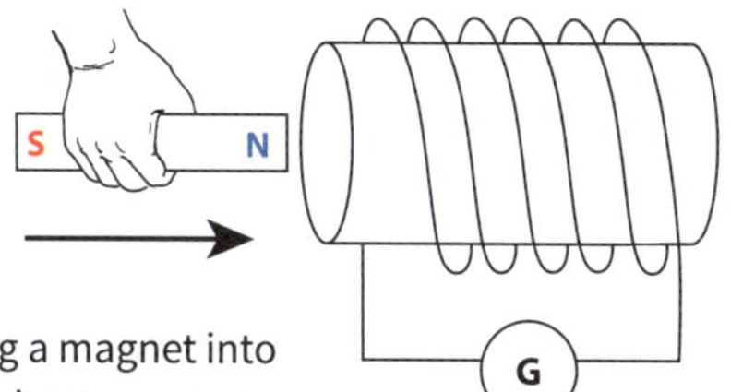

Figure 6.4 Moving a magnet into a solenoid to produce a current

Students should investigate the relationship between the amount of current produced and the relative speed, the number of turns on the solenoid and the strength of the magnetic field used (i.e. by using a stronger bar magnet). Care should be taken to change only one variable at a time while keeping the other variables constant.

Students may be able to apply the right-hand palm rule to charges in the wire as they are moved relative to the magnetic field to predict the direction of the induced current in the solenoid.

Adjacent solenoids

To investigate whether the production of current is due to the motion of a wire in the magnetic field or to changing the magnetic field or the magnetic flux in the coil, students should conduct experiments with adjacent solenoids. As shown in Figure 6.5, connecting one solenoid to an external power supply will produce a constant magnetic field that will be linked to the second solenoid.

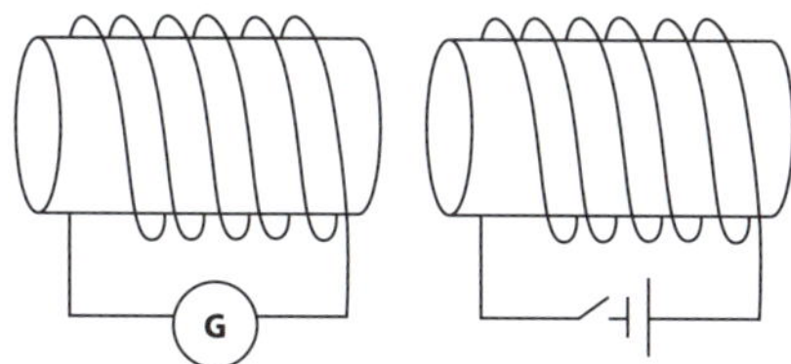

Figure 6.5 Two adjacent solenoids; one connected to a galvanometer and the other connected to a battery via a switch

Students will observe that current is induced in the second solenoid for an instant only when the switch is opened or closed on the other solenoid. That is, current is induced in the coil connected to the galvanometer only when the magnetic field changes.

After students notice that a constant DC current in the first coil does not produce electricity in the second coil, they should connect the first coil to an AC source rather than a DC source. This will produce a continually changing magnetic field. By connecting the second coil to an oscilloscope or an AC ammeter, students will observe that an AC current is induced in the second coil. Students may also investigate the effect of increasing the magnetic field by placing a soft iron rod through the coils when using an AC voltage.

- Producing a current in a loop by changing the magnetic environment of the loop is called **electromagnetic induction**.
- **Faraday's law** of electromagnetic induction relates the change in the magnetic environment of a circuit to the electrical potential difference induced in a circuit. The law states that the **electromotive force** (emf or ε) induced in a circuit is proportional to the negative rate of change of the magnetic flux linking the circuit, which can be expressed as:

$$\varepsilon = -N\frac{\Delta \Phi}{\Delta t}$$

where ε is the electromotive force
$\Delta \Phi$ is the change in magnetic flux
Δt is the time over which the change occurred and
N is the number of turns on the circuit

electromagnetic induction: the production of an electromotive force across a conductor produced by a changing magnetic flux

Faraday's law: the induced emf in a circuit is equal to the negative rate of change of magnetic flux in the circuit

electromotive force: the potential difference produced by a source of electricity (e.g. a battery or generator) measured in volts when the source is not connected to a circuit; an emf can be used to produce a current by connecting the source to a circuit

- The negative sign in the equation relates to the polarity of the induced electromotive force and the resulting current direction in the circuit. We will examine this in more detail in the next section when we introduce Lenz's law.

- ➔ Electromotive force (emf) is the name given to the electrical potential difference produced by a battery or generated by electromagnetic induction. The emf refers to the voltage across a power source when no current is being drawn from the source. For example, the emf of a battery is the potential difference across the battery when it is not connected to a circuit. When you buy a 1.5 V battery you are buying a battery with an emf of 1.5 V.
- ➔ Faraday's law tells us that if the magnetic flux through a circuit changes, an electrical potential difference will be induced in the circuit while the flux is changing. Thus changing the strength of the applied magnetic field, the area of the circuit in the magnetic field, and/or the angle between the area vector and the magnetic field will all induce an emf.

➔ KEY QUESTIONS

3 **How can a magnetic field be used to produce an emf in a circuit?**

4 **Why does having more turns on a coil in a changing magnetic field produce a greater emf than a coil with fewer turns?**

Answers ➲ p. 90

EXAMPLE 2

A circular coil has an area of 0.4 m^2 and consists of 100 turns of wire. The coil is placed perpendicular to a uniform magnetic field of 0.05 T, as shown in Figure 6.6.

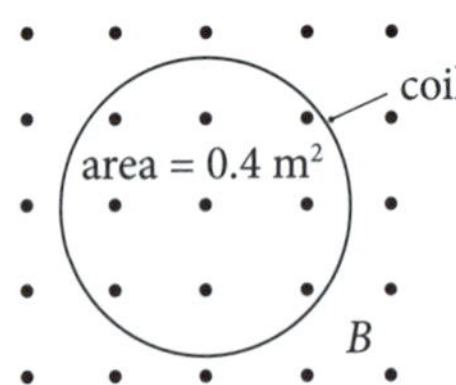

Figure 6.6 Circular coil in a magnetic field

If the magnetic field was reduced at a constant rate to zero over a period of 0.2 s, what electromotive force would be induced in the coil?

> The rate of change of a quantity is the change divided by the time over which the change occurred

Answer:

Because there are 100 turns of wire on the coil and the angle between the magnetic field and the area vector is 90°, the initial magnetic flux linking the coil would be:

$\Phi = BA\cos\theta = 0.05 \times 0.4 \times \cos 0° = 0.02$ Wb

The change in magnetic flux is therefore:

$\Delta\Phi = \Phi_{final} - \Phi_{initial} = 0 - 0.02 = -0.02$ Wb

Applying Faraday's law:

$$\varepsilon = -N\frac{\Delta\Phi}{\Delta t} = \frac{-(100 \times (-0.02))}{0.2} = 10 \text{ V}$$

EXAMPLE 3

An aluminium rod moves along a pair of parallel, conducting rails 40 cm apart at 10 ms^{-1} perpendicular to a 0.1 T magnetic field, as shown in Figure 6.7.

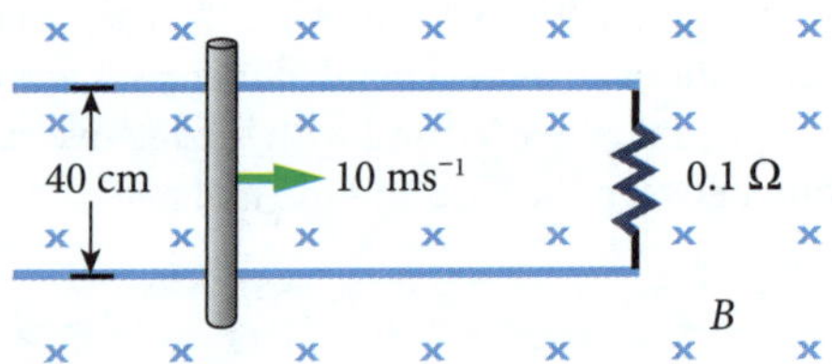

Figure 6.7 Aluminium rod moving along a pair of parallel rails at constant velocity

Find the magnitude of the current that would be induced in the circuit if the circuit had a resistance of 0.1 Ω.

> The rate of change of the area is how much the area changes each second

Answer:

If the rod moved at 10 ms^{-1}, after 1 s the rod would have moved by 10 m to the right and the area of the field enclosed by the circuit would have decreased by 10 × 0.4 = 4 m^2. Thus the rate of change of magnetic flux linking the circuit would be:

$$\frac{\Delta\Phi}{\Delta t} = 0.1 \times \frac{4}{1} = 0.4 \text{ Wbs}^{-1}$$

The induced electromotive force will be given by:

$$\varepsilon = -N\frac{\Delta\Phi}{\Delta t} = -0.4 \text{ V}$$

And applying Ohm's laws gives:

$$I = V/R = \frac{0.4}{0.1} = 4 \text{ A}$$

EXAMPLE 4

Figure 6.8 shows a rectangular loop of wire with an area of 0.6 m^2 aligned such that the plane of the loop is parallel to a magnetic field of strength 0.2 T.

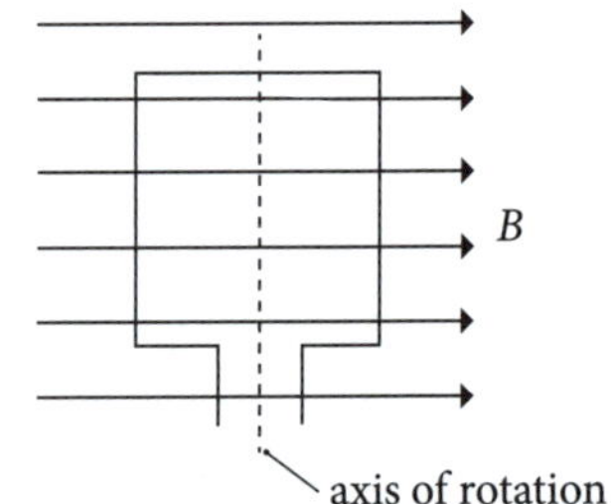

Figure 6.8 Rectangular loop of wire parallel in a magnetic field

If the loop illustrated in Figure 6.8 was rotated about the axis of rotation by 30° from its initial position in 0.1 s, what emf would be generated on the loop during the rotation?

If no lines of force pass through an area, the magnetic flux linking the area is zero

Answer:

In this example the component of the area of the loop perpendicular to the magnetic field changes as the loop is rotated and hence the change in magnetic flux is:

$$\Delta\Phi = \Phi_{\text{final}} - \Phi_{\text{initial}} = BA\cos\theta - 0 = 0.2 \times 0.6 \times \cos 60° = 0.06 \text{ Wb}$$

Note that $\theta = 60°$ because the area vector moves 30° from its initial orientation of 90°.

The induced electromotive force is therefore:

$$\varepsilon = -N\frac{\Delta\Phi}{\Delta t} = \frac{0.06}{0.1} = 0.6 \text{ V}$$

Lenz's law

- An **electrical load** is a circuit component that uses electrical power. When we connect a resistor, motor, light globe or any other circuit component to an electrical energy supply, we place a load on the supply that will transform electrical energy into other forms of energy such as heat, light and energy of rotation. Provided the source voltage is constant, the electrical power leaving the voltage source is proportional to current supplied to the circuit. Hence when a circuit (i.e. a load) draws more current, it will consume more electrical power.

electrical load: an electrical appliance or circuit that uses electrical power

- When electromagnetic induction produces an emf in a circuit no work is done but if a load is connected, current will flow and electrical energy will be transferred to the load where it will do work.
- Because energy is conserved, work must be done when electromagnetic induction produces a current (i.e. provides power to a load) because energy cannot be created; it can only be transferred or transformed. That is, work must be done by an external agent (us) when we produce electrical energy by changing the magnetic flux in a circuit.
- Early experiments confirmed that the more electrical work that was done (i.e. the more current drawn) the more work had to be done to produce the electricity.
- Lenz showed that for energy to be conserved in electromagnetic induction, the induced current had to produce a magnetic field that opposed the change in magnetic flux that caused the current to be induced.
- **Lenz's law** uses conservation of energy to predict the direction of induced current in a circuit. The law states that the direction of the induced current is always such that it produces a magnetic field that opposes the change in magnetic flux that produced the current.

Lenz's law: the direction of the induced current is such that it will produce a magnetic field that will oppose the change in magnetic flux that created the current

- To see how Lenz's law can be used, consider the rod sliding on the rails shown in Figure 6.9. In this case we can find the direction of the induced current using the right-hand palm rule because the force on a positive charge on the rod moving to the right will be upwards, resulting in a clockwise induced current. Note that it is actually the electrons that are moving (anticlockwise) in the wire, but conventional current is defined as the direction that positive charges would move if they could move.

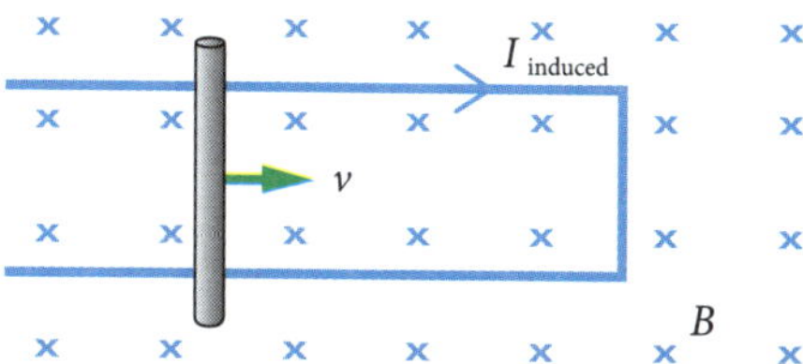

Figure 6.9 Direction of induced current in a moving rod

- Now if we apply Lenz's law to the situation shown in Figure 6.9, the current direction must oppose the change that produced the current; that is, the induced current must oppose the rod moving to the right. The current in the rod must therefore be upwards because only this current direction will produce a motor effect force on the rod that would oppose the rod's velocity. Thus to move the rod at constant velocity requires a force to be applied to balance the force exerted on the rod by the induced current. Applying a force over a distance means work is done and, because energy must be conserved, the work done to move the rod is equal to the electrical energy used in the circuit.
- Another example of Lenz's law is shown in Figure 6.10. Moving a bar magnet towards solenoid A will induce a current in the solenoid that will be in a direction that will produce a magnetic field to oppose the magnet being moved towards the solenoid. If the north pole is closest to the solenoid, the induced current in the solenoid will produce a north pole to oppose the magnet. If the magnet is moved away from the solenoid B the induced current will be in a direction that produces the opposite pole to try to prevent the magnet being moved away from the solenoid. We can determine the induced current direction in the solenoid by choosing the current direction that will produce a magnetic field that will oppose the change that produced the current.

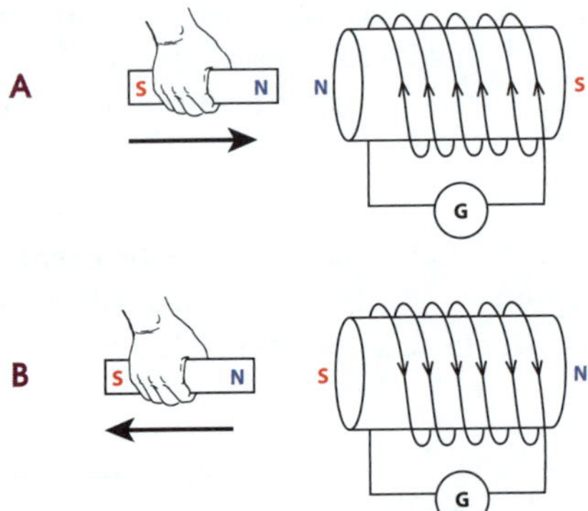

Figure 6.10 Direction of induced current when a bar magnet is moved towards solenoid A and away from solenoid B

➔ An interesting example of Lenz's law is seen when the magnetic field in the region of a solid conductor is changed. The changing field causes **eddy currents** to be induced in the conductor that produce their own magnetic fields that oppose the change that caused the eddy currents. For example, consider the spinning aluminium disc shown in Figure 6.11. If a magnet is brought close to the disc it will cause eddy currents to be induced in the disc that will oppose the motion of the disc. In Figure 6.11 we see that an upwards current, by the right-hand palm rule, will produce a force in the opposite direction to the rotation of the disc. This force would slow the disk rapidly because the disc's mechanical energy of rotation is converted into electrical energy (i.e. heat) in the disc.

eddy currents: loops of current that form in solid conductors exposed to a changing magnetic environment; for example, a changing magnetic field or a moving magnetic field

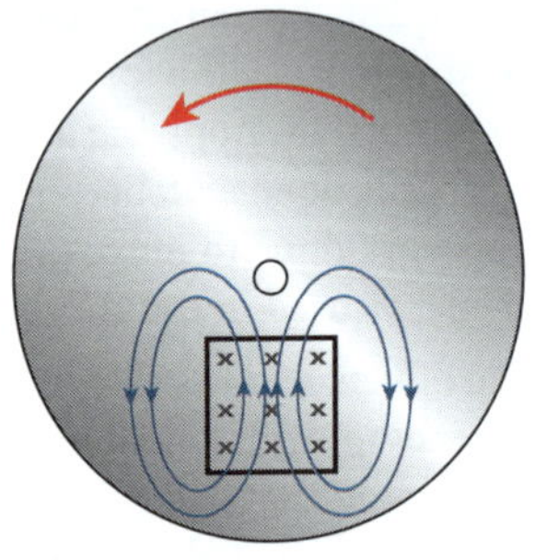

Figure 6.11 Eddy currents induced in an aluminium disc as it is rotated through an external magnetic field

EXAMPLE 5

A magnet is dropped through an aluminium ring, as shown in Figure 6.12. Explain how the induced current in the ring will change as the magnet falls through the ring. You should give the direction of any currents induced in the ring.

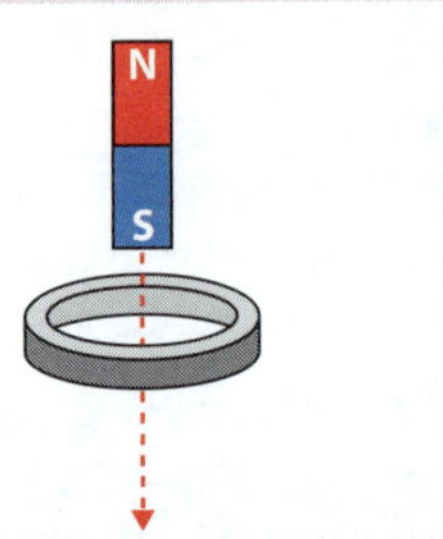

Figure 6.12 Magnet dropped through an aluminium ring

The induced current direction always opposes the change in magnetic flux that produced the current

Answer:

Because the magnetic flux in the ring changes as the magnet falls through it, a current will be induced in the ring. As the magnet approaches, the induced current in the ring will be clockwise (viewed from above) as this will produce a magnetic field downwards in the ring, which would make the top of the ring a south pole and oppose the approaching magnet. The induced current would be anticlockwise just after the magnet has passed through the ring, temporarily making the bottom of the ring a south pole, which would oppose the north pole of the magnet moving away from the ring.

➔ KEY QUESTIONS

5 **What is Lenz's law?**

6 **How is Lenz's law related to conservation of energy?**

7 **Why do eddy currents form in conductors?**

Answers ➲ p. 90

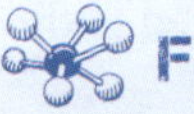

FIRSTHAND INVESTIGATION 2

Lenz's law

Students should revisit some of the experiments they conducted earlier to demonstrate electromagnetic induction, predict the induced current direction using Lenz's law and then confirm their prediction experimentally.

For example, consider the right-hand sketch shown Figure 6.3, which shows a magnet moving near a stationary wire. Lenz's law states that for energy to be conserved, the induced current direction must oppose the change that induced the current. Therefore moving the magnet upwards should induce a current that would exert an upwards force on the wire to move it with the magnet and hence oppose the change in the magnetic field. Applying the right-hand palm rule shows that the induced current between the poles of the magnet must be into the page (clockwise around the circuit) as this current direction would exert an upwards force ($F = ILB$) on the wire in the magnetic field.

Using the same technique to find the current direction when the wire is moved upwards, as shown in the left-hand sketch in Figure 6.3, gives the induced current direction as out of the page between the magnetic poles. Current in this direction would produce a downwards force on the wire, which would oppose the change that produced the current (i.e. the wire moving upwards).

Non-ferrous conducting plate moving in a magnetic field

A common experiment used to demonstrate Lenz's law and the production of eddy currents is shown in Figure 6.13.

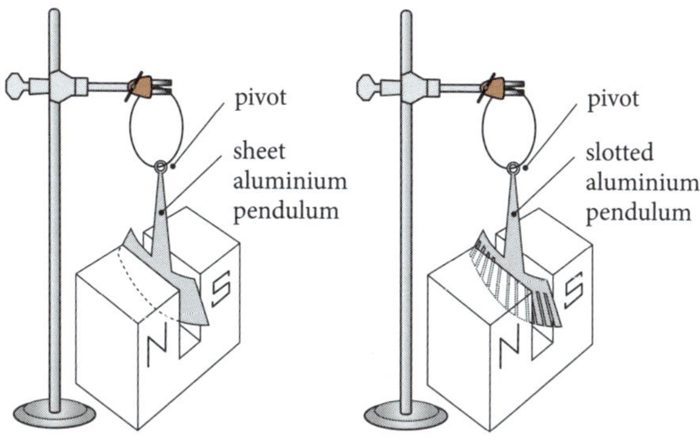

Figure 6.13 Pendulums swinging in a magnetic field

Students will observe that the sheet aluminium pendulum is rapidly brought to rest when it swings through the magnetic field due to the eddy currents induced in the sheet. Lenz's law tells us that the direction of the current in the sheet will oppose the change that produced the current and hence the force exerted on the sheet will oppose its motion through the magnetic field. Students should try to work out the direction of these currents as the sheet enters and leaves the field.
The slotted sheet swings for much longer because the slots prevent some of the eddy currents forming and hence there will be less force opposing the motion of the pendulum.

Rare earth magnet falling in a non-ferrous conducting tube

Students should also observe a rare earth magnet falling through aluminium and copper tubes. Students will find that magnets fall through the tubes much more slowly than a free-falling magnet due to the eddy currents induced in the tube exerting a downwards force on the tube and an upwards force on the falling magnet.

Students should try to work out the direction of the eddy currents around the tube above and below the magnet. (A tip here is to remember that above the magnet the magnetic field produced by the eddy currents will try to attract the falling magnet and below the magnet the eddy current will try to repel the falling magnet.)

Students may also try to measure the reaction force exerted on the tube by the falling magnet.

3 The transformer

» Students analyse quantitatively the operation of ideal transformers through the application of:
- $\frac{V_p}{V_s} = \frac{N_p}{N_s}$
- $V_p I_p = V_s I_s$.

➔ A **transformer** is a device that uses electromagnetic induction to increase (step up) or decrease (step down) the AC voltage. Basic transformers consist of two coils of insulted copper wire wrapped around a laminated, **soft iron core**, as shown in Figure 6.14.

transformer: a device used to increase or decrease AC voltage
soft iron core: structure used to intensify, contain and link the magnetic field of the primary coil to the secondary coil in a transformer

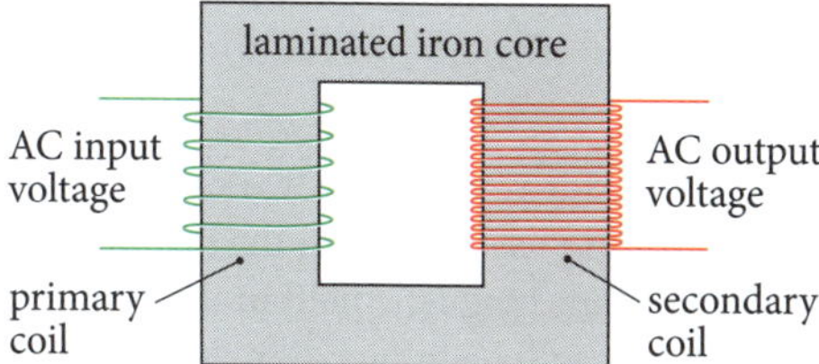

Figure 6.14 Step-up transformer

➔ Referring to Figure 6.14, an AC input voltage is placed on the **primary** (input) **coil** of the transformer. The resulting AC current in the primary coil produces a constantly changing magnetic field, which is intensified and linked to the **secondary** (output) **coil** by the iron core. The changing magnetic flux in the secondary coil causes an AC voltage to be induced on the secondary coil.

primary coil: a coil connected to an AC voltage that produces a changing magnetic field
secondary coil: a coil on which an AC voltage is induced by the changing magnetic flux produced by the primary coil

- As the magnetic flux in the iron core is proportional to the number of turns on the primary coil and the induced voltage is proportional to the number of turns on the secondary coil, the input and output voltages will be related by:

$$\frac{V_P}{V_S} = \frac{N_P}{N_S}$$

where N_P and N_S are the number of turns on the primary and secondary coils respectively and V_P and V_S are the input (primary) and output (secondary) voltages respectively

- Step-up transformers increase AC voltage and consequently always have more turns on the secondary coil than the primary coil, while step-down transformers reduce AC voltage and have fewer turns on the secondary coil than the primary coil.
- An ideal transformer would transform all the input energy into useful output energy with no energy wasted. For an ideal transformer:

power in = power out

$$V_P I_P = V_S I_S$$

where V_P and V_S are the input (primary) and output (secondary) voltages respectively and I_P and I_S are the currents in the primary and secondary coils respectively

ideal transformer: a transformer that transforms all the input power into useful output power (i.e. an ideal transformer has an efficiency of 100%)

- Note that a transformer will not work using DC because if a direct current passes through the primary coil it will produce a constant magnetic field in the iron core and hence no voltage will be induced on the secondary coil.

KEY QUESTIONS

8 **How does a transformer work?**

9 **What is the difference between a step-up and step-down transformer?**

Answers p. 90

EXAMPLE 6

An ideal transformer has a primary coil with 480 turns and transforms 240 V AC to 20 V AC.

a **Is this a step-up or step-down transformer?**

b **How many turns are on the secondary coil?**

c **If the input current was 2 A, what would the maximum current in the secondary be?**

d **How would the output change if the input voltage was DC rather than AC?**

Ideal transformers operate with 100% efficiency and a changing magnetic flux is required to induce a current

Answer:

a Because the transformer reduces the input voltage it is a step-down transformer.

b Applying the transformer equation:

$$\frac{V_P}{V_S} = \frac{N_P}{N_S}$$

$$N_S = N_P V_S / V_P = \frac{480}{240} \times 20 = 40 \text{ turns}$$

c As the transformer is ideal:

$$V_P I_P = V_S I_S$$

$$I_S = V_P I_P / V_S = 240 \times \frac{2}{20} = 24 \text{ A}$$

d The output voltage would be zero because if DC rather than AC was used as an input voltage, the magnetic flux in the iron core would not change and no voltage would be induced on the secondary coil.

4 Transformer efficiency

» Students evaluate qualitatively the limitations of the ideal transformer model and the strategies used to improve transformer efficiency, including but not limited to:

- incomplete flux linkage
- resistive heat production and eddy currents.

- Transformers are very efficient devices but not ideal. The efficiency of real transformers is generally between 95% and 98.5%.

efficiency: the ratio of useful power out to input power, expressed as a percentage (i.e. efficiency = (useful power out/input power) × 100%)

- Transformers are not perfectly efficient because of resistance heating of the windings, incomplete flux linkage, the production of eddy currents in the core or nearby metal objects and because changing the polarity of the magnetic domains in the core is not perfectly efficient.
- The energy lost in resistance heating of the windings is given by:

  Power loss = I^2R

  Clearly, to minimise resistance heating we must keep the resistance of the windings low. This may require cooling the transformer as the resistance of metals increases with temperature. Note that resistance heating is proportional to the square of the current.

- Eddy currents in the core heat the core and reduce the magnetic flux change in the core; in accordance with Lenz's law the currents produce magnetic fields that oppose the change in flux in the core. Eddy currents are reduced by using a **laminated iron core**. Laminating involves cutting the iron core into thin slices perpendicular to the plane of the windings and then rejoining the slices with insulating material between each slice. A transformer with a laminated iron core is shown in Figure 6.15.

laminated iron core: an iron core that consists of thin slices with insulating material between the slices to reduce the formation of eddy currents

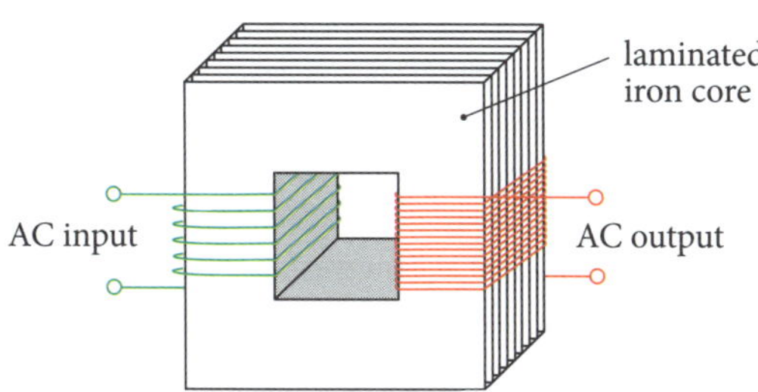

Figure 6.15 Transformer with laminated iron core

- Care must be taken when designing the transformer to ensure maximum flux linkage between the primary and secondary windings to provide maximum efficiency. Any magnetic flux not contained in the core will reduce the voltage induced by the secondary coil and may induce eddy currents in nearby metal objects that would reduce the efficiency further.
- Changing the polarity of the magnetic domains in the iron core also causes some heating and loss of efficiency. We call this a *hysteresis loss*. Special materials are chosen for transformer cores (such as silicon steel) to minimise hysteresis losses.

FIRSTHAND INVESTIGATION 3

Students should construct and investigate the efficiency of a simple transformer. If a demonstration transformer is not available students could use two adjacent solenoids, as shown in Figure 6.5. An AC voltage could be placed on one solenoid and an AC meter or oscilloscope connected to the second solenoid could be used to determine the induced voltage on the secondary coil. Students could investigate the effect of inserting a soft iron rod and a laminated iron rod into the solenoids.

Students will find the transformer efficiency improves greatly when a soft iron core is used to intensify and better link the magnetic flux between the coils. Students will also find laminating the iron core and using a core that confines the magnetic field, as shown in Figure 6.15, also significantly improves the transformer efficiency.

KEY QUESTIONS

10 **With respect to efficiency, what is an ideal transformer?**

11 **Why are real transformers not ideal?**

12 **How can the efficiency of transformers be improved?**

Answers p. 90

EXAMPLE 7

A student constructs a transformer using two solenoids and a solid, u-shaped piece of soft iron, as shown in Figure 6.16.

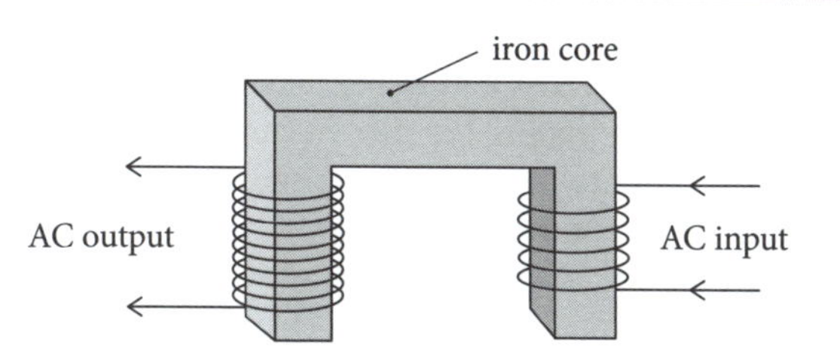

Figure 6.16 Student transformer

a **Is this a step-up or step-down transformer? Justify your answer.**

b **Outline two changes the student could make to improve the transformer's efficiency and explain how these changes would improve the efficiency.**

Transformer efficiency is reduced by incomplete flux linkage and resistance heating due to eddy currents in the core

Answer:

a This is a step-up transformer because it has more turns on the secondary (output) coil than on the input coil. As $\frac{V_P}{V_S} = \frac{N_P}{N_S}$, having more turns on the secondary coil means the output voltage will be greater than the input voltage.

b i Laminating the iron core would reduce eddy currents and resistance heating in the core and thus increase the efficiency.

ii Placing a laminated soft iron bar across the bottom of the iron core will confine and intensify the magnetic field and will improve the efficiency by increasing the flux linkage between the coils.

5 Applications of transformers

» Students analyse applications of step-up and step-down transformers, including but not limited to the distribution of energy using high-voltage transmission lines.

- Because transformers can efficiently step AC voltages up or down they are used in a wide range of industrial and domestic applications.
- Perhaps the most important use of transformers is in the efficient distribution of electrical power. During transmission, energy is lost through the resistance heating in the transmission lines. This **line loss** is related to the resistance of the line and the amount of current the line carries by:

  $P_{\text{loss}} = I^2R$

line loss: power lost due to resistance heating in transmission lines; line loss is proportional to the resistance of the line and the square of the current in the line (i.e. $P_{\text{loss}} = I^2R$)

To minimise line loss we need to reduce the resistance of the lines and/or reduce the current carried in the lines. There is a limit to reducing the resistance, as thicker copper or aluminium wires cost more and are more difficult to work with. We can use a lower current but to transmit the same power at lower current we must use a higher voltage (as $P = IV$).

- To minimise line loss, power distribution systems use transformers to transmit AC electricity through transmission lines at low current and high voltage.
- Figure 6.17 shows how step-up and step-down transformers are used in an AC power distribution system. Note that the voltage from the generator is stepped up to 330 kV to reduce the current in the long-distance transmission line. After transmission the voltage is stepped down to 11 kV by a transformer in the substation and near the consumer the voltage is again stepped down by a transformer on a street pole to 240 V before reaching the consumer.
- Modern AC distribution systems have operating efficiencies in the range of 8% to 15%. Power loss is predominantly due to heating of the transmission lines and losses stepping up and stepping down the voltage.
- Transformers are used in homes and industry to step-up and step-down the supply voltage for specific applications. For example, most electronic devices operate on a few volts DC and hence the AC supply voltage must be stepped down using a transformer and then converted from AC to DC (rectified). Step-down transformers are used in computers that are connected to the mains and in laptop, tablet and mobile phone chargers.
- Sometimes the mains voltage needs to be stepped up for domestic or industry applications. For example, the magnetron that produces the microwaves in a microwave oven operates at high voltage and every microwave oven uses a step-up transformer to convert the mains voltage to approximately 2500 V. (Interested students might like to investigate how this AC voltage is converted to DC and then the voltage doubled to power the magnetron in a microwave oven.) Step-up transformers are also used in voltage adaptors, which enable us to use our 240 V appliances in countries with a mains voltage of 110 V or 120 V AC.

→ KEY QUESTIONS

13 **How are transformers used to transfer electrical power efficiently over long distances?**

14 **Why are transformers used in homes and industry?**

Answers ⊃ p. 90

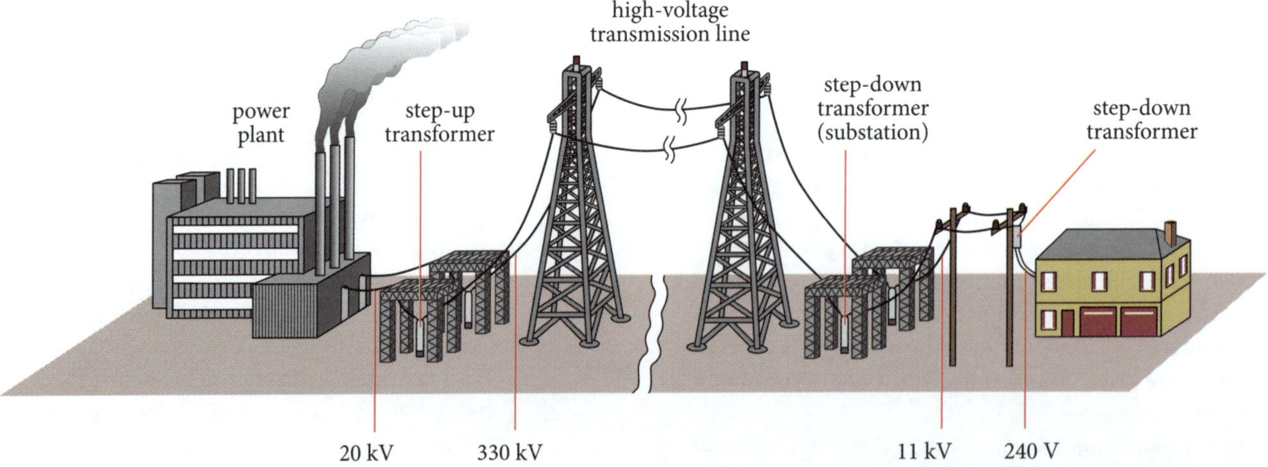

Figure 6.17 Typical AC power distribution system

CHAPTER SYLLABUS CHECKLIST

Are you able to answer these questions from the syllabus for this chapter? Tick each question as you go through the checklist if you are able to answer it. If you cannot answer a question, turn to the relevant page in the study guide to find the answer. For NESA key word meanings, go to www.educationstandards.nsw.edu.au and search 'key words'.

	FOR A COMPLETE UNDERSTANDING OF THIS TOPIC:	PAGE NO.	✓
1	Can I define magnetic flux and magnetic flux density and recall the SI units used for these quantities?	77	
2	Can I find the angle between the area vector and the magnetic field?	77	
3	Can I calculate the magnetic flux linking a loop?	77–78	
4	Can I outline the ways in which the magnetic flux linking a circuit can be changed?	78	
5	Can I describe a firsthand investigation to demonstrate electromagnetic induction?	78–79	
6	Can I use Faraday's law to calculate the electromotive force that is induced when the magnetic flux changes in a circuit?	79–81	
7	Can I explain how Lenz's law is related to the conservation of energy?	81–82	
8	Can I use Lenz's law and the right-hand palm rule to determine the direction of induced current in a range of situations?	81–82	
9	Can I explain why eddy currents are produced in conductors?	82	
10	Can I explain how transformers use electromagnetic induction to step up and step down AC voltages?	83–84	
11	Can I calculate the voltage and current output of an ideal, zero-loss transformer?	84	
12	Can I perform calculations using the transformer equation that relates the ratio of turns on the primary and secondary coils to the input and output voltages?	84–85	
13	Can I explain how design features can be used to increase the operating efficiency of transformers?	85	
14	Can I explain how and why transformers are used in electricity distribution systems?	86	
15	Can I describe applications of transformers, other than in electrical energy distribution systems?	86	

HSC EXAM-TYPE QUESTIONS

Objective-response questions (1 mark each)

1 **Which of the following statements best describes Faraday's law of electromagnetic induction?**

A The induced current is always in a direction that will produce a magnetic field that opposes the change in magnetic flux that produced the current.

B The induced electromotive force is proportional to the negative rate of change of the magnetic flux density.

C The induced current in a circuit is equal to the negative rate of change of the magnetic flux.

D The induced electromotive force is proportional to the negative rate of change of the magnetic flux.

2 **Figure 6.18 shows some circular loops of wire in magnetic fields. Note that in all the examples shown, the plane of the coil is perpendicular to the magnetic field. Which loop has the greatest magnetic flux?**

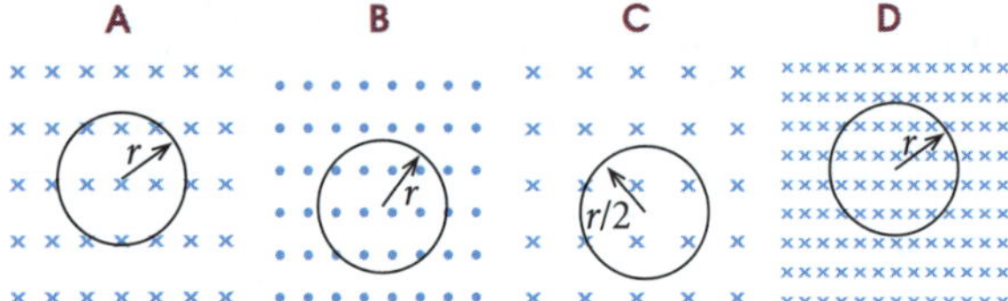

Figure 6.18 Circular loop perpendicular to a magnetic field

3 **A copper rod sliding along frictionless rails was travelling with a velocity *v* when it entered a uniform magnetic field, as shown in Figure 6.19.**

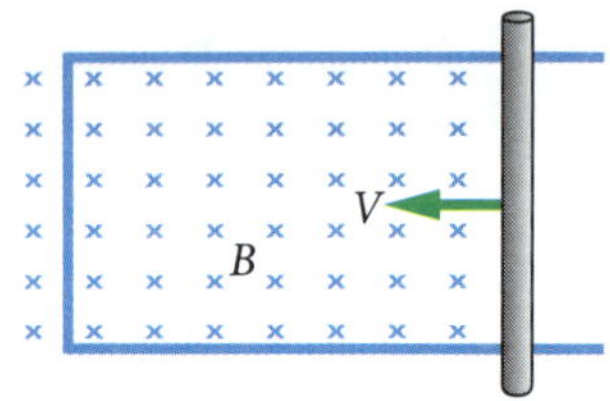

Figure 6.19 Copper rod sliding along a pair of frictionless, parallel rails

Which of the following best describes what will happen when the rod enters the magnetic field?

A A clockwise current will be induced in the circuit and the rod will continue to move at constant velocity.

B A clockwise current will be induced in the circuit and the rod will slow down.

C An anticlockwise current will be induced in the circuit and the rod will continue to move at constant velocity.

D An anticlockwise current will be induced in the circuit and the rod will slow down.

4 **Why do transformers use laminated iron cores?**

A To minimise the eddy currents that form in the core due to the motor effect.

B To minimise the eddy currents that are induced in the core that form magnetic fields that are perpendicular to the magnetic fields produced by the windings.

C To minimise the eddy currents that are induced in the core that produce magnetic fields that are in the opposite direction to the magnetic fields produced by the windings.

D To reduce the flux linking the primary and secondary coils.

5 **Two 50 km AC transmission lines A and B have the same resistance and transmit the same amount of power. Line A uses a current *I* and line B uses a current of 5*I*. What is the ratio of power lost due to resistance heating in transmission line A to power lost in transmission line B?**

A 1:25

B 25:1

C 5:1

D 1:5

Extended-response questions

6 **Consider a rectangular loop of wire moving at velocity *v*, travelling through a magnetic field directed into the page, as shown in Figure 6.20.**

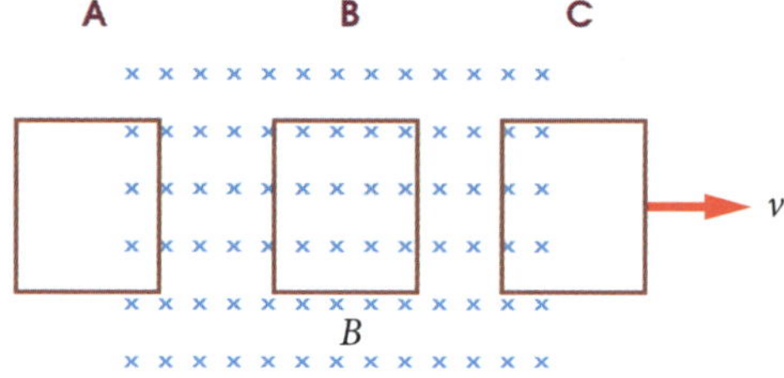

Figure 6.20 Rectangular loop of wire moving through a magnetic field

a **In what direction would induced current flow in the loop at positions A, B and C?** (3 marks)

b **Explain why current is induced in the loop at positions A, B and C.** (3 marks)

c **Compare the forces acting on the loop at positions A and B.** (1 mark)

7 **State two transformer design features that are used to increase transformer efficiency and explain how they increase efficiency.** (4 marks)

8 A transformer at a power station is used to step up an AC voltage from 20 000 V to 500 000 V. The transformer has 200 turns on the primary coil and when the current in the primary coil is 100 A the current in the secondary coil is 3.9 A.

a Find the number of turns on the secondary coil. (1 mark)

b Is this an ideal transformer? Justify your answer. (2 marks)

c Explain why large transformers like this need to be oil cooled. (1 mark)

9 A square coil with sides 10 cm is made up of 150 turns of copper wire. The coil was initially perpendicular to a magnetic field of 0.02 T, as shown in Figure 6.21. The coil was then rotated around a vertical axis by 90° until the plane of the coil was parallel with the magnetic field. The rotation took 0.05 s to complete.

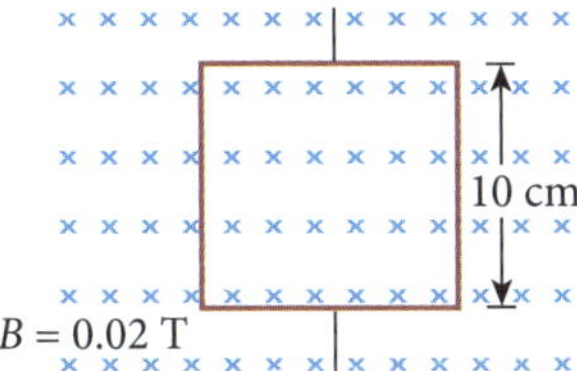

Figure 6.21 Square coil of 100 turns in a magnetic field

a Find the initial and final magnetic flux in the coil. (2 marks)

b Determine the emf that would be induced in the coil during its rotation. (1 mark)

c Explain why the coil would be more difficult to turn if the ends of the coil were joined together to make a closed circuit (rather than slightly apart, making an incomplete circuit). (2 marks)

10 A circular aluminium disc is rotated below a freely suspended bar magnet, as shown in Figure 6.22a. When the disc was turned the magnet also began to rotate.

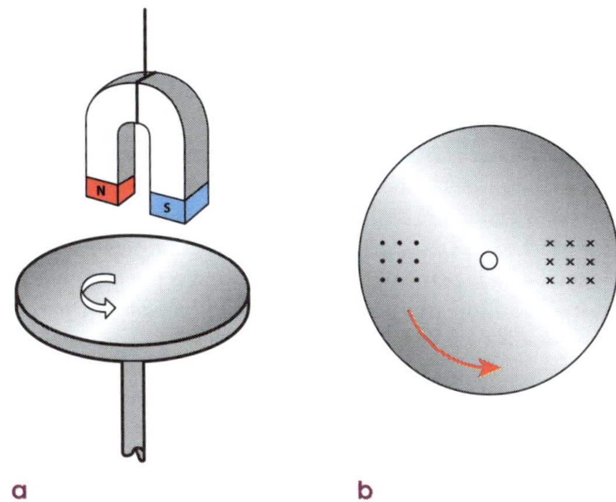

Figure 6.22 a A freely suspended magnet above a spinning aluminium disc and b magnetic fields on the spinning aluminium disc

a Redraw Figure 6.22a and include any eddy currents that would be formed as the disc is rotated. (2 marks)

b State the direction of rotation of the magnet and explain why the magnet rotates in this direction. (3 marks)

ANSWERS

KEY QUESTIONS

Key questions ➲ p. 78

1 *Magnetic flux density* is a term used to describe the magnetic field (B), which is a vector measured in tesla. *Magnetic flux* (Φ), in contrast, is a measure of the magnetic flux density perpendicular to a surface and is a scalar quantity measured in webers. If the magnetic field is perpendicular to the surface, the magnetic flux density linking an area A of the surface is related to the magnetic flux by $\Phi = BA$.

2 To find the magnetic flux for a magnetic field that is not perpendicular to the surface, we multiply the area by the component of the field that is perpendicular to the surface. We can do this by using $\Phi = BA \cos \theta$, where θ is the angle between the magnetic field and the area vector (a vector perpendicular to the surface).

Key questions ➲ p. 80

3 If the magnetic flux linking the circuit changes in any way a current will be induced in the circuit.

4 Changing the magnetic flux in a single loop of wire generates an emf in the loop. Each time an extra turn is added to the coil it adds this amount of emf to the total on the coil. Thus the emf is proportional to the number of turns on the coil.

Key questions ➲ p. 82

5 Lenz's law states that an induced current will always be in a direction that produces a magnetic field that opposes the change in magnetic flux that produced the current.

6 Lenz's law is a direct consequence of conservation of energy. The opposing magnetic field produced by the induced current means a force is required to change the magnetic flux. This force does work that is transformed into electrical energy. If the circuit draws more current, more electrical energy is consumed and the opposing magnetic field produced by the current is greater and hence more work must be done to produce the current. Thus the work done is equal to the electrical energy used in the circuit and the energy of the system is conserved.

7 If the magnetic flux linking a solid conductor changes, eddy currents will be induced in accordance with Faraday's law. The direction of these currents will be such that they produce a magnetic field that will oppose the change in magnetic flux that caused them, in accordance with Lenz's law.

Key questions ➲ p. 84

8 By connecting the primary coil to an AC source, a constantly changing magnetic field is produced in the iron core. This iron core links the changing magnetic field to the secondary coil and an AC voltage is induced on the secondary coil because it experiences a constantly changing magnetic flux. By changing the ratio of primary turns to secondary turns we can use a transformer to increase or decrease an AC voltage.

9 A step-up transformer has more turns on the secondary coil than the primary coil and hence increases the AC input voltage. A step-down transformer has fewer turns on the secondary coil than on the primary coil and hence it is used to reduce the input AC voltage.

Key questions ➲ p. 85

10 An ideal transformer is one that would operate at 100% efficiency.

11 Real transformers fail to convert all the input power to useful output power because some energy goes into resistance heating in the windings and resistance heating in the core (because of induced eddy currents), some of the magnetic flux might not be totally linked to the secondary coil and changing the magnetic state of the core causes some losses.

12 Transformer efficiency is improved by effective design to maximise flux linkage, laminating the iron core to reduce eddy currents, keeping the transformer cool to reduce the increase in resistance with temperature, and using low-resistance wire in the coils.

Key questions ➲ p. 86

13 As line loss is proportional to the square of the current in the lines, power is most effectively transmitted through transmission lines by using low currents and high voltages. Step-up transformers are used to raise the voltage for transmission and step-down transformers are used to lower the voltage after transmission to make it more convenient and safer to use.

14 Mains electricity is delivered at 240 V, which may be too high or too low for some applications. Transformers are used to change the mains AC voltage to a voltage suitable for use in specific appliances in homes and in industry.

HSC EXAM-TYPE QUESTIONS

Objective-response questions

1 **D**. This is the only answer that relates induced emf to the rate of change of magnetic flux. **A** is incorrect as it is a statement of Lenz's law. **B** is incorrect as it wrongly states that induced emf is proportional to the magnetic flux density rather than the magnetic flux. **C** is incorrect because it refers to the induced current being equal to the rate of change of magnetic flux rather than the induced emf.

2 **D**. As the loop is perpendicular to the magnetic field, the magnetic flux is $\Phi = BA$. The strength of the magnetic field is related to the density of dots or crosses and hence **D** has the strongest magnetic field and an area equal to or greater than the other loops. **A** and **B** are incorrect as they have the same area as **D** but lower magnetic field strength. **C** is incorrect as it has a comparable magnetic field strength to **D** but a smaller area.

3 **B**. We can use the right-hand palm rule or Lenz's law to show that a current down the page will be induced in the rod. A current in this direction will exert a force to the right on the rod (opposing the change in magnetic flux), which will cause the rod to slow down. **A** is incorrect as Lenz's law tells us that the induced current will be in a direction that will oppose the change that creates it, which in this case is the motion of the rod. Hence the rod will slow down. **C** and **D** are incorrect because if the induced current flowed in this direction the rod would speed up, violating the law of conservation of energy.

4 **C**. By Lenz's law the induced eddy currents will produce magnetic fields that oppose the change in magnetic flux that created the current. In this case, because an increasing and decreasing field is the cause of the change, the current will make a field parallel to

the changing field to oppose the change in the field intensity. **A** is incorrect as induced eddy currents are not caused by the motor effect. **B** is incorrect because perpendicular fields would not oppose the change that caused the current to be induced. **D** is incorrect because laminating the core would increase the magnetic flux linking the secondary coil rather than decrease it.

5 **A**. The line loss is proportional to the square of the current. Hence the ratio of the loss in A to the line loss in B = $I^2/(5I)^2 = 1:25$. **B, C** and **D** are incorrect as they give the wrong ratio.

Extended-response questions

6 EM This question tests students' understanding of electromagnetic induction, magnetic flux and Lenz's law.

a In position A the current will be anticlockwise. ✓ In position B no current will flow. ✓ In position C the current will be clockwise. ✓

b Using the right-hand palm rule or Lenz's law to predict the current direction that will oppose the change in flux, we see that when one side of the coil is in the magnetic field, a force up the page will be exerted on the positive charges on the side of the coil in the magnetic field. ✓ As the other side of the coil is not in the field in positions A and C, the current will flow anticlockwise and clockwise respectively in positions A and C. ✓ Because there is no change in magnetic flux in position B, no current will flow around the circuit. ✓

Note that you could also give an answer in terms of the increasing magnetic flux at position A and the decreasing magnetic flux at position C, resulting in induced currents in the given directions due to Lenz's law.

c If the loop continues to move at constant velocity, the induced current and force will be the same at positions A and C. In both cases the current will produce a force to the left, ✓ which will oppose the motion of the coil (i.e. oppose the change in magnetic flux).

7 EM This question tests students' understanding of the factors that reduce the efficiency of transformers and how these effects can be minimised.

Transformers use a soft iron core ✓ to intensify and contain the magnetic field. The iron core increases the efficiency as it leads to a greater rate of change of magnetic flux in the secondary coil. ✓ Transformers also use laminated iron cores, ✓ which consist of thin sheets of soft iron electrically insulated from one another. The laminations are perpendicular to the magnetic field and hence greatly reduce the eddy currents ✓ formed by the changing magnetic field in the iron core. This increases efficiency as eddy currents reduce the magnetic flux in the core and the waste energy by resistance-heating in the core.

Note that you could also talk about cooling the transformer or using low-resistance windings to increase the efficiency.

8 EM This question tests students' ability to apply the transformer equations and their understanding of ideal and non-ideal transformers.

a Applying the transformer equation:

$$\frac{V_P}{V_S} = \frac{N_P}{N_S}$$

$$N_S = \frac{V_S N_P}{V_P} = 500\,000 \times \frac{200}{20\,000} = 5000 \text{ turns} ✓$$

b An ideal transformer has no loss of power; that is, $IV_{input} = IV_{output}$.

In this case the input power = $IV_{input} = 20\,000 \times 100 = 2 \times 10^6$ W, and the output power $IV_{output} = 3.9 \times 500\,000 = 1.95 \times 10^6$ W. Hence the output power is less than the input power ✓ and therefore the transformer is not ideal. ✓

c The power not transformed to useful output power (i.e. $2 \times 10^6 - 1.95 \times 10^6 = 5 \times 10^4$ W) is dissipated as heat in the transformer. The transformer is cooled to remove this heat energy. If this was not done, the transformer would be less efficient because the winding resistance would increase as the temperature of the windings increased. ✓

9 EM This question tests students' understanding of rate of change of magnetic flux, Lenz's law and Faraday's law.

a Magnetic flux is given by:

$\Phi = B_{||}A = BA\cos\theta$

In the initial position $\theta = 0$ and hence $\Phi = (0.02) \times (0.1 \times 0.1) = 2 \times 10^{-4}$ Wb ✓

In the final position the angle between the magnetic field and the area vector, $\theta = 90°$ and hence $\Phi = 0$ ✓

b $\Delta\Phi = \Phi_{final} - \Phi_{initial} = 0 - 2 \times 10^{-4} = -2 \times 10^{-4}$ Wb

The emf is given by $\varepsilon = -N\frac{\Delta\Phi}{\Delta t} = \frac{-150 \times (-2 \times 10^{-4})}{0.05} = 0.6$ V ✓

c If the ends were not joined, no current would flow and hence there would be no magnetic field opposing the rotation (i.e. opposing the change in magnetic flux). ✓ However, if a circuit was formed and current flowed, Lenz's law tells us that the current will flow in a direction that will oppose the change in magnetic flux and hence the coil will be harder to turn. ✓

Note that you could also answer this question in terms of conservation of energy as electrical work is done only when current flows and hence energy will need to be put into turning the coil only when current is flowing.

10 EM This question tests students' understanding of Lenz's law and eddy currents and their ability to apply Newton's third law to electromagnetic induction.

a Figure A6.1 shows the direction of the eddy currents in the rotating disc. The induced current must flow to the left in the magnetic field as this would produce a force on the disc that would oppose the rotation of the disc, as required by Lenz's law. We can determine the current direction by applying Lenz's law or by applying the right-hand palm rule to a positive charge on the disc as it moves through each magnetic field.

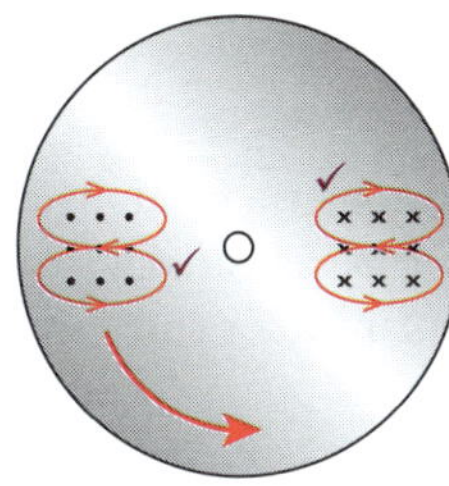

Figure A6.1 Eddy currents on rotating aluminium disc

b By Lenz's law, the induced eddy currents on the disc as it moves in the magnetic field will oppose the rotation of the disc. ✓ By Newton's third law, an equal and opposite force will be exerted on the magnet ✓ in the direction of the disc's rotation. Thus the magnet will begin to turn in the same direction as the disc. ✓

We can understand the motion of the magnet in terms of Lenz's law because if the magnet moves in the same direction as the disc it will oppose the change that induced the eddy currents (i.e. the relative motion of the magnet and the disc). Indeed, if the magnet rotated at the same rate as the disc, no eddy currents would be induced in the disc.

CHAPTER 7 APPLICATIONS OF THE MOTOR EFFECT

INQUIRY QUESTION:

How has knowledge about the motor effect been applied to technological advances?

An understanding of the motor effect led to the development of a variety of important technologies based on the effect, including the voltmeter, ammeter and electric motor. Similarly, an understanding of electromagnetic induction enabled the development of a wide range of technologies and devices such as electric generators, transformers, electromagnetic brakes and the AC induction motor.

1 The DC motor

» Students investigate the operation of a simple DC motor to analyse:
- the functions of its components
- its production of a torque $\tau = nIA_{\perp}B = nIAB\sin\theta$
- the effects of back emf.

Simple electric motors

➔ Students should construct and investigate some simple electric motors, such as Faraday's rotor, a rare-earth magnet homopolar motor and a simple single coil motor.

FIRSTHAND INVESTIGATION 1

Faraday's rotor

The device illustrated in Figure 7.1 is a copy of the world's first electric motor. The original was constructed by Michael Faraday in 1822 and used mercury rather than an ionic solution as the liquid conductor.

Students should use the right-hand palm rule to predict the direction of rotation and investigate the effect of reversing the current and inverting the magnet (to place the south pole upwards). Drawing the magnetic field lines around the magnet will help students apply the right-hand palm rule to the rod.

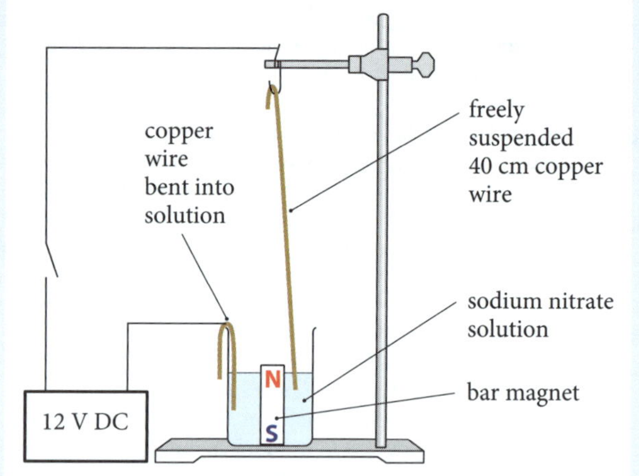

Figure 7.1 First electric motor

This type of DC motor is called a *homopolar motor* as the current in the rotating part (the armature) remains in one direction. Another homopolar motor that students can construct is shown in Figure 7.2. In this motor a screw acts as a conducting axle attracted to a rare earth magnet. Current running into or out of the centre of the conductive coating on the rare earth magnet causes it and the attached screw to rotate at high speed.

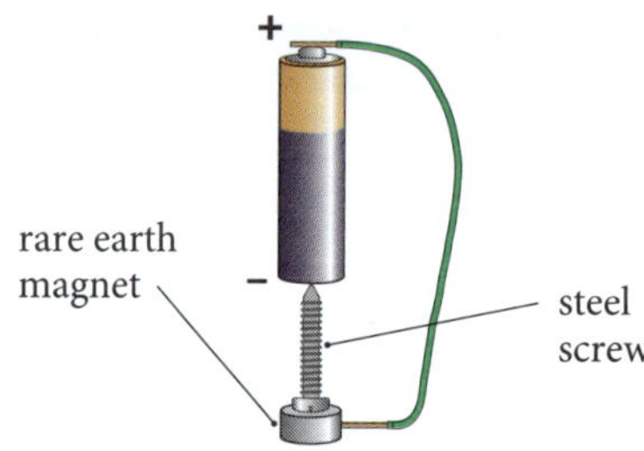

Figure 7.2 Simple homopolar DC motor

This is one of many ways a homopolar motor of this type can be configured and students could investigate and construct other simple motors of this type.

Simple DC motor

A simple DC motor can be constructed using a piece of enamelled (insulated) copper wire bent into a 2 cm diameter coil and suspended above a rare earth magnet, as shown in Figure 7.3. By cleaning off the enamel insulation on the sides of coil arms placed at 90° to the plane of the coil, as shown in Figure 7.4, a torque will be applied to the coil for a short time each cycle and the coil will rotate rapidly.

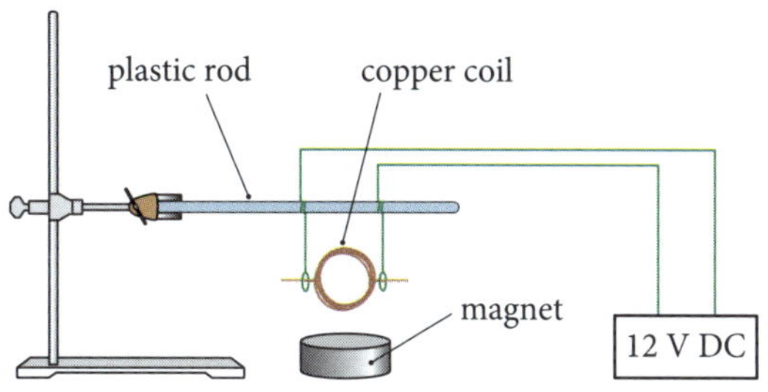

Figure 7.3 Simple DC electric motor

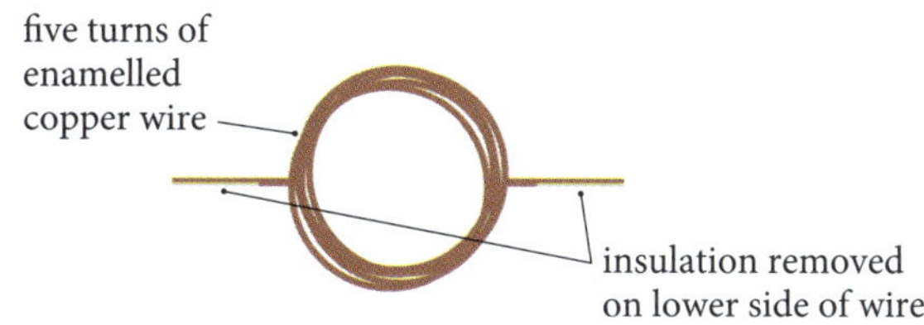

Figure 7.4 A simple copper coil rotor

Students should use the right-hand palm rule to predict the direction of rotation and investigate the effect of changing the number of turns on the coil and the current direction. Students should also investigate why removing all the insulation from the coil arms prevents the coil from continually rotating.

armature: the rotating coil in a motor

split-ring commutator: a device that connects the armature to the external power supply and reverses the connection between the coil and the power supply each half-cycle of rotation to ensure the current in the armature only produces a torque in one direction as it rotates

brushes: component that electrically connects a moving part to a stationary part; in a DC motor the brushes connect the rotating split-ring to the external power supply. Brushes eventually wear out and need to be replaced. Brushes are generally made of graphite or graphite composites

FIRSTHAND INVESTIGATION 2

Split-ring commutator DC motor

Students should also investigate DC motors with split-ring commutators. A simple demonstration motor with a single coil is shown in Figure 7.5. Students should examine the structure and function of the split-ring commutator and use the right-hand palm rule to predict the direction of the forces on each side of the coil throughout the rotation.

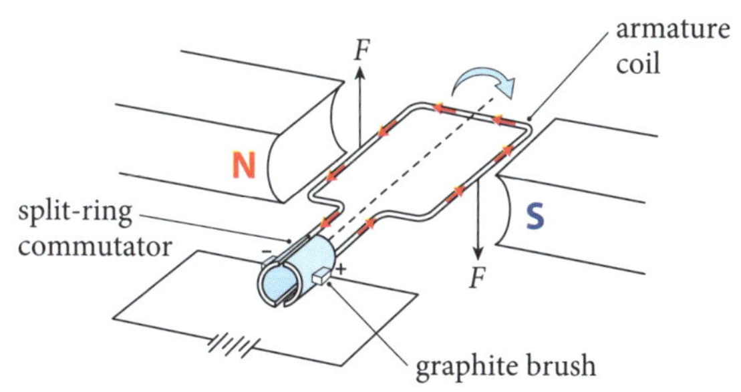

Figure 7.5 Simple DC motor with split-ring commutator

Motor force and the DC electric motor

- Figure 7.5 shows a simple DC motor and illustrates how the motor force on each side of the **armature** coil produces a torque on the coil. This torque would cause the armature to rotate 90° from the initial position and come to rest unless the direction of the forces on the sides of the armature could be reversed.
- The external magnetic field applied to the coil can be provided by permanent magnets or electromagnets.
- To ensure the armature coil continues to rotate in one direction, the DC motor shown changes the direction of the current in the armature coil each half-cycle by using a device called a **split-ring commutator**. As shown in Figure 7.5, the commutator is a cylinder split into two halves and re-joined with insulating material between the halves. The commutator is connected to a DC power source by two graphite **brushes**. Graphite is used because it is a conductor, has a low coefficient of friction and has a much higher melting point than a metal.
- We can calculate the torque on an armature coil by examining the motor force exerted on the coil. When the plane of the coil is parallel to the applied magnetic field, as shown in Figure 7.6, the motor force will produce a force on each side of the coil and, as these sides are perpendicular to the applied magnetic field, each force will be given by:

  $F = nILB$

  where n is the number of turns of wire on the coil
- The torque produced by the force on each side of the coil will be:

  $\tau = r_{\perp}F = r_{\perp}BnIl$

  And as the forces on each side produces a torque in the same direction, the total toque on the coil will be:

  $\tau = 2r_{\perp}BnIl$

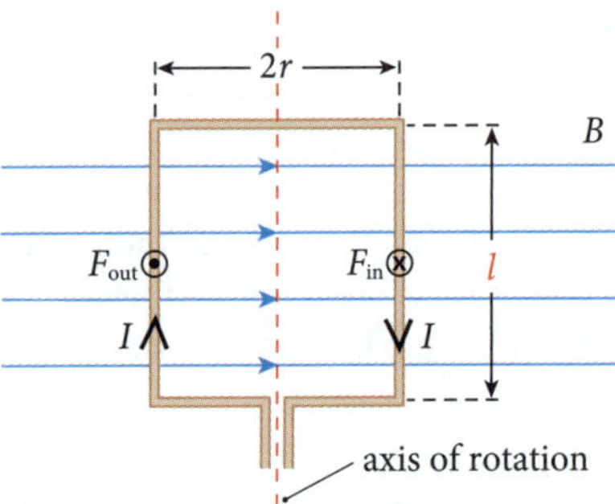

Figure 7.6 Force on coil with *n* turns of wire with the plane of the coil parallel a magnetic field

Now as $2r \times l = A$ = area of the coil, the torque on the coil is given by:

$$\tau = nIA_{\perp}B = nIAB\sin\theta$$

where n is the number of turns on the coil
I is the current in the coil
A is the area of the coil
B is the magnetic field and
θ is the angle between the area vector and the magnetic field

area vector: a vector with magnitude equal to the area of the coil and a direction perpendicular to the plane of the coil

- You will recall from the last chapter that the area vector has a magnitude equal to the area of the coil and is directed at 90° to the plane of the coil. The above equation tells us that the torque on the coil will be maximum when the area vector is perpendicular to the applied magnetic field, and zero when the area vector is parallel to the applied magnetic field. We can understand this by recalling from the Year 11 course that the torque (turning force) is the product of the applied force and the perpendicular distance between the axis of rotation and the line of action of the force. When the plane of the coil is parallel to the magnetic field, as in Figure 7.6, the torque is maximum because the perpendicular distance between the axis and the force is maximum.
- If the coil was rotated 90°, the perpendicular distance between the line of action of the force and the axis would be zero and the torque would be zero as the line of action of the force would pass through the axis. At this position, as shown in Figure 7.7, the forces are colinear (in the same line) and hence do not apply a turning force on the coil.
- To increase efficiency, modern DC motors normally use three armature coils at different angles, a commutator with multiple splits to power each of the coils sequentially, curved pole faces on the magnet to shape the magnetic field and a laminated iron core in the armature. Students may wish in investigate how each of these design features increases the efficiency of the motor.

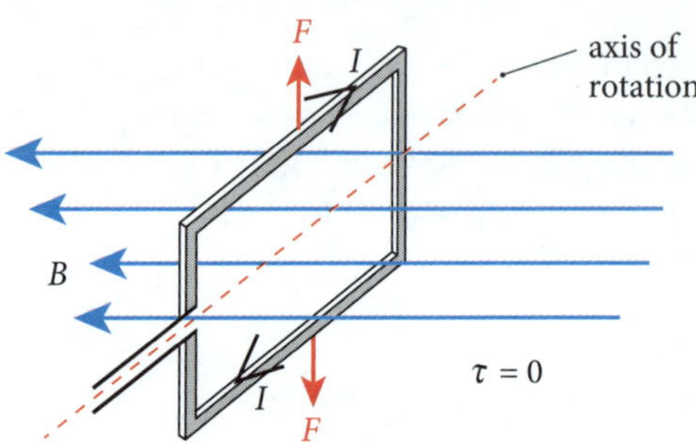

Figure 7.7 Force on a coil when the plane of the coil is perpendicular to the magnetic field

- We can also understand the operation of a DC motor by considering the interaction of magnetic fields rather than the torque produced by the motor effect. Students may wish to investigate how the rotation of the armature can be explained by treating the armature coil as an electromagnet placed to an external magnetic field.

KEY QUESTIONS

1. **Why do DC motors need a split-ring commutator?**
2. **Why do DC motors require an external magnetic field?**
3. **Explain why the net force on the rotor coil of a DC motor in a linear magnetic field is zero and why the torque on the motor changes as it rotates.**

Answers ➲ p. 106

EXAMPLE 1

Figure 7.8 shows a simple electric motor. The rotor coil has 200 turns and has an area of 4 cm^2. The magnetic flux density is 0.02 T in the region of the coil and 5 A of current flows in the rotor coil.

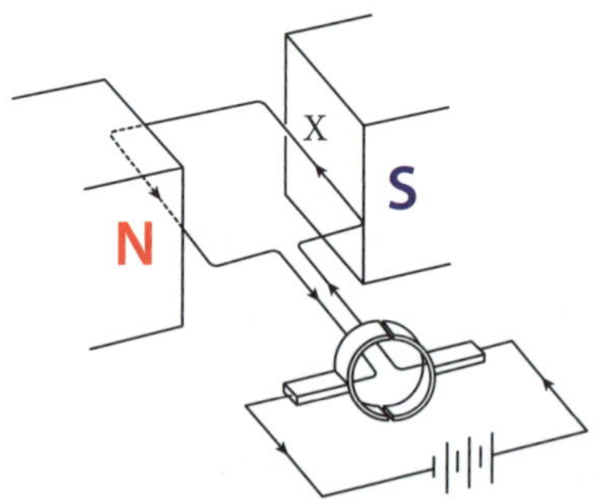

Figure 7.8 Simple DC electric motor

a **In which direction will the motor rotate?**
b **Find the initial torque on the rotor coil.**
c **After the coil had rotated 20°, what would the torque on the coil be?**

d **Sketch a graph to show how the force on the side of the coil marked X changes throughout one rotation (numerical values not required).**

e **Sketch a graph to show how the torque on the coil varies over one complete rotation (numerical values not required).**

Recall that the torque is proportional to the perpendicular distance between the line of action of the force and the axis of rotation

Answer:

a Using the right-hand palm rule on the side of the coil marked X shows the force on this side will be downwards and hence the motor will rotate clockwise.

b Note that 4 $cm^2 = 4 \times 10^{-4}$ m^2 and in the initial position the angle between the area vector and the magnetic field (θ) is 90°. Thus the torque is:

$\tau = nIAB \sin\theta$
$= 200 \times 5 \times 4 \times 10^{-4} \times 0.02$
$= 8 \times 10^{-3}$ Nm clockwise

c The angle between the area vector and the magnetic field (θ) is now 70°and the torque will be:

$\tau = nIAB \sin\theta$
$= 200 \times 5 \times 4 \times 10^{-4} \times 0.02 \times \sin 70°$
$= 7.52 \times 10^{-3}$ Nm clockwise

d Using the right-hand grip rule, the force on side X is upwards for the first 90° of rotation then downwards for the next 180° and finally upwards for the final 90°. A graph of force on side X against the angle of rotation is shown in Figure 7.9. Note that the angle in this case is the angle of rotation from when the plane of the coil was parallel to the magnetic field (i.e. not the angle between the area vector and the magnetic field).

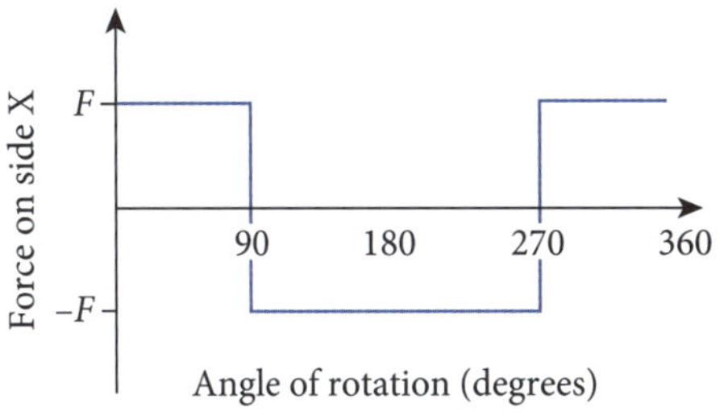

Figure 7.9 Graph of force on side X against the angle of rotation

e Without a commutator the torque would vary as a cosine function from the position shown (i.e. it starts at a maximum value) but as the split-ring commutator ensures the torque is in one direction, the torque will vary as the absolute value of a cosine function. Figure 7.10 shows the torque on the coil as a function of the angle of rotation from when the plane of the coil was parallel with the magnetic field.

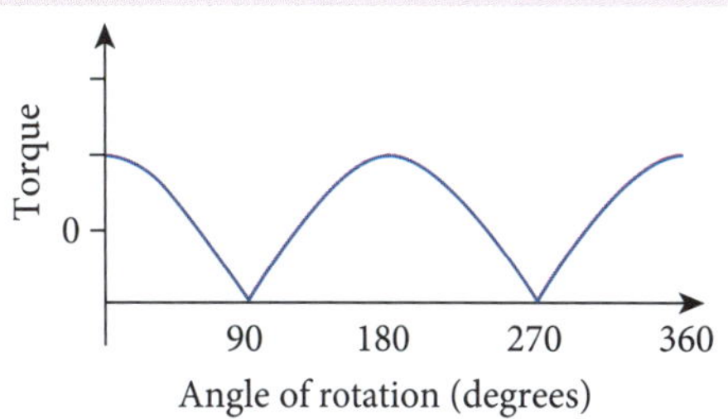

Figure 7.10 Torque on the rotor coil against the angle of rotation

Back emf

- Because the magnetic flux linking the armature coil changes as the armature rotates, an emf is induced on the rotating coil. Lenz's law tells us that the current produced by this induced emf will oppose the change in magnetic flux that produced the current and hence the induced current would be in the opposite direction of the current flowing through the coil. Because the induced emf opposes the emf applied to the motor, we call the induced emf the back emf. The emf across the armature coil would be related to the applied emf and back emf by:

  emf on rotor coil = applied emf − back emf

back emf: the emf induced in the armature of a motor when it rotates; it is called a back emf because it always opposes the emf applied by the power supply to the motor (Lenz's law)

- Because the rate of change of magnetic flux linking the armature coil increases with the speed of rotation, the back emf also increases with the speed of rotation. This increase in the back emf reduces the potential difference across the coil, which reduces the current in the armature and the torque produced by the armature.
- When a motor is first switched on, before it begins to rotate, there is no back emf and a large current would flow in the armature coil. This large current produces a large torque, which increases the rate of rotation of the motor. As the motor turns faster the back emf increases, reducing the current and reducing the torque until an equilibrium is reached when the rate of rotation produces just enough back emf to produce a motor torque that balances the load torque on the motor. When this occurs there is no longer a net torque on the armature and its rate of rotation remains constant.
- If we were to place a greater mechanical load on the motor, the torques would no longer be balanced and the rate of rotation of the motor would decrease, reducing the back emf and increasing the current in the armature coil. This increase in current would increase the torque

supplied by the motor force, until a new equilibrium was reached between the motor and the load, at a lower rate of rotation.

- If we were to prevent a motor from turning completely there would be no back emf and a large current would flow in the armature coil, which could melt the insulating enamel between the wires in the armature coil and **burn out** the motor.

burn out: refers to the overheating damage caused to the armature coils in an electric motor when too much current passes through the coils and melts the insulation between the wires

- In an ideal, frictionless motor (no such thing exists in practice), the back emf would equal the applied emf and there would be no current in the armature as no torque would be required to turn the motor. However, if a load was attached to the motor, the motor would have to supply a torque and the back emf would have to be less than the supply emf to provide a current in the armature so it could exert a torque to balance the load torque.

→ KEY QUESTIONS

4 **Why is a back emf induced when a DC motor operates?**

5 **How does back emf affect the current flowing in a DC motor?**

6 **Why is there a danger of burning a motor out if the motor is prevented from rotating?**

Answers ➲ p. 106

EXAMPLE 2

A student measured the time it took for a DC motor to lift a mass from the floor to a tabletop when a constant voltage was applied to the motor (as shown in Figure 7.11). The student found that when the mass being lifted was increased, it took longer for the motor to lift the mass and the motor drew more current. Explain the student's observations in terms of the back emf produced by the motor.

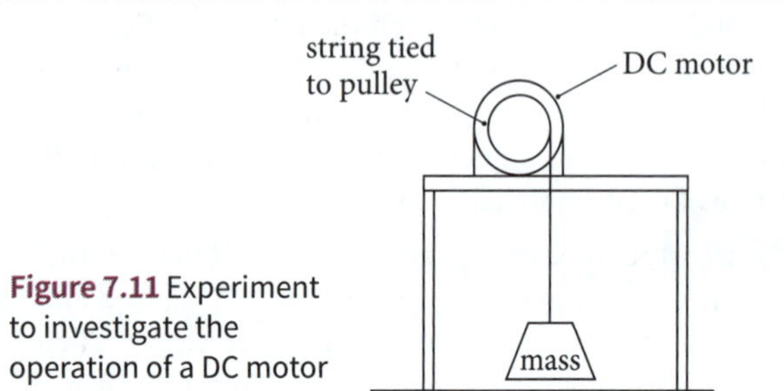

Figure 7.11 Experiment to investigate the operation of a DC motor

Recall that the back emf increases and the current in the armature decreases as the rotational speed of a motor increases

Answer:

As the speed of rotation of the motor increases the back emf increases, reducing the emf across and the current in the motor. If there is a net toque on the motor, the rate of rotation of the armature will change. The large initial torque supplied by the motor causes the rate of rotation to increase, but as it does so the torque supplied by the motor decreases (due to an increase in back emf) until the torque produced by the current in the armature balances the toque exerted by the load. At this point the motor turns at a constant rate.

Because a greater mass requires a greater torque to lift it up, the motor torque and load torque would come into equilibrium at a lower rate of rotation and, as the motor is turning more slowly, it will take longer to lift the heavier mass. That is, if the motor operated at a lower rotational speed, the back emf would be lower, more current would flow in the armature and the motor would produce a greater torque. Hence more current is supplied to the motor by its power source when the mass is added and the speed of rotation is reduced.

2 Generators and the AC induction motor

» Students analyse the operation of simple DC and AC generators and AC induction motors.

Generators

- As we saw in the previous section, the changing magnetic flux in the rotating coil of a DC motor induces a back emf on the coil. If we were to use an external source of mechanical energy to turn a DC motor, the changing flux in the coil would produce an emf that could be used to convert the mechanical energy of rotation to electrical energy.
- Students should conduct firsthand investigations to examine how electric motors can be used to convert mechanical energy into electricity (i.e. to be used as electric generators).

FIRSTHAND INVESTIGATION 3

The DC generator

Students should connect a simple DC motor to a galvanometer, data logger or cathode ray oscilloscope and investigate the output voltage that is produced when the DC motor is turned. The relationship between the output voltage and the rate and direction of rotation should be investigated. For a single-rotor coil DC motor they will find that a pulsed DC voltage is induced, with two pulses of DC per rotation. If a DC motor with multiple armature coils is used they will find a more consistent DC voltage is generated. If the direction of rotation is reversed, the polarity of the output will be reversed, and if the generator is turned faster, the peak induced voltage will increase and it will pulse at a higher frequency.

The AC generator

If a simple slip-ring commutator AC generator is available, students could investigate the output that is produced and compare the output from a DC generator and an AC generator. Students could also construct a simple AC generator by rotating a magnet near a stationary coil and investigate the voltage output using an oscilloscope, a centre-reading galvanometer or a sensitive AC voltmeter.

- A **generator** is a device that converts mechanical energy into electrical energy. A DC generator has the same structure as a DC motor but it has the opposite function. While a DC motor changes electrical energy into mechanical energy, a DC generator changes mechanical energy into electrical energy.
- The split-ring commutator in the DC generator ensures the output voltage is DC. By replacing the split-ring commutator with a **slip-ring commutator**, a DC generator can be converted into an AC generator. A slip-ring commutator connects the armature directly to the external circuit and, unlike a split-ring commutator, it does not reverse the connection as the armature coil rotates. Figure 7.12 shows a simple AC generator with a slip-ring commutator. Figure 7.13 shows the output voltage from a DC generator with a split-ring commutator and the output voltage from a simple AC generator with a slip-ring commutator.

generator: a device that uses electromagnetic induction to convert mechanical energy into electrical energy
slip-ring commutator: a device that connects the armature coils of a motor or generator to the external circuit; unlike the split-ring commutator the slip-ring commutator does not reverse the connection to the external circuit each half-cycle

- Modern AC generators often feed a DC current through the slip-ring commutators into the armature coil, turning the coil into a rotating electromagnet. The emf in this type of generator is produced by stationary coils (**stator coils**) placed around the rotating magnet. As the electromagnet moves past each stator coil the magnetic flux in the coil changes and an emf is induced in the stator coil. Because

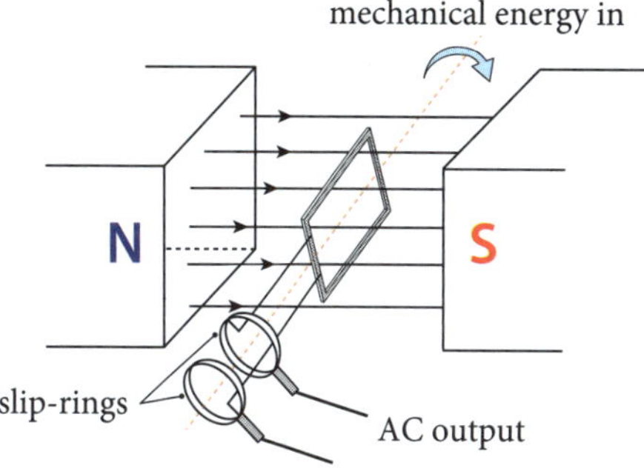

Figure 7.12 Simple AC generator with slip-ring commutator

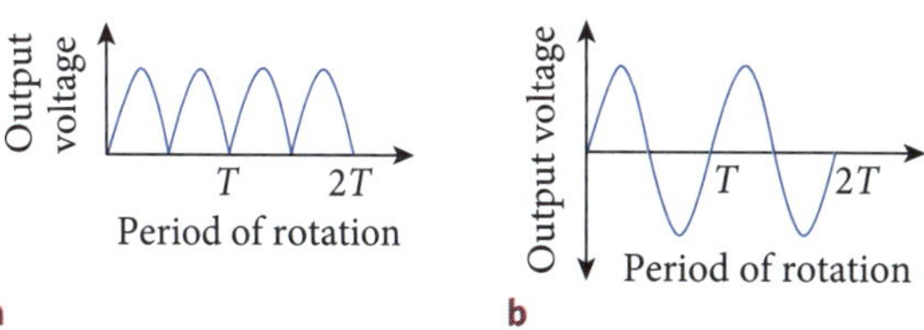

Figure 7.13 Output emf from **a** a DC generator and **b** an AC generator

the stator coils see the north and south pole of the magnet pass alternatively, an AC emf is induced on the coils. Power station AC generators use three stator coils placed around the rotating magnet (armature) to produce AC with three different AC output voltages, each one-third of a cycle (120°) out of phase from the other. Figure 7.14 shows the phase relationships between the three output voltages from a three-phase AC generator.

stator or stator coils: the stationary (non-rotating) coils in an electric motor or generator

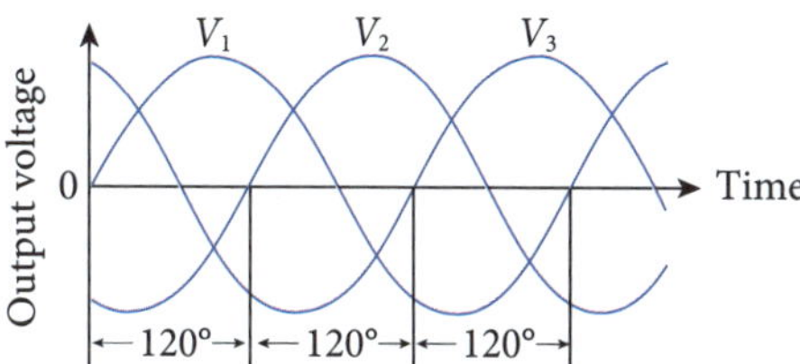

Figure 7.14 Output voltages from a three-phase AC generator

- If we place a resistance across the output terminals of a generator, current will flow and electrical power will be transferred from the generator to the resistor ($P = I^2R$). When we do this we say a load had been placed across the output. Because energy must be conserved, placing a load on a generator will make the generator more difficult to turn. This is because more mechanical input power must be provided for the generator to supply more electrical output power.
- We can understand how this occurs by thinking about the current in the armature. As more electrical energy is drawn from the generator more current flows in the rotor coils. But Lenz's law tells us that the induced current is

in a direction that opposes the change that produced the current. That is, the increased current in the armature will produce magnetic fields that will make the generator harder to turn and hence ensure that more mechanical work is done to produce the extra electrical energy.

➔ Hence if no load is connected to a generator (i.e. it is open circuited), an emf will be induced on the armature coils but no current will flow and hence no opposing magnetic field will be produced and the generator will be easy to turn. But when a load is connected, current flows in the armature coils and a magnetic field is produced that opposes the rotation, making the generator more difficult to turn. This is how energy is conserved in an electric generator.

➔ KEY QUESTIONS

7 **How does the structure and function of a DC generator compare with the structure and function of a DC motor?**

8 **How does the structure and voltage output of a DC generator compare with the structure and output voltage of an AC generator?**

9 **Why is a generator more difficult to turn when it is connected to a circuit that draws more electrical power (i.e. connected to a larger load)?**

Answers ➲ p. 106

EXAMPLE 3

A slip-ring AC generator has a single coil rotor with one turn of wire. The rotor coil was initially oriented at 90° to the applied magnetic field and in this position the magnetic flux linking the coil was 0.1 Wb. At time $t = 0$ the coil was rotated with a frequency of 10 Hz and the maximum rate of change of magnetic flux was found to be 6.3 Wbs^{-1}.

a **Sketch a graph of the magnetic flux linking the rotor against time for one complete oscillation.**

b **Sketch a graph of the emf induced by the generator against time for one complete oscillation.**

c **Explain how the graph of the induced emf would change if the frequency of rotation was halved.**

Recall that emf is the negative rate of change of the magnetic flux, which is the negative gradient of the graph of magnetic flux as a function of time

Answer:

a Because the plane of the coil is initially perpendicular to the magnetic field, the initial magnetic flux will be maximum. We can also see this from the magnetic flux equation. The graph will therefore be a cosine function, as shown in Figure 7.15:

$\Phi = BA \cos \theta$

In this case, the initial angle between the magnetic field and the area vector is zero and the initial flux is therefore:

$\Phi = BA \cos 0° = BA$

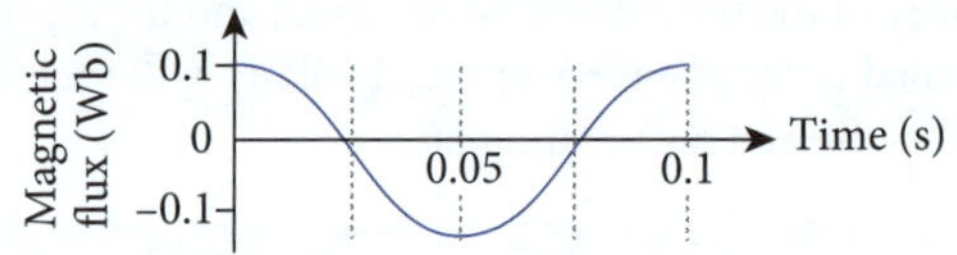

Figure 7.15 Magnetic flux as a function of time

b Because the induced emf is given by:

$$\varepsilon = -N\frac{\Delta\Phi}{\Delta t}$$

the emf (ε) will be the negative gradient of the magnetic flux graph, as shown in Figure 7.16.

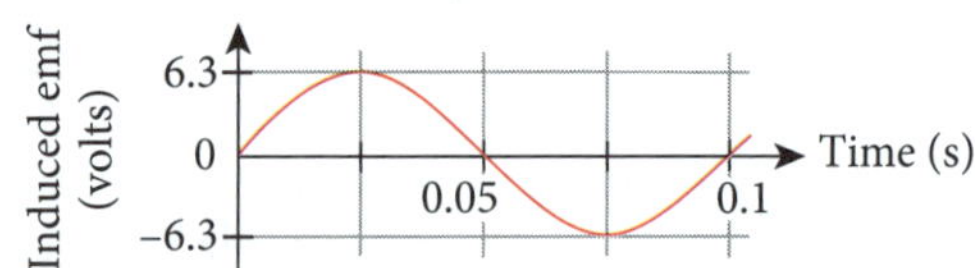

Figure 7.16 Electromotive force induced on a rotating coil as a function of time

c Halving the frequency of rotation will double the period but halve the peak emf as the rate of change of magnetic flux would be halved.

AC induction motor

➔ It would be possible to place an AC power supply across the slip-ring commutator of an AC generator to convert it into a motor, but such a motor could only run at the rate at which the AC input voltage oscillated. We call this type of motor an *AC synchronous motor* because its rotation is synchronised to the AC input voltage frequency (i.e. 50 Hz in Australia).

➔ A far more useful and efficient AC motor is the *AC induction motor.* This type of motor uses the force produced when eddy currents are induced in a conductor by a moving magnetic field. Students should conduct a firsthand investigation to examine how moving magnets induce eddy currents and forces on nearby non-magnetic conductors.

FIRSTHAND INVESTIGATION 4

Induction motor principle

To understand how the AC induction motor operates it is instructive for students to investigate the key principle involved. They can do this with an experiment such as the one shown in Figure 7.17. When a pair of strong rare earth magnets is rotated below an aluminium (non-magnetic) can suspended by a light string, the can begins to rotate in the same direction as the magnets. Students should investigate some of the factors that influence the rate at which the aluminium can rotates and try to explain the motion in terms of induced eddy currents and Lenz's law.

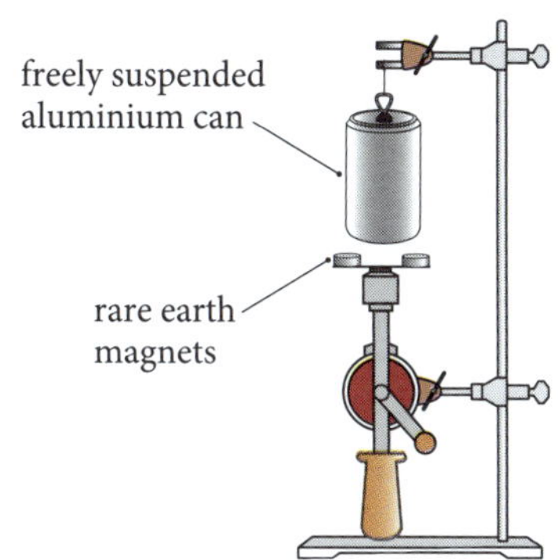

Figure 7.17 Experiment to investigate the principle of the AC induction motor

- The AC induction motor works on the principle illustrated in Figure 7.18. If a magnet is spun around a non-magnetic conductor, eddy currents will be induced in the conductor and these will produce magnetic fields that will oppose the motion of the magnets. These induced magnetic fields would exert a force on the magnets in the opposite direction of their motion and an equal and opposite force on the conductor in the same direction that the magnets were moving. Hence the conductor would spin around following the motion of the magnets.

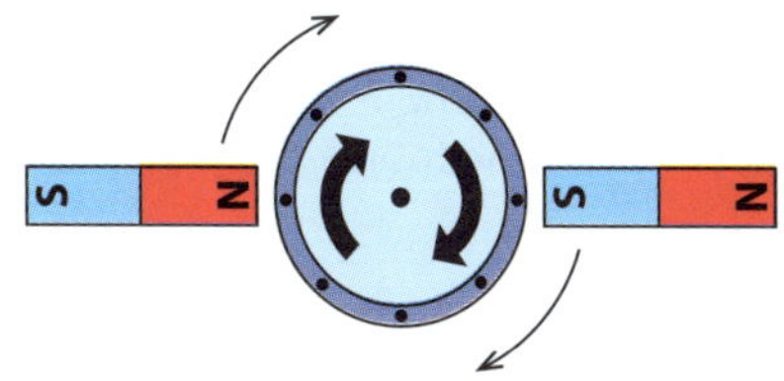

Figure 7.18 Magnets moved around a copper cylinder would cause it to rotate with the magnets

- By replacing the permanent magnets shown in Figure 7.18 with electromagnets and connecting each to AC power to ensure that each magnet was driven by AC at a slightly different phase, the central cylinder would experience a magnetic field that moved like the rotating magnets shown in Figure 7.18. An AC induction motor of this type is shown in Figure 7.19. This is a three-phase induction motor in which an AC voltage with a different phase is placed on each of the stator (stationary) coils. The phase of the AC voltage on the B coils is one-third of a cycle (i.e. 120°) behind the AC on the A coils, and the AC on the C coil is one-third of a cycle behind the AC on the B coils. This causes the north magnetic pole shown in the diagram to move anticlockwise from coil A to B to C in each cycle. This motion induces a current in the rotor that produces magnetic fields which cause the rotor to turn in the same direction as the moving magnetic field (i.e. anticlockwise). Three-phase AC from the power station could be used to power the motor directly or phase-changing circuits could be used to convert single-phase AC into three phases to operate the motor.

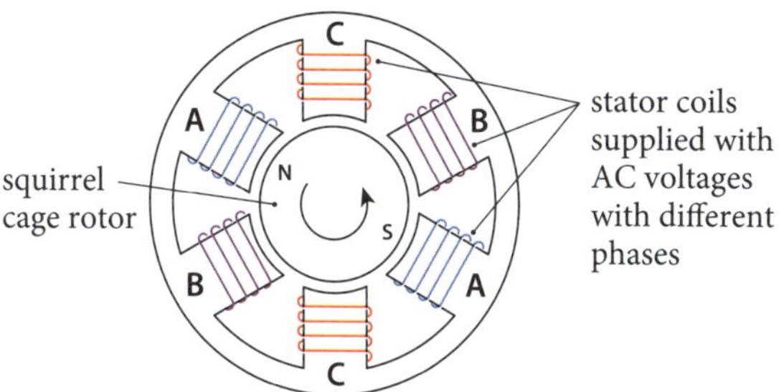

Figure 7.19 AC electromagnets and motion of the rotor in an induction motor

- To intensify the induced currents in the rotor, the rotor in modern induction motors consists of a cage of thick conducting copper or aluminium wires, as shown in Figure 7.20**a**. This type of armature is called a **squirrel cage rotor**. To improve efficiency further, the squirrel cage is generally filled with a laminated iron core and the conducting rods are skewed slightly, as shown in Figure 7.20**b**. The iron core increases the magnetic fields in the motor, and the skewed conducting rods smooth out the torque produced by the motor as it rotates and ensure that the conducting rods will produce a torque when the motor is switched on.

squirrel cage rotor: the rotating part of an AC induction motor that is made up of thick conducting rods designed to increase the current induced in the rods by a changing magnetic flux

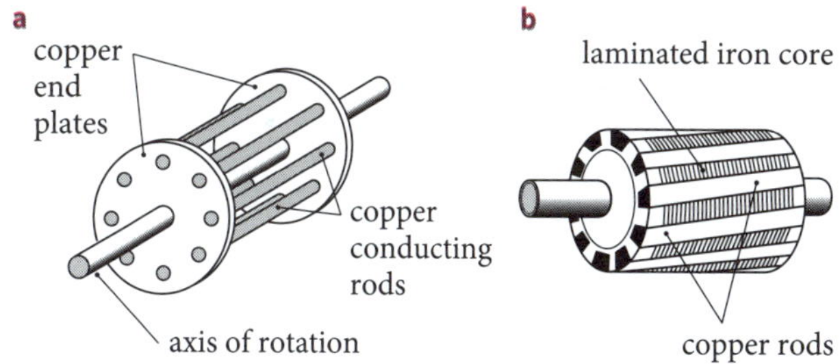

Figure 7.20 Squirrel cage rotor **a** without iron laminations and **b** with iron laminations and skewed conducting rods

- In an ideal (frictionless) induction motor the rotor would revolve at the same speed as the magnetic field produced by the stator coils because no torque would be required to turn the motor. If the rotor turned at the same rate as the magnetic field, no torque would be produced because the magnets would not be moving with respect to the rotor. Therefore there would be no change in the magnetic flux in the rotor, no currents induced and no torque generated.

- In a real motor, friction and the addition of mechanical loads slow the rotor and, as the rotor is no longer rotating at the same rate as the magnetic field, the rotor experiences a changing magnetic flux, currents are induced in the rotor and a torque is produced.
- When an induction motor is first switched on, the rotor is initially not moving and the high rate of magnetic flux change induces a large current in the rotor. This large current will produce a large torque, which will increase the rate of rotation of the rotor. As the rotor turns faster it experiences a lower rate of change of magnetic flux (as it catches up on the rotating magnetic field) and less torque is produced. The motor stops turning faster and revolves at a constant rate when the lag between the rotor and the revolving magnetic field is just enough to provide the torque needed to balance the torques produced by friction and the load. If the load is increased the rotor would slow, producing more torque until a new equilibrium is reached at a lower speed. If the load was decreased the motor would speed up until a new equilibrium was reached at a higher rate of rotation.
- This change in rotational speed ensures energy is conserved because the rotational power is equal to the product of the torque and the angular velocity (rate of rotation). Thus if the input power is constant and a greater load is placed on the motor, it can only provide a greater torque by running at a slower rate.
- AC induction motors are used widely because they are efficient and require little maintenance as, unlike DC motors, they do not have brushes that can wear out.

→ KEY QUESTIONS

10 **How can the laws of Faraday and Lenz be used to explain the motion of the squirrel cage rotor in an AC induction motor?**

11 **How can AC voltage be used to move a magnetic field around a squirrel cage rotor?**

Answers ⊃ p. 106

EXAMPLE 4

Use your knowledge of electromagnetic induction to explain the following observations:

a **The current in the squirrel cage rotor decreases as it rotates faster.**

b **If an AC induction motor is prevented from turning it gets very hot.**

c **If the frequency of the AC voltage placed on the stator coils of an induction motor is increased, the motor turns faster.**

Remember that the amount of current induced in the squirrel cage rotor will be determined by the relative motion of the magnetic field moving around the rotor

Answer:

a Current is induced in the squirrel cage rotor of an induction motor because the rotor is surrounded by AC stator coils that produce a rapidly rotating magnetic field. This causes a changing magnetic flux in the rotor, which by Faraday's law induces a current in the rotor. As the speed of rotation of the rotor increases, the apparent speed of rotation of the magnetic field decreases with respect to the rotor. This will decrease the rate of change of magnetic flux experienced by the rotor and hence decrease the emf induced on the rotor and the current in the rotor.

b If the squirrel cage rotor is prevented from turning when current is flowing in the stator coils the rotor will experience a large rate of change of magnetic flux and hence a large current will be induced in the rotor. This will cause resistance heating in the rotor, which will increase its temperature.

c If a higher frequency AC voltage is placed on the stator coils the frequency of rotation of the magnetic field produced by the stator coils will also increase. This will cause a greater rate of change of magnetic flux in the rotor, more induced current and an increased torque on the rotor, leading to a greater rate of rotation.

3 Energy conservation in DC motors and electromagnetic braking

» Students relate Lenz's law to the law of conservation of energy and apply the law of conservation of energy to DC motors and magnetic braking.

DC motor and conservation of energy

- Energy is conserved in a DC motor because the electrical power used by the motor always equals the power supplied to the load.
- As we saw earlier in the discussion of the DC motor, the speed of the motor changes until the armature coils draw just enough current to produce a torque that is equal to the load torque. This occurs because the rate of rotation will increase or decrease if there is a net torque on the motor. If the motor supplies more torque that the load torque, the rate of rotation of the motor will increase. This will increase the back emf induced in the armature, which by Lenz's law is in a direction that opposes the supply emf. Thus as the armature spins faster, the current

in the armature decreases and the torque produced by the motor also decreases until the torque supplied by the motor balances the load torque. When there is no net torque on the armature it will turn at a constant rate. Whether the rate of rotation is changing or is constant, the electrical power supplied to the motor is equal to the power drawn from the motor by the load.

➔ Increasing the voltage supplied to a DC motor with the same load will require the motor to turn faster to induce enough back emf to produce the current required to generate the torque required to balance the load torque. Changing the voltage supplied to the motor therefore changes the operating speed of the motor. If the motor operates at a higher voltage with the same load it will turn faster and, as rotational power is related to the product of the torque and the operating speed, the increased input power (higher applied voltage with the same current) will result in an equal increase in output power. Again we see that energy is conserved.

➔ We have seen that when the speed of rotation of the motor changes, the current in the armature, the torque and the power provided by the motor also change but the electrical power going into the motor is always equal to the power transferred to the load. That is, energy is always conserved in the DC motor/load system when the torques have reached equilibrium and when the speed of the motor is changing before it has reached equilibrium.

➔ KEY QUESTION

12 **Why does placing a greater load on a motor make it run slower?**

Answers ➲ p. 106

EXAMPLE 5

An ideal 240 V DC motor is used to lift a 50 kg mass, 10 m vertically in 2 s.

a **Find the current in the motor when it is lifting the mass.**

b **Explain using conservation of energy how the answer to part b would change if a real motor was used rather than an ideal motor.**

c **Explain why the current in the motor was greater when it was first switched on than when it was rotating.**

Recall that an ideal machine is 100% efficient and power is the rate at which energy is transferred or transformed

Answer:

a If the motor is ideal all the electrical power will be converted into mechanical power. The power required to lift the mass is given by:

Power = energy/time = mgh/t

$= \frac{50 \times 9.8 \times 10}{2} = 2450$ W

Equating this power to the electrical power:

$P = IV = 2450$ W

$I = 2450/V = \frac{2450}{240} = 10.2$ A

b A non-ideal motor is not 100% efficient and therefore not all the input energy is converted into useful output energy. Because energy is conserved, some of the input energy is converted into heat. The motor would therefore use more electrical power and hence require a greater current to lift the mass in the same time.

c The motor acts as a generator when it rotates, inducing a back emf that opposes the supply emf and hence reducing the current in the armature. When the motor is first switched on, it is not rotating and therefore the magnetic flux is not changing and no back emf is induced to oppose the supply emf. Thus the initial emf across the armature coil is large and a large current will flow in the armature coil. However, when the armature coil is rotating in the magnetic field, it will induce a back emf that will reduce the emf across the armature coil. This will reduce the current in the armature when it is rotating.

Eddy current brakes

➔ **Eddy current brakes** use the magnetic fields produced by eddy currents to apply a braking force. A simple eddy current brake is shown in Figure 7.21. You will recall that we looked at several other examples of the retarding force produced by eddy currents in the previous chapter when we investigated Lenz's law, including an aluminium plate swinging though a magnetic field and a magnet dropped through an aluminium tube.

eddy current brake: a brake that uses induced eddy currents produced by the relative movement between a conductor and a magnetic field to slow the conductor by converting kinetic energy to heat

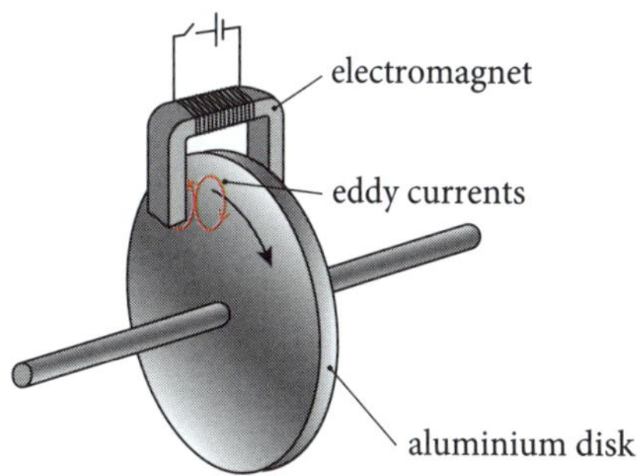

Figure 7.21 Eddy current brake

➔ Eddy current brakes are used in industry on machines where movement needs to be damped, on some trains and even in some amusement park rides. Unlike a traditional friction brake, eddy current brakes do not have parts in contact that will wear out. However, eddy current brakes cannot replace friction brakes entirely because eddy current brakes cannot be used to lock a moving conductor and they produce very little braking power at low speeds. For this reason, eddy current brakes are often used together with friction brakes.

- Eddy current brakes, like friction brakes, convert mechanical energy into heat energy. To do this, traditional brakes use friction, while eddy current brakes use resistance heating. If a friction brake and an eddy current brake were to provide the same braking power, the same amount of heat would be generated in each type of brake.

→ KEY QUESTION

13 **How can eddy currents be used to slow a moving object?**

Answers ⊃ p. 106

Regenerative electromagnetic braking

- **Regenerative electromagnetic braking** involves using an electric motor as a generator during braking to return energy to a battery (or capacitor) for later use. You will recall that conservation of energy requires that the more current that is drawn from a generator, the harder it is to turn the generator. This is because the current is proportional to the electrical power drawn from the generator and hence to the mechanical power required to turn the generator. Therefore the more current drawn from the generator, the greater will be its braking effect.

regenerative electromagnetic braking: a braking system that converts kinetic energy to electrical energy to be stored for later use

- The use of regenerative braking in electric cars and hybrid cars significantly increases the car's efficiency. Rather than all the car's kinetic energy being wasted as heat during braking, some of the kinetic energy is converted to electrical energy and stored for later use.
- Large aircraft also use regenerative braking. Aircraft use electric motors on each wheel to accelerate the wheels to a high speed of rotation before landing, to reduce tyre damage during landing. When the tyres touch the ground, the electric motors on each wheel are used to generate electricity and help to slow the plane.
- Regenerative braking cannot recover all the kinetic energy because some energy is always converted to heat; but with careful design, a significant proportion of the kinetic energy can be stored for later use during braking.

→ KEY QUESTION

14 **Why is regenerative braking more effective when it produces more current?**

Answers ⊃ p. 106

CHAPTER SYLLABUS CHECKLIST

Are you able to answer these questions from the syllabus for this chapter? Tick each question as you go through the checklist if you are able to answer it. If you cannot answer a question, turn to the relevant page in the study guide to find the answer. For NESA key word meanings, go to www.educationstandards.nsw.edu.au and search 'key words'.

	FOR A COMPLETE UNDERSTANDING OF THIS TOPIC:	PAGE NO.	✓
1	Can I explain how a DC motor operates?	93–95	
2	Can I describe the structure and function of the principle components in a DC motor?	93–95	
3	Can I relate the torque produced by a DC motor to the angle between the applied magnetic field and the area vector of the rotor coil?	94	
4	Can I solve quantitative problems involving the torque produced by a DC motor?	95	
5	Can I explain how a DC motor produces back emf and why back emf increases with rotational speed?	95	
6	Can I explain how back emf affects the current if the armature?	95–96	
7	Can I explain how DC and AC generators operate?	96–97	
8	Can I differentiate between the structure and output voltage of a DC and AC generator?	97	
9	Can I relate the rate of change of magnetic flux and the emf induced to the angle of rotation of the rotor coil in an AC and DC generator?	97–98	
10	Can I explain why drawing more power from a generator makes the generator harder to turn?	97–98	
11	Can I describe the structure of an AC induction motor and explain how an induction motor operates?	98–99	
12	Can I apply Lenz's law to the operation of DC motors and electromagnetic brakes to explain how energy is conserved in these devices?	100–102	
13	Can I explain how an eddy current brake operates?	101–102	
14	Can I explain how regenerative electromagnetic braking operates?	102	

HSC EXAM-TYPE QUESTIONS

Objective-response questions

(1 mark each)

1 **If the split-ring commutator in the motor shown in Figure 7.22 was replaced with a slip-ring commutator, what would happen to the rotor coil?**

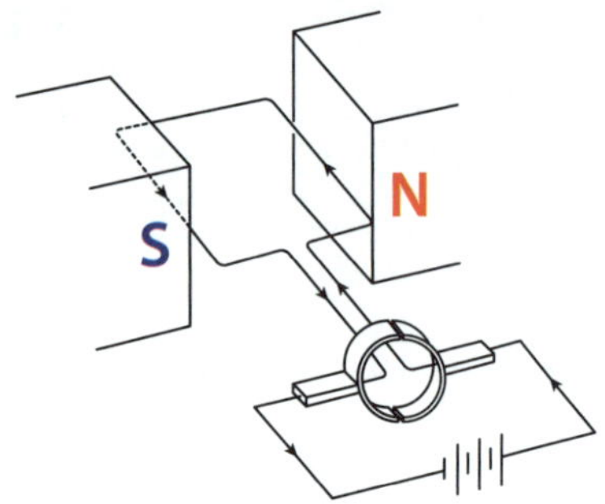

Figure 7.22 Electric motor

- A The coil would rotate clockwise continually.
- B The coil would rotate clockwise 90° and then come to rest.
- C The coil would rotate anticlockwise continually.
- D The coil would rotate anticlockwise 90° and then come to rest.

2 **Why does the rotor coil of a DC motor rotate when current flows in the coil?**

- A because the motor force exerts a net force on the rotor coil
- B because the motor force produces opposing forces on opposite sides of the rotor coil
- C because the back emf induces a torque on the rotor coil
- D because the motor force exerts a force in the same direction on each side of the rotor coil

3 **If the magnetic field in an AC generator was doubled and the period of rotation was halved, how would the peak voltage and frequency of the output change?**

- A The frequency and peak voltage would halve.
- B The frequency and peak voltage would double.
- C The frequency would double and the peak voltage would increase by a factor of four.
- D The frequency would halve and the peak voltage would double.

4 **Why does the squirrel cage rotor rotate in an induction motor?**

- A The stator coils produce a magnetic field that attracts the ferro-magnetic rotor.
- B The stator coils induce a current in the rotor that repels the stator coils.
- C The stator coils induce a current in the rotor that opposes the changing magnetic flux in the rotor.
- D The current fed into the rotor via the commutator produces a torque on the rotor.

5 **Table 7.1 compares the advantages and disadvantages of an eddy current brake over a friction brake. Which of the answers is most correct?**

Table 7.1 Advantages and disadvantages of eddy current brakes over friction brakes

Answer	Advantage	Disadvantage
A	Work well at low relative speed	Produce more heat than a friction brake
B	Low maintenance	Provide little braking force at low speed
C	Produce less heat than a friction brake	Cannot lock the moving wheel or disc
D	Can be controlled electronically	Back EMF can burn out the disc

Extended-response questions

6 **Figure 7.23 shows a component from an electric motor. The wires are connected to a power supply, which is not shown in the diagram.**

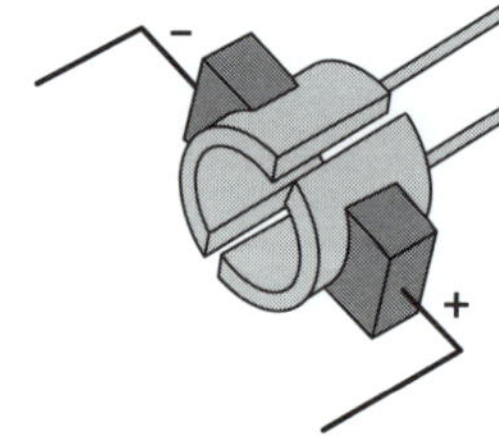

Figure 7.23 Component used in an electric motor

- a **Name the type of electric motor and the component shown in Figure 7.23.** (2 marks)
- b **Explain the function of this component.** (2 marks)
- c **When the component is in the position shown in Figure 7.22 the motor produces a clockwise torque of 8 Nm. What torque would be produced by the motor when the drum of the component was rotated 150° clockwise from the position shown?** (2 marks)

7 Figure 7.24 shows the current passing through a DC motor after it is switched on until it reaches operating speed.

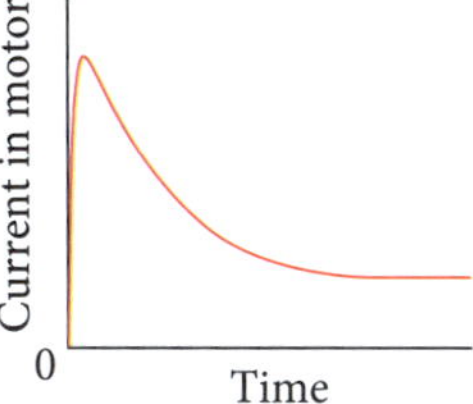

Figure 7.24 Current in armature coil of DC motor after it is switched on

a Explain why the current changes in this way? (4 marks)

b After the motor had reached operating speed the mechanical load on the motor was suddenly increased. Explain how and why the current would change when the load was increased. (2 marks)

8 These questions refer to the simple generator shown in Figure 7.25.

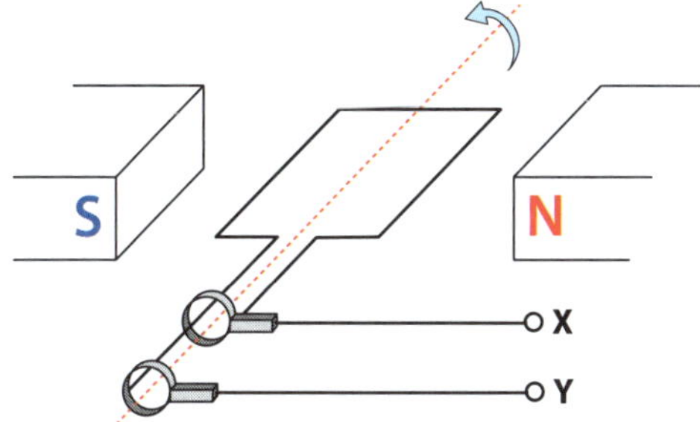

Figure 7.25 Electric generator turned anticlockwise

a Is this an AC or DC generator? Justify your answer. (2 marks)

b If the rotor coil was turned anticlockwise from the position shown, which of the terminals X or Y would have the higher electrical potential? (2 marks)

c When the coil was rotating, at what point in the cycle would the induced emf on the coil be zero? (2 marks)

9 Explain why the AC voltage placed on adjacent stator coils in an induction motion must have a different phase. (3 marks)

10 When an ideal (100% efficient) generator is used to provide an emf of 240 V to a specific circuit, it requires 3.6 kW of power to turn the generator at a specific rate.

a What is the current in the circuit and the total resistance of the circuit? (2 marks)

b If the total resistance of the circuit was doubled, what power would be required to turn the generator to maintain a 240 V potential across the circuit? (1 mark)

c Use the law of conservation of energy and Lenz's law to explain why increasing the total resistance of the circuit changes the torque required to turn the generator at a specific rate. (3 marks)

ANSWERS

KEY QUESTIONS

Key questions ➲ p. 94

1 DC motors require a split-ring commutator to reverse the connection between the power supply and the rotor coil each half-cycle of rotation. This ensures the torque on the rotor coil remains in the one direction and the coil continues to rotate. Without a split-ring commutator, the coil would come to rest when the plane of the coil was perpendicular to the applied magnetic field.

2 DC motors use the motor force (the force on a current-carrying wire in a magnetic field) to produce a torque on the rotor. If there was no magnetic field, there would be no motor force to produce a torque on the coil and the armature would not rotate.

3 The force on each side of the rotor coil is equal, but because the forces operate in opposite directions, the net force on the rotor coil is zero. There is a net torque on the rotor because the forces on each side of the coil produce a torque in the same direction of rotation. However, because the perpendicular distance between the line of action of the force and the axis of rotation changes as the coil rotates, the magnitude of this net torque changes as the coil rotates.

Key questions ➲ p. 96

4 When the coil in a DC motor rotates, the magnetic flux linking the coil continually changes and this induces an emf on the coil in accordance with Faraday's law of electromagnetic induction. This induced emf is called a back emf because it opposes the change that produced it (by Lenz's law), which in this case is the supply emf.

5 The net emf on the rotor coil of a DC motor is the supply emf minus the back emf. Because the back emf increases with the speed of rotation, the current in the rotor coil (and the torque it produces) will decrease as the motor turns faster.

6 If the motor is prevented from rotating there will be no change in magnetic flux to generate back emf. Because there is no back emf to reduce the supply emf, the coil will have the entire supply emf across it, which will result in a large current that may cause enough resistance heating in the rotor coil to melt the enamel insulation between the windings and burn out the motor.

Key questions ➲ p. 98

7 DC motors and generators have the same structure but have the opposite function. DC motors transform electrical energy into mechanical energy of rotation, while DC generators transform mechanical energy of rotation into electrical energy.

8 The structure of AC and DC generators differ only in the way the rotor coil is connected to the external circuit. To ensure the output current from a DC generator will flow in only one direction, DC generators use a split-ring commutator to reverse the connection between the rotor coil and the external circuit each half-cycle. In contrast, AC generators use a slip-ring commutator to connect the rotor coil to the external circuit because they require the connection between the external circuit and the rotor coil to be changed as the coil rotates. AC generators produce a sine-wave voltage output, while single-coil DC generators produce a voltage that does not reverse its polarity (i.e. a |sine| function).

9 For energy to be conserved, a greater amount of mechanical work must be done to turn the generator when it is providing more electrical energy, as the output energy must be equal to the input energy. Another way to think about this is to consider Lenz's law. Because the induced current opposes the change that caused it, the generator will be more difficult to turn when more current flows in the generator (i.e. when it is supplying more electrical power).

Key questions ➲ p. 100

10 The magnetic field moving around the squirrel cage will produce a changing magnetic flux in the cage, which, by Faraday's law, will induce an emf on the coil. By Lenz's law, the induced emf will produce a current in the squirrel cage rotor, which will oppose the change in magnetic flux and cause the cage to rotate with the magnetic field.

11 A rotating magnetic field can be produced by placing AC electromagnets around the squirrel cage with an AC voltage on each successive electromagnet that has a slightly different phase to that on the previous magnet. The different phases ensure that the pole on each electromagnet facing the squirrel cage rotor will become a north pole in sequence around the squirrel cage each AC cycle.

Key question ➲ p. 101

12 Placing a load on a motor increases the torque the motor must supply. The torque a motor can supply increases at slower rotation speed and hence the increased load will slow the motor until the motor can supply the torque to turn the load. In a DC motor, reducing the rotation speed decreases the back emf and hence increases the current in the rotor coil and the torque it produces. In an AC induction motor, slowing the motor increases the lag between the rotor and the rotating magnetic field, which increases the rate of change of magnetic flux linking the rotor, increasing the current in the rotor and the torque it produces. Energy is conserved in both types of motor and hence the electrical power used by the motor is equal to the output power of the motor. Note that in a non-ideal motor some of this output power goes into heat.

Key question ➲ p. 102

13 When eddy currents are induced in a conductor when it moves in a magnetic field, the eddy currents are always in a direction (by Lenz's law) that oppose the relative motion between the conductor and the magnetic field. Thus eddy currents can be used to provide a braking force on a conductor when it moves relative to a magnetic field.

14 Regenerative braking uses an electric motor as a generator. The greater the load on a generator, the more difficult it is to turn (by Lenz's law and conservation of energy). Because drawing a greater current from the generator means it is supplying more power (as $P = IV$), drawing a greater current will make the generator harder to turn and hence it will provide a greater braking force.

HSC EXAM-TYPE QUESTIONS

Objective-response questions

1 **D.** The motor force on the right-hand side of the coil is upwards and on the left-hand side of the coil is downwards (by the right-hand palm rule). With a slip-ring commutator these forces will remain constant in magnitude and direction, which will cause the coil to rotate 90° anticlockwise and come to rest. **A** is incorrect as the direction of rotation is wrong and continual rotation would

require a split-ring commutator. **B** is incorrect as the direction of rotation is wrong. **C** is incorrect as continual rotation would require a split-ring commutator.

2 **B.** The motor force on each side of the coil is in the opposite direction, but both forces produce a torque in the same direction, which causes the coil to rotate. **A** is incorrect because there is no net force on the coil as the forces on each side are equal and operate in opposite directions and hence sum to zero. **C** is incorrect as the back emf opposes the rotation of the rotor and no back emf is induced until the coil is rotating. **D** is incorrect because the current is in opposite directions on each side of the rotor coil and therefore the motor force on each side must operate in opposite directions.

3 **C.** If the period is halved, the frequency would double ($f = 1/T$), and doubling the frequency and the magnetic field would increase the rate of change of magnetic flux by a factor of 4. This would also increase the peak induced emf by a factor of 4. **A** and **D** are incorrect as the frequency must double. **B** is incorrect because the rate of change of magnetic flux is doubled by the increased magnetic field and doubled again by the increased frequency of rotation.

4 **C.** Induced current in the rotor opposes the change in magnetic flux that caused it (Lenz's law). This will move the rotor with the rotating magnetic field produced by the stator coils. **A** and **B** are incorrect because a repelling or attracting force between the stator coils and the squirrel cage rotor would not produce a torque on the rotor. **D** is incorrect as induction motors do not have commutators.

5 **B.** Friction brakes require friction pads to be replaced regularly and hence require more maintenance than eddy current brakes, and eddy current brakes are not efficient at low speed because less current is induced. **A** is incorrect as eddy current brakes are not effective at low speed. **C** is incorrect because both types of brake produce the same amount of heat. **D** is incorrect as there is no significant back emf induced in the disc.

Extended-response questions

6 EM This question tests students' understanding of the split-ring commutator and the torque produced by a DC motor.

- a This is a DC motor ✓ and the component shown is a split-ring commutator. ✓
- b The split-ring commutator connects the armature to the external power supply in a DC motor. The split-ring commutator reverses the connection between the coil and the power supply each half-cycle of rotation ✓ to ensure the current in the armature produces a torque in one direction as it rotates. ✓
- c In the initial position, the angle (θ) between the magnetic field and area vector is 90°. Rotating the coil by 150° places the plane of the coil at an angle of 30° to the magnetic field, which means the angle between the magnetic field and the area vector $\theta = 60°$. ✓

  In the initial position:

  $\tau = 8 \text{ Nm} = nIAB \sin\theta = nIAB$

  as $\theta = 90°$ and $\sin\theta = 1$

  In the new position:

  $\tau = nIAB \sin\theta = nIAB \sin 60°$

  $= 8 \times \sin 60° = 6.93$ Nm clockwise ✓

7 EM This question tests students' understanding of back emf and the operation of a DC motor under load.

- a The current is initially large because the emf on the rotor coil is equal to the supply emf and the current in the coil is proportional to the emf on the coil. ✓ As the rotational speed of the motor increases, an increasing amount of back emf is induced on the rotor coil. ✓ By Lenz's law this back emf opposes the supply emf and hence the emf over the rotor coil decreases and the current in the coil decreases as the rotational speed increases. ✓ When the motor reaches operating speed the back emf, the emf across the rotor and the current in the coil remain constant and therefore the graph levels out. ✓ (Note the current takes a little time to reach its maximum value because the magnetic field in the rotor changes when the current is switched on and this induces an emf in the rotor that opposes the current being switched on.)
- b The load torque would initially be greater than the torque being supplied by the motor, which would reduce the rotational speed of the motor. ✓ As the rotational speed decreased the back emf would also decrease, increasing the emf across the rotor coil and the current in the rotor coil. Therefore, as the rate of rotation decreased, the current in the coil and the torque produced by the motor would increase until the motor torque balanced the load torque. When this occurred there would be no net torque on the motor and the speed of revolution (angular velocity) would remain constant. ✓

8 EM This question tests students' understanding of the AC generator and the relationship between the angle of rotation of the coil and the rate of change of magnetic flux.

- a This is an AC generator ✓ as it has a slip-ring commutator. ✓
- b X would become positively charged and would therefore be at a higher electrical potential than Y. ✓ We can prove this by using Lenz's law as a current (positive charge flow) would have to move towards X on the right-hand side of the coil to oppose the rotation. ✓ Or we could consider the force on a free positively charged particle on the right-hand side of the coil. As the wire moves upwards, a force is exerted out of the page on the positively charged particle by the right-hand palm rule. Thus positive charges would be moved towards X, making it more positive than Y.
- c When the coil is perpendicular to the magnetic field, the induced emf is momentarily zero. ✓ We can see this by considering the force on a free charged particle on the coil at that point. Imagine a positively charged particle on the top arm of the coil when the coil is perpendicular to the magnetic field. The side of the coil and the charge would be moving parallel to the magnetic field and hence there would be no force on the charges to produce a current. ✓

  Note that we could also justify our choice by considering the rate of change of magnetic flux in the coil. The rate of change of the magnetic flux is zero when the plane of the coil is perpendicular to the magnetic field and hence the induced emf is also zero at this point. This is because the magnetic flux is given by $BA\cos\theta$, where θ is the angle between the magnetic field and the area vector ($\theta = 0$ when the plane of the coil is perpendicular to the magnetic field), and the gradient of $BA\cos\theta$ against θ is zero at $\theta = 0$.

9 EM This question tests students' understanding of the principle of operation of the AC induction motor.

The AC induction motor uses a magnetic field that moves around a squirrel cage rotor to induce current in the rotor that, by Lenz's law, will cause the rotor to rotate with the magnetic field. ✓ To produce the rotating magnetic field the induction motor uses several electromagnets placed around the rotor and connected to AC at different phases. The AC phase of each successive electromagnet around the rotor has a phase lag from the preceding magnet. ✓ In this way one magnet has its north pole pointing inwards and a short time later the adjacent magnet has its north pole inwards, and so on. Thus the north pole is moved around the squirrel cage rotor, producing a changing magnetic flux in the rotor that induces currents in the rotor that cause it to oppose the change in magnetic flux ✓ by rotating with the revolving magnetic field.

10 EM This question tests students' ability to apply Lenz's law and the conservation of energy to an electric generator.

a As the generator is ideal, the output power must equal the input power and hence the power provided to the circuit is 3.6 kW. Now:

$P = IV$ and hence

$I = P/V = \dfrac{3600}{240} = 15$ A ✓

The total resistance of the circuit is therefore:

$I = V/R = \dfrac{240}{15} = 16\ \Omega$ ✓

b If the total circuit resistance was 32 Ω:

$P = IV$ and $I = V/R$

Therefore $P = V^2/R = \dfrac{240^2}{32} = 1800$ W = 1.8 kW ✓

c Increasing the resistance decreases the current and the power dissipated in the circuit. ✓ Energy is conserved when power is generated. As this is an ideal system, the input power is equal to the useful output power and hence less mechanical power will be needed to turn the generator as it is supplying less power to the circuit. As rotational power is related to the product of the torque and the rate of rotation, if the rate of rotation remains constant and less power is required, a smaller torque will be required to turn the generator. ✓

We can also explain this in terms of Lenz's law, which states that the induced current is in a direction that will oppose the change that produced it initially. For the generator, this means that the induced current will produce magnetic fields that oppose the generator being turned. Increasing the resistance of the circuit decreases the current required and hence a smaller current is flowing in the generator. A smaller induced current means that less torque is opposing the generator being turned (provided the generator is rotated at the same rate) and hence the torque required to turn the generator will be lower. ✓

CHAPTER 8

ELECTROMAGNETIC SPECTRUM

INQUIRY QUESTION:

What is light?

Light is an electromagnetic wave that can be detected by the human eye. Electromagnetic waves can have a wide range of frequencies, from radio wave frequencies to gamma wave frequencies. We call this range of electromagnetic waves the electromagnetic spectrum. Because the human eye can detect only a small range of electromagnetic wave frequencies, visible light makes up only a small portion of the electromagnetic spectrum.

1 Predicting the existence of electromagnetic waves

» Students investigate Maxwell's contribution to the classical theory of electromagnetism, including the unification of electricity and magnetism, the prediction of electromagnetic waves and the prediction of velocity.

- Michael Faraday (1791–1867) introduced the field concept to help explain the action of electric and magnetic forces. Faraday imagined that electric and magnetic lines of force permeated all space and in 1846 he boldly speculated that light might be a vibration of these electric and magnetic lines of force. This idea was treated as pure speculation by the scientific community at the time because Faraday could not provide theoretical or experimental evidence to support his hypothesis.
- In the 1860s James Clerk Maxwell (1831–1879) combined all the existing equations and observation of electricity and magnetism into a single coherent theory of electromagnetism based on Faraday's field concept. Maxwell replaced Faraday's lines of force with electric and magnetic fields but the field concept remained central to the theory. The unification of electricity and magnetism by Maxwell was a major step forward in the development of physics.
- Maxwell's theory initially consisted of 20 equations but was later simplified to four equations that still form the foundation of the **classical theory** of electromagnetism. **Maxwell's equations** use advanced mathematics (vector calculus) beyond the scope of the HSC course but we can paraphrase the content of his equations as follows:
    - The first equation tells us that electric field lines originate from positive charges and end at negative charges, and that the force exerted by the electric field is related to a constant (ε_0) called the *permittivity of free space.*
    - The second equation tells us that magnetic field lines are continuous and therefore magnetic monopoles (isolated north or south poles) cannot exist. The equation also shows that the strength of the magnetic force is related to a constant (μ_0) called the *permeability of free space.*
    - The third equation states that a changing magnetic field will produce an electric field that can generate an emf in a direction that opposes the change. This law combines Faraday's law of electromagnetic induction and Lenz's law.
    - The forth equation states that magnetic fields can be produced by moving charges (i.e. Oersted's law) or by changing electric fields.

classical theory: refers to a theory in physics that does not incorporate quantum mechanics

Maxwell's equations: four equations that describe electromagnetic forces in terms of the relationship between charge particles, and magnetic and electric fields; the classical theory of electromagnetism is based on Maxwell's equations

- The fourth law is testament to Maxwell's remarkable insight. There was no proof at the time that a changing electric field would induce a magnetic field but Maxwell argued from the point of view of symmetry that if a changing magnetic field could produce an electric field (i.e. Faraday's law), then a changing electric field should produce a magnetic field.
- In 1864 Maxwell presented some new work to the Royal Institute in London that was later published in

1865. Maxwell showed that his equations involving the rate of change of electric fields and the rate of change of magnetic fields could be combined to produce a wave equation and that the equation gave the wave speed as:

$$v = \frac{1}{\sqrt{\mu_0 \varepsilon_0}} = \frac{1}{\sqrt{(1.2566^{-6})(8.8542^{-12})}} = 3 \times 10^8 \text{ ms}^{-1}$$

As this is the speed of light, Maxwell suggested that light must be an **electromagnetic wave**.

electromagnetic waves: mutually perpendicular, changing electric and magnetic fields that form a self-sustaining, transverse wave that travels at the speed of light

- Maxwell's theory showed that oscillating charges would produce changing electric and magnetic fields that would move away from the charge at the speed of light. The waves would have the same frequency as the oscillating charge and hence electromagnetic waves could have a wide range of frequencies. The visible light spectrum would make up only a small portion of the possible electromagnetic wave frequencies.
- By developing the electromagnetic wave theory of light, Maxwell demonstrated that light, magnetism and electricity were interrelated rather than independent phenomena.
- Maxwell did not show how his equations could describe wave properties such as reflection and refraction or how such waves could be produced in the laboratory, but he did suggest that electromagnetic waves might exist over a much wider range of frequencies than the frequencies of visible light.
- Maxwell died at the age of 48 in 1879, about a decade before his theoretical prediction of electromagnetic waves was experimentally confirmed by Heinrich Hertz. Hertz produced, detected and investigated the behaviour of electromagnetic waves in the frequency range we call today *radio waves*. He showed that electromagnetic waves travelled at the speed of light and could be polarised, reflected, refracted and diffracted, just like light waves.

EXAMPLE 1

Maxwell condensed all the known observations on electricity and magnetism into four equations and suggested that changes in electric and magnetic fields would propagate as electromagnetic waves.

a How did Maxwell explain the propagation of electromagnetic waves?

b Why did Maxwell think light was an electromagnetic wave?

Recall the relationship between changing electric and magnetic fields expressed in Maxwell's equations

Answer:

a Maxwell's equations showed that a changing magnetic field would produce a changing electric field and that a changing magnetic field would produce a changing electric field. Hence a change in one field induces a change in the other and vice versa continually. These changes produce a self-sustaining wave that travels at the speed of light and which does not require a medium in which to travel.

b Maxwell's wave equation related the speed of the waves to known constants in electricity and magnetism, which gave the speed of the wave as 3×10^8 ms^{-1}. This was very close to the measured speed of light at the time and Maxwell therefore proposed light was an electromagnetic wave.

KEY QUESTIONS

1 What was Maxwell's contribution to electromagnetism?

2 What led Maxwell to suggest light was related to electromagnetism?

Answers p. 122

2 Electromagnetic wave properties

» Students describe the production and propagation of electromagnetic waves and relate these processes qualitatively to the predictions made by Maxwell's electromagnetic theory.

- Today we call the wide range of electromagnetic frequencies predicted by Maxwell the **electromagnetic spectrum**. As shown in Figure 8.1, the electromagnetic spectrum ranges from low frequency (long wavelength) radio waves to incredibly high frequency (short wavelength) gamma rays. Visible light makes up only a small portion of electromagnetic wave frequencies, near the middle of the electromagnetic spectrum.

electromagnetic spectrum: the range of frequencies that electromagnetic radiation can have

- All electromagnetic waves are produced by accelerating charged particles. Continuous electromagnetic waves can be produced by oscillating charged particles because they accelerate continually as they move back and forth. The electromagnetic waves produced would have the

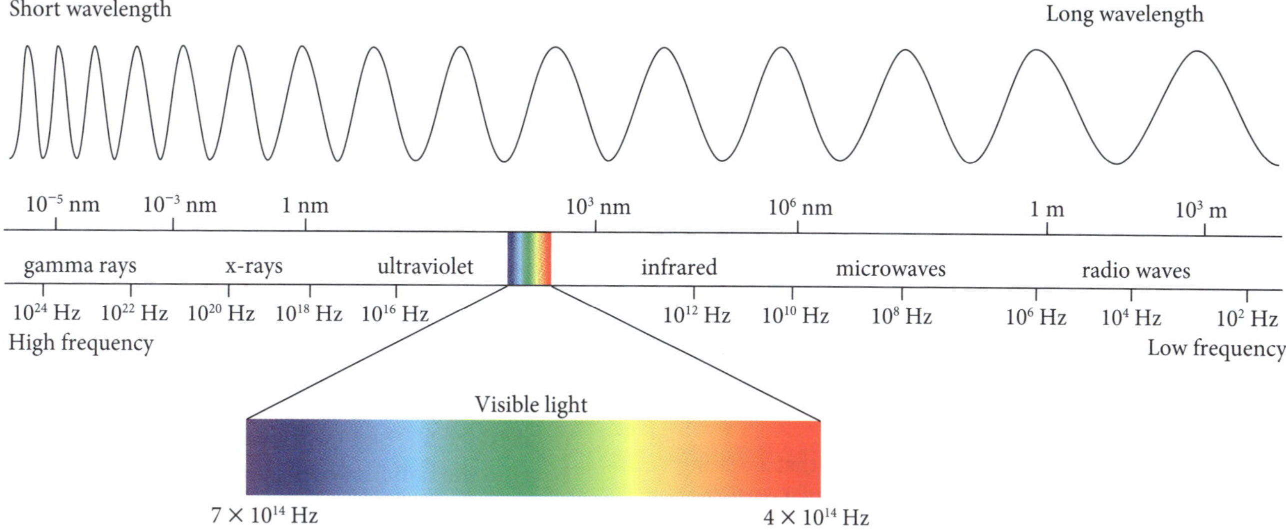

Figure 8.1 The electromagnetic spectrum

same frequency as the oscillating charged particle. Thus the electromagnetic spectrum is produced by charges oscillating at different rates. Gamma rays are produced by charges oscillating in the nucleus of atoms, visible light by charges oscillating in atoms, infrared by charge oscillation in molecules and radio waves by charge oscillation in metal wires (antennas).

➔ We can understand the production of electromagnetic wave in terms of Maxwell's theory by considering what would happen to the electric and magnetic fields around a charge as it oscillated. We saw in the Year 11 course that a current-carrying wire produces a magnetic field in concentric circles around the wire. An oscillating charge would therefore produce a circular magnetic field around the charge that continually changed its intensity and reversed its direction each half-cycle. The oscillating charge would also produce a changing electric field perpendicular to the changing magnetic field. As predicted by Maxwell, the changing magnetic field would induce a changing electric field, and the changing electric field would induce a changing magnetic field. The changing magnetic and electric fields would be in phase and the change would move away from the charge at the speed of light. Because the fields oscillate in a direction perpendicular to the velocity, electromagnetic waves are transverse waves. Figure 8.2 shows an electromagnetic wave produced by an oscillating charged particle. Note that electromagnetic waves would actually be emitted in all directions around the particle, except in the direction parallel to the charge's motion.

➔ Another way to understand electromagnetic waves is to consider the electric field that surrounds every charged particle. The electric field extends to infinity around a charged particle and influences other charged particles in the field. If a charged particle is

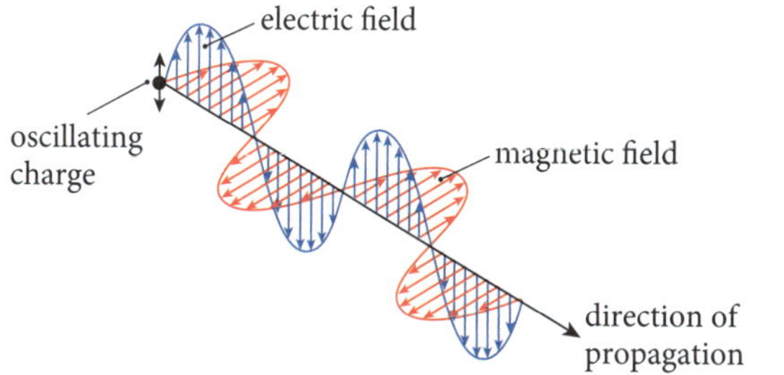

Figure 8.2 Electromagnetic wave produced by an oscillating negatively charged particle

accelerated (or oscillated), the electric field around the particle will change but this change is not immediately experienced by surrounding charges. The change in the electric field travels at the speed of light away from the oscillating charged particle and therefore the change takes time to reach and influence the surrounding charged particles. We can think of electromagnetic waves as a change in the electric field moving through the field. Because changing an electric field produces a magnetic field and vice versa, these travelling changes are electromagnetic waves.

➔ Later we will see that the speed of light is the maximum speed that any information or object can travel. If we accept that this is the case, we can appreciate why different-frequency electromagnetic radiation comes from such different sources. Observations show that radio waves can be produced by charges oscillating in a wire (antenna), while high frequency electromagnetic waves are produced by charges that oscillate over much smaller distances. Because charged particles cannot travel faster than the speed of light the maximum distance moved by a charge when it oscillates is determined by its frequency. For example, gamma rays, which originate from the nucleus of atoms, have a frequency of about 10^{22} Hz.

One-half a cycle would take $0.5 \times \frac{1}{10^{22}} = 5 \times 10^{-23}$ seconds and hence, even if the charged particle was travelling at the speed of light, it could only travel a distance of:

$s = vt = 3 \times 10^8 \times 5 \times 10^{-23} = 1.5 \times 10^{-14}$ m

Therefore to oscillate at 10^{22} Hz the charged particle would have to be confined to a region with a maximum diameter of about 10^{-14} m, which is about the diameter of an atomic nucleus.

- Another consequence of Maxwell's theory is that charged particles oscillating in the one direction would produce polarised waves (i.e. waves in which the electric field is in one plane).
- Maxwell's theory also correctly predicted that a passing electromagnetic wave would cause a charged particle to oscillate in the same way that a passing water wave would cause a duck on a pond to oscillate. The electric field in the passing electromagnetic wave exerts a changing force on the charge that will attempt to oscillate the charge parallel to the electric field (i.e. perpendicular to the electromagnetic wave velocity). This oscillating charge would then generate electromagnetic waves at the same frequency. Light travels through transparent materials by exciting electron oscillations that cause the wave to be re-emitting. This transmission via excitation and re-emission slows the light and results in the light travelling more slowly through transparent materials than in a vacuum.
- Maxwell's theory of electromagnetism was extremely successful in describing the propagation of electromagnetic waves but it failed to explain some observations about the way light interacts with matter on the atomic scale. One of the difficulties with Maxwell's theory, that would eventually lead to the development of quantum physics, was that it could not correctly predict the black-body radiation curves we investigated in the Year 11 course. However, Maxwell's theory did predict that the vibrating charges in hot bodies would emit electromagnetic radiation and that more radiation would be emitted from such a body if the temperature of the body was increased. Maxwell's theory also supported the observation that more radiation was emitted at higher frequencies because when the temperature of a body was increased, higher temperatures would cause charges to vibrate at higher frequency.

EXAMPLE 2

Figure 8.3 shows a proton moving perpendicular to a 0.02 T magnetic field with a velocity of 200 ms^{-1}. Note that a proton has a charge of 1.602×10^{-19} C and a mass of 1.673×10^{-27} kg.

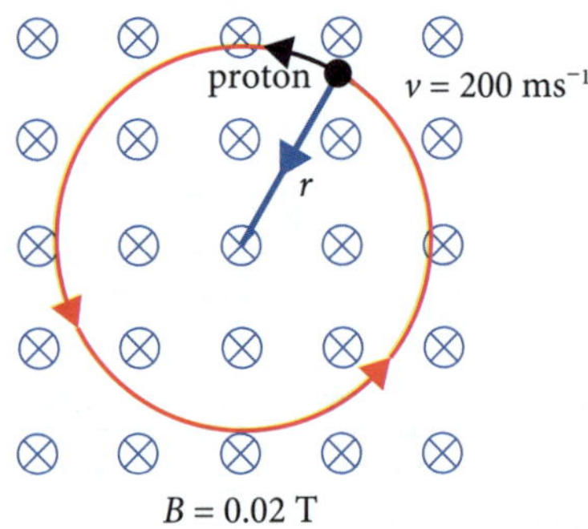

Figure 8.3 Proton moving in a magnetic field

a **Find the force on the proton.**

b **Find the radius of curvature of the circular path taken by the electron.**

c **Find the frequency of the electromagnetic waves emitted by the proton.**

d **In which direction would polarised electromagnetic waves be emitted from the proton? Justify your answer.**

A particle undergoing circular motion appears to be oscillating back and forth when viewed from the side

Answer:

a The force on the proton due to its motion in the magnetic field will be given by:

$F_B = qvB = 1.602 \times 10^{-19} \times 200 \times 0.02 = 6.408 \times 10^{-19}$ N towards the centre of the circle

b The force on the charge ($F_B = F_c$) is a centripetal force and hence:

$F_c = mv^2/r$ or

$r = mv^2/F_c = \frac{(1.673 \times 10^{-27}) \times (200)^2}{6.408 \times 10^{-19}} = 1.044 \times 10^{-4}$ m

c The emitted waves will have the same frequency as the charge:

$v = 2\pi/T = 2\pi rf$ and hence

$f = v/2\pi r = \frac{200}{2\pi \times 1.044 \times 10^{-4}} = 3.048 \times 10^5$ Hz

d Polarised electromagnetic waves would be emitted from the charge in the plane of the circular motion (plane of the page) because when the charge is viewed from this plane it will appear to be oscillating in a line and hence will produce a changing electric field in only one plane.

KEY QUESTIONS

3 **How are electromagnetic waves produced?**

4 **Why do electromagnetic waves always have both an electric field and magnetic field component?**

5 **How do electromagnetic waves influence other charged particles?**

Answers p. 122

3 The speed of light

» Students conduct investigations of historical and contemporary methods used to determine the speed of light and its current relationship to the measurement of time and distance.

➔ The first successful measurements of the speed of light used astronomical observations. Later, mechanical experiments involving high-speed, spinning-toothed wheels and mirrors were used and today lasers and electronic measuring techniques are used. Table 8.1 summarises the key milestones in measuring the speed of light.

Table 8.1 History of the measurement of the speed of light

Year	Scientist	Method	Result (kms^{-1})
1676	Romer	Eclipse of Jupiter's moon Io	220 000
1728	Bradley	Aberration of star light	301 000
1849	Fizeau	Rotating toothed wheel	313 300
1862	Foucault	Rotating mirror	309 000
1909	Rosa and Dorsay	Accurate measurements of electric and magnetic constants	299 788
1926	Michelson	Rotating mirror	299 796 ± 4
1958	Froome	Radio interferometry	299 792.5 ± 0.1
1976	Woods (and others)	Frequency and wavelength of a laser	299 792.459 ± 0.001
From 1983		Value fixed and metre defined in terms of the speed of light	299 792.458

➔ By the early 1970s the speed of light was known to within 1 ms^{-1}, which was more accurate than the metre could be measured.

➔ In 1983, through international agreement, the speed of light in a vacuum was fixed at 299 792 458 ms^{-1}. This speed was then used to define the standard metre as the distance light travels in $\frac{1}{299\,792\,458}$ seconds.

➔ Astronomers also measure distances using the speed of light. A light year is the distance travelled by light in a vacuum in one Earth year.

➔ The standard second is not measured using the speed of light but is set as a particular number of oscillations of the radiation emitted from a specific electron transition in a caesium-133 atom.

➔ Note that the definition of the metre depends on the time standard and the fixed speed of light.

SECONDARY-SOURCED INVESTIGATION

Measuring the speed of light

Students should conduct a secondary investigation of how the accuracy of our measurement of the speed of light has improved over time. Typical finding would include:

- In the time of Galileo, scientists argued about whether the speed of light was infinite or finite. Galileo tried to measure the speed of light by direct measurement but concluded that the speed of light was instantaneous or too fast to measure with the simple method he used.
- Proof that the speed of light was finite came in 1676 when the Danish astronomer, Ole Romer, notices that Jupiter's moon Io was eclipsed by Jupiter at a different time as the Earth moved in its orbit. He noted that Io disappeared behind Jupiter 22 minutes later when the Earth was at the far side of its orbit and he concluded that light must take 22 minutes to travel the diameter of the Earth's orbit. Romer concluded that the speed of light was therefore finite. Christian Huygens used Romer's result to calculate that the light was 2.2×10^8 ms^{-1}, which is about 25% lower than the value we use today.
- Another astronomical method to find the speed of light was used by the English astronomer James Bradley in 1728. He used the shift in the position of stars caused by the Earth's orbital velocity, stellar aberration, to calculate a value of 3.01×10^8 ms^{-1}.
- The first experimental measurement of the speed of light was made by the French physicist Fizeau in 1849. His experiment is shown in Figure 8.4. Fizeau passed light through a gap in a spinning-toothed wheel and, after reflecting the light from a mirror 8 km away, he adjusted the speed of his wheel until the reflected light passed through the next gap in the wheel. Fizeau obtained a velocity of 3.133×10^8 ms^{-1}, which is about 1.5% above the correct value.

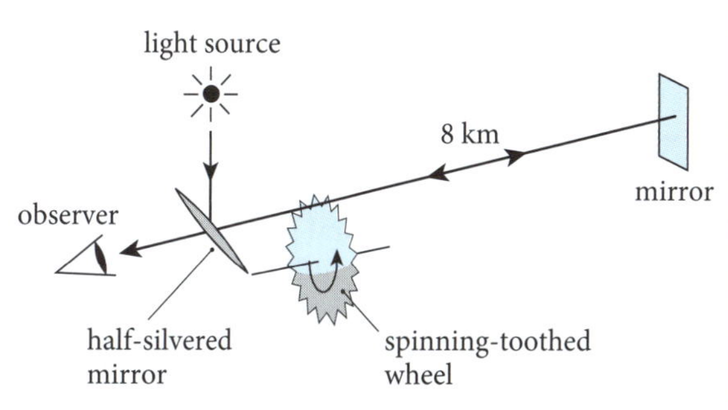

Figure 8.4 Fizeau's experiment to measure the speed of light

- In the period 1850 to 1862, Foucault obtained a more accurate value for the speed of light by replacing Fizeau's spinning-toothed wheel with a spinning mirror and relating the shift in the reflected image from the rotating mirror to the distance covered by the beam. Foucault's value of 3.09×10^8 was within 1.5% of the accepted value. Foucault's experiment was later refined by the American physicist Albert Michelson, who used a longer reflecting path and eventually obtained a value of $2.997\,96 \times 10^8$ ms^{-1} in 1926. The experiment that Michelson used to measure the speed of light is illustrated in Figure 8.5.

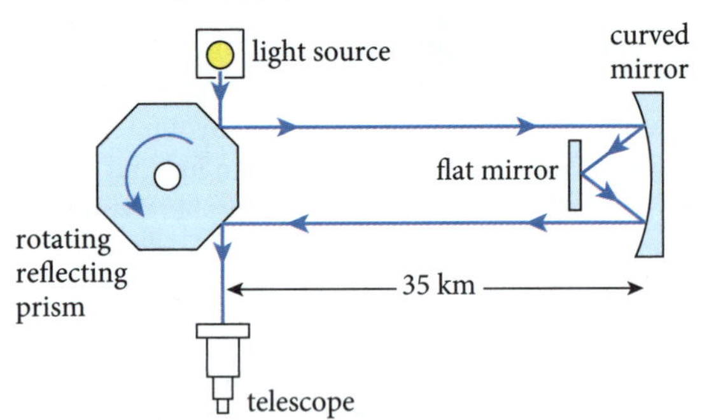

Figure 8.5 The beam will be reflected into the telescope if the prism turns one-eighth of a turn while the light is travelling to the distant mirror and back

- In 1909, Rosa and Dorsay used accurate measurements of the magnetic permeability and electric permittivity constants and Maxwell's relationship between these constants and the speed of light to obtain a value of $2.997\,88 \times 10^8$ ms^{-1}.
- In 1958, Froome used radio interferometry (interference) to measure the speed at $2.997\,925 \times 10^8$ ms^{-1}.
- In the 1970s several groups measured the frequency and wavelength of lasers and then used the wave equation to obtain even more accurate values for the speed of light. A team in 1976 found the speed to be 299 792.4574 kms^{-1}.
- In 1983 the value was fixed at 299 792.458 kms^{-1}.

EXAMPLE 3

Explain why many early natural philosophers thought the speed of light might be infinite.

Recall that the speed of light is incredibly fast

Answer:

The speed of light is so fast that it could not be measured by any of the techniques available to early natural philosophers and, without proof of its finite speed, philosophers could only speculate about its speed.

FIRSTHAND INVESTIGATION 1

Using a microwave oven to measure the speed of light

Students can measure the speed of light in the school laboratory by using a simple microwave oven. Microwave ovens work by setting up standing waves in the oven. Students should remove the revolving turntable and replace it with a piece of ovenproof paper covered in chocolate buttons. After running the microwave for a short time, the buttons should be examined. The distance between the antinodes in the oven can be determined by measuring the distance between the most-melted chocolate buttons. Then, using the frequency from the compliance plate on the oven and recalling that the distance between adjacent antinodes is one half-wavelength, the wave equation can be used to calculate the velocity of the microwaves.

→ KEY QUESTIONS

6 **How was the speed of light first shown to be finite rather than infinite?**

7 **Explain why the speed of light will no longer change if it is measured more accurately?**

8 **How is the speed of light used to define the metre?**

Answers ➲ p. 122

4 Spectra

» Students conduct an investigation to examine a variety of spectra produced by discharge tubes, reflected sunlight or incandescent filaments.

» Students investigate how spectroscopy can be used to provide information about the identification of elements.

Types of spectra

→ A spectrum is produced when light is passed through a dispersive element, such as a prism, which separates the different wavelength components of the light. The light emitted by an incandescent lamp is an example of a continuous spectrum because it contains all the visible wavelengths of light. In contrast, a monochromatic laser beam contains only one wavelength of light and will produce a single line spectrum if it is passed through a prism.

FIRSTHAND INVESTIGATION 2

Continuous spectra

Students should examine **continuous spectra** by using a prism or grating **spectrometer** to view the light produced by an incandescent light globe. As shown in Figure 8.6, students will find the hot element in the globe will emit all visible wavelengths and produce a continuous visible light spectrum when viewed through the spectrometer. Students will also find that when the globe is operated at lower voltages, less light is emitted at the blue end of the spectrum.

continuous spectra: electromagnetic radiation with a continuous range of frequencies

spectrometer: a general term for a device that separates the different wavelength components in a sample of light; a *spectroscope* is a spectrometer that is used for viewing spectra directly, and a *spectrograph* is a spectrometer that records spectra photographically or electronically

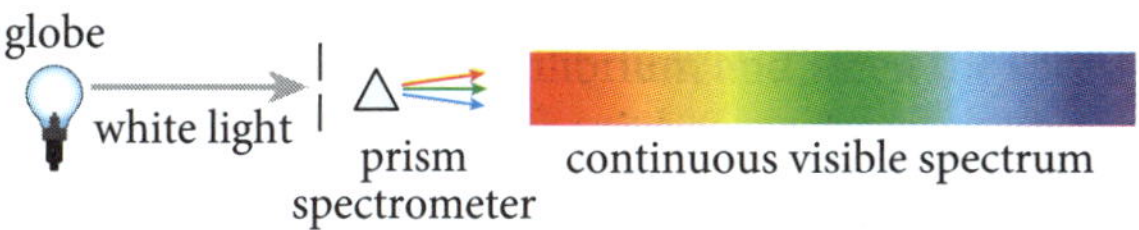

Figure 8.6 Continuous spectrum emitted from an incandescent light globe

Students will recall from the Year 11 course that the electromagnetic radiation emitted from the hot substance, like a globe filament, is called *black-body radiation*. The intensity of the electromagnetic radiation emitted increases and the wavelength of maximum emission decreases as the temperature of the **black body** increases. This is why the blue end of the spectrum is not present in the spectra when the globe is operated at low voltage, but the entire visible spectrum is emitted when the globe is operated at high voltage.

black body: a body that absorbs all the light that falls on it and reflects none; the radiation emitted from a black body is therefore due to the temperature of the body alone

Atomic spectroscopy

- When gaseous atoms are excited (given energy) by an electric discharge or a flame, they emit electromagnetic radiation at specific wavelengths that are characteristic of the type of element being excited. We call this type of spectra, atomic **emission spectra** or line spectra.
- Because each element emits different wavelengths, atomic emission spectra can be used to identify the elements in a sample. The elements in a gas or vapour sample can be identified by examining the spectral lines emitted when gases or vapour are excited by an electric discharge or flame. This technique is used in **flame spectroscopy**, which is an import tool in chemical analysis.

emission spectra: specific wavelengths (lines) emitted from excited atomic gases; the frequencies of radiation emitted from an elemental gas are discontinuous rather than continuous and the set of wavelengths emitted from an element are specific to that element

flame spectroscopy: the study of the light emitted from elements excited in a flame

- Fireworks utilise the emission spectra from excited metal atoms to produce specific colours of light in firework displays. Table 8.2 shows the principle emission wavelength and colour emitted from some metallic elements.

Table 8.2 Principle wavelengths emitted from and colour of some excited metal vapours

Element	Principle emission wavelength (nm)	Colour in a flame
Copper (Cu)	510 and 578	yellow-green
Barium (Ba)	554	lime green
Sodium (Na)	589	yellow
Calcium (Ca)	662	orange
Lithium (Li)	670	red

- When a continuous spectrum is sent through an unexcited gas or vapour the gas absorbs the same unique wavelengths that it emits when it is excited, leaving dark lines in the spectra. We call this type of spectra atomic **absorption spectra**. Because the absorbed lines act as a signature of the element, absorption spectra can also be used to identify the elements in a sample. The advantage of using absorption spectra is that the sample does not need to be excited in a discharge or flame to examine the spectra. Chemists regularly use atomic absorption spectroscopy to identify atoms and molecules and in a sample.

absorption spectra: the specific wavelengths absorbed by an atomic gas when a continuous spectrum is passed through the gas; an absorption spectrum is a continuous spectrum with some wavelengths missing (i.e. some dark lines in the spectrum)

EXAMPLE 4

Explain how spectroscopy can be used to identify the elements present in a gas.

> Recall that the atoms in an excited elemental gas emit specific wavelengths of light that are characteristic of the element

Answer:

Absorption spectrometry could be used to identify the elements in the gas. This technique uses a spectrometer to examine the absorption spectrum produced after a continuous light has been passed through the gas. Because elements in the gas absorb specific wavelengths of the light, dark lines at specific frequencies will be present in the light spectrum after it has passed through the gas. Because each element absorbs specific known wavelengths, comparing the absorption lines in the spectrum with the known spectral lines from the elements will enable the elements in the gas to be identified.

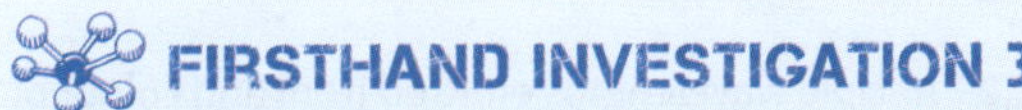

FIRSTHAND INVESTIGATION 3

Atomic emission spectra

An investigation of the light emitted from a low-pressure gas discharge tube excited by a high voltage can be used to demonstrate atomic line spectra. As shown in Figure 8.7, when the light emitted from a gas discharge tube is passed through a prism spectrometer, it splits into specific wavelengths that appear as different coloured lines of light in the spectrometer. We call this type of spectrum a *line spectrum*. Different gases emit different wavelengths when they are excited in a discharge tube and the wavelengths emitted are specific to the element being excited in the tube. Indeed, the presence of a specific element in a discharge tube can be confirmed by the emission of the wavelengths that are characteristic of that element. For example, excited copper atoms emit a green line with a wavelength of 510.6 nm and a yellow line with a wavelength of 578.2 nm, while sodium emits two closely spaced yellow lines with wavelengths of 589.0 nm and 589.6 nm.

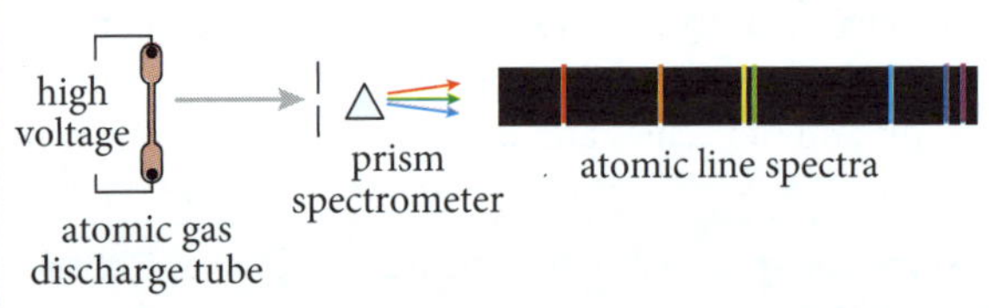

Figure 8.7 Atomic emission line spectra emitted from an excited atomic gas

Students should also use a spectrometer to examine the light emitted from some elements when they are placed in a flame. For example, placing a copper wire in a flame will change the colour of the flame to a yellow-green colour due to the emission of copper atoms vaporised from the metal. Metal salts such as sodium chloride and strontium chloride could also be investigated using flame spectroscopy.

Atomic absorption spectra

Examining reflected light from the Sun using a hand spectroscope produces a continuous spectrum, as one would expect from a hot body like the Sun. But if students have access to a more accurate prism or grating spectrometer, careful examination of reflected sunlight will reveal dark lines in the spectra. These dark lines in the Sun's spectra were first studied in detail by Joseph von Fraunhofer, who identified 570 lines. The dark lines in the spectra of the Sun and other stars are still known as *Fraunhofer lines*. Later studies showed the lines were produced by some of the wavelengths emitted from the Sun's surface being preferentially absorbed by atoms and ions in the Sun's less dense outer layers, the photosphere and chromosphere. Spectra that exhibit missing wavelengths (dark lines) are called *absorption spectra*. Some of the Fraunhofer absorption lines that students might be able to distinguish and the elements responsible for the lines are shown in Figure 8.8.

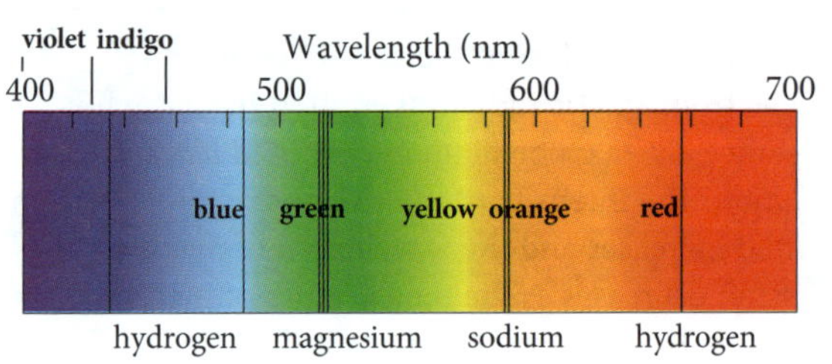

Figure 8.8 Some of the stronger Fraunhofer absorption lines in the Sun's visible spectra

→ KEY QUESTIONS

9 **What is the difference between continuous spectra and line spectra?**

10 **What is the difference between atomic emission spectra and atomic absorption spectra?**

11 **How can spectra be used to identify the elements in a sample of material?**

Answers ➲ p. 122

5 Stellar spectra

» Students investigate how the spectra of stars can provide information on surface temperature, rotational and translational velocity, density and chemical composition.

- Spectroscopy can tell us a great deal about the motion and composition of stars. Stellar spectra are obtained by passing the light from a telescope through a grating or prism spectrograph. Usually the intensity of the light is sampled electronically at each wavelength and digitised so computers can be used to analyse the captured spectral data.
- We investigated black-body radiation in the Year 11 course and you will recall that the wavelength of maximum emission from a hot body (like a star) is related to the temperature of the body by **Wien's law**:

$$\lambda_{max} = b/T$$

where λ_{max} is the wavelength where the emitted light intensity is greatest
T is the absolute temperature of the surface of the body and
$b = 2.898 \times 10^{-3}$ mK is a constant called *Wien's displacement constant*

Thus by using a spectrograph to measure the peak of the black-body radiation curve of the radiation emitted by the star we can determine the surface temperature of the star.

Wien's law: a relationship between the temperature of a black body and the wavelength at which the radiant emission from the body is greatest

- We also examined the Doppler Effect in the Year 11 course where the observed frequency of sound changed when there was relative motion between the source of the sound and the observer. A similar effect occurs for light waves. This **optical Doppler Effect** causes light waves to shift to longer frequencies (i.e. towards the red end of the spectrum) when the source of the light is moving away from the observer. When this occurs we say the observed light has been *redshifted*. If the source is moving towards the observer, the observed frequency is greater than the emitted frequency (i.e. moves towards the blue end of the spectrum) and we say the light has been *blueshifted*.

optical Doppler Effect: refers to the Doppler Effect for electromagnetic radiation; that is, the difference between the emitted frequency and observed frequency of light when the source is moving relative to the observer

- Stellar spectra always contain dark lines produced by specific wavelengths being absorbed by elements in the photosphere and chromosphere (outer layers) of the star. We can determine the amount a star's light has been red- or blueshifted by comparing the absorption lines in the star's spectrum with the known spectral lines emitted from elements on Earth. Once the red- or blueshift has been measured, the following optical Doppler Effect equation can be used to determine the velocity of the star with respect to the Earth:

$$v = c(\Delta\lambda/\lambda_0)$$

where v is the relative velocity of the source with respect to the observer
$\Delta\lambda$ is the observed change in wavelength
λ_0 is the wavelength emitted at the source and
c is the speed of light

Note that this equation is presented for completeness and students will not be required to use this equation in the HSC Exam.

- The optical Doppler Effect can also be used to find the rate of rotation of a star because, as a star rotates, one side of the star will be moving towards us and the other side will be moving away from us. Because each point of the surface of a rotating star is moving at a different velocity with respect to an observer on the Earth, the wavelengths of the emitted and absorbed lines will be broadened by the red- and blueshift caused by the rotation. By measuring the amount of broadening of the absorption lines, astronomers can calculate the rate at which a star is rotating.
- The density of the outer layers of a star will also affect the depth and width of absorption lines. Because in dense gases the particles are closer together and have a higher rate of atomic collisions, the absorption lines are broadened. We call this *pressure broadening* and it can be used to gain information about the density of the star's outer layers.
- Stars are made up of the same elements as the Earth but in very different proportions. By examining the absorption lines in the spectra of stars and comparing them with the spectral signatures of elements on Earth, the chemical composition of the outer layers of the star can be determined. Further, by examining the relative intensity of absorption lines the relative composition of each element contained in the star can be estimated. The element helium was actually discovered because solar spectra was found to contain some absorption lines that did not correspond to any known elements on Earth. Helium was first found on Earth in mineral water, when spectral studies of mineral water showed the same dark lines observed in the Sun's spectra.

The Sun, like most stars, is made up of approximately 71% hydrogen, 27% helium, 1% oxygen, 0.4% carbon and small traces of many other elements.

EXAMPLE 5

The intensity of the light emitted from a star at different wavelengths is shown in Figure 8.9.

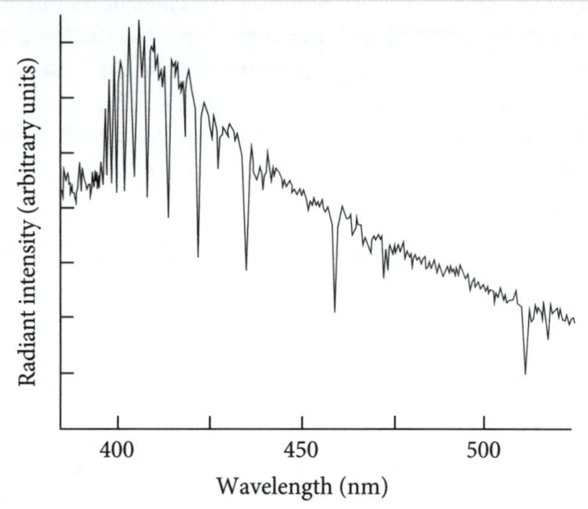

Figure 8.9 Radiant light intensity emitted from a star as a function of wavelength

a **Estimate the wavelength of maximum radiation emission from the star.**

b **Use the stellar spectra to determine the surface temperature of the star.**

c **Explain why the curve has dips rather than being a smooth line.**

d **If the temperature of the surface of the star suddenly increased, how would a graph of radiant intensity change?**

Recall that stars emit a black-body radiation spectrum that is related to the surface temperature of the star by Wien's law

Answer:

a From Figure 8.9, the wavelength at which the maximum radiant emission occurs is 410 nm.

b The temperature of the surface can be determined using Wien's law:

$\lambda_{max} = b/T$ and hence

$$T = b/\lambda_{max} = \frac{2.898 \times 10^{-3}}{410 \times 10^{-9}} = 7068 \text{ K}$$

c The sharp dips in the graph are produced by atoms in the outer layers of the star (photosphere and chromosphere) absorbing specific wavelengths of light as it passes outwards from the hot surface below.

d If the star's surface was hotter, the star would emit more radiation and the wavelength of maximum emission would decrease (we saw in the Year 11 course that all black-body radiators behave in this way).

→ KEY QUESTIONS

12 **How can stellar spectra be used to find the translational and rotational speed of a star?**

13 **How can stellar spectra be used to find the surface temperature and composition of the outer atmosphere of a star?**

14 **How does the density of a star affect the stellar spectra of a star?**

Answers ➲ p. 122

CHAPTER SYLLABUS CHECKLIST

Are you able to answer these questions from the syllabus for this chapter? Tick each question as you go through the checklist if you are able to answer it. If you cannot answer a question, turn to the relevant page in the study guide to find the answer. For NESA key word meanings, go to www.educationstandards.nsw.edu.au and search 'key words'.

	FOR A COMPLETE UNDERSTANDING OF THIS TOPIC:	PAGE NO.	✓
1	Can I outline Maxwell's contribution to our understanding of electromagnetism?	109–110	
2	Can I describe how electromagnetic waves propagate in terms of electric and magnetic fields?	110–111	
3	Can I recall that Maxwell related the speed of electromagnetic waves to the permeability constant of magnetism and the permittivity constant of electrostatics?	110	
4	Can I explain how electromagnetic waves are produced and how they affect other charges?	111–112	
5	Can I describe the frequency bands of the electromagnetic spectrum?	111	
6	Can I describe how the speed of light was measured using astronomical observations, mechanical devices and electrical devices?	113	
7	Can I explain why the speed of light was fixed and how the speed of light and the second are now used to define the metre?	113	
8	Can I describe continuous spectra, emission line spectra and absorption line spectra?	115–116	
9	Can I explain how spectroscopy can be used to identify elements?	115	
10	Can I use Wien's law and stellar spectra to determine the surface temperature of a star?	117	
11	Can I explain how stellar spectra can be used to find the density, relative velocity and rotational period of stars?	117	
12	Can I explain how stellar spectra can be used to identify the chemical composition of stars?	117	

HSC EXAM-TYPE QUESTIONS

Objective-response questions (1 mark each)

1 **What characteristic of electromagnetic waves led Maxwell to believe that light was an electromagnetic wave?**

A Electromagnetic waves were transverse waves.

B Electromagnetic waves travelled at the same speed as light.

C Electromagnetic waves could have a wide range of wavelengths.

D Electromagnetic waves were non-mechanical waves.

2 **If a polarised electromagnetic wave was incident upon an electron that was free to move, what would happen to the electron?**

A It would rotate at the same frequency as the wave.

B It would oscillate parallel to the wave velocity.

C It would oscillate parallel to the changing magnetic field.

D It would oscillate perpendicular to the changing magnetic field.

3 **What can we conclude about a star from the spectra shown in Figure 8.10?**

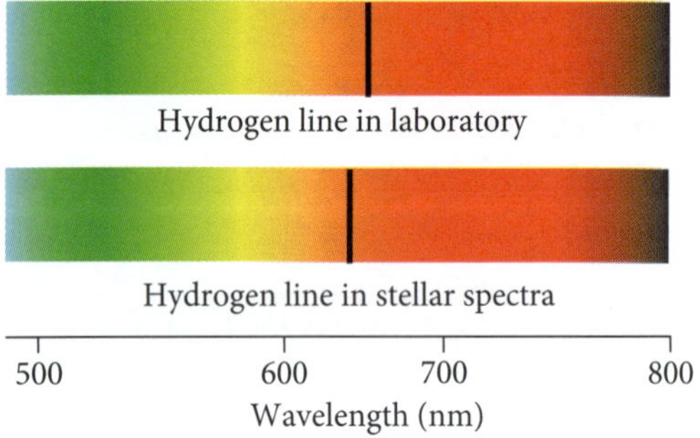

Figure 8.10 A specific hydrogen line observed on Earth and in the spectra of a star

A The star is moving rapidly away from the Earth.

B The star is moving rapidly towards the Earth.

C The star does not contain hydrogen in its atmosphere.

D The star is rotating rapidly.

4 **Why does a Bunsen flame change from a blue colour to a yellow-green colour when a copper wire is placed in the flame?**

A Vaporised copper atoms absorb the blue wavelengths from the flame and emit a broad band of green frequencies.

B Vaporised copper atoms are excited by the flame and emit a broad band of yellow and green frequencies.

C Vaporised copper atoms are excited by the flame and emit specific frequencies of light that turn the flame yellow-green.

D Vaporised copper atoms absorb specific wavelengths, which changes the flame's light to yellow-green.

5 **If the wavelength of maximum light emission from two stars was equal, what could we conclude about the two stars?**

A The stars have the same surface temperature.

B The stars are rotating at the same rate.

C The stars have the same core temperatures.

D The stars have the same chemical composition.

Extended-response questions

6 **Figure 8.11 shows the emission spectrum of sodium.**

Figure 8.11 Visible emission spectrum of sodium

a **Describe an experiment that could be used to examine the emission spectrum of an element.** (3 marks)

b **Explain how emission spectra can be used to identify the elements in a sample?** (2 marks)

c **If sodium was present in the atmosphere of the Sun, how would it affect the emission spectra of the Sun?** (2 marks)

7 **Outline one method that has been used to measure the speed of light.** (3 marks)

8 **About half of the visible stars in the sky are binary star systems, which consist of two stars orbiting each other. Consider two stars of similar mass in a binary system that has its orbital plane almost parallel to the line of sight from the Earth. Each of the hydrogen atomic absorption lines in the stellar spectrum of the binary star system are observed to split into two lines of slightly different frequency and recombine to a single line every four days.**

a **Explain why the motion of the stars would cause the atomic absorption lines in the stellar spectra to change in this way.** (3 marks)

b **Determine the orbital period of this binary star system.** (1 mark)

c **Explain why some non-hydrogen absorption lines from the binary system move back and forth around a specific frequency periodically but do not split and recombine like the hydrogen lines.** (2 marks)

9 Maxwell used his equations of electromagnetism to predict the existence of electromagnetic waves.

a Explain how an electromagnetic wave can be produced. (2 marks)

b Describe the structure of electromagnetic waves in terms of electric and magnetic fields. (3 marks)

c If the source of an electromagnetic wave was removed, would the existing waves produced by the source continue to propagate? Justify your answer. (2 marks)

10 Explain how an examination of the light from a star can be used to provide information on THREE physical properties of the star. (6 marks)

ANSWERS

KEY QUESTIONS

Key questions ➲ p. 110

1 Maxwell consolidated the known laws of electricity and magnetism into a single theory of electromagnetism based on the field concept. He also showed that a consequence of his theory was that changes in the electric field would travel at the speed of light as self-propagating electromagnetic waves and these waves could have a wide range of possible frequencies. Maxwell's four equations of electromagnetism still form the foundation of classical electromagnetism.

2 Maxwell used his theory to predict the existence of electromagnetic waves. Because the calculated speed of these waves was close to the measured value for the speed of light at the time, Maxwell suggested that light itself must be an electromagnetic wave.

Key questions ➲ p. 112

3 Electromagnetic waves are produced by oscillating (i.e. accelerating) charges. An oscillating charge produces changing, perpendicular electric and magnetic fields that oscillate at the same speed as the charge and move away from the charge at the speed of light.

4 Electromagnetic waves involve changing electric and magnetic fields. Maxwell's equations tell us that a changing magnetic field produces a changing electric field, and a changing magnetic field produces a changing electric field. Neither field can change without producing the other field and hence electromagnetic waves always consist of electric and magnetic fields.

5 A charged particle experiences a force when placed in an electric field ($F = qE$). Therefore, when an electromagnetic wave interacts with a charged particle, it exerts a changing force on the charged particle that would cause the particle to oscillate if it was free to do so. Note that in classical physics, if the particle does oscillate it will produce its own electromagnetic waves.

Key questions ➲ p. 114

6 The first successful measurement of the speed of light was achieved by Romer, who used observations of the eclipse of one of the moons of Jupiter (Io) at different times of the year to show that light took 22 minutes to cross the Earth's orbital diameter. Prior to this measurement scientists were not sure if light travelled infinitely fast or at a finite speed.

7 The speed of light was fixed in 1983 and the metre defined in terms of the speed of light and the second. Any new measurement of the speed of light is circular because the measurement requires using the metre, which itself is defined in term of the speed of light.

8 When the speed of light was fixed at 299 792.458 ms^{-1} in 1983, it was used to define the metre as the distance travelled by light in $\frac{1}{299\ 792.458}$ seconds. Hence the metre is defined in terms of the speed of light (a constant) and the second, which is defined in terms of a certain number of cycles of a specific atomic frequency.

Key questions ➲ p. 116

9 Continuous spectra contain all wavelengths (i.e. frequencies) of light, while line spectra contain only specific wavelengths of light. Incandescent lightbulbs emit continuous visible spectra and excited elemental gases in electric discharge tubes (or in a flame) emit line spectra. Each element emits a specific line spectrum, which acts as a spectral signature of the element.

10 Atomic emission spectra are emitted from excited elemental gases and consist of spectral lines of specific wavelengths. An absorption spectrum consists of specific lines missing from a continuous spectrum sent through an unexcited elemental gas. Many of the emission lines of an element are seen in the absorption spectrum of the element, although the two spectra are not identical and may vary if the gas is excited. For example, the absorption spectra from a very hot, ionised gas would show few atomic absorption lines as the density of the unionised atoms in the gas would be low.

11 The wavelengths emitted or absorbed by an element are specific to that element. These spectral signatures have been studied in the laboratory and are available to use to identify elements. By vaporising the sample and exciting it in an electric discharge, the emission spectra could be compared with the known elemental wavelengths. Alternatively, a continuous spectrum could be passed through the vapour and the absorption spectra examined.

Key questions ➲ p. 118

12 By comparing the spectral lines in a stellar spectrum with those from excited elements in the laboratory we will find that, if the star is moving away from us, the spectral lines will be shifted to longer wavelengths (redshifted), and if the star is moving towards us, the lines will be shifted to shorter wavelengths (blueshifted). The change in the observed wavelengths can then be used with the optical Doppler Effect equation to calculate the relative velocity of the star.

Because each side of a rotating star is moving, one side will show a different optical Doppler Effect to the other side. This results in a broadening of the spectral lines that can be measured and used to estimate the rotation rate.

13 Stars emit a radiation curve that is close to that of a black body. By measuring the wavelength at which the star emits most radiation we can use Wien's law to calculate the surface temperature of the star.

The elements in a star can be determined from the star's absorption spectrum. This spectrum is produced by atoms in the star's photosphere and chromosphere absorbing specific wavelengths from the continuous black-body radiation emitted from the surface of the star.

14 The density of the star's atmosphere influences the width of spectral lines because higher density gases have many more atomic and ionic collisions, which tend to spread the frequency of radiation emitted and absorbed by the atom.

HSC EXAM-TYPE QUESTIONS

Objective-response questions

1 **B.** Maxwell used known constants from electricity and magnetism to show that electromagnetic waves would travel at the speed of light. The link between the constants of electricity and magnetism to the speed of light provided strong circumstantial evidence that light was an electromagnetic wave. **A**, **C** and **D** are incorrect because, while light has similar properties, none of these properties provided strong enough evidence for Maxwell to suggest that light was an electromagnetic wave.

2 **D.** The electric field would exert a force on the charge parallel to the electric field and hence perpendicular to the magnetic field. **A** is incorrect as the electric field is in one plane as the incident electromagnetic wave was polarised. **B** is incorrect because the electric field is perpendicular to the velocity and hence the charge will oscillate perpendicular to the velocity. **C** is incorrect as the electric field is perpendicular to the magnetic field and the charge therefore must oscillate perpendicular to the magnetic field.

3 **B.** Because the line is shifted towards a shorter wavelength (blueshifted), the star must be moving towards the Earth. **A** is incorrect as this would shift the line to a longer wavelength (redshifted). **C** is incorrect as the absorption line for hydrogen would not appear in the spectra unless the star contained hydrogen in its atmosphere. **D** is incorrect as a rotating star would have a broader absorption line due to the red- and blueshift on each side of the star.

4 **C.** Copper atoms will vaporise from the wire and emit an atomic line spectrum when they are excited (given energy) by the flame. **A** and **B** are incorrect because elemental atoms in a flame emit line spectra not broad band spectra. **D** is incorrect because if the copper absorbed light it would only absorb specific wavelengths, which would not change the colour of the flame.

5 **A.** The surface temperature of a star determines the wavelength of maximum emission from the star (Wien's law). **B** is incorrect as rotational speed only affects the breadth of spectral lines, not the wavelength of maximum emission. **C** is incorrect as black-body radiation (Wien's law) relates to the surface temperature of the radiating body, not the core temperature. The stars could be different sizes, for example, and have the same surface temperature but different core temperatures. **D** is incorrect as the chemical composition of the star changes the absorption lines in the spectra but it is not related to the wavelength of maximum emission.

Extended-response questions

6 EM This question tests students' understanding of the relationship between terrestrial and stellar emission and absorption spectra.

a This question could be answered by referring to excitation by a flame or in a low-pressure gas discharge tube. Using a discharge tube, as shown in Figure A8.1, connect a low-pressure gas discharge tube containing the elemental gas ✓ to be investigated to a high-voltage power supply (e.g. an induction coil). ✓ Keeping a safe distance from the high-voltage wires, use a prism spectrometer to examine the wavelength of lines emitted from the excited atoms in the tube. ✓

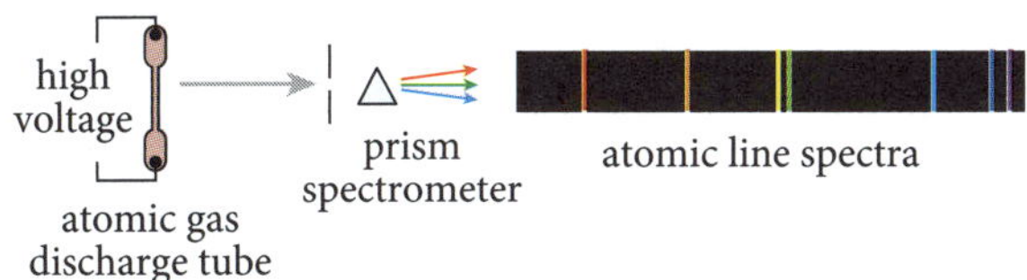

Figure A8.1 Experiment to examine line spectra of an elemental gas

b Each element emits a characteristic set of wavelengths that act as a spectral signature of the element. ✓ By comparing the emission or absorption spectrum of a sample with the known wavelength signatures of elements, the elements in the sample can be identified. ✓

c The sodium atoms in the Sun's atmosphere would absorb those wavelengths that are emitted by excited sodium atoms ✓ and hence dark lines at 589 nm and 590 nm would appear in the stellar spectra ✓ reaching the Earth.

7 EM This question test students' understanding of how the speed of light can be measured.

Any of the methods outlined in the chapter could be used. For example, Fizeau passed light through a gap in a spinning-toothed wheel ✓ and then, after reflecting the light from a mirror 8 km away, he adjusted the speed of his wheel until the reflected light passed through the next gap in the wheel. ✓ Using the rotational speed of the wheel, Fizeau was able to calculate the time the light beam took to complete the 16 km round trip and hence calculate the speed of light. ✓

8 EM This question test students' ability to apply their understanding of the optical Doppler Effect.

a Because binary stars orbit around each other, the stars would move perpendicular to the line of sight to the Earth twice each orbit and hence the light would not be red- or blueshifted. At these times an absorption line present in the spectra from both of the stars, such as hydrogen, would appear as a single line. ✓ But twice per cycle, one star would be moving away from the Earth and one star would be moving towards the Earth. At these times the absorption line from one star would be redshifted and the same line in the spectra of the other star would be blueshifted. ✓ This would cause the observed absorption line to split into two lines twice per orbit. ✓

b As the absorption line would split and recombine twice every orbital period, the orbital period must be $2 \times 4 = 8$ days. ✓

c If the low density, outer layers of one star contained an element that was missing from the outer layer of the other star, the line would simply move back and forth as the star orbited. ✓ The line would not split to higher and lower frequencies but would move to lower and higher frequencies periodically because the line would be periodically red- and blueshifted as the star orbited. ✓

9 EM This question tests students' understanding of the nature of electromagnetic waves, how they are produced and how they propagate.

a Electromagnetic waves are produced by oscillating charged particles. ✓ The oscillating charge will change the electric field parallel to the charge movement and produce a changing magnetic field that is perpendicular to the charge movement. ✓ The electromagnetic waves produced will have the same frequency as the oscillating charged particle.

b Electromagnetic waves are transverse waves that consist of mutually perpendicular electric and magnetic fields. ✓ Figure A8.2 shows how the changing electric and magnetic fields are perpendicular to one another ✓ and to the velocity of the wave. ✓

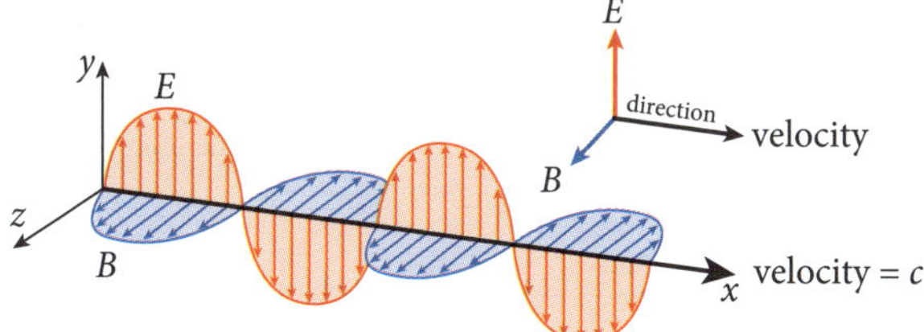

Figure A8.2 An electromagnetic wave

c Maxwell's theory showed that a changing electric field produces a changing magnetic field and a changing magnetic field produces a changing electric field. ✓ Hence, once changing electric and magnetic fields have been produced, the change is self-propagating ✓ and continues to propagate at the speed of light.

10 EM This question tests students' understanding of how the properties of a star can be deduced from the star's spectrum.

Astronomers pass the light from stars gathered by their telescopes into a spectrograph, which records the spectra of the star electronically. The stellar spectrum can tell us a great deal about the composition, structure and motion of the star. ✓ For example:

i Stars produce black-body radiation that is characteristic of a body with a temperature of the surface of the star. By measuring the wavelength of maximum emission of a star's spectrum and applying Wien's law for black-body radiators we can find the temperature of the surface of the star. ✓

ii Elements emit and absorb specific wavelengths that form a spectral signature of the element. ✓ By comparing the absorption lines in the stellar spectrum with the absorption/emission lines gathered from elements in the laboratory on Earth we can find which elements are present in the outer layers (photosphere and chromosphere) above the denser surface of the star. ✓

iii By comparing the shift of all the absorption lines in a star's spectrum to the lines on Earth ✓ we can measure the red- or blueshift of the star and hence determine its velocity relative to Earth. ✓

CHAPTER 9 LIGHT WAVE MODEL

MODULE 7 THE NATURE OF LIGHT

INQUIRY QUESTION:

What evidence supports the classical wave model of light and what predictions can be made using this model?

Light was thought be a wave because it was observed to diffract, reflect and refract in a similar manner to sound waves and water waves. The early longitudinal wave model of light was used to derive laws of refraction and reflection and, when the model was later changed to a transverse wave model, it could also be used to explain polarisation and interference.

1 Diffraction of light

» Students conduct investigations to analyse qualitatively the diffraction of light.

- You will recall from the Year 11 course that when a wave encounters a small object (of the same order as the wavelength) or a small gap or edge, the wave diffracts (spreads out) behind the object, edge or gap. You will also recall, as shown in Figure 9.1, that the amount of **diffraction** increases as the gap becomes smaller or the wavelength increases. Diffraction is optimum when the gap is equal to the wavelength of the incident light.

diffraction: the spreading out of a wave when it encounters a gap, edge or small object

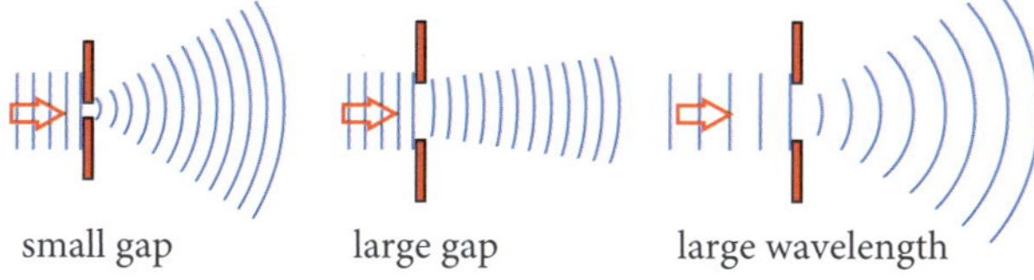

Figure 9.1 Diffraction increases with wavelength but decreases with gap size

- Light, like other waves, diffracts around corners, as shown in Figure 9.2.
- Diffraction of light was first observed by Francesco Grimaldi in 1665 and led scientists at the time to suggest that light might be some form of wave.

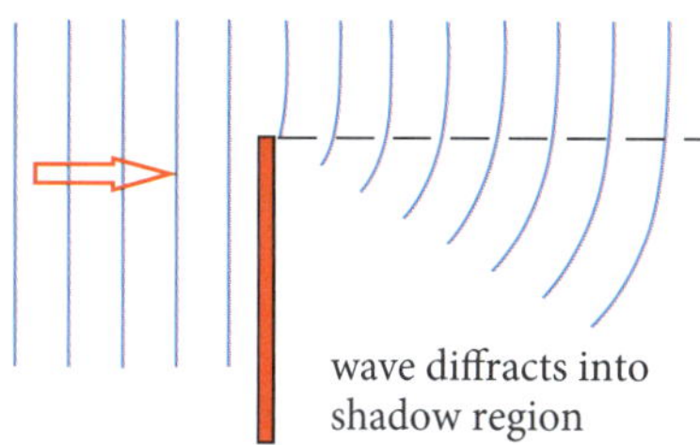

Figure 9.2 Diffraction of wavefronts at an edge

- Christiaan Huygens used the work of Grimaldi and others to propose a wave model of light. Huygens believed that light was a longitudinal pressure wave like sound and that the light waves moved through a substance, called *the ether*, which filled all of space. Huygens believed that the ether would support light waves but not sound waves.
- Huygens proposed that each point on a wavefront acted as an individual point source of secondary wavelets that moved forwards in the direction of the wave velocity. As the wave propagates, the new wave front forms at a tangent to the secondary wavelets. Figure 9.3 illustrates how these individual point sources can produce straight and circular wavefronts. This concept of wave propagation through secondary wavelets is now called **Huygens' principle**.

Huygens' principle: a principle that states that each point on a wavefront acts as a point source of secondary wavelets that propagate forwards in the direction of the velocity

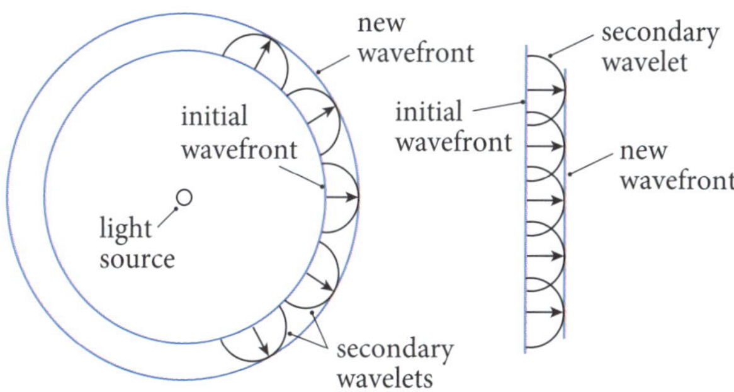

Figure 9.3 Propagation of a curved and straight wavefronts via secondary wavelets

- Huygens' principle can be used to explain diffraction. For example, if light passed through a slit comparable in size to the wavelength of the light, the gap would act as a point source of waves that would produce a circular wavefront. Waves passing through a slit that was large compared with the wavelength would tend to continue to propagate as a straight wavefront as only the points on each side of the slit would diffract. The diffraction patterns predicted by Huygens' principle for large and small slits are illustrated in Figure 9.4.

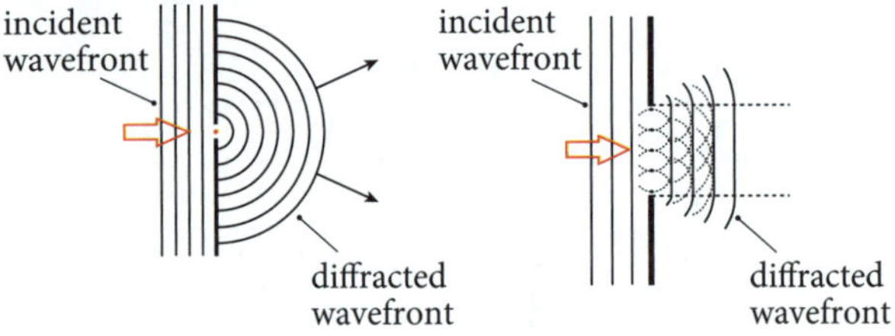

Figure 9.4 Using Huygens' principle to explain diffraction at a narrow and wide slit

➔ Later investigations of wave diffraction found that diffraction is also accompanied by **interference** effects. See the firsthand investigations in this section for more details.

interference: the superposition of two waves to form a single wave of greater, smaller or equal amplitude to the initial waves

FIRSTHAND INVESTIGATION 1

Diffraction of light

Warning: Lasers can cause eye damage. When using lasers in experiments, use the lowest-power laser available. Never look directly into the beam or have the beam accidently reflect into your eye.

Diffraction refers to the way light spreads out when it encounters a small object or a small gap (a slit). Both types of diffraction can be demonstrated using a laser pointer.

Diffraction by a human hair

In this experiment a human hair is taped across the output lens of a laser pointer and projected onto a screen, as shown in Figure 9.5.

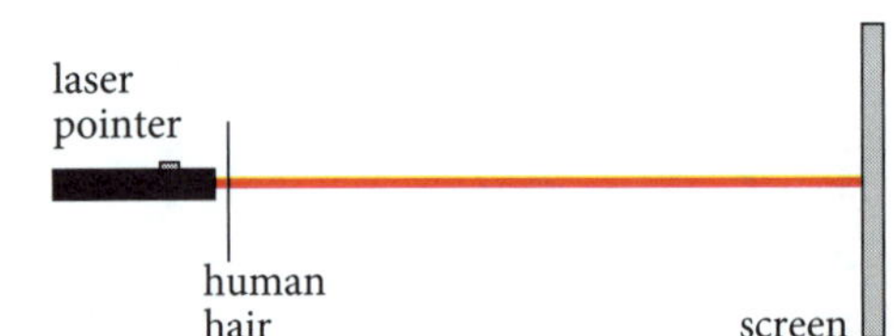

Figure 9.5 Experiment to investigate diffraction by a small object

The laser beam will be diffracted by the hair and the light on the screen will be spread out in a direction perpendicular to the hair. Students may note that, as shown in Figure 9.6, there are dark bands in the diffracted light on each side of the hair.

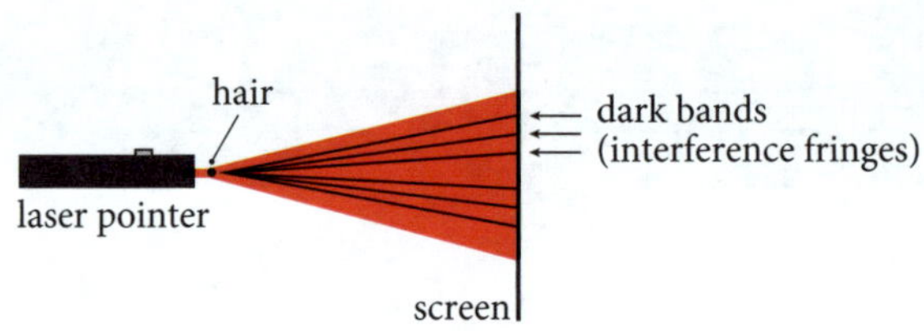

Figure 9.6 Diffraction pattern produced by a human hair in a laser beam

Although it is beyond the syllabus, interested students might like to use the diffraction pattern to estimate the diameter of the human hair used in the experiment. To do this they must measure the angle (θ) between the centre of the pattern and the first dark band; that is, the angle by which the first dark band has been diffracted. The diameter of the hair (w) is related to the wavelength (λ) of the light used and the angle of diffraction (θ) for the first minimum by:

$w = \lambda/\sin\theta$

Diffraction by a small slit

Students should also examine the diffraction of light when it passes through a narrow slit. Again, a laser pointer can be used to show how light is diffracted (spreads out) when it passes through a narrow slit. A typical experiment of this type is shown in Figure 9.7.

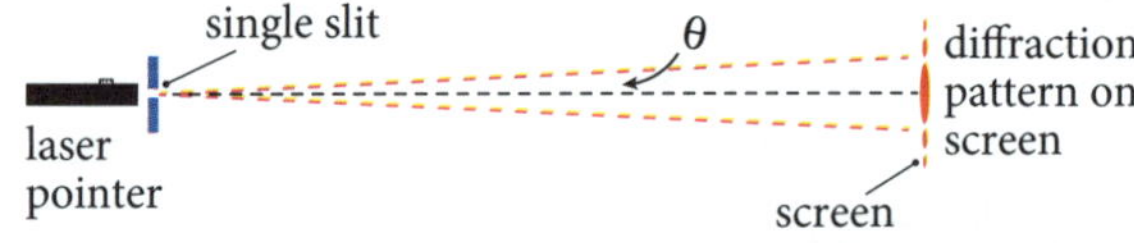

Figure 9.7 Simple experiment to demonstrate the diffraction of light by a single narrow slit

If possible, students should investigate how the diffraction pattern changes when the slit width is changed and when the wavelength (colour) of the incident light is changed. Students will find that the amount of diffraction increases when the slit width is decreases and when the wavelength of the incident light is increased.

Interested students may wish to investigate the relationship between the angle of diffraction of the first dark fringe (θ) and the slit width (a) and wavelength (λ) used:

$a = \lambda/\sin\theta$

Note that this is the same equation that was used in the previous investigation but the width of the human hair has been replaced with the width of the slit. Although single slit interference is beyond the syllabus, interested students may wish to experimentally verify this equation and investigate how it could be derived.

→ KEY QUESTIONS

1 What is diffraction?

2 How is diffraction related to wavelength?

3 How does Huygens' principle explain the propagation of light waves?

Answers ⊃ p. 138

2 Interference of light

» Students conduct investigations to analyse quantitatively the interference of light using double-slit apparatus and diffraction gratings $d\sin\theta = m\lambda$.

Double-slit interference

→ A light source that emits light waves of only one wavelength is called a **monochromatic light** source, and a light source that emits waves that have their crests and troughs aligned across the wavefront is called a **coherent light** source. Interference effects are easier to demonstrate and analyse when a coherent, monochromatic light source is used.

monochromatic light: light of one wavelength
coherent light: light waves that are in phase with one another

→ In early experiments on interference, coherent light was produced by passing light through a narrow slit. The diffracted light is essentially coherent because it originates from a point in accordance with Huygens' principle. However, this method of producing coherent light is problematic because the diffracted light from the narrow slit has very low intensity. Today we often use laser light, which is both monochromatic and coherent for interference experiments, as it is much more intense than the diffracted light from a single slit.

→ When coherent monochromatic light passes through two closely placed slits the diffracted light from the slits overlaps and produces an interference pattern, as shown in Figure 9.8. At some places the superimposed waves from each slit will be one half-cycle out of phase and produce *destructive interference*. At other places the waves from each slit will be in phase and produce *constructive interference*. A screen placed in front of the slits will therefore show bright and dark bands, respectively representing the positions where constructive and destructive interference occurs.

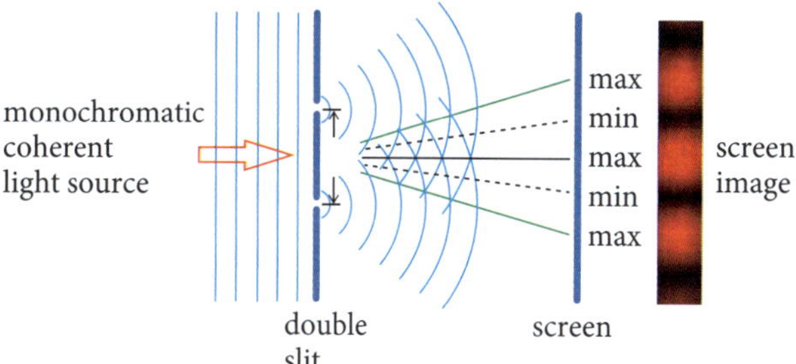

Figure 9.8 Interference pattern produced by the light diffracted from a double slit (note that the diagram is not drawn to scale)

→ We can quantitatively analyse the interference pattern produced by a double slit by recalling that, to produce a maximum of interference at the screen, the light from both slits must arrive at the screen with the same phase.

→ A central **interference maximum** is produced because the light from both slits travels an equal distance from each slit to the screen. If the light waves are in phase at the slits then they will still be in **phase** at the screen as they have travelled the same length (i.e. the same number of wavelengths).

interference maximum: a point of maximum intensity in an interference pattern produced when waves with the same phase arrive at the point
phase: a point in time of a wave in its wave cycle

→ The first **interference minima** on each side of the screen occurs at the point where the light from each slit is one half-wavelength out of phase and hence interferes destructively. This will happen when the light from one slit travels one half-wavelength further to the screen than the light from the other slit. We say the beams have a path length difference (Δ) of one half-wavelength at this point (i.e. $\Delta = \lambda/2$). The next minima will occur when the path difference is three half-wavelengths (i.e. when $\Delta = 3\lambda/2$), and so on.

interference minimum: a point of minimum intensity in an interference pattern produced by waves arriving at the point one half-cycle out of phase

→ If the distance to the screen is very much larger than the slit separation we can calculate the path difference (Δ) for the beams from each slit by assuming that the rays travelling from the slits to the screen are parallel. Consider the two rays travelling from the slits to point *P* on a distant screen, as shown in Figure 9.9.

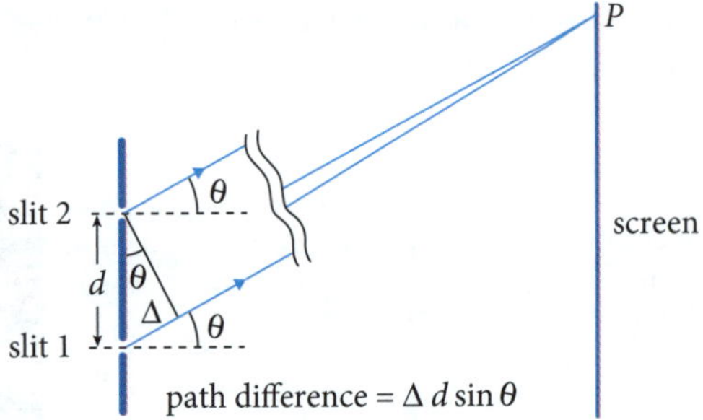

Figure 9.9 Path difference between two rays travelling from a pair of double slits to a distant screen

We can now use trigonometry to relate the path difference (Δ) to the slit separation (d) and the angle of diffraction (θ) to give:

Path difference = $\Delta = d\sin\theta$

➔ The first maximum of interference will occur when the path difference is equal to one wavelength and the second maximum will occur when the path difference equals two wavelengths, and so on. If we call the order of interference m, maxima will occur when:

$$d\sin\theta = m\lambda$$

where d is the slit separation
λ in the wavelength of light incident on the slits
θ is the angle of diffraction for the mth maximum and
m is the order of interference (i.e. the number of maxima from the central maximum)

order of interference: the number of whole wavelengths in the path difference between the rays from adjacent slits

EXAMPLE 1

When coherent light of wavelength 520 nm is incident on a double slit it produces an interference pattern on a screen. The second bright maximum in the interference pattern occurs at an angle of 0.4° from the central maximum.

a Find the slit separation.

b If the slit separation was halved, how would the diffraction pattern change?

Recall that when the distance to the screen is very large compared with the slit separation, $\sin\theta = x/L$

Answer:

a Applying the double-slit equation for interference maxima:

$d\sin\theta = m\lambda$

Or $d = \dfrac{m\lambda}{\sin\theta}$

And as this is the second order maxim $m = 2$:

$d = \dfrac{2 \times 520 \times 10^{-9}}{\sin(0.5)}$

$d = 1.2 \times 10^{-4}$ m

b As the separation between the central maximum and mth order maximum in the diffraction pattern is inversely proportional to the slit width (i.e. $x = 2\lambda y/d$), the diffraction pattern would spread out and the separation between the central maximum and the other maxima would double.

FIRSTHAND INVESTIGATION 2

Young's double-slit experiment

Students should use a laser or another suitable monochromatic light source to investigate the interference pattern produced by a double slit and to determine the wavelength of the light used. A schematic diagram of the experimental set up and the interference pattern produced is shown in Figure 9.10.

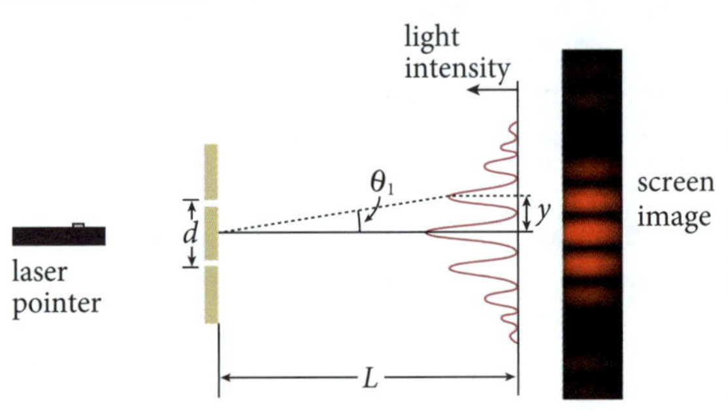

Figure 9.10 Illustration of Young's double-slit interference experiment; note that the angle to the first order maximum of interference is labelled θ_1

Provided the distance to the screen is large compared with the slit separation, the wavelength is related to the slit width (d) and angle (θ) between the centre maximum and mth maximum by:

$d\sin\theta = m\lambda$

Students will observe that the distance between the interference maxima on the screen increases when the separation between the slits (d) is decreased and when the wavelength (λ) of the light used is increased.

To experimentally determine the wavelength of light used in the experiment, students should measure the angle between the central maximum and the first interference maximum and, as $m = 1$ for the first maximum, the wavelength will be given by:

$y = d\sin\theta$

Note that, because θ is a very small angle:

$\sin\theta = \tan\theta = y/L$

and hence students can determine sin θ simply by measuring the distance to the screen (L) and the distance between the central maxima and the first interference maxima (y).

Although it is beyond the syllabus, students may also wish to investigate the interference pattern produced by a single slit and how it affects the intensity of the double-slit interference pattern. Interestingly the minima of the interference pattern produced by a single slit of width (a) occur at:

$a \sin \theta = m\lambda$

Note that this is the same equation that gives the interference maxima of a double slit (but with the slit separation replaced by the slit width).

→ KEY QUESTIONS

4 **How do double slits cause interference?**

5 **How is path length related to constructive and destructive interference in a double slit?**

Answers ➲ p. 138

The diffraction grating

→ If more than two slits are used, the maxima in the interference pattern produced are brighter and sharper. This is because the maxima are produced by the superposition of the light from many slits rather than from two slits and because there are multiple minima rather than one minima between the maxima. We can see why this occurs by examining the interference produced when coherent light passes through four slits, as shown in Figure 9.11.

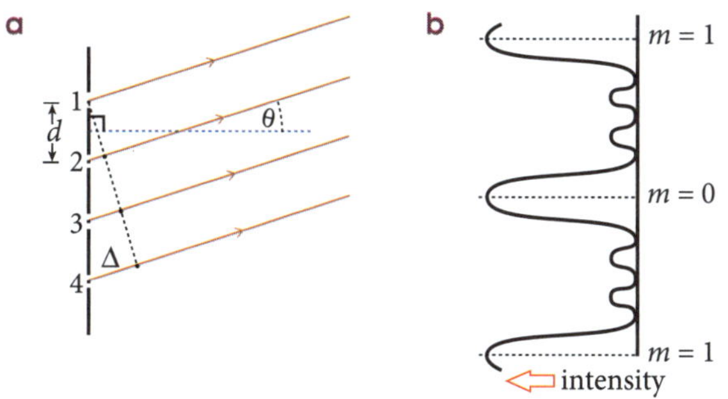

Figure 9.11 **a** Light passing through four slits and **b** the resulting interference pattern

→ With reference to Figure 9.11**a**, when the path difference for the rays travelling to the screen from adjacent slits is equal to a whole number of wavelengths, the rays will arrive at the screen in phase and a maximum will appear on the screen. This is the same condition that is obeyed for double-slit maxima. A central minimum will occur when the path difference between adjacent slits is an odd number of half-wavelengths (i.e. when $\Delta = \lambda/2$, $3\lambda/2$, $5\lambda/2$, and so on), and again this is also the case for double-slit interference. But because there are four slits, minima also occur when the path difference between adjacent slits is an odd number of quarter-wavelengths (i.e. $\Delta = \lambda/4$, $3\lambda/4$, $5\lambda/4$, etc.). This will mean, as shown in Figure 9.11**b**, that between the central maximum and the first-order maximum there will be three minima: one corresponding to a path difference of $\lambda/4$ between adjacent slits, one corresponding to a path difference of $\lambda/2$, and one corresponding to a path difference of $3\lambda/4$.

→ To check if this makes sense we can examine 9.11**a** to see what happens when the path difference between adjacent slits is $\lambda/4$. If this was the case, the path difference between slits one and three would be $\lambda/2$ and the light from this pair would be one half-cycle out of phase at the screen and cancel out. Similarly the path difference between the rays from slits two and four would be one half-wavelength different, resulting in the waves cancelling out at the screen. Similarly, a path difference of $3\lambda/4$ would also result in light from the two pairs of slits cancelling out to produce a minimum at the screen.

→ Thus if four slits are used there will be three minima between each pair of interference maxima. Indeed it can be shown that if there are n slits then there will be $n - 1$ minima between each pair of maxima.

→ **Diffraction gratings** are glass slides containing thousands of closely placed slits and consequently they will produce interference patterns with a huge number of minima between each pair of maxima. The net effect of this, as shown in Figure 9. 12, is to produce very sharp interference maxima.

diffraction grating: a periodic structure that diffracts light to produce interference; diffraction gratings are usually made by drawing or etching a large number of closely spaced parallel lines on a glass or metal plate

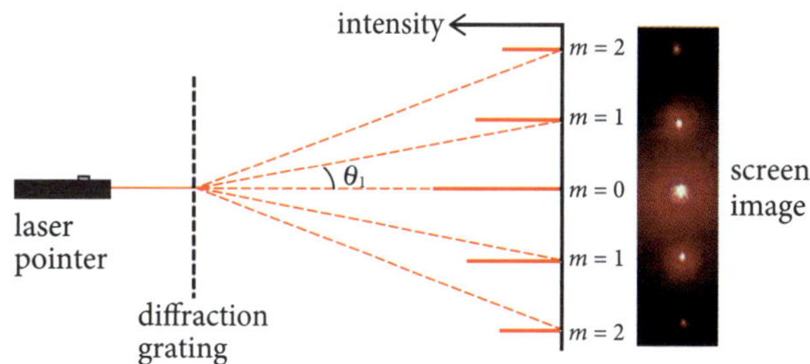

Figure 9.12 Experimental set up and interference pattern produced when monochromatic laser light passes through a diffraction grating; note that the angle to the first order maximum of interference is labelled θ_1

→ Because the interference maxima consist of the superposition of the light from hundreds of slits, the maxima are extremely bright. We can also use the law of conservation of energy to explain why the maxima are so bright. The energy of the light passing through the slits cannot disappear; it must still reach the screen. Because it cannot appear between the maxima, it must appear in the maxima, resulting in very bright maxima.

→ The interference maxima produced by diffraction gratings occur at the same angles as they do for the double slit. That is, maxima occur when:

$d \sin\theta = m\lambda$

where d is the separation between adjacent slits (or between adjacent lines) and
m is the order of interference

→ Because light of different wavelengths produces interference maxima at different angles (different positions on the screen), diffraction gratings are often used as **dispersive elements** to examine the spectral distribution of light. Spectrometers can use glass prisms or diffraction gratings to disperse light.

dispersive element: a device that separates a wave into its constituent components (wavelengths) such as a glass prism or diffraction grating

→ If white light is passed through a diffraction grating the central interference maxima will be white. This is because the path difference from each pair of slits either side of the middle would be equal for all wavelengths, resulting in a central maximum for all wavelengths. However, the other maxima would appear as continuous spectra because the maxima for each colour (wavelength) would occur at a slightly different angle for each order of interference. Long wavelength maxima occur at a greater angle (θ) and hence, unlike the spectrum produced by a prism, diffraction gratings produce spectra with the red end of the spectrum deflected by the greatest angle. The interference pattern produced when white light is passed through a diffraction grating is shown in Figure 9.13.

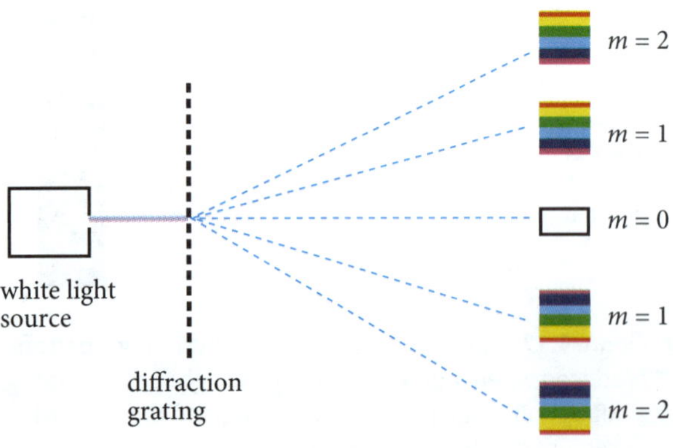

Figure 9.13 Interference pattern produced when white light is passed through a diffraction grating

EXAMPLE 2

After passing through a single slit, light from a hydrogen discharge tube is incident on a diffraction grating with 2000 lines per cm. The visible hydrogen spectrum consists of two violet lines with wavelengths of 410 nm and 434 nm, a blue-green line with a wavelength of 486 nm and a red line with a wavelength of 656 nm.

a Find the separation between adjacent slits in the grating.

b Find the angular spread of the first-order interference spectrum when the light from the hydrogen lamp is passed through the grating.

c How many of orders of interference maxima will be produced when light from a hydrogen lamp is passed through the grating?

Recall that the maxima for a diffraction grating occur at the same angles as the maxima for a double slit

Answer:

a If there are 2000 lines per cm, the distance between each pair of lines will be:

$$d = \frac{0.01}{2000} = 5 \times 10^{-6}\text{ m} = 5\ \mu\text{m}$$

b Using the equation for a first-order ($m = 1$) maximum to find the angle for the largest and smallest wavelengths gives:

$d \sin\theta = m\lambda$ and hence $\theta = \sin^{-1}\left(\frac{\lambda}{d}\right)$

$$\theta_{red} = \sin^{-1}\left(\frac{656 \times 10^{-9}}{5 \times 10^{-6}}\right) = 7.54°$$

$$\theta_{violet} = \sin^{-1}\left(\frac{410 \times 10^{-9}}{5 \times 10^{-6}}\right) = 4.70°$$

The angular spread of first-order spectrum is therefore from 4.70° to 7.54°.

c Red light has the longest wavelength and will have the largest angle for a maximum for each order of interference. The maximum path difference between adjacent slits would occur at an angle of 90° and hence the maximum path difference possible is d, the slit separation. The maximum order of interference will be equal to the maximum number of whole red wavelengths less than the slit separation d.

$$\text{Number of wavelengths} = d/\lambda_{red} = \frac{5 \times 10^{-6}}{656 \times 10^{-9}} = 7.62$$

and hence seven orders of interference would be produced by the grating.

FIRSTHAND INVESTIGATION 3

The diffraction grating

Students should also investigate the interference pattern produced when laser light is incident on a diffraction grating. A diffraction grating consists of many closely spaced slits produced by drawing or engraving equally spaced parallel lines on a glass slide. The interference maxima produced by a diffraction grating occur at the same place as for a double slit and hence for a diffraction grating:

$d \sin \theta = m\lambda$

A schematic diagram of an experimental set up that could be used to examine the interference pattern produced by a diffraction grating and the interference pattern produced are shown in Figure 9.12.

Students will observe, as shown in Figure 9.12, that the interference maxima produced by the grating are much sharper and brighter than the maxima produced by a double slit.

Again, as the angle between the central maxima and the first maxima (θ_1) is very small, we can use the approximation:

$\sin \theta = \tan \theta = y/L$

Hence by measuring the distance to the screen (L) and the separation between the central and first interference maxima (y), students will again be able to determine the wavelength of the laser light using:

$\lambda = d \sin \theta = d\left(\frac{y}{L}\right)$

Students will observe that the interference pattern produced by the grating spreads out when the number of lines per unit length on the grating is increased (i.e. slit separation decreased) and when the wavelength of the incident light is increased.

Students should also replace the incident monochromatic light with a white light source. When white light is used students will observe that each interference maximum is replaced by a white light spectrum, as shown in Figure 9.13. Students should also note that the central maximum is white and that the red light maxima are further from the central maxima than the blue light maxima in each spectrum.

➔ KEY QUESTIONS

6 **How can a diffraction grating be used to find the wavelength of light?**

7 **How does the interference pattern produced by a double slit compare with the interference pattern produced by a diffraction grating?**

8 **Why does white light produce multiple white light spectra when it is used to illuminate a diffraction grating?**

Answers ➲ p. 138

3 Wave and particle models of light

» Students analyse the experimental evidence that supported the models of light that were proposed by Newton and Huygens.

Early scientific models of light

➔ By 1700 two quite different models were being used to explain the nature of light: a compressional wave model proposed by Christiaan Huygens (1629–1695) and a **corpuscular** (particle) model proposed by Isaac Newton (1643–1727). At the time a less developed, transverse wave model was also proposed by Robert Hooke.

corpuscle: a minute particle

➔ Neither of the models explained all observed wave phenomena satisfactorily. For example, light beams were observed to pass through one another without being scattered by the interaction. This observation would seem to be at odds with a particle model. Light also could travel in straight lines as rays, unlike other types of waves, which tend to spread out.

➔ Double refraction (**birefringence**) in calcite crystals proved difficult for either models to explain. Calcite splits incident light into two rays that are diffracted by different angles, which implies there were two types of light. Newton suggested light particles must be 'sided' rather than spherical to explain the double refraction, while Huygens proposed that forces in the calcite caused two types of secondary wavelets to be produced. Today we attribute double refraction to the perpendicular components of the electric field of the incident ray being refracted by different angles. That is, double refraction is related to the polarisation of light, a phenomenon that was unknown at the times of Newton and Huygens.

birefringence or double refraction: the ability of some materials such as calcite to refract an incident ray of light at two different angles; birefringence results from the material refracting perpendicular components of polarisation in the incident light at different angles

- The wave models of Huygens and Hooke provided simpler explanations for many wave phenomena, but largely because of Newton's standing in the scientific community his particle model was the dominant model throughout the 18th century.
- In the early 19th century scientists turned against the particle model when Thomas Young and Augustin-Jean Fresnel independently demonstrated that interference phenomena could only be explained by using the wave model of light.
- Observations of polarisation at that time led Young and Fresnel to return to Hooke's concept of light being transverse wave rather than a longitudinal wave. The transverse wave model of light developed by Fresnel could be used to explain double refraction and polarisation as well as all the other properties of light that had been observed at the time.
- The final blow to Newton's particle model was delivered by Foucault in 1850. A major difference between the wave model and particle model was the way each explained refraction. Newton proposed that light particles moved faster in a transparent medium than in air, while Huygens's model required light to travel more slowly in a transparent medium than in air. Foucault measured the speed of light in water and confirmed Huygens' prediction that light would travel slower in the water than in air.

Evidence supporting Newton's particle model of light

- The success of geometric optics in explaining reflection and refraction led Newton to propose a corpuscular (particle) model of light. Newton first suggested light was a stream of particles in 1672 and went on to develop his corpuscular theory of light over the next 30 years.
- Newton saw light rays as fast-moving streams of elastic particles and proposed that different-coloured light corresponded to particles of different mass.
- The particle model developed by Newton could be used to explain reflection, dispersion and refraction (although his model erroneously predicted that light travelled faster in denser materials).
- To explain double refraction Newton had to modify his theory and proposed that light particles had to be 'sided' rather than spherical. Newton also had difficulty applying his model to thin-film interference and had to resort to proposing that light particles had 'fits' of easy transmission and easy reflection.
- Table 9.1 sets out the light phenomena explained by Newton's corpuscular (particle) model.
- There was no direct evidence to prove that light was a wave or a particle in Newton's time but the particle model could be used to explain the rectilinear propagation of light, reflection, refraction, dispersion, thin-film interference and double refraction. However, the model could not explain diffraction, partial reflection or the observation that two rays of light can cross each other without scattering. The particle model also had difficulty explaining interference phenomena. To explain refraction with his model Newton had to assume that light particles must travel faster in dense transparent materials than in a vacuum. Many scientists at the time thought this defied common sense.

Table 9.1 Evidence supporting Newton's particle model of light

Light phenomena	Newton's particle model explanation
Light rays travel in straight lines	Light rays are beams of very high-speed particles that will travel in a straight line due to their inertia.
Laws of reflection	Light particles are repelled by the reflecting surface and have collisions with the surface.
Laws of refraction	Light particles are attracted to refracting materials, which makes them travel faster in the material than in a vacuum and causes rays of light to refract.
Dispersion	Different-coloured light particles must have different masses.
Double refraction	Light particles are not spherical but must have 'sides' that cause them to be refracted at two different angles.

- Because of Newton's standing in the scientific community his particle model was accepted by most scientists throughout the 18th century, even though the evidence supporting it was arguably no stronger than the evidence supporting the wave model.

Evidence supporting Huygens' wave model of light

- Christiaan Huygens published a detailed wave model of light in 1678. Huygens modelled light as a longitudinal pressure wave that travelled in an elastic medium called the ether that he believed filled all of space. He proposed that light was propagated by each point on the wavefront, acting as a source of secondary wavelets. Huygens used his wave propagation principle to also explain reflection, diffraction and refraction. Table 9.2 sets out the key light phenomena that could be explained with Huygens' wave model.

- There was no direct evidence that light was a wave in Huygens' time, but the ability of the model to explain a wide range of light phenomena and the similarity of the behaviour of light to other waves, such as sound, provided strong support for the model.
- The evidence for the wave model became overwhelming in the early 1800s when the model was used to explain interference experiments. Modifying the model to a transverse wave model enabled polarisation and double refraction to be explained, and Foucault experimentally confirmed the model's prediction about the speed of light being slower in denser materials.

Table 9.2 Evidence supporting Huygens's wave model of light

Light phenomena	Explanation using the wave model
Light travels in straight lines.	Explained using Huygens' principle of secondary wavelets.
Light emitted from a point source produces spherical wave fronts.	Explained using Huygens' principle of secondary wavelets.
Reflection	Laws of reflection could be derived using Huygens' principle.
Refraction	Laws of refraction could be derived using Huygens' principle and by assuming that light travelled slower in denser materials.
Diffraction	Could be explained qualitatively using Huygens' principle.
Double refraction (birefringence)	Explained (erroneously) by Huygens as two types of refraction making secondary wavelengths of different shapes and hence causing rays to be refracted at two different angles.
Interference by thin films, double slits and periodic structures	Not explained by Huygens' theory. First explained using a wave model by Young in 1801, who proposed that different-coloured light consisted of periodic waves with different wavelengths.

- Maxwell's theoretical prediction in the 1860s of the existence of transverse electromagnetic waves and the production and detection of electromagnetic waves by Hertz in the 1880s appeared to provide conclusive evidence for the wave model of light.

KEY QUESTIONS

9 **What evidence was used to support Huygens' wave model?**

10 **What evidence was used to support the particle model of light?**

Answers p. 138

EXAMPLE 3

Newton and Huygens proposed quite different models to explain the behaviour of light. Compare the evidence used by Newton and Huygens to support their models.

A good scientific model is based on reliable observations and can be used to explain characteristics of the observations used to formulate it; a good model should also be able to explain other observations and make predictions that can be tested

Answer:

Newton believed light consisted of particles because of the geometric nature of light. He believed the success of geometric optics (the ray model) in explaining reflection and refraction was due to rays of light being streams of particles. Newton could not see how waves could travel in straight lines like rays. Newton's model could also explain dispersion and diffraction, which he saw as a type of refraction. He had to modify his model to include sidedness in the particles to explain double refraction and could not satisfactorily explain thin-film interference or the way light beams could cross one another without scattering. To explain refraction, Newton had to assume that light particles travelled faster in transparent materials than in a vacuum. Many scientists thought this simply did not make sense.

Huygens believed light was a wave because it exhibited many of the same characteristics of other waves like sound. Huygens' principle could be used to explain the propagation of light, diffraction, reflection, refraction and double refraction (although this explanation was later shown to be incorrect). To some degree Huygens' wave theory was inadequate because it was based on pulses rather than periodic waves and because Huygens thought the waves would be compressional waves like sound. Huygens' model predicted that light would travel slower in transparent materials than in a vacuum, which seemed a reasonable assumption by most scientists at the time.

The evidence for each model was inconclusive and indirect and the technology was not available at the time to test the predictions made by the models about the speed of light in materials. Both models could be used to explain many of the observations of properties of light but neither model could explain all the observations of the behaviour of light. The prominence of the particle model during the 18th century was more about Newton's standing in the scientific community than the strength of evidence supporting his model.

➔ KEY QUESTIONS

11 Why were scientists unable to prove that one model or the other was correct?

12 Why was Newton's particle model of light accepted by most scientists in the 18th century?

13 Why did scientists abandon Newton's particle model of light?

Answers ➲ p. 138

4 Polarisation of light

» Students conduct investigations quantitatively using the relationship of Malus's law $I = I_{max}\cos^2\theta$ for plane polarisation of light, to evaluate the significance of polarisation in developing a model for light.

Malus's law

➔ You will recall from our earlier investigation of electromagnetic waves that **polarised light** is light that has a changing electric field in one plane only.

polarised light: light made up of electromagnetic waves which have their electric fields oscillating in only one plane

➔ We can produce polarised light by using a plastic polarising filter (polaroid sheet) or by scattering or reflection. Although it is beyond the syllabus requirements, interested students may wish to investigate why light scattered at 90° or reflected at the Brewster angle is polarised.

➔ As shown in Figure 9.14, when plane-polarised light is passed through a polarising filter, the filter only transmits the component of the incident electric field that is parallel to the polarisation axis of the filter. That is, if the incident electric field E_0 is at an angle θ to the polarisation axis of the film, then the transmitted electric field strength will be:

$E = E_0 \cos\theta$

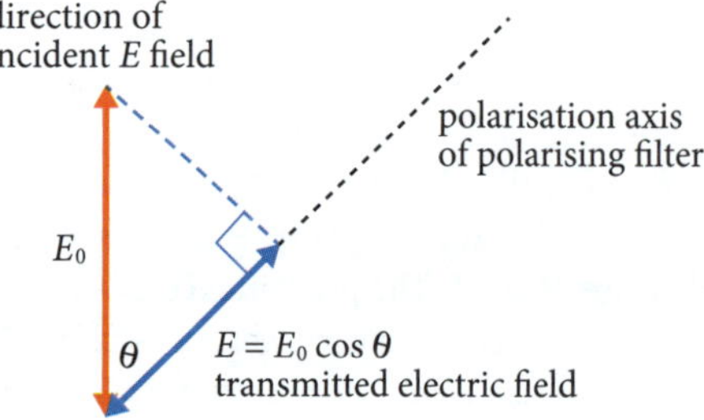

Figure 9.14 Electric field transmitted by a polarising filter

➔ Because the **intensity** of a wave is proportional to the square of the **amplitude**, the intensity of the light transmitted through a polaroid filter is given by:

$I = E^2 = E_0^2\cos^2\theta$

intensity: the power carried by a wave per unit area perpendicular to the wave velocity measured in Wm^{-2}

amplitude: maximum displacement of a wave from its equilibrium position; in the case of an electromagnetic wave the electric field amplitude is the maximum value of the electric field in each cycle

The maximum transmitted intensity ($I_{max} = E_0^2$) will occur when $\theta = 0$ and we can therefore write the transmitted intensity in terms of the maximum possible transmission as:

$$I = I_{max}\cos^2\theta$$

where I is the transmitted intensity
I_{max} is the maximum possible transmitted intensity (i.e. the incident light intensity) and
θ is the angle between the incident electric field and the polarisation axis of the polarising filter

This relationship is called **Malus's law** after Etienne Louis Malus (1775–1812), who discovered polarisation by reflection in 1808.

Malus's law: relates the light intensity transmitted through a perfect polarising filter to the angle between the electric field of the incident polarised light and the polarisation axis of the polarising filter; Malus's law can be written as $I = I_{max}\cos^2\theta$

➔ Malus's law is seldom absolutely adhered to in practice because it assumes the polariser is perfect. Real polarising filters always absorb and reflect some of the incident light.

➔ If unpolarised light is incident on a perfect polariser, half the light intensity will be absorbed and half transmitted. This is because if the electric field of the incident light exists equally in all directions, breaking the electric field into two mutually perpendicular components should always result in half the intensity going into each component.

Polarisation and the wave and particle models of light

➔ Malus's investigation of polarisation seemed initially to provide evidence against the Huygens' wave model because compressional waves cannot be polarised.

However, at about the same time, Thomas Young's work with double-slit interference provided strong evidence in support of the wave model. Double-slit interference could be explained by the wave model but Young at the time could not see how the wave model could explain polarisation. He wrote to Malus in 1801 and said 'Your experiments show the insufficiency of the theory which I have adopted (i.e. the wave model), but they do not prove it false.'

➔ Young initially could not understand how a wave passing through the ether could have 'sides' that would enable it to become polarised, and it was not until 1817 that he proposed that light waves were transverse rather than longitudinal. Augustin-Jean Fresnel in France independently suggested the same change to the wave model as a solution to the polarisation problem.

EXAMPLE 4

Unpolarised light of intensity 500 Wm^{-2} is incident on a polarising filter. The plane-polarised light that emerges from the first filter is then passed through a second polarising filter with its axis of polarisation at 45° to the axis of polarisation of the first polarising filter. You may assume the polarising filters are perfect polarisers.

a **Find the intensity of the light passing through the first polarising filter.**

b **Find the intensity of the light passing through the second polarising filter.**

Malus's law enables us to determine the intensity of light that is transmitted by a polarising filter

Answer:

a When unpolarised light is passed through a polarising filter, half the incident intensity is transmitted and hence the transmitted light will have an intensity of 250 Wm^{-2}.

b Applying Malus's equation:

$I = I_{max}\cos^2\theta = 250 \times \cos^2 45°$
$= 125$ Wm^{-2}

EXAMPLE 5

Two polarising filters are aligned so they have their axes of polarisation at 90° to each other. Unpolarised light of intensity 2000 Wm^{-2} is incident on the crossed polarising filters. You may assume the filters are perfect polarisers.

a **What will be the intensity of the light that emerges from the crossed polarising filters?**

b **A third polarising filter is placed between the other filters with its optic axis at 45° to the polarisation axes of the other two filters. Find the intensity of the light that will be transmitted through this combination of filters.**

Recall that a polarising filter only transmits the component of the incident electric field that is parallel to the optic axis

Answer:

a Because there is no component of the incident light parallel to the polarisation axis of the second filter no light will be transmitted through the crossed filters.

We can also see this from Malus's law because $\theta = 90°$ and $\cos 90° = 0$.

b We must calculate the intensity of light that passes through each filter.

Because the incident light was unpolarised, half of the incident light will pass through the first filter.

Hence $\frac{2000}{2} = 1000$ Wm^{-2} will be transmitted.

Now this incident light passes through the next filter, which is at 45° to the first, and hence the transmitted intensity will be:

$I = I_{max}\cos^2\theta = 1000\cos^2 45° = 500$ Wm^{-2}

This intensity is incident on the final filter and the intensity of the light transmitted will be given by:

$I = I_{max}\cos^2\theta = 500\cos^2 45° = 250$ Wm^{-2}

Thus placing a filter oriented at 45° between the other two increases the transmitted light intensity from zero to 250 Wm^{-2}.

FIRSTHAND INVESTIGATION 4

Polarised light

Students should use a light meter to measure the intensity of plane-polarised light that passes through a sheet of polaroid film rotated at different angles to the incoming polarised light. Figure 9.15 shows an experimental arrangement that could be used to conduct this type of experiment.

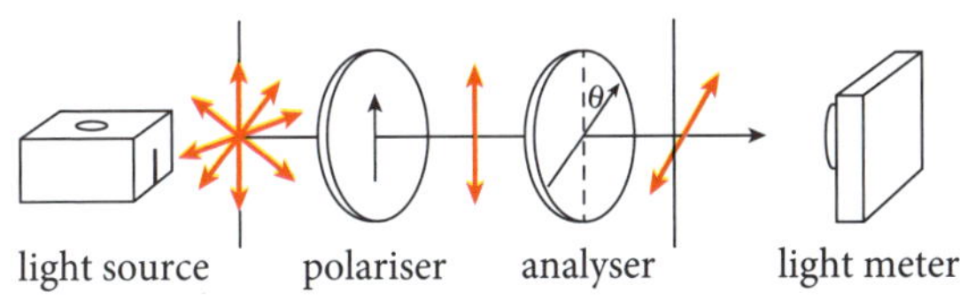

Figure 9.15 Experiment to examine the polarisation of light

As shown in Figure 9.15, the experiment involves two sheets of polaroid film. The first sheet (the polariser) converts the incident unpolarised light to plane-polarised light, and the second sheet (the analyser) has it polarisation axis rotated by an angle θ with respect to the polarising axis of the first sheet.

Students should measure the intensity of the light transmitted through the analyser as a function of the angle θ. Students can then confirm Malus's law by showing that a graph of intensity (I) against $\cos^2\theta$ is a straight line.

➔ KEY QUESTIONS

14 **How can Malus's law be used to determine the change in intensity of polarised light when it passes through a polarising filter?**

15 **How did the observations of polarisation influence the development of the wave model of light?**

Answers ➲ p. 138

CHAPTER SYLLABUS CHECKLIST

Are you able to answer these questions from the syllabus for this chapter? Tick each question as you go through the checklist if you are able to answer it. If you cannot answer a question, turn to the relevant page in the study guide to find the answer. For NESA key word meanings, go to www.educationstandards.nsw.edu.au and search 'key words'.

	FOR A COMPLETE UNDERSTANDING OF THIS TOPIC:	PAGE NO.	✓
1	Can I relate the amount of diffraction to the wavelength of the light being diffracted and the size of the gap or object that produced the diffraction?	125	
2	Can I describe the diffraction of light by small objects, gaps and edges?	125–126	
3	Can I describe Huygens' principle of secondary wavelets and use it to describe how light is propagated and diffracted?	125–126	
4	Can I explain how double slits and diffraction gratings produce interference patterns?	127–128	
5	Can I relate the order of interference to the number of wavelengths in the path difference between the rays that produce interference maxima?	127	
6	Can I calculate the angle and position on a screen of the interference maxima produced by a double slit or diffraction grating?	128	
7	Can I explain why the interference maxima produced by a diffraction grating are much sharper than those produced by a double slit?	129	
8	Can I explain why multiple spectra are present in the interference pattern produced by a diffraction grating when white light passes through the grating?	130	
9	Can I describe the models of light proposed by Newton and Huygens?	131–132	
10	Can I outline the evidence used to support the models proposed by Newton and Huygens?	132–133	
11	Can I recall shortcomings and predictions made by the models proposed by Newton and Huygens?	132–133	
12	Can I describe polarisation and calculate the intensity of polarised and unpolarised light transmitted by a polarising filter?	134–135	
13	Can I derive and solve problems using Malus's law?	134	
14	Can I evaluate the significance of polarisation to the development of the wave model of light?	134–135	
15	Can I evaluate the significance of double-slit interference to the acceptance of the wave model of light?	133 & 135	

HSC EXAM-TYPE QUESTIONS

Objective-response questions (1 mark each)

1 **A light beam made up of red and blue light is incident on a hole of diameter 40 μm in a screen. Which of the following statements best describes the effect of the hole on the light ray?**

A The red and blue light will be diffracted by the same amount by the hole.
B The red light will be diffracted more than the blue light.
C The blue light will be diffracted more than the red light.
D The light will form a parallel beam with the diameter of the hole.

2 **A double-slit interference experiment produces an interference pattern on a screen. What condition is required to produce the second maximum from the central maximum?**

A The rays from both slits must travel an equal path length before superimposing at the screen.
B The ray from one slit must travel one wavelength further than the light from the other slit before superimposing at the screen.
C The ray from one slit must travel two wavelengths further than the light from the other slit before superimposing at the screen.
D The rays from both slits must travel two wavelengths further than the rays that produce the central maximum before superimposing at the screen.

3 **Why did the particle model of light become the preferred model in the 18th century?**

A The particle model was supported by direct evidence.
B The wave model could not explain refraction.
C The particle model successfully explained more wave phenomena than the wave model.
D Newton had great standing in the scientific community.

4 **Why was the wave model of light modified by Young and Fresnel in the early 19th century?**

A because Huygens' wave model could not be used to explain polarisation
B because Huygens' wave model could not be used to explain interference
C because Maxwell predicted that electromagnetic waves would be transverse waves
D because the speed of light was found to be slower in denser materials

5 **How did Newton use the particle model to explain dispersion?**

A Newton suggested that different colours were made up of different-shaped particles.
B Newton suggested that light particles were 'sided' rather than spherical.
C Newton suggested that different colours were made up of particles with different mass.
D Newton's theory could not explain dispersion.

Extended-response questions

6 **Explain why light is refracted, using:**

a **Huygens' wave model.** (2 marks)
b **Newton's particle model.** (2 marks)

7 **In an experiment shown in Figure 9.16, unpolarised light is incident upon a pair of polaroid filters which have their polarisation axes at an angle θ with respect to each other. The transmitted light was found to have an intensity of 10% of the incident light intensity.**

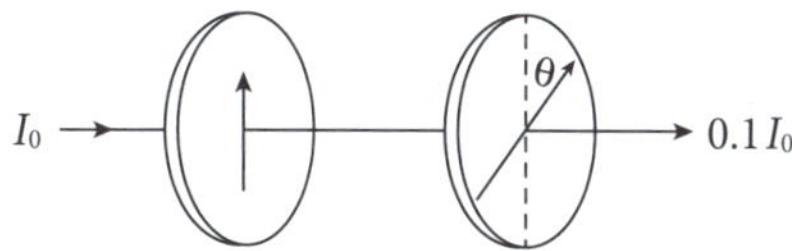

Figure 9.16 Unpolarised light incident on a pair or polaroid filters

a **Explain the difference between unpolarised and plane-polarised electromagnetic waves.** (2 marks)
b **Find the angle (θ) between the axes of polarisation of the two polaroid filters.** (2 marks)

8 **When laser light of wavelength 630 nm is passed through a pair of slits, the first maxima on each side of the central maxima are found to be 12 cm apart from each other on a screen 4 m from the slits.**

a **Why is laser light rather than light from an incandescent lamp used for interference experiments like this?** (2 marks)
b **What is the separation between the slits?** (2 marks)
c **If the double slits were replaced with a diffraction grating with a slit separation one-quarter of the double-slit separation, how would the interference pattern change?** (3 marks)

9 **Evaluate the importance of the polarisation of light to the development of the wave model of light.** (4 marks)

10 **A diffraction grating is used to investigate an atomic spectrum which is known to contain a wavelength of 430 nm and one other longer wavelength. In the interference pattern produced by the grating, the fourth-order maxima of the 430 nm line is observed to fall at the same place as the third maxima of the longer wavelength line.**

a **Sketch a qualitative diagram to show the intensity of the interference pattern that would be produced by the grating. Include the first three maxima of the longer wavelength and the first four maxima of the shorter wavelength. Label the short wavelength maxima and long wavelength maxima.** (3 marks)
b **Find the wavelength of the second line emitted from the atomic spectrum.** (2 marks)

ANSWERS

KEY QUESTIONS

Key questions ➲ p. 127

1 Diffraction is the spreading out of the wave front that occurs when the wave encounters a small object, gap or edge. Diffracted light also exhibits interference.

2 Light is diffracted most when it has a wavelength that is similar to the gap size. If the wavelength is much smaller than the gap, less diffraction will occur.

3 Each point on the wavefront acts as an independent source of secondary wavelets, according to Huygens' principle. A tangent drawn across the secondary wavelets a short time after the wavelets are emitted gives the position of the wavefront at that time. The wavefront propagates continually in this way.

Key questions ➲ p. 129

4 Light is diffracted from each slit and is superimposed with the light from the other slit. At some angles this superposition will result in constructive interference and at other angles the superposition will result in destructive interference.

5 Maxima occur when the path difference of the rays from each slit is a whole number of wavelengths when the waves reach the screen.

Key questions ➲ p. 131

6 Because the maxima from a diffraction grating occur when $d \sin \theta = m\lambda$, we can find the wavelength of the incident light by measuring the angular displacement (θ) of the mth maxima after the light has been passed through the grating. If the distance to the screen is much larger than the slit separation we can replace $\sin \theta$ with x/L, where x is the distance between central maxima and the mth maxima and L is the distance between the grating and the screen.

7 Each interference maxima on the interference pattern produced by a diffraction grating is sharper and brighter than the maxima produced by a double slit. This is because the double slit has one minima between each pair of maxima, but the diffraction grating has a very large number of minima between each pair of maxima.

8 The positions of the maxima depend on the wavelength of the light and hence the maximum corresponding to each colour occurs at a slightly different angle, which produce a white light spectrum at each order of interference.

Key questions ➲ p. 133

9 Light appeared to have many of the same characteristics as sound, which implied it was a wave, like sound. Huygens' principle can be applied to light to explain diffraction and derive the laws of reflection and refraction. The model predicts light will move slower in a transparent material than in a vacuum. Huygens' principle can also be used to explain double refraction (though this explanation is erroneous).

10 Newton believed that waves would not travel in straight lines and that light rays were therefore best represented as a stream of particles. By assuming that light particles would collide elastically with surfaces, Newton could use his model to derive the law of reflection. By assuming transparent materials attracted light particles and made them travel faster, Newton could also explain refraction.

Key questions ➲ p. 134

11 The evidence for both models was indirect and could be interpreted in different ways. In addition, neither model made predictions that could be tested with the technologies available at the time.

12 The evidence for the particle model was not significantly stronger than the evidence for the wave model, but Newton's standing in the scientific community led to the particle model being accepted as the better model for a century.

13 Interference experiments conducted in the early 19th century could only be explained with the wave theory. In addition, when the theory was modified from a compressional wave to a transverse wave theory, the model could also explain polarisation and double refraction. Most scientists had already abandoned the particle model by the time the speed of light in water was measured and found to confirm the prediction of the wave model rather than that of the particle model.

Key questions ➲ p. 136

14 Malus's law is based on the transverse wave model and quantitatively relates the intensity of the light that passes through a polaroid filter to the intensity of the incident polarised light and the angle between the electric field of the incident light and the polarisation axis of the filter. The transmitted intensity can be calculated using Malus equation, $I = I_{max} \cos^2 \theta$.

15 Investigations of the effects of polarisation led to Huygens' compressional wave model being replaced with a transverse wave model. The transverse wave model could be used to explain double refraction and polarisation.

HSC EXAM-TYPE QUESTIONS

Objective-response questions

1 **B**. Long wavelength light is diffracted more readily than short wavelength light. **A** is incorrect because diffraction is wavelength dependent. **C** is incorrect as blue light has a shorter wavelength than red light and hence will be diffracted less. **D** is incorrect because the hole is small and will result in the beam spreading out due to diffraction.

2 **C**. The first maximum requires a path difference of one wavelength and the second maximum requires a path difference of two wavelengths. This ensures the waves from each slit are in phase and produce constructive interference when they superimpose at the screen. **A** is incorrect because this is the condition that applies to the central ($m = 0$) maximum. **B** is incorrect because it is the condition required for the first ($m = 1$) maximum. **D** is incorrect as it would imply an equal path length from both slits, which is the condition for the central ($m = 0$) maxima.

3 **D** is correct as both models were supported by about the same amount of (inconclusive) evidence, but Newton was hugely respected within the scientific community of the time. **A** is incorrect because neither the wave nor particle models were supported by direct evidence. **B** is incorrect because the wave model could explain refraction. **C** is incorrect because both models explained around the same amount of wave behaviour.

4 **A**. To explain polarisation, Huygens' compressional wave model had to be modified to a transverse wave model. **B** is incorrect as Young initially explained interference using the compressional wave

theory. **C** is incorrect because Young and Fresnel modified the wave model before Maxwell predicted the existence of electromagnetic waves. **D** is incorrect because this measurement was made 30 years after the model was changed to a transverse wave model and the measurement was not related to the type of wave model used.

5 **C**. Newton believed different-coloured light had different masses and therefore the attractive forces that he believed produced refraction affected each colour differently. **A** is incorrect because different shapes would not be attracted differently and hence would refract by the same amount. **B** is incorrect as Newton made this suggestion to explain double refraction (polarisation) rather than refraction. **D** is incorrect because the particle model could be used to explain dispersion by assuming different-coloured particles had different masses.

Extended-response questions

6 EM This question tests students' understanding of the wave and particle models and the contradictory predictions they make about the speed of light in transparent materials.

a Huygens showed that if light waves travelled slower ✓ in a transparent medium than in air a ray would be refracted towards the normal to the surface when it passed from air into the transparent medium. Figure A9.1 shows how secondary wavelets entering a material and moving slower led to the wavefront changing direction (i.e. the light being refracted). ✓

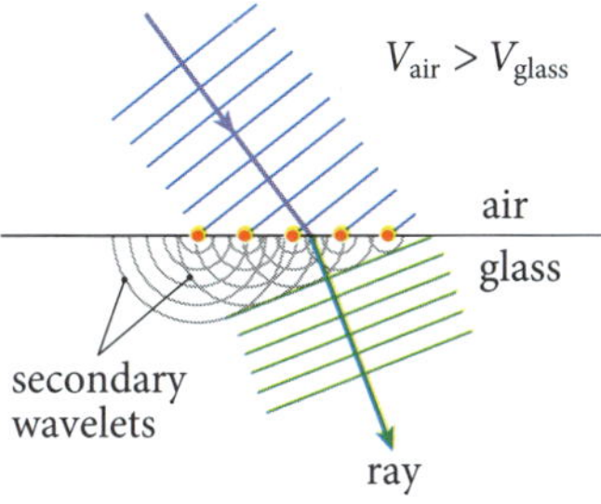

Figure A9.1 Huygens' principle used to explain refraction

b Newton showed that a particle of light would change its direction of travel when it passed from air into a transparent medium if the medium attracted the particle. This would cause a beam of particles (a ray) to be refracted towards the normal ✓ and would require the particles to travel faster in the material than in air. ✓

7 EM This question tests students' understanding of polarisation and their ability to calculate quantities using Malus's law.

a Unpolarised waves have the electric field oscillating in all planes ✓ perpendicular to the wave velocity, but polarised electromagnetic waves have their oscillating electric field confined to one plane, ✓ perpendicular to the velocity.

b The intensity will halve when the polarised light passes through the first polarising filter. Now if we call the initial intensity 100 arbitrary units, 50 units will be incident on the second filter ✓ and 10 units will be transmitted through the filter. Now by applying Malus's law to the second filter:

$I = I_{max} \cos^2\theta$ or

$\theta = \cos^{-1}\sqrt{I/I_{max}} = \cos^{-1}\sqrt{10/50} = 63.4°$ ✓

8 EM This question tests students' understanding of coherence and interference produced by double slits and diffraction gratings.

a The double-slit interference is clearest when monochromatic (one wavelength), coherent (in phase) light is used. Light from an incandescent light is not monochromatic or coherent. In contrast, laser light is both monochromatic ✓ and coherent. ✓

b If the separation between the two first maxima is 12 cm, the distance between the central maximum and the first maximum must be $\frac{0.12}{2} = 0.06$ m. Now applying the double-slit equation, with the simplification $\sin\theta = x/L$ = distance to first maxima/distance to screen and setting $m = 1$: ✓

$d = \frac{m\lambda}{\sin\theta} = \frac{m\lambda L}{x} = \frac{(630 \times 10^{-9} \times 4)}{0.06} = 4.2 \times 10^{-5}$ m = 42 μm ✓

c A diffraction grating produces much sharper ✓ and more intense ✓ maxima than a double slit. In addition, because the separation between maxima is inversely proportional to the slit separation, the interference maxima will be four times further from the central maxima with the grating than they were with the double slit. ✓

9 EM This question tests students' understanding of how studies of polarisation led to the development of the transverse model of light.

Generally light is unpolarised, which means its electric field can be oriented on all planes perpendicular to the wave velocity. In contrast, polarised light has an electric field in only one plane. ✓ Early investigations of polarised light by Malus and others had a major impact on the wave model of light. ✓

The original wave model of light proposed by Huygens was a compressional wave model. Compressional waves are symmetric and cannot be polarised. ✓ Hence advocates of Newton's particle model could point to observations of double refraction and polarisation as proof that light could not be a compressional wave. In the early 1800s Young and Fresnel independently proposed that light must be a transverse wave. ✓ This modification to the wave model enabled polarisation, interference and all the other properties of light known at the time to be explained using the wave theory.

10 EM This question tests students' understanding of diffraction gratings and order of interference.

a See Figure A9.2.

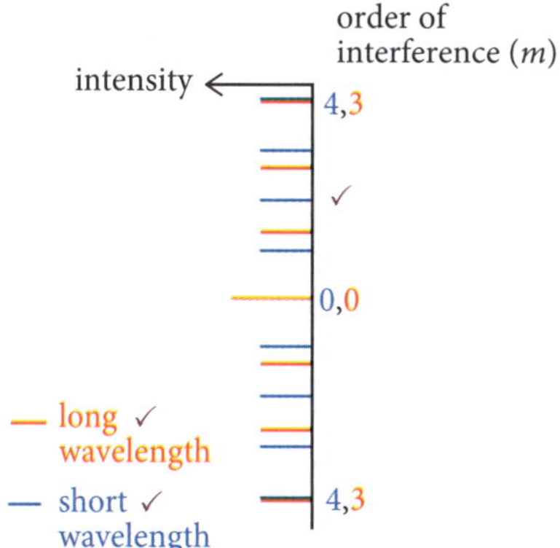

Figure A9.2 Interference pattern produced by diffraction grating

b The path difference between each ray for the fourth-order maximum at 430 nm must be $4\lambda_{short}$. The path difference between each ray for the third-order maximum at the longer wavelength will be $3\lambda_{long}$.

Now as the maxima occur at the same place, these path lengths must be equal ✓ and we can write:

$4 \times 430 \times 10^{-9} = 3 \times \lambda_{long}$

Or $\lambda_{long} = \frac{(4 \times 430 \times 10^{-9})}{3} = 573.3 \times 10^{-9}$ m = 573.3 nm ✓

CHAPTER 10 LIGHT QUANTUM MODEL

MODULE 7
THE NATURE OF LIGHT

INQUIRY QUESTION:

What evidence supports the particle model of light and what are the implications of this evidence for the development of the quantum model of light?

By the late 19th century the electromagnetic wave model of light was well established as it explained almost all observed light phenomena. However, the classical theory of electromagnetism could not account for the spectra emitted from black bodies or some aspect of the photoelectric effect. In 1899 Max Planck introduced energy quanta to explain the radiation emitted by black bodies and in 1905 Albert Einstein extended Planck's energy quanta to light itself. Einstein used this new model of light to explain a range of problematic observations, including those related to the photoelectric effect.

Einstein showed that to explain the interaction of light with matter on the atomic scale, light had to be treated as spatially quantised packets of energy (photons) rather than waves. While the modern quantum theory of light is quite different to Newton's particle model, we can think of the photon model as a particle model. The evidence for the existence of photons today is overwhelming and the modern theory of light and matter, Quantum Electrodynamics, is based on the photon concept.

1 Black-body radiation and quanta

» Students analyse the experimental evidence gathered about black-body radiation, including Wien's law related to Planck's contribution to a changed model of light: $\lambda_{max} = b/T$

➔ You will recall that you investigated **black-body radiation** in the Year 11 course and also when you looked at the radiation emitted from stars in Chapter 8 of this book.

black-body radiation: the electromagnetic radiation emitted from a black body

➔ Recall that a black body is defined in physics as a body that absorbs all the radiation that falls on it but reflects none. The radiation emitted from a black body at different temperatures is shown in Figure 10.1. Note that the total radiation emitted (the area under the graph) increases when the temperature of the black body is increased. We saw in the Year 11 course that the Stefan–Boltzmann law shows that the power emitted per unit area of a black body is proportional to the fourth power of the absolute temperature:

$P/A = \sigma T^4$

where P is the power
A the area
T the absolute temperature and
$\sigma = 5.67 \times 10^{-8}\ \text{Wm}^{-2}\text{K}^{-4}$ is a constant

This equation explains why the power emitted from a black body increases so rapidly with temperature. Note this equation is included for completeness; you will not be required to use it in the HSC Exam.

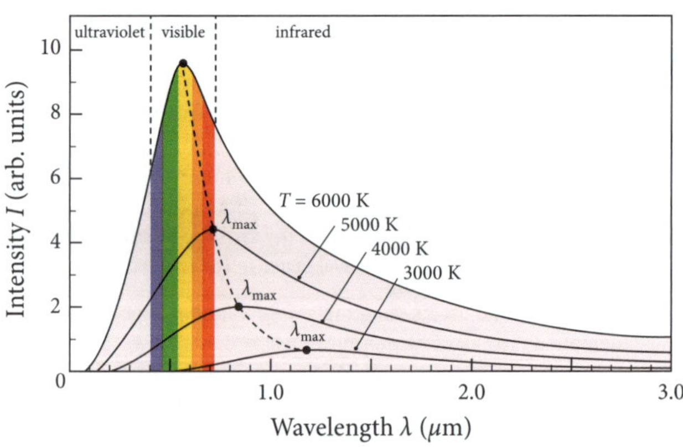

Figure 10.1 Intensity of radiation emitted at different wavelengths from a black body at different temperatures

➔ You will also recall that the wavelength of maximum emission from a hot body is related to the temperature of the body by **Wien's law:**

$$\lambda_{max} = b/T$$

where λ_{max} is the wavelength where the emitted light intensity is greatest
T is the absolute temperature of the surface of the body and
$b = 2.898 \times 10^{-3}$ mK is a constant called *Wien's displacement constant*

Thus by using a spectrograph to determine the wavelength at which the most intense black-body radiation is emitted, we can remotely determine the surface temperature of the body. This is how we can measure the surface temperature of distant stars or the filament temperature of an incandescent lamp.

Wien's law: the wavelength at which the most intense electromagnetic radiation is emitted from a black body is inversely proportional to the absolute temperature of the body

- In the late 19th century attempts to reproduce the black-body radiation curves using classical electromagnetic theory failed. Classical theory predicted that the radiation emitted at high frequencies should be much more intense than the radiation emitted at low frequencies. However, as we have seen, the peak emission is closer to the middle of the frequency (or wavelength) range. Indeed, classical theory predicted that an infinite amount of power would be radiated at very high frequencies; this problem became known as the 'ultraviolet catastrophe'.
- In 1896 Wilhelm Wien suggested an empirical equation that at first appeared to reproduce the black-body radiation curves. Max Planck showed how this equation could be derived from thermodynamics and this Wien–Planck law equation initially fitted the experimental data well. However, when more accurate experimental data became available in the late 1890s, the equation was shown to overestimate the emission at low frequencies.
- When the experimentalists pointed out the shortcomings in the Wien–Planck equation, Planck found a way to change the equation to enable it to reproduce the radiation curves exactly. He then worked backwards from the answer to see what assumptions he would have to make to derive the equation from known physical principles.
- In 1900, in what he called 'an act of desperation', Planck found a way to derive the empirical equation that he had earlier shown could reproduce the black-body radiation curves.
- Scientists had previously assumed that electromagnetic radiation was produced by the oscillating charges on the surface of the black body and that these charges could oscillate with any amount of energy. To derive the black-body equation Planck had to make two radical assumptions:
  1. The first assumption was that the energy of the oscillators was quantised. The oscillators could not have any energy; they could only have a whole-number multiple of a specific minimum amount of energy. This minimum energy was related to the frequency of oscillation by:

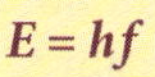

$$E = hf$$

where E is the minimum quanta of energy that an oscillator can possess
f is the frequency of the oscillator and
$h = 6.626 \times 10^{-34}$ Js is a constant called *Planck's constant*

  2. Planck's second assumption was that if the oscillators could only emit or absorb whole-number multiples of the minimum energy, light could therefore only be emitted or absorbed from the surface of the black body as chunks or **quanta** of energy.

quanta: a discrete amount of energy

- Planck still considered light to be an electromagnetic wave at the time and thought only the oscillators themselves and the light being absorbed and emitted at the oscillator was quantised. He did not think the radiation field (i.e. the light) was quantised. Planck initially believed that his assumptions were mathematical tricks to get the correct answer, rather than revolutionary insights into the true nature of energy and light.
- It is difficult to understand why Planck's quantisation gives the correct answer for black-body radiation without delving into complex mathematics. However, we can get a feel for why Planck's assumptions work by thinking about light at equilibrium inside a hollow black body at a specific temperature. At equilibrium the walls would absorb light energy at the same rate as they emit light energy. Electromagnetic waves would form standing waves in such a cavity. The four longest-wavelength (lowest frequency) modes of vibration that could form in such a cavity are shown in Figure 10.2. Now clearly there are many more modes of vibration at higher frequencies that could exist in the cavity. Classical physics implies that each mode contains a specific amount of energy ($E = kT$) and, as there are an infinite number of high-frequency modes, the cavity would contain an infinite amount of energy (this was later called the 'ultraviolet catastrophe'). But if we apply Planck's quantisation condition, the oscillators in the wall can only emit or absorb energy in whole-number multiples of a specific minimum energy related to their frequency (i.e. hf).

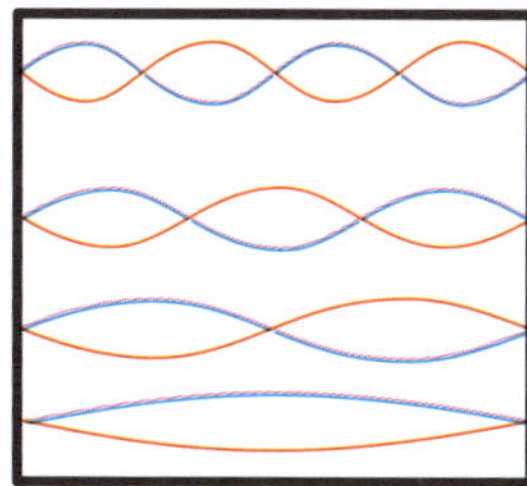

Figure 10.2 Electromagnetic standing waves in a black-body cavity

As the frequency increases, so does the minimum energy until the minimum energy that can be exchanged with the oscillator becomes greater than the energy in the mode (i.e. $hf > kT$). At this point the mode no longer has enough energy to excite the oscillator and the oscillator has too much energy to excite the mode. It is as if you went to an ATM to withdraw $20 and the smallest denomination held in the machine was $50. You simply could not withdraw the $20 just as the high-frequency mode could not obtain the energy to exist in the cavity. This is why the black-body curves turn over at high frequencies and why the ultraviolet catastrophe predicted by classical physics does not occur.

➔ Because energy quanta are so small, physicists often express their energy in **electron volts** (eV) rather than joules. An electron volt is the energy gained by an electron when it is accelerated over a potential difference of one volt. Hence the energy of an electron volt is:

$E = qV = 1.6012 \times 10^{-19} \times 1 = 1.602 \times 10^{-19}$ J

electron volt: energy gained by an electron when it is accelerated across an electrical potential difference of 1 volt (i.e. 1 eV = 1.602×10^{-19} J)

➔ Planck did not know it at the time but his 'act of desperation' would lead to a deeper understanding of nature and a new physics.

EXAMPLE 1

Planck assumed that energy was quantised and that the quanta of an oscillator was related to its frequency of oscillation.

a Find the minimum energy in electron volts of an electron oscillating at 4×10^{14} Hz.

b Why do we not notice energy quantisation in everyday situations like pushing a child on a swing?

Recall that $E = hf$ and that an electron volt = 1.602×10^{-19} joules

Answer:

a The energy of the quanta will be given by:

$E = hf = 6.626 \times 10^{-34} \times 4 \times 10^{14}$

$= 2.65 \times 10^{-19}$ J

$= \dfrac{2.65 \times 10^{-19}}{1.602 \times 10^{-19}} = 1.65$ eV

b Planck's constant is incredibly small, which means the energy quanta are also incredibly small. Even for electrons oscillating at 10^{14} Hz, the energy quanta are only of the order of 10^{-19} J. For a child on a swing with a frequency of 0.4 Hz, the energy quanta would be:

$E = hf = 6.626 \times 10^{-34} \times 0.4 = 2.65 \times 10^{-34}$ J

This energy difference is too small to measure and hence we see the energy of the swing as continuous rather than quantised.

➔ KEY QUESTIONS

1 What is a black body?

2 What is the black-body emission spectrum and why could it not be reproduced from classical electromagnetic wave theory?

3 How does the radiation emitted from a black body change as the temperature of the black body is increased?

4 How was Planck able to reproduce black-body radiation spectra?

Answers ➲ p. 151

2 Photoelectric effect and the wave model of light

» Students investigate the evidence from photoelectric effect investigations that demonstrated inconsistency with the wave model for light.

➔ The photelectric effect shown schematically in Figure 10.3 refers to the emission of electrons from clean metal surfaces when high-frequency light is incident on the surface. The effect was first reported in 1887 by Hertz, who noticed it when he was conducting experiments to confirm the existence of electromagnetic waves.

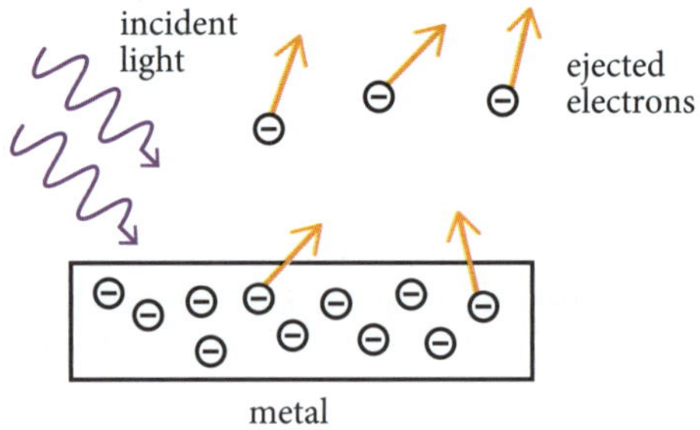

Figure 10.3 Photoelectric effect

➔ Like the radiation emitted from black bodies, the photoelectric effect could not be fully explained by the classical wave model of light. By the end of the 19th century several observations made about the effect puzzled physicists, three of which are described below.

Existence of a threshold light frequency

➔ No **photoelectrons** are emitted from the metal surface if the frequency of the incident light is below a specific **threshold frequency** (f_0). This threshold frequency is characteristic of the metal used in the experiment. Even when very high-intensity light is incident on the surface, no electrons are emitted if the incident frequency is less than the threshold frequency.

photoelectron: an electron emitted from a surface by light incident on the surface
threshold frequency: the minimum frequency of incident light required to eject electrons from the surface of a specific metal

- This observation cannot be explained by the wave theory of light because the wave theory relates the intensity of the wave to the amplitude of the wave and not the frequency. The wave theory wrongly predicts that, provided it is intense enough, incident light of any frequency would eject electrons from the surface.
- The threshold frequency and wavelength for a range of materials are shown in Table 10.1. Note that most metals have threshold frequencies in the ultraviolet region ($\lambda < 400$ nm) of the electromagnetic spectrum.

Table 10.1 Threshold frequencies and wavelengths for a range of materials

Metals	Threshold frequency ($\times 10^{15}$ Hz)	Threshold wavelength (nm)
Copper	1.376	218
Zinc	1.293	232
Calcium	0.9615	312
Sodium	0.8174	367

Instantaneous emission of electrons at low light intensities

- Provided the frequency of the incident light is above the threshold frequency, electrons are immediately ejected from the surface even when very low intensity light is incident on the surface.
- This observation could not be explained by classical electromagnetic wave theory because the theory assumes that the energy carried by light is spread evenly across the wave front. At very low light intensities, wave theory predicts that it would take some time before an electron could absorb enough energy from the wave to be ejected from the surface.

Maximum kinetic energy of ejected electrons

- Measurements of the kinetic energy of the electrons ejected from the surface showed that the maximum kinetic energy of the electrons was proportional to the frequency of the incident light. Classical electromagnetic wave theory could not explain this observation because it relates the intensity of the wave to the square of the amplitude of the wave. Wave theory predicts, incorrectly, that the energy of the electrons ejected from the surface would be proportional to the intensity of the light used and not be related to the frequency.

EXAMPLE 2

Explain how classical electromagnetic wave theory explains the transfer of energy from light to an electron in the surface of a metal.

Remember that classical wave theory treats the electron as an oscillator that can be driven by a changing electric field

Answer:

Classical electromagnetic wave theory treats the electron as an oscillator that would be pushed back and forth by the electric field of the electromagnetic wave as it moved past the charge. The electromagnetic wave would cause the oscillator amplitude and energy to increase until the electron had enough energy to escape from the surface. Note that this would mean it would take some time before electrons were emitted when very low light intensity was used, which contradicts experimental observations.

FIRSTHAND INVESTIGATION 1

Demonstrating the photoelectric effect

Students can demonstrate the photelectric effect by using an electroscope and a UV-C lamp.

SAFETY: Avoid exposing your skin and eyes to the short-wavelength light emitted from the UV-C lamp.

Students use emery paper to clean a sheet of zinc and place it on top of a negatively charged electroscope. Shining the UV-C light on the zinc sheet will cause the electroscope to rapidly discharge because the photoelectric effect causes electrons to be ejected from the zinc sheet. The experiment is illustrated in Figure 10.4.

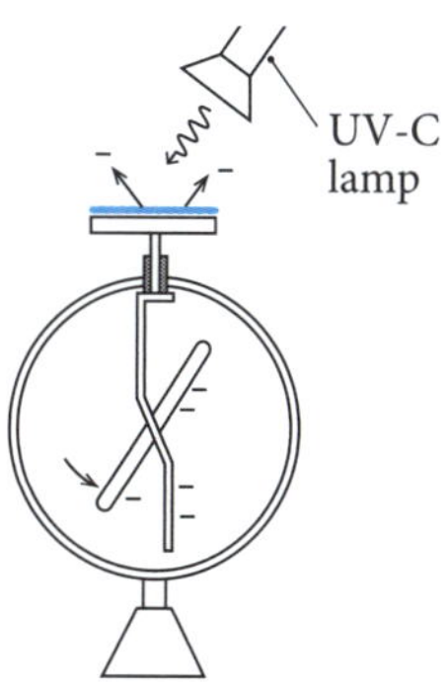

Figure 10.4 Simple experiment to demonstrate how the photoelectric effect can discharge a negatively charged electroscope

Note that if an electroscope is not available there are several videos of this type of experiment available online.

FIRSTHAND INVESTIGATION 2

Investigating the photoelectric effect

If the apparatus is available, students can conduct a firsthand investigation using a photocell and coloured light filters to measure the threshold frequency, maximum photocurrent and stopping voltage. They could then use the stopping voltage to calculate the maximum kinetic energy of the photoelectrons and plot it against the incident light frequency. Figure 10.5 shows a photocell and the circuit required to make these measurements.

If the photoelectric equipment is not available, students should conduct a virtual photoelectric effect experiment using one of the applets available online (e.g. https://phet.colorado.edu/en/simulation/photoelectric).

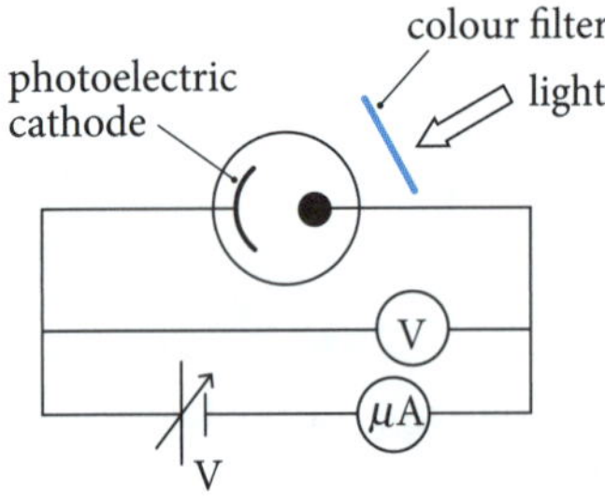

Figure 10.5 Photoelectric experiment

➔ KEY QUESTIONS

5 **What observations about the photoelectric effect could not be explained by classical electromagnetic wave theory?**

6 **What incorrect predictions does the classical theory of electromagnetism make about the photoelectric effect?**

Answers ➲ p. 151

3 Photoelectric effect and the photon model of light

» Students analyse the photoelectric effect $K_{max} = hf - \phi$ as it occurs in metallic elements by applying the law of conservation of energy and the photon model of light.

➔ In 1905, five years after Planck solved the black-body radiation problem, Einstein extended Planck's idea to light itself. By treating light as a particle, Einstein was able to produce a more rigorous derivation of Planck's black-body radiation equation and explain the problematic photoelectric effect.

➔ Einstein suggested that electromagnetic radiation was spatially quantised and made up of a stream of particles (later named **photons**) that had an energy given by Planck's equation:

$E = hf$

photon: quanta of electromagnetic radiation with energy $E = hf$

By using the wave equation for light waves, we can express the photon energy in terms of the light's wavelength. As $c = f\lambda$, we can write $\lambda = c/f$ and hence:

$$E = hc/\lambda$$

where E is the energy of the photon
$h = 6.626 \times 10^{-34}$ Js is Planck's constant
$c = 3 \times 10^{8}$ ms^{-1} is the speed of light and
λ is the wavelength

➔ In Einstein's model the intensity of a beam of light is determined not by the amplitude of a wave but by the number of photons that pass through a unit area per second multiplied by the photon energy.

➔ Einstein was aware of the overwhelming evidence supporting the wave nature of light and was careful to qualify his hypothesis about free radiation being quantised as 'provisional'. It took many years before the scientific community accepted that light itself was quantised.

➔ Scientists now believe that light is made up of photons and that it sometimes exhibits wave-like behaviour and at other times exhibits particle-like behaviour. This strange, **wave-particle duality** is today a fundamental tenet of quantum physics.

wave-particle duality: the ability of photons or small particles to exhibit both wave properties and particle properties

➔ To explain the photoelectric effect, Einstein made three assumptions:

1. Light consisted of a stream of photons with energy $E = hf$.
2. One photon would interact with one electron on the surface of the metal.
3. A specific amount of energy, called the **work function** (ϕ) was required to remove an electron from the surface of a metal. Because the electrons are held more tightly in some crystal lattices than others, different metals have different work functions.

work function: the minimum energy required for a photon to liberate an electron from the surface of a specific metal

➔ By applying conservation of energy to the interaction between an incident photon and a surface electron, Einstein was able to use his photon model to explain the photoelectric effect, as outlined below.

A. The existence of a threshold frequency

If an electron on the surface requires a specific amount of energy (ϕ) to escape the surface, only photons with this energy or more will liberate electrons from the surface. The threshold frequency (f_0) is hence related to the work function of the material by:

$hf_0 = \phi$

and electrons will only be ejected from the surface when the photon energy is greater than the work function for the surface. That is, electrons will only be ejected when:

$hf > \phi$

B. The instantaneous emission of electrons at low incident light intensity

If we assume that light is spatially quantised, the energy of the light will be concentrated at specific points in space rather than spread out across the wavefront. If the incident light intensity was very low, the wave model would predict that it would take a long time for an electron in the surface to gain enough energy to escape from the surface. But if light was quantised, some photons would reach the surface as soon as the light was switched on and electrons would be released from the surface instantly (provided the photon energy was greater than the work function for the metal).

C. The maximum kinetic energy of ejected electrons

If we apply the law of conservation of energy to the interaction between an incident photon and an electron on the surface of a metal we can write:

Energy of incident photon	=	Energy to release electron	+	Maximum kinetic energy of released electron

Hence the maximum energy of the photoelectron will be given by:

$$K_{max} = hf - \phi$$

where K_{max} is the maximum kinetic energy of the photoelectron
h is Planck's constant
f is the frequency of the incident light and
ϕ is the work function for the metal

Thus, as shown in Figure 10.6, a plot of the maximum kinetic energy of photoelectrons against the incident light frequency (f) for any metal should produce a straight line with a gradient equal to Planck's constant (h). This theoretical prediction was confirmed experimentally by Robert Millikan in 1916.

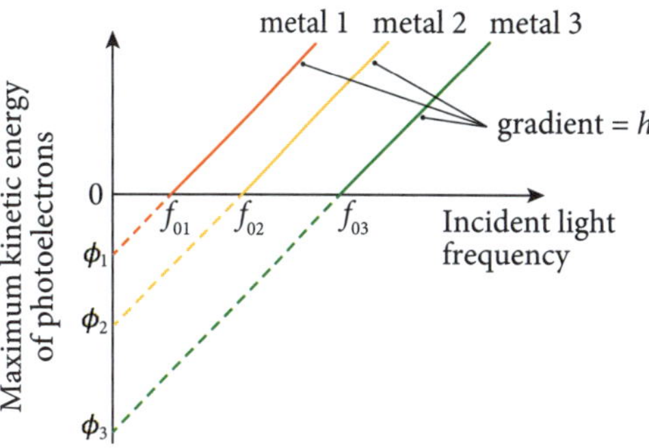

Figure 10.6 Maximum kinetic energy of photoelectron as a function of incident light frequency for different metals

➔ KEY QUESTIONS

7 **What was Einstein's quantum hypothesis about the nature of light?**

8 **How did Einstein use the photon concept to explain those features of the photoelectric effect that could not be explained by classical electromagnetic wave theory?**

Answers ➲ p. 151

➔ The circuit used by Millikan to measure the maximum energy of the photoelectrons is shown in Figure 10.7. Note that a quartz window is used because glass absorbs ultraviolet light.

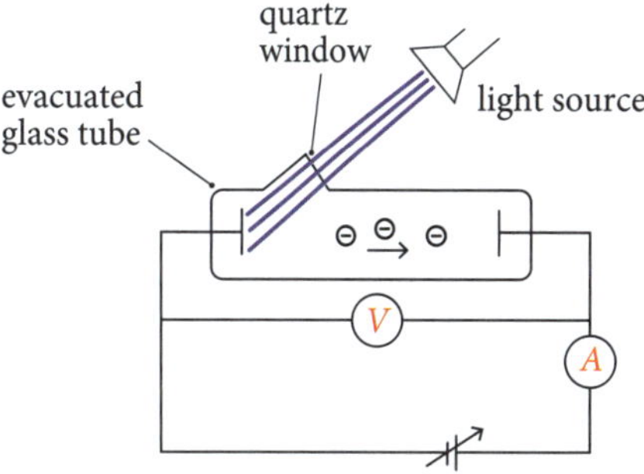

Figure 10.7 Apparatus used to investigate the photoelectric effect

If the voltage is adjusted to ensure the right-hand (collector) electrode is positive, all the photoelectrons will be attracted to the positive electrode and the ammeter will record the maximum **photocurrent**. If the voltage on the collector electrode is made negative and increased, the current in the circuit is reduced because the photoelectrons are repelled from the collector electrode by the electric field between the plates. Eventually this voltage will become large enough so that even the most energetic photoelectron will not quite

photocurrent: the current produced by the electrons emitted from the surface of a metal by the photoelectric effect

reach the collector electrode and the current in the circuit will be reduced to zero. At this point the initial kinetic energy of the most energetic photoelectron would be equal to the work done by the electric field to stop the electron and we can write:

$K_{max} = qV_{stop}$

where K_{max} is the maximum kinetic energy of the ejected electrons

q is the charge of the electron and

V_{stop} is the minimum applied voltage that will reduce the current to zero

Hence we can find the maximum kinetic energy of the photoelectrons by measuring the stopping voltage when the current in the circuit drops to zero.

➔ Figure 10.8 shows the circuit current as a function of the voltage between the electrodes for a photoelectric experiment that used three different light intensities at the same frequency. Note that the photocurrent increases with the incident light intensity because more photons would be incident on the metal surface per second. Note also that the stopping voltage for each light intensity is equal because the stopping voltage is proportional to frequency of the incident light, which is constant in this case.

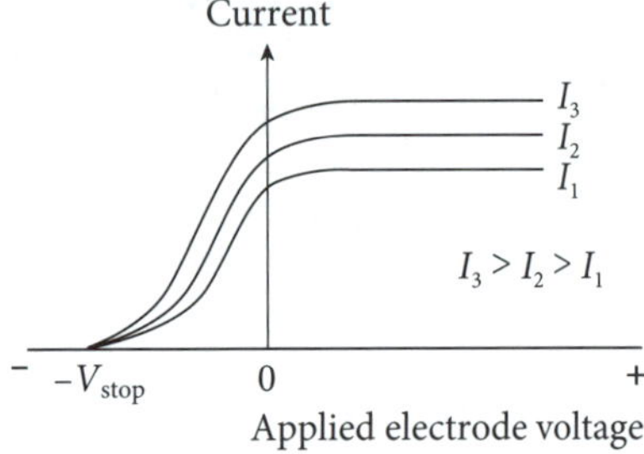

Figure 10.8 Current as a function of electrode voltage

➔ To obtain the experimental data plotted in Figure 10.6, the maximum kinetic energy at each frequency is determined by measuring the stopping voltage for each metal at that frequency. Note that photoelectric effect graphs, like the one shown in Figure 10.6, sometimes plot qV_{stop} on the vertical axis rather than K_{max} because the two quantities are equivalent.

➔ KEY QUESTIONS

9 **How is the threshold frequency related to the photocurrent?**

10 **What determines the maximum kinetic energy of a photoelectron?**

11 **What determines the maximum photocurrent emitted from a surface?**

12 **How does the photon model account for the intensity of light?**

Answers ➲ p. 151

EXAMPLE 3

Two lasers are used in an experiment. Laser A emits 1 mW of light at a wavelength of 630 nm. Laser B emits 2 mW of light with a wavelength of 400 nm.

a **Find the energy in electron volts of the photons emitted by each laser.**

b **Determine the number of photons emitted per second by each laser.**

c **Compare the effect of shining each of the lasers on a metal with a work function of $\phi = 2.8$ eV.**

Recall that the intensity of light is related to the number of photons per second in the beam and the energy of the photons

Answer:

a Photon energy is given by:

$E = hf = hc/\lambda$

For laser A

$$E = \frac{6.626 \times 10^{-34} \times 3 \times 10^{8}}{630 \times 10^{-9}} = 3.16 \times 10^{-19}\text{ J}$$

$$= \frac{3.16 \times 10^{-19}}{1.602 \times 10^{-19}} = 1.97\text{ eV}$$

For laser B

$$E = \frac{6.626 \times 10^{-34} \times 3 \times 10^{8}}{400 \times 10^{-9}} = 4.97 \times 10^{-19}\text{ J}$$

$$= \frac{4.97 \times 10^{-19}}{1.602 \times 10^{-1}} = 3.10\text{ eV}$$

b The power is related to the photon energy by:

Power = energy/time = photon energy × number of photons per second

Hence the number of photons/second = power/photon energy

For laser A

Number of photons emitted per second

$$= \frac{1 \times 10^{-3}}{3.16 \times 10^{-19}} = 3.16 \times 10^{15}\text{ s}^{-1}$$

For laser B

Number of photons emitted per second

$$= \frac{2 \times 10^{-3}}{4.97 \times 10^{-19}} = 4.02 \times 10^{15}\text{ s}^{-1}$$

c Because the photons emitted by laser A have less energy than the work function they will not have enough energy to eject an electron from the surface of the metal.

The photons emitted by laser B have more energy than the work function and hence will eject electrons from the surface of the metal. If every photon emitted an electron, laser B would cause 4.02×10^{15} electrons to be emitted from the surface per second. The electrons would have a maximum kinetic energy of:

$K_{max} = hf - \phi = 3.10 - 2.8 = 0.3$ eV

We can also calculate the photocurrent as current = number of electrons per second × charge on each electron:

Photocurrent $= I = 4.02 \times 10^{15} \times 1.602 \times 10^{-19}$

$= 6.44 \times 10^{-4}$ A

EXAMPLE 4

The results of a photoelectric effect experiment for a specific metal are plotted on the graph in Figure 10.9.

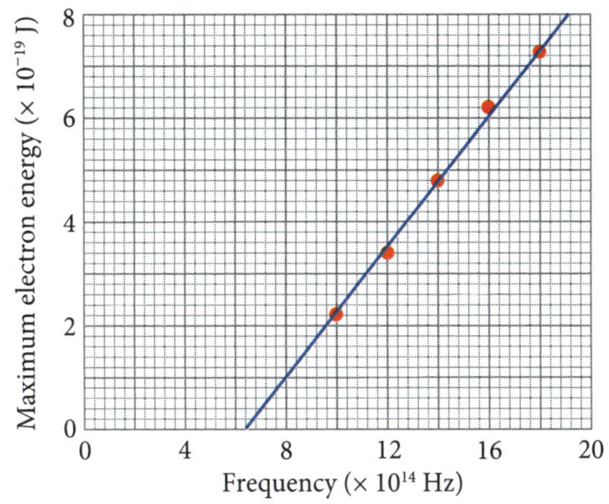

Figure 10.9 Maximum photoelectron energy as a function of incident light frequency

a **Determine the work function of the metal.**

b **Find the gradient of the graph and comment on this experimental result.**

c **Outline how the graph would change if a different metal was used.**

Recall that the work function is related to the threshold frequency and that different metals have different threshold frequencies

Answer:

a The work function is the minimum energy required to remove an electron from the surface. The threshold frequency is the minimum incident photon frequency required to remove an electron from the metal. Hence the x-intercept on the graph will be the threshold frequency. Reading from the graph, the threshold frequency is:

$f_0 = 6.4 \times 10^{14}$ Hz

The work function is therefore:

$\phi = hf_0 = 6.626 \times 10^{-34} \times 6.4 \times 10^{14} = 4.24 \times 10^{-19}$ J

Or $\dfrac{4.24 \times 10^{-19}}{1.602 \times 10^{-19}} = 2.65$ eV

b Gradient = rise/run = $\dfrac{8 \times 10^{-19}}{12.8 \times 10^{14}} = 6.25 \times 10^{-34}$ Js

We would expect the gradient to be Planck's constant from theory but there is some experimental error here as the gradient of the graph is about 5% lower than the accepted value of Planck's constant.

c If a different metal was investigated we would still expect the gradient to be equal to Planck's constant but the second metal might require a different amount of energy to eject an electron from the surface and hence would cross the frequency axis at a different point on the axis.

CHAPTER SYLLABUS CHECKLIST

Are you able to answer these questions from the syllabus for this chapter? Tick each question as you go through the checklist if you are able to answer it. If you cannot answer a question, turn to the relevant page in the study guide to find the answer. For NESA key word meanings, go to www.educationstandards.nsw.edu.au and search 'key words'.

	FOR A COMPLETE UNDERSTANDING OF THIS TOPIC:	PAGE NO.	✓
1	Can I define black-body radiation?	140	
2	Can I explain how black-body radiation changes with the temperature of the body?	140	
3	Can I use Wien's law to relate the peak emission from a black body to the temperature of the body?	140–141	
4	Can I recall that the spectral distribution of black-body radiation could not be reproduced by classical electromagnetic wave theory?	141	
5	Can I outline the assumptions Planck made to derive the equation that correctly reproduced the black-body radiation spectrum at different temperatures?	141	
6	Can I describe the photoelectric effect?	142–143	
7	Can I recall that photoelectrons are emitted instantly, even when the incident light frequency is very low?	143	
8	Can I recall the observations about the photoelectric effect that could not be explained by classical electromagnetic wave theory and outline why they could not be explained?	142–145	
9	Can I relate the threshold frequency of the incident light to the work function of a metal?	144–145	
10	Can I describe the relationship between the incident light frequency and the maximum kinetic energy of photoelectrons?	145	
11	Can I outline how Einstein extended Planck's quantum hypothesis to light?	144	
12	Can I recall how Einstein was able to explain the photoelectric effect using conservation of energy and the photon model of light?	144–145	
13	Can I recall that Einstein predicted that a plot of the maximum energy of photoelectrons against the incident light frequency would produce a straight line with a gradient equal to Planck's constant?	145	
14	Can I describe how 'stopping voltage' can be measured and used to determine the maximum kinetic energy of photoelectrons?	145–146	

HSC EXAM-TYPE QUESTIONS

Objective-response questions (1 mark each)

1 **How does the radiation from a black body change when the temperature of the black body is increased?**

A The wavelength of maximum emission and the total radiated energy decrease.

B The wavelength of maximum emission and the total radiated energy increase.

C The wavelength of maximum emission increases but the total radiated energy decreases.

D The wavelength of maximum emission decreases but the total radiated energy increases.

2 **What assumptions did Planck make to derive an equation that could reproduce the spectral distribution of black-body radiation?**

A Light was quantised and electromagnetic waves did not exist.

B The energy of the oscillators was quantised but the oscillators could emit or absorb any amount of energy.

C The energy of the oscillators was quantised and oscillators could only emit and absorb energy in discrete packets.

D Light is made up of a stream of photons and one photon interacts with one electron in the black body.

3 **In a photoelectric effect experiment, what determines the maximum kinetic energy of the photoelectrons?**

A the incident light intensity and wavelength

B the incident light intensity and the work function of the surface

C the incident light wavelength and the work function for the surface

D the stopping voltage and the incident light wavelength

4 **Which of the following changes to a photoelectric effect experiment would result in an increase in the photocurrent? (You may assume that the incident photon energy is always greater than the work function for the surface.)**

A The incident light intensity remains constant but the wavelength of the light is increased.

B The incident light intensity remains constant but the wavelength of the light is reduced.

C A metal with a lower work function is used.

D A metal with a higher work function is used.

5 **Which of the following observations about the photoelectric effect could not be explained by the electromagnetic wave model?**

A the existence of a threshold wavelength for the emission of electrons

B the observation that electrons are ejected from a metal surface by incident light

C the existence of a work function for metals

D the observation that increasing the light intensity increases the photocurrent

Extended-response questions

6 **The Sun has a surface temperature of 5778 K, yet to a good approximation it can be treated as a perfect black body.**

a **What do physicists mean when they talk about a 'perfect black body'?** (1 mark)

b **Determine the wavelength and frequency at which the Sun emits the greatest intensity electromagnetic radiation.** (2 marks)

7 **This question refers to Figure 10.10.**

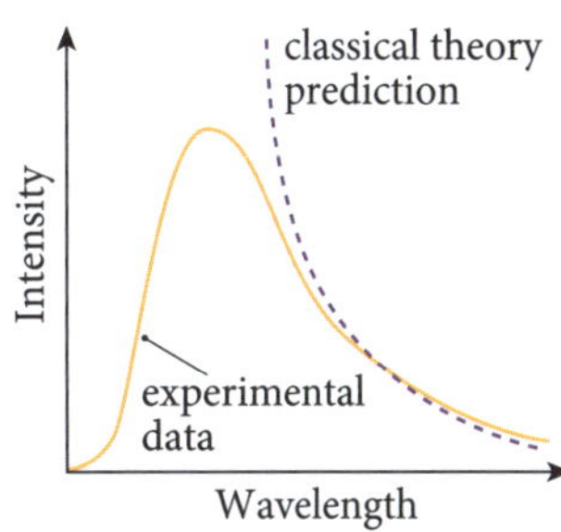

Figure 10.10 Black-body radiation emitted at specific temperatures and the classical wave theory prediction for the emission at the same temperature

a **Explain why the data shown in Figure 10.10 caused concern to physicists in the late 19th century.** (2 marks)

b **Planck solved the black-body problem by making assumptions that Einstein built upon to produce a new model of light. Compare Einstein's model of light with the classical electromagnetic wave model.** (4 marks)

8 **Experiments with the photoelectric effect show that electrons are only emitted from a metal surface when the incident light has a frequency greater than a specific threshold frequency.**

a **Outline why this observation could not be explained by the classical wave model of light.** (1 mark)

b **Outline how the photon model can be used to explain this observation.** (2 marks)

c **If a metal had a work function of 4 eV, what would be its threshold frequency?** (2 marks)

9 This question refers to the apparatus illustrated in Figure 10.11.

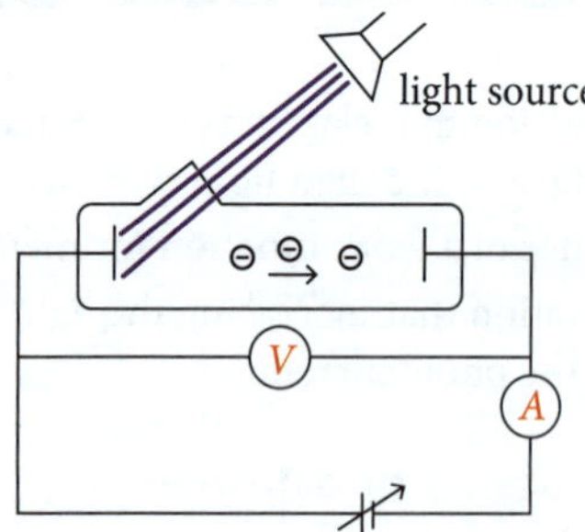

Figure 10.11 Photoelectric effect experiment

a Explain how the apparatus in Figure 10.11 can be used to determine the total photocurrent for a specific light intensity and wavelength. (2 marks)

b Explain how the maximum kinetic energy of the photoelectrons can be determined using the apparatus. (3 marks)

c Explain why ultraviolet light is often used in photoelectric experiments. (2 marks)

10 In a photelectric effect experiment an incident light wavelength of 350 nm is found to produce photoelectrons with a maximum kinetic energy of 1.2 eV.

a Find the energy of the incident photons. (1 mark)

b Find the work function of the metal used in the experiment. (2 marks)

c Determine the threshold frequency for this metal. (1 mark)

ANSWERS

KEY QUESTIONS

Key questions ➲ p. 142

1 A black body is a perfect emitter and a perfect absorber of radiation. It absorbs all the light that falls on it and reflects none. The electromagnetic radiation emitted by a black body is therefore due to the temperature of the body alone.

2 A black-body emission spectrum refers to the range of electromagnetic waves emitted from the black body. Classical electromagnetic wave theory was unable to reproduce the black-body spectrum because when light interacts with matter on the atomic scale it behaves like a particle rather than a wave. The black-body radiation spectrum could only be predicted by *quantising* the interaction between light and the oscillators (atoms or molecules) in the black body.

3 The total amount of electromagnetic wave energy emitted from a black body increases rapidly with temperature. However, the wavelength at which the maximum emission occurs decreases when the temperature of the black body in increased.

4 Planck quantised the energy of the oscillators in the black body, specifying that they could only have a whole number multiple of minimum energy $E = hf$. In addition, Planck had to assume that radiation could only be emitted or absorbed by the oscillators in discrete quanta. With these assumptions Planck could derive an equation that reproduced the black-body radiation spectra for a black body of any temperature.

Key questions ➲ p. 144

5 Classical electromagnetic wave theory could not explain:

- a the existence of a threshold frequency for the incident light. Only light with a frequency above the threshold frequency of the metal could liberate electrons from the surface of the metal.
- b the immediate emission of electrons from a metal when it was irradiated with very weak electromagnetic radiation (with a frequency above the threshold frequency).
- c that the maximum kinetic energy of the photoelectrons was found to be proportional to the frequency of the incident light and unrelated to the intensity.

6 The classical theory erroneously predicted that:

- a there would be no threshold frequency and any frequency of light should liberate electrons from the surface provided the light had sufficient intensity.
- b weak light would take some time to transfer enough energy to electrons in the surface for the electrons to be liberated from the surface.
- c the maximum kinetic energy of the photoelectrons should be related to the intensity of the incident light.

Key questions ➲ p. 145

7 Einstein proposed that light was spatially quantised into points of energy called *photons*. These photons would have an energy related to the frequency of the light by $E = hf$.

8 Einstein assumed that one photon would interact with one electron on the metal's surface. He then applied conservation of energy to show that the maximum kinetic energy of the photoelectrons would be given by $K_{max} = hf - \phi$, where ϕ is the energy to remove an electron from the metal surface (i.e. the work function of the metal). This equation shows that the maximum kinetic energy of photoelectrons is proportional to the incident light frequency. Clearly the photons must have at least as much energy as the work function to remove an electron from the surface, which explains the existence of a threshold frequency. Finally, if light consists of a stream of particles, energy will be concentrated at points on the wavefront rather than being spread evenly over the wavefront. Electrons will therefore be emitted as soon as light reaches the surface (provided the photon energy is equal to or greater than the work function).

Key questions ➲ p. 146

9 No electrons will leave the surface (i.e. photocurrent will be zero) until the incident light frequency is above the threshold frequency. Increases in frequency beyond this point do not affect photocurrent.

10 The maximum kinetic energy of the photoelectrons is determined by the work function of the material and the incident wave frequency (i.e. $K_{max} = hf - \phi$).

11 Provided the incident light has a frequency above the threshold frequency for the metal, the photocurrent will be proportional to the incident light intensity (i.e. the number of photons incident on the surface per second).

12 In the wave model, intensity = wave energy per second passing through a unit area. In the photon model, light intensity = (photon energy) × (number of photons that pass through a unit area per second).

HSC EXAM-TYPE QUESTIONS

Objective-response questions

1 **D.** The increased temperature increases the vibrational energy of the atoms in the body, resulting in more energy being radiated (power radiated $\alpha\ T^4$) and the maximum emission wavelength will decrease in accordance with Wien's law. **A** and **C** are incorrect as the radiated energy must increase with temperature. **B** is incorrect because by Wien's law the wavelength of maximum emission must decrease when the temperature of the radiating body is increased.

2 **C.** Both assumptions are required to derive an equation that reproduces the black-body radiation spectrum. **A** is incorrect because Planck believed that light was an electromagnetic wave once it left the surface of the black body. **B** is incorrect as the derivation requires the energy emitted or absorbed by the oscillators to be quantised. **D** is incorrect because Planck did not propose that light in free space was quantised.

3 **C.** Because of conservation of energy, the maximum kinetic energy of the photoelectron is equal to the difference between the incident photon energy and the work function (i.e. $K_{max} = \frac{hc}{\lambda} - \phi$).

A and **B** are incorrect as the maximum kinetic energy of the photoelectron is not related to the incident light intensity. **D** is incorrect as the work function of the metal affects the maximum kinetic energy of the photoelectrons.

4 **A.** The photocurrent is determined by the number of photons reaching the metal per second. If the intensity of the light remains the same but the wavelength increases (photon energy reduced), more photons per second will be in the beam and hence more electrons will be ejected per second. **B** is incorrect because if the photon energy is increased there would be fewer photons in the

beam per second. **C** and **D** are incorrect because, provided the incident photon energy is greater than the work function, the work function does not affect the photocurrent.

5 **A**. The wave model predicts that all wavelengths would liberate electrons from the surface. **B** is incorrect as the wave model can explain the ejection of electrons. **C** is incorrect as the work function is a property of the material, independent of the model of light. **D** is incorrect as the wave model and particle model predict that the photocurrent will increase with light intensity (provided $hf > \phi$).

Extended-response questions

6 EM This question tests students' understanding of the black-body concept and their ability to apply Wien's law.

a A black body is a perfect absorber ✓ and emitter of radiation. It absorbs all the light that falls on it and reflects none.

b Applying Wien's law:

$$\lambda_{max} = \frac{b}{T} = \frac{2.898 \times 10^{-3}}{5778} = 501 \text{ nm} ✓$$

$$f = c/\lambda = \frac{3 \times 10^8}{501 \times 10^{-9}} = 5.99 \times 10^{14} \text{ Hz} ✓$$

7 EM This questions tests students' understanding of the unsuccessful application of classical wave theory to explain black-body radiation and their understanding of the wave and particle models of light.

a The classical electromagnetic wave model of light was supported by a large amount of evidence but when it was applied to black-body radiation, as shown in Figure 10.10, it predicted that black bodies should emit an infinite amount of energy at high frequencies. ✓ As shown in Figure 10.10, the wave model prediction did not match experimental observations. ✓ The failure of the wave model to reproduce the black-body radiation curves caused considerable concern to physicists at the time.

b The classical electromagnetic wave model considers light to be a transverse wave made up of mutually perpendicular, oscillating electric and magnetic fields. ✓ In the wave model the energy of the wave spreads evenly across the wavefront and the intensity of the wave was related to the wave amplitude. ✓ In contrast, Einstein saw light as made of a stream of particles (photons) with an energy $E = hf$. ✓ In this model the energy of the wave was spatially quantised rather than spread evenly across the wavefront. Light intensity according to Einstein was related to the number of photons in the beam and the energy of the photons. ✓

8 EM This questions tests students' understanding of why the classical wave model cannot be used to explain observations related to the photoelectric effect and why the photon model can be used to explain these observations.

a The classical theory predicts that the power of the wave is spread evenly over the wavefront and is independent of the frequency. Hence the model predicts that, provided the intensity is great enough, electrons will be liberated from the surface at all incident wave frequencies. ✓

b The photon model assumes one photon transfers its energy to one electron. An electron will only be ejected from the surface if the photon has more energy than the energy required to remove the electron from the surface (the work function of the surface). ✓ As the photon energy is determined by the frequency of the light ($E = hf$) only frequencies above a threshold frequency will liberate electrons from the surface. ✓

c First converting the work function to joules:

$\phi = 4 \times 1.602 \times 10^{-19} = 6.408 \times 10^{-19}$ J ✓

To find the threshold frequency we must equate the photon energy and the work function:

$hf = \phi$

Hence $f = \frac{\phi}{h} = \frac{6.408 \times 10^{-19}}{6.626 \times 10^{-34}} = 9.67 \times 10^{14}$ Hz ✓

9 EM This question tests students' understanding of the photoelectric effect experiment.

a The voltage is adjusted ✓ to make the right-hand electrode (the collector) more positive until the current shown on the ammeter is maximum. This maximum current flow is the total photocurrent. ✓

b The voltage on the collector electrode is made progressively more negative until the current in the circuit drops to zero. ✓ At this point the potential across the plates (V_{stop}) ✓ is just large enough to stop the most energetic photoelectron from reaching the collector plate and the maximum kinetic energy will be given by $K_{max} = qV_{stop}$. ✓

c The threshold frequency of many metals is high ✓ and therefore requires ultraviolet light (which has a higher frequency than visible light) to liberate electrons from the surface. ✓

10 EM This question tests students' ability to use equations related to the photoelectric effect to calculate a range of quantities.

a The energy of the photons will be given by:

$E = hf = hc/\lambda$

$= \frac{6.626 \times 10^{-34} \times 3 \times 10^8}{350 \times 10^{-9}} = 5.68 \times 10^{-19}$ J ✓

b First converting the work function into joules:

$\phi = 1.2 \times 1.602 \times 10^{-19} = 1.92 \times 10^{-19}$ J ✓

Applying Einstein's equation for the photoelectric effect:

$K_{max} = hf - \phi$

Hence $\phi = hf - K_{max}$

$= 5.68 \times 10^{-19} - 1.92 \times 10^{-19}$

$= 3.76 \times 10^{-19}$ J ✓

c The work function is related to the threshold photon energy:

$\phi = hf_0$

Hence $f_0 = \phi/h$

$= \frac{3.76 \times 10^{-19}}{6.626 \times 10^{-34}}$

$= 5.67 \times 10^{14}$ Hz ✓

CHAPTER 11 LIGHT AND SPECIAL RELATIVITY

MODULE 7 THE NATURE OF LIGHT

INQUIRY QUESTION:

How does the behaviour of light affect concepts of time, space and matter?

Einstein showed that if the speed of light in a vacuum is an absolute constant and all inertial frames are equivalent then time, length and momentum must be relative rather than absolute quantities. That is, these quantities will be different when viewed by observers in different frames of reference moving with respect to one another. Einstein's postulates also enabled him to discover a relationship between mass and energy, and to predict that mass could be converted to energy and energy into mass.

1 Postulates of special relativity

» Students analyse and evaluate the evidence confirming or denying Einstein's two postulates:

- All inertial frames of reference are equivalent.
- The speed of light in a vacuum is an absolute constant.

➔ In 1905 Albert Einstein published a theory, now called the **theory of special relativity**, that changed our understanding of mass, length and time. The theory showed that mass, space and time were not absolute but relative quantities.

➔ Einstein based his theory on two **postulates**:

1 All inertial frames of reference are equivalent.

2 The speed of light in a vacuum is an absolute constant.

theory of special relativity: an axiomatic theory (i.e. a theory based on postulates) that proposes that mass, space and time are relative rather than absolute quantities
postulate: a basic principle that is assumed to be true in order to develop further arguments or theories

➔ Scientific theories are usually based on postulates or assumptions. Postulates like these can never be absolutely proven but they can be disproved. If an experiment or observation shows that a postulate is not true, the postulate must be changed or abandoned. While we can never absolutely prove a postulate, we can assume a postulate is true if it is supported by all the available observations and experiments. The success of a theory or theories developed from postulates also acts as evidence for the veracity of the postulates.

➔ The first postulate, which Einstein called the *principle of relativity*, deals with the equivalence of inertial reference frames. You will recall from the Year 11 course that a frame of reference is a set of coordinates (axes) that can be used to measure the position and velocity of objects in the frame. An **inertial reference frame** is a frame in which a free body (one without forces acting on it) does not accelerate. This means that an inertial frame of reference is one in which Newton's first law is obeyed. Any frame of reference that moves at constant velocity with respect to another inertial frame of reference will itself be an inertial frame.

inertial reference frame: a frame of reference in which a free body does not accelerate (i.e. Newton's first law is obeyed)

➔ Einstein's first postulate was an extension of **Galilean–Newtonian relativity**. To explain motion, Galileo and Newton proposed that all inertial frames were equivalent for the laws of mechanics. Einstein's relativity extends Galilean–Newtonian relativity to all the laws of physics. That is, not only are inertial frames equivalent for the laws of mechanics but they are also equivalent for all other areas of physics, such as thermodynamics, optics and electromagnetism.

Galilean–Newtonian relativity: a relativity theory that asserts that all inertial frames are equivalent for the laws of mechanics

➔ The equivalence of inertial frames means there is no privileged frame and hence there is no such thing as an absolute frame of rest. Another way to think about this is to say there is no experiment that could be done inside an inertial reference frame that could tell us if the frame is moving or at rest. We experience this every day when we drive in cars or fly in planes at constant velocity. In each case, it feels like we are at rest because we in an inertial frame. Whether we are sitting at home or flying at constant velocity in a plane, it will feel the same and any experiment we conduct in either frame will produce the same result.

→ The **invariance** of the speed of light expressed in the second postulate is a consequence of the first postulate. This is because if all inertial frames are equivalent then the speed of light in all inertial frames must be the same, otherwise the frames would not be equivalent.

invariant: remaining constant

→ The second postulate is hard to understand at first because it seems to contradict common sense. It tells us that light travels at a constant rate in a vacuum, independent of the speed of the source or observer. We saw in the Year 11 course that the velocity of one object relative to another can be determined by performing a vector subtraction:

$$v_{\text{relative}} = v_{\text{object}} - v_{\text{observer}}$$

Note that if the observer and object are approaching or receding from each other the velocity vectors will be in opposite directions and the magnitude of relative velocity will be the sum of the magnitudes of the two velocities.

→ If Einstein's postulate about light is correct then this equation does not hold for light because no matter how fast an observer is travelling, they will still measure the same velocity for light. Consider the space ships shown in Figure 11.1. Ship A is travelling at $0.6c$ towards the Earth when it observes a laser beam sent from a station on the surface of the Earth. According to Galileo the observer in the spaceship would see the beam travelling at $1.6c$, but by Einstein's postulate the observer in the spaceship and the observer on the Earth would both measure the speed of the laser beam to be the same speed, the speed of light (c). Similarly for spaceship B, which shines a laser back to the Earth while moving away from the Earth at $0.6c$. Galileo predicts the Earth-bound observer would measure a speed of $(1 - 0.6)c = 0.4c$, but Einstein's postulate tells us that the observer on the Earth and the observer in spacecraft B would measure the same speed of light (c).

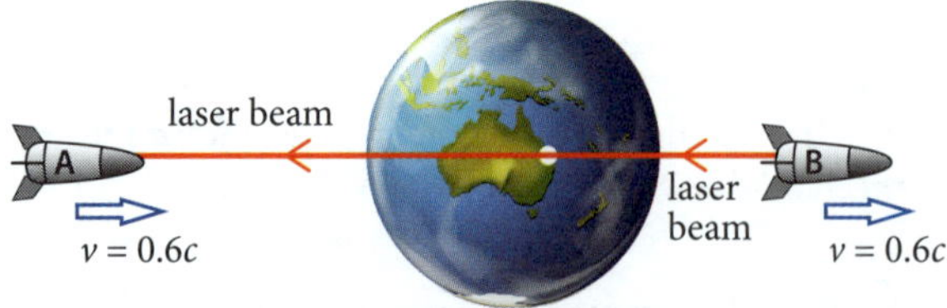

Figure 11.1 Measuring the speed of light in different frames of reference

→ Einstein demonstrated the difference between Galilean relativity and his relativity by a simple example. He imagined an observer travelling at the speed of light holding a mirror in front of him and looking at his reflection in a mirror. Galileo predicts that light could not reach the mirror from the observer and therefore the observer would not see their image reflected from the mirror. But Einstein predicts that the observer would see their refection because the speed of light will not be affected by the motion of the observer or the mirror. That is, the observer would still see light move away from them at the speed of light.

→ Although it is beyond the syllabus, students may wish to investigate how Einstein's theory led to a new relative velocity equation. Einstein showed that the magnitude of relative velocity when the observer and object approach or recede from each other is given by:

$$v_{\text{relative}} = \frac{v + u}{1 + uv/c^2}$$

where v is the velocity of the object and
u is the velocity of the observer

Note this equation reduces to Galileo's relative motion equation when $uv/c^2 \ll 1$.

→ Now that we have a better understanding of Einstein's postulates we can examine some of the evidence confirming or denying the two postulates.

→ Einstein initially argued from logic and his deep understanding of Maxwell's theory of electromagnetism that the postulates must be correct. He also showed that Galileo's transformation did not make sense for light. This is because if an observer was travelling at the speed of light, according to Galileo they would see a stationary electromagnetic wave. But this appears to contradict Maxwell's theory, which shows that electromagnetic waves consist of changing electric and magnetic fields. Einstein argued that Maxwell's equations had no preferred frame of reference and that they implied the speed of light would be the same for all observers.

→ Einstein also noted that attempts to find the absolute motion of the Earth with respect to the ether had been unsuccessful. You will recall that physicists at the time assumed light waves travelled in a substance called the 'ether'. The ether was believed to permeate all space and matter, be transparent, be incompressible and have zero viscosity.

→ Neither of the postulates was consistent with the concept of an ether because the ether would constitute an absolute rest frame, and frames moving at different speeds relative to the ether would experience different physics. For example, the speed of light would be different for observers in frames moving at different velocities with respect to the ether. If the ether existed it would invalidate both postulates because inertial frames would not be equivalent and the speed of light would be different in different frames.

→ While the existence of the ether was widely accepted at the time, there was no experimental evidence to support the concept and Einstein simply dismissed the ether because it was 'superfluous' to his new theory.

- Although Einstein did not refer to their work specifically, an important experiment had been conducted by Albert Michelson and Edward Morley in 1887 that provided strong support for his postulates.
- Michelson and Morley believed that light travelled in the ether. They knew the Earth was moving at 30 kms^{-1} in its orbit through the ether and set out to design an experiment that was sensitive enough to detect the ether wind that would be produced by the Earth's motion. Their experiment shown in Figure 11.2 used the interference of light that was produced when a light beam was split into two beams, sent on perpendicular paths and recombined. They calculated that the beam that travelled with and against the ether wind would take longer to travel the distance than the beam that passed back and forth across the ether wind. This change in time is due to the speed of light being different in each direction due to the Earth's motion through the stationary ether. Michelson and Morley showed that if the apparatus was rotated by 90° the time lag between the paths would change enough to produce a measurable change in the interference pattern (i.e. the interference fringes would shift). This type of device is called an **interferometer** because it uses interference to make precise measurements.

interferometer: a device that uses the interference of light to make precise measurement

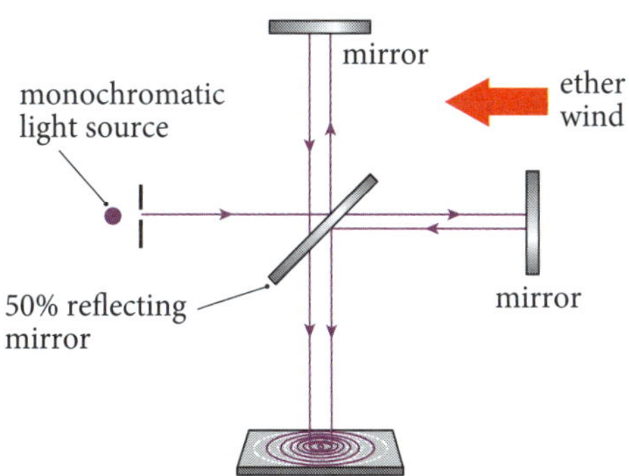

Figure 11.2 The Michelson–Morley experiment that failed to detect an ether wind

- When Michelson and Morley rotated their interferometer, they were shocked to find that the interference pattern did not change. Although their experiment was sensitive enough to measure the expected ether wind, it found no ether wind. Einstein's first postulate is supported by this **null result** because it is exactly what you would expect if all inertial frames were equivalent. It also supports the second postulate because the time to travel each path would be identical if the speed of light was not affected by the speed of the Earth's motion.

null result: an experimental result that does not support the initial hypothesis

- More recent experiments also provide strong support for Einstein's postulates.
- Modern versions of the Michelson–Morley experiment (e.g. Chen 1997) have shown, to an accuracy of one part in 10^{17}, that the speed of light parallel and perpendicular to the Earth's orbital motion is equal.
- Experiments that involve measuring the speed of gamma rays emitted from the decay of subatomic particles show that, even when the subatomic particles are travelling near the speed of light, the gamma rays travel at the speed of light with respect to an observer at rest in the laboratory frame.
- A vast number of astronomical observations also support the postulates. For example, the light from each star in a binary pair orbiting each other reaches the Earth travelling at the speed of light. The relative motion of the star towards or away from the Earth changes the frequency of the light but does not affect the velocity of the light. Supernovae have also been used to test Einstein's postulates. The decay products of supernovae explosions can travel at 3% of the speed of light, yet the light reaches the Earth from matter ejected towards the Earth and perpendicular to the line of sight from the Earth at the same time. This is true for supernovae experiments at a wide range of distances from the Earth. If the velocity of the source changed the speed of the light, the light emitted from particles moving towards the Earth would reach to the Earth faster.
- There have been many preliminary reports about observations that contradict Einstein's predictions but further study has, to date, always revealed problems with the analysis of the date or errors that were underestimated.
- The theory of special relativity that was based on the postulates has been tested over a wide range of conditions to ever-increasing levels of precision and to date its predictions have always been verified. The success of the theory of special relativity provides strong evidence for the postulates upon which the theory is based.
- Physicists today agree that there are no observations that directly contradict either postulate. However, like all current scientific knowledge, the theory will continue to be supported and used by scientists unless some scientifically verifiable evidence against the postulates comes to light.

KEY QUESTIONS

1. **What were the two postulates Einstein proposed to justify his special theory of relativity?**
2. **What was the purpose and outcome of the Michelson–Morley experiment?**
3. **What evidence do we have to support or deny Einstein's postulates?**

Answers p. 166

2 Time dilation and length contraction

» Students investigate the evidence, from Einstein's thought experiments and subsequent experimental validation, for time dilation $t = \frac{t_0}{\sqrt{\left(1 - \frac{v^2}{c^2}\right)}}$ and length contraction $l = l_0\sqrt{\left(1 - \frac{v^2}{c^2}\right)}$, and analyse quantitatively situations in which these are observed, for example:

- observations of cosmic-origin muons at the Earth's surface
- atomic clocks (Hafele–Keating experiment)
- evidence from particle accelerators
- evidence from cosmological studies.

Time dilation

➔ Einstein's two postulates were inconsistent with the accepted notion that time and distance were absolute quantities. Einstein showed that observers in reference frames moving with respect to one another experienced different realms of space and time. Only by measuring different lengths and times could the observers measure light to be travelling with the same velocity in each frame. Einstein used **thought experiments** to derive quantitative relationships between the time and lengths measured by observers in different frames of reference.

thought experiment: a theoretical rather than a physical experiment designed to extend and investigate the implications of a theory; often used when the technology is not available to carry out the proposed experiments

➔ To relate a time interval in one reference frame to a time interval in another frame we can revisit one of Einstein's famous though experiments. Consider the train carriage shown in Figure 11.3. A clock is set up in the carriage that measures time using a pulse of light going from the bottom to the top of the carriage and back. An observer in the train sees the light go up and back as shown in Figure 11.3**a** and measures the time t_0 for the beam to make the round trip. If the height of the carriage is h, the observer in the carriage will find:

$h = ct_0/2$

If the train was travelling horizontally with a constant velocity v, the observer in the train would be at rest with respect to the clock and the relationship above would still be correct. This is because moving at constant velocity will be identical to being at rest as both are inertial reference frames.

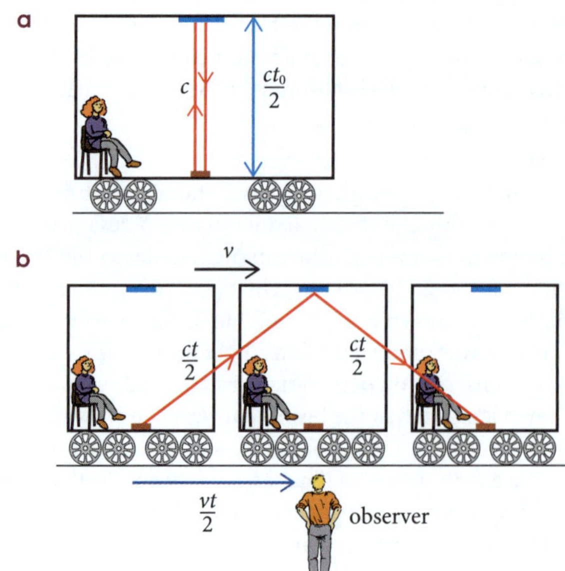

Figure 11.3 Light clock as seen by observer **a** at rest with respect to the clock and **b** light clock moving with respect to the observer

➔ Now consider an observer outside the train, as shown in Figure 11.3**b**. This observer sees the beam move through a longer distance during the time interval than the observer in the train. But both must agree on the speed of light and hence the time interval (t) for the observer outside the train will be different to the time interval (t_0) for the observer inside the train. For the observer outside the train, the length of the path taken by the light to reach the top of the carriage will be:

$Length = ct/2$

Figure 11.4 shows how we can relate the clocks in each reference frame to the distance travelled by the train in the time t measured by the outside observer (the inside observer does not know they are moving).

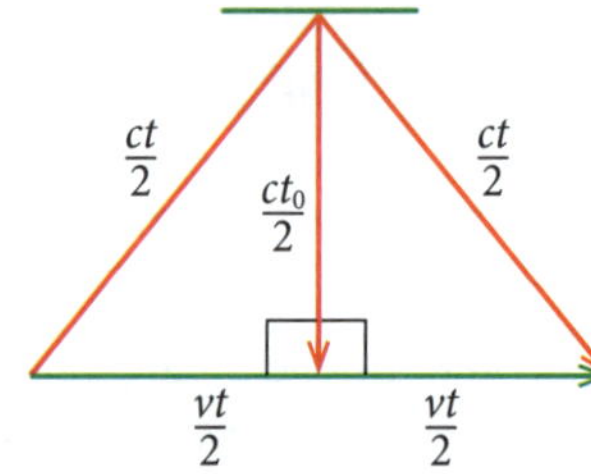

Figure 11.4 Relationship between the light path observed by the observer **a** in the train and the observer **b** outside the train

Now applying Pythagoras's theorem to either of the right-angled triangles shown in Figure 11.4:

$(ct/2)^2 = (ct_0/2)^2 + (vt/2)^2$

$c^2t^2 = c^2{t_0}^2 + v^2t^2$

Rearranging we get:

$$t = \frac{t_0}{\sqrt{\left(1 - \frac{v^2}{c^2}\right)}}$$

As the velocity of the train will always be less than the speed of light, the denominator will always be less than 1 and hence t will always be greater than t_0. When one second passes for the observer at rest outside the train, less than one second passes for the person in the train. This means that a person at rest outside the train will see time passing more slowly for the person on the moving train. The equation tells us that when a clock moves with respect to an observer, the moving clock will always run slower than the observer's clock. But it is not just the clocks that are different, it is time itself. We call the effect **time dilation**.

time dilation: the reduction in the rate at which time passes in a frame moving with respect to an observer

- Now, because Einstein said there was no such thing as an absolute rest frame, the person travelling on the train would consider themselves to be at rest and the observer on the ground outside would be moving with respect to them. The observer inside the train therefore sees time running slower for the observer outside the train. The observer outside the train also sees time running slower for the observer inside the train. This at first seemed crazy and was highlighted by a famous thought experiment called the *twin paradox*.
- In this thought experiment, shown in Figure 11.5, one twin stays on Earth while the other twin goes on a long, very high-speed journey into space. They both see each other's clocks run slowly as they both see the other twin's clock moving with respect to them. But when the flying twin returns to Earth, would he be younger or older than the twin who stayed at home? Which twin's clock has been running slowly? The answer is that the twin that has been to space will return as a younger man than the twin who stayed at home. The reason is beyond special relativity and is to do with the acceleration experienced by the twin that went into space. The twin that accelerated is the one that moved into a different realm of space-time. The full explanation of the twin paradox was eventually provided by Einstein when he later generalised the special theory of relativity to accelerating (non-inertial) reference frames.
- When Einstein proposed the theory of time dilation in 1905 there was no experimental evidence for the theory but since that time a large amount of experimental evidence has been collected that supports the concept. For example:
  - Subatomic particles called **muons** are created by cosmic ray collisions in the Earth's upper atmosphere. In the laboratory, muons are found to decay spontaneously with a half-life of 1.5 μs. The muons created in the upper atmosphere travel at $0.98c$, but with a half-life of 1.5 μs they would not be expected to reach the Earth's surface. However, because the muon is moving so fast, its time is running slower than it is for an earthbound observer by a factor of five and for this reason most muons reach the surface.
  - In 1971 two American physicists, Hafele and Keating, conducted a direct test of time dilation. They took four caesium-beam atomic clocks aboard an airliner and flew around the world twice from west to east and then from east to west. When the clocks aboard the plane were compared with the Earth-bound clocks after the journey, the clocks on the plane read a different time to the clocks left in Washington. The time difference agreed with the predictions for such a journey from time dilation and the general theory of relativity (gravity slows time). This experiment was repeated with more accurate atomic clocks in 2010 and the results were again consistent with the predictions of special and general relativity.
  - With the development of high-speed particle accelerators, physicists have been able to test time dilation by measuring the increase in the lifetimes of subatomic particles as a function of the speed of the particles. Again these experiments have confirmed the predictions of the theory of special relativity.

muon: a short-lived subatomic particle that can be created in the upper atmosphere by cosmic ray collisions

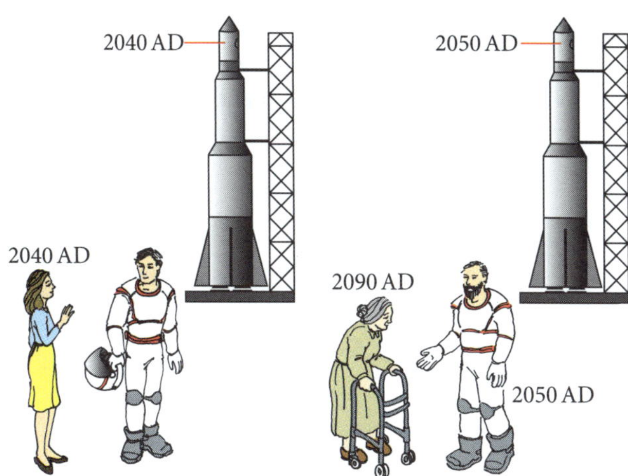

Figure 11.5 An astronaut leaving and returning from space

Length contraction

- To reconcile time dilation with distance travelled, space must contract parallel to the direction of the motion.
- Consider a spacecraft travelling to a star at a distance L_0 as measured from the Earth, illustrated in Figure 11.6.

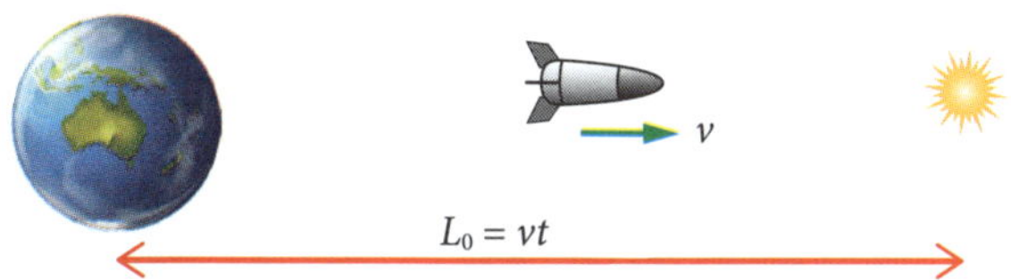

Figure 11.6 Spacecraft travelling from the Earth to a star a distance L_0 from the Earth

➔ If it takes a time t to reach the star as measured by an observer on the Earth, the velocity of the craft would be $v = L_0/t$. But because of time dilation, time runs slower for the space travellers and they reach the star in the lesser time t_0, where:

$$t_0 = t\sqrt{\left(1-\frac{v^2}{c^2}\right)}$$

The distance travelled by the space traveller (L) to the star is therefore given by:

$L = vt_0$

But as $v = L_0/t$:

$L = L_0 t_0/t$ and as $t_0/t = \sqrt{\left(1-\frac{v^2}{c^2}\right)}$

$$L = L_0\sqrt{\left(1-\frac{v^2}{c^2}\right)}$$

where L_0 is the length measured for an observer at rest and
L is the contracted length in the direction of motion in a frame moving with a velocity v with respect to the observer

➔ From the space traveller's frame of reference, the stars are moving rapidly and length contraction causes the distance between the stars to shrink in the direction of the motion. The star map they brought from Earth is no longer relevant as the motion of the stars past them causes space to contract in the direction they are moving. We call this effect **length contraction**. It applies to space and everything in space. For example, if the spaceship had a length L_0 when measured at rest on the launch pad it would have a shorter length L if it was measured from Earth when it was travelling at a velocity v with respect to the Earth. Figure 11.7 illustrates this idea. Because motion is relative, astronauts on the spaceship would still measure the length of the ship to be L_0 and would see the observer moving with respect to them and hence the width of the observer would decease as the craft went faster.

length contraction: the contraction of length in the direction of motion in a frame moving with respect to an observer

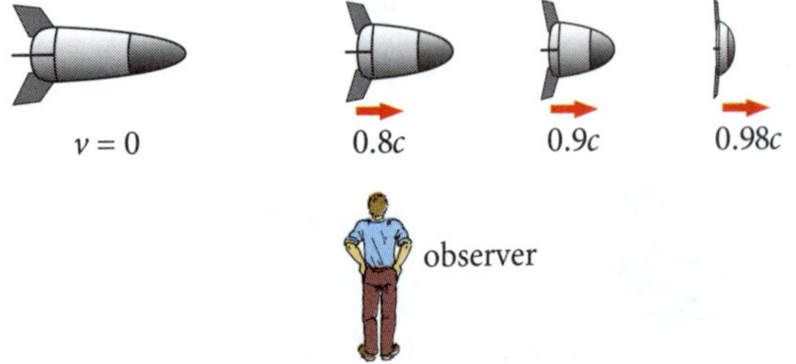

Figure 11.7 Length contraction of a spacecraft travelling at different speeds

➔ We have seen that length contraction and time dilation occur simultaneously in each frame of reference. Muons reaching the surface of the Earth also illustrate this idea. From the Earth-observer's point of view, time dilation increases the half-life of the muon, allowing it to reach the Earth's surface. But from the muon's frame of reference the depth of the atmosphere is reduced by length contraction as the atmosphere rushes past the muon.

EXAMPLE 1

Muons are produced in cosmic ray collision 1.5 km above the Earth's surface. Cosmic rays have very high energies and muons created by cosmic ray collisions travel with a velocity of 0.99c. Experiments with muons in the laboratory on Earth show that they decay into an electron and two neutrinos with a half-life of 2.2 μs.

a What is the half-life of a muon produced in the upper atmosphere and how far could a muon travel through the atmosphere:
  i as measured in the muon's frame of reference?
  ii as measured by an observer at rest on the Earth's surface?

b Explain how an observer on the ground and an observer in the reference frame of the muon account for muons reaching the surface of the Earth.

When a frame of reference moves with respect to an observer, time runs slower and lengths contract in the moving frame

Answer:

a i In the muon's frame of reference, the muon is at rest and the half-life is 2.2 μs.

The distance travelled by the muon would be:

Distance = velocity × time

$= 0.99(3 \times 10^8) \times 2.2 \times 10^{-6} = 653$ m

ii The muon is moving with respect to the Earth observer and hence when 2.2 μs passes for the muon, the Earth observer will experience more time passing. Apply the time dilation equation:

$$t = \frac{t_0}{\sqrt{\left(1-\frac{v^2}{c^2}\right)}} = \frac{2.2}{\sqrt{\left(1-\frac{(0.99c)^2}{c^2}\right)}} = 15.6\ \mu s$$

The distance travelled by the muon as measured by the Earth observer will be:

Distance = velocity × time

$= (0.99)(3 \times 10^8) \times 15.6 \times 10^{-6} = 4633$ m

b An observer on Earth sees the muon travelling very fast and time dilation causes the muon's half-life to be increased to 15.6 μs, as measured by the Earth-bound observer. The muon will easily travel the 1.5 km to the Earth's surface in this time.

An observer travelling with the muon sees themselves at rest and the Earth and atmosphere rushing towards them at 0.99*c*. This causes space to contract in the direction of motion. Applying the length-contraction equation:

$$L = L_0 = \sqrt{\left(1 - \frac{v^2}{c^2}\right)} = 1500\sqrt{\left(1 - \frac{(0.99c)^2}{c^2}\right)} = 212 \text{ m}$$

As the muon can travel 653 m in its frame of reference, it will easily reach the Earth's surface.

EXAMPLE 2

A spacecraft has a length of 40 m on the launch pad. The spacecraft travels at 0.95*c* to a planet that orbits a star, 10 light years from the Earth.

a **Find the length of the craft when it is travelling at 0.95c as measured by:**
  i **an astronaut in the ship**
  ii **an observer on the Earth.**

b **How much time would have passed for an observer on the Earth when the spacecraft reached the planet?**

c **How much time would have passed for an observer in the spaceship when it reaches the planet?**

d **How far is the planet from the Earth as measured by an observer in the spaceship?**

When an object moves with respect to an observer, time passes more slowly for the moving object than for the observer and the moving object contracts in its direction of motion

Answer:

a **i** An observer in the spacecraft is at rest with respect to the ship and hence will measure the length to be 40 m.

  ii An observer on the Earth will see the ship contracted in the direction of its motion and measure the length to be:

$$L = L_0 = \sqrt{\left(1 - \frac{v^2}{c^2}\right)} = 40\sqrt{\left(1 - \frac{(0.95c)^2}{c^2}\right)} = 12.5 \text{ m}$$

b For an observer on Earth:

Time = distance/velocity = $\frac{10}{0.95}$ = 10.53 years

c Because the spacecraft is moving at 0.95*c* with respect to the Earth, less time will have passed on the spacecraft than on the Earth. The elapsed time on the craft will be:

$$t_0 = t\sqrt{\left(1 - \frac{v^2}{c^2}\right)} = 10.53\sqrt{\left(1 - \frac{(0.95c)^2}{c^2}\right)} = 3.29 \text{ years}$$

d The observers in the spacecraft will see the universe rushing towards them at 0.95*c* and hence space will contract in their direction of movement by:

$$L = L_0\sqrt{\left(1 - \frac{v^2}{c^2}\right)} = 10\sqrt{\left(1 - \frac{(0.95c)^2}{c^2}\right)} = 3.12 \text{ light years}$$

➔ KEY QUESTIONS

4 **Why did Einstein use thought experiments?**

5 **What is length contraction and time dilation?**

6 **Explain why an observer does not see length contraction and time dilation in their own frame of reference.**

7 **What observations and experiments have supported Einstein's prediction of length contraction and time dilation?**

8 **How would a photon perceive time and space?**

Answers ➲ p. 166

3 Relativistic momentum

» Students describe the consequences and applications of relativistic momentum with reference to:

- $p_v = \frac{m_0 v}{\sqrt{\left(1 - \frac{v^2}{c^2}\right)}}$
- the limitation on the maximum velocity of a particle imposed by special relativity.

➔ Consider the thought experiment shown in Figure 11.8. Here two flat, smooth carriages, each carrying a mass of 1 kg, approach each other at a relative speed of *v*. As they pass each other a weightless spring is attached to the masses for a moment and then discarded. An equal force is applied to each mass by the spring for the same time and therefore each mass gains the same momentum, perpendicular to the carriages' velocity. After the spring is removed, the observer on each carriage measures the velocity of their 1 kg mass and both agree that their mass moves 1 m across the carriage in 1 s. Hence both find that the momentum of the mass is $p = 1 \text{ kgms}^{-1}$.

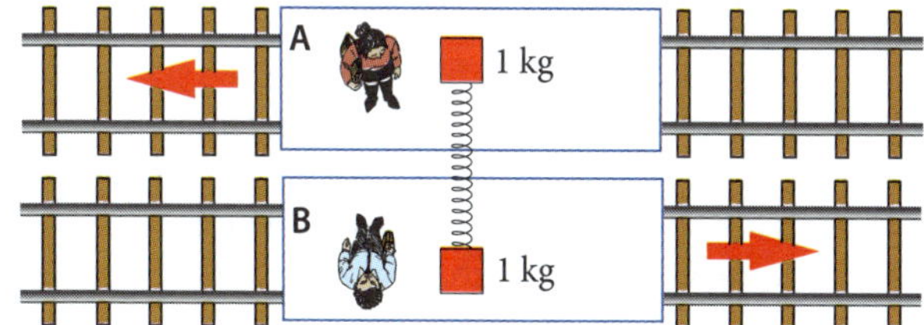

Figure 11.8 Thought experiment to determine relative velocity

But when observer A looks at B making the measurement they notice that, because of time dilation, B's clock is running slowly and hence when B measures 1 s, more than 1 s has passed for observer A. Observer A concludes that time dilation due to the relative motion of the carriages causes the velocity measured by observer B not to be 1 ms^{-1} but the smaller velocity:

$$v = 1 \times \sqrt{\left(1 - \frac{v^2}{c^2}\right)} ms^{-1}$$

Both observers agree that momentum should always be conserved and conclude that the momentum must be affected by the relative motion of the observers. They reason that if the motion caused the mass to increase by the same factor as the velocity decreased, both observers would agree that momentum was conserved. The observers conclude that the relative motion must increase the mass by:

$$m_v = \frac{m_0}{\sqrt{\left(1 - \frac{v^2}{c^2}\right)}}$$

where m_0 is the mass measured in the frame of the body (**rest mass or invariant mass**) and
m_v is the mass of the object moving with velocity v with respect to the observer

rest mass or invariant mass: the mass as measured in the rest frame of the object

This change in mass means that momentum is a relative, rather than an absolute, quantity. The momentum in the moving frame of reference will be related to the momentum in the rest frame (m_0v) by:

$$p_v = m_v v = \frac{m_0 v}{\sqrt{\left(1 - \frac{v^2}{c^2}\right)}}$$

where p_v is the momentum of an object moving at velocity v with respect to an observer and m_0 is the mass measured in the frame of the object (i.e. the rest mass)

We call this momentum (p_v) the **relativistic momentum** to distinguish it from the Newtonian momentum (m_0v).

relativistic momentum: the actual momentum of a moving body, as opposed to Newtonian momentum, which is a low-speed approximation of a body's momentum; as a body approaches the speed of light, the relativistic momentum approaches infinity because the effective mass approaches infinity

- We see from the relationship above that when an object moves at speeds much smaller than the speed of light it will have a momentum $p = mv$, but if it moves at an appreciable fraction of the speed of light (relativistic speeds) it will have a greater momentum. Newtonian mechanics works well at Earthly speeds but at extremely high velocities we must modify Newtonian mechanics to take into account the relative nature of space, time and momentum.
- Note that as the speed of an object approaches the speed of light, the denominator of the relativistic momentum equation approaches zero and the momentum of the object would approach infinity. As we saw in the Year 11 course, $\Delta p = Ft$, and an infinite amount of force would therefore be required to accelerate an object to the speed of light. This means it is impossible for a material object to travel at the speed of light with respect to an observer. The speed of light is therefore the speed limit of the universe. (The only exception to this rule occurs when the space between objects expands. Curved space is beyond the HSC syllabus but interested students may wish to pursue this idea.)
- Length contraction and time dilation also show us that the speed of light is a natural speed limit. If we watched an object reaching the speed of light, its time would slow to zero and its length in the direction of motion would contract to a plane. From the object's frame of reference, the universe would be rushing towards the object so fast that time in the universe would approach zero and the universe would shrink to a plane in the direction of the motion. It does not make sense to travel faster than light, because you would have no space to travel into. This is how a photon travelling at the speed of light would experience the universe. For the photon, time has no meaning and the universe has shrunk to a plane in the direction of the photon's motion. The photon is at the start of its journey and at the end at the same time. We see the photon being created and travelling at the speed of light until it reaches its destination and disappearing as it gives its energy to atom. But for the photon, no time passes as it is simultaneously created and annihilated at one point in time and space.
- The relativistic momentum equation is confirmed every time there is a collision between subatomic particles in a particle accelerator. The large hadron collider can accelerate protons to 0.999 999 991c, which is 3 ms^{-1} slower than the speed of light. The relativistic momentum of a proton travelling at this speed is almost 8000 times greater than the momentum predicted by Newton's momentum equation, and when they collide with another particle, it is as if they had a mass 8000 times greater than a proton at rest.

EXAMPLE 3

An electron in a particle accelerator is accelerated to a velocity of 0.98c.

a Find the relativistic momentum of the electron.
b Find the ratio of the relativistic momentum to the Newtonian momentum.

Answer:

Newtonian momentum (m_0v) can only be applied to objects moving much slower than light

a Applying the relativistic momentum equation:

$$p_v = \frac{m_0 v}{\sqrt{\left(1 - \frac{v^2}{c^2}\right)}}$$

And hence:

$$p_v = \frac{m_0 v}{\sqrt{\left(1 - \frac{v^2}{c^2}\right)}} = \frac{9.109 \times 10^{-31} \times 0.98 \times (3 \times 10^8)}{\sqrt{(1 - 0.98^2)}}$$

$= 1.35 \times 10^{-21}$ kgms^{-1}

b The Newtonian momentum is given by:

$p = m_0 v = 9.109 \times 10^{-31} \times 0.98 \times 3 \times 10^8$

$= 2.68 \times 10^{-22}$ kgms^{-1}

Hence the ratio of relativistic momentum/Newtonian momentum is:

$$\frac{1.35 \times 10^{-21}}{2.68 \times 10^{-22}} = 5.0$$

Thus the relativistic momentum for an object travelling at $0.98c$ is five times greater than the Newtonian momentum.

➔ KEY QUESTIONS

9 **Compare the Newtonian and relativistic momentum of a moving object.**

10 **Explain why the speed of light is the limiting velocity for objects with mass.**

11 **What experimental evidence supports the concept of relativistic momentum?**

Answers ➲ p. 166

4 Mass–energy equivalence

» Students use Einstein's mass–energy equivalence relationship $E = mc^2$ to calculate the energy released by processes in which mass is converted to energy, for example:
- the production of energy by the Sun
- particle–antiparticle interactions, e.g. positron–electron annihilation
- the combustion of conventional fuel.

➔ We have seen that the speed of light is the maximum speed an object can achieve and that the relativistic mass increases as an object approaches the speed of light. With this knowledge we can use a thought experiment to find a relationship between mass and energy.

➔ Consider a spacecraft that is pushed with a constant force F. The force moves the craft over a distance and hence does work on the spacecraft that increases the craft's kinetic energy. At non-relativistic speeds the kinetic energy increases because the speed of the spacecraft increases. But when the spacecraft approaches the speed of light, the craft can no longer go faster and the only way the kinetic energy can increase is by increasing the mass of the spacecraft. Hence when we get close to the speed of light the energy transferred to the craft is transformed into mass.

➔ Imagine the spacecraft has almost reached the speed of light but the engines are still applying a force to the craft. To simplify the calculations, consider what happens to the craft in unit time intervals of one second. The energy put into the craft will be equal to the work done on the craft in one second. Now, as the craft is essentially travelling at the speed of light it will travel a distance $s = ct = c$ (as $t = 1$ s) each second and the energy transferred to the craft will be:

Energy transferred per second = work done in one second

$$E = Fs = Fc$$

Now the change of momentum of the spacecraft would be given by:

$\Delta p = m_f v - m_i u = Ft$

And as the velocity stays constant, $v = u = c$ and the change in the momentum per second ($t = 1$ s) will be:

$\Delta p = m_f c - m_i c = \Delta mc = F$

where m_i is the initial mass and

m_f is the final mass

Now $m_f - m_i = \Delta m$ is the amount of energy converted to mass and by substituting this expression for force (i.e. $F = \Delta mc$) into the equation for energy above we obtain:

$E = (\Delta mc)c = \Delta mc^2$

This is often simplified to:

$$E = mc^2$$

where E is the energy
m is the mass and
c is the speed of light

This is Einstein's famous **mass–energy equivalence** equation. Relativity tells us that mass is a form of energy and when mass is converted to energy or energy to mass the relationship derived above is obeyed. For example, this equation tells us that 1 kg of mass is equivalent to 9×10^{16} J of energy.

mass–energy equivalence: a relationship that shows that mass is a form of energy and that the energy contained in a body is proportional to the mass of the body; the SI unit of proportionality is the square of the speed of light

- In 1905, when Einstein first published his famous mass–energy equation, there was no direct evidence available to support the proposal. Einstein suggested radioactivity might be an example of mass being converted to energy and this was later verified by experiment.
- A very large number of experimental observations have now been made with increasing precision and all have verified Einstein's mass–energy equivalence equation. Today we accept that all chemical and nuclear reactions are accompanied by the release or absorption of energy and involve mass being converted to energy or energy to mass, respectively.
- The change in mass involved in chemical reactions is tiny because chemical reactions do not produce very much energy per kilogram of reactant. Nuclear reactions produce millions of times more energy per kilogram than chemical reactions and the change in mass can approach 1% of the mass of the reactants.
- When a chemical or nuclear reaction releases energy, mass is lost, and when a reaction stores energy, mass increases. Some examples of mass–energy equivalence include:
  - Our Sun and all other stars convert mass to energy in their core by fusing small nuclei together to form larger nuclei.
  - Nuclear power reactors use the fission of large nuclei to convert mass to energy.
  - The combustion of chemical fuels converts mass to energy by rearranging chemical bonds.
  - A charged battery is slightly more massive than an uncharged battery because of the energy stored in chemical bonds in the battery.
  - When a spring is compressed or an elastic band stretched, potential energy is stored in the stretched chemical bonds and the spring or band is slightly more massive than when it is relaxed.
  - Increasing the temperature of an object increases the internal energy of the object and consequently mass increases slightly with temperature.
  - We have already seen that the measured mass of a body increases with velocity and hence an increase in kinetic energy is accompanied by an increase in mass.
  - A spinning body has kinetic energy and hence an increased mass. The faster the body is spun the more massive the body becomes.
  - In high-speed particle accelerators, collisions between particles often result in the creation of more and different subatomic particles as energy is converted into mass.
- Because of the c^2 term in Einstein's equation, many of the mass changes listed above are minute, but modern measuring techniques have enabled all these predictions to be tested and to date there is no credible evidence that contradicts Einstein's equation of mass–energy equivalence.
- We will see in the following examples that the energy released per kilogram of fuel (i.e. the **specific energy**) is determined by the fraction of the mass that is converted to energy when the fuel is used. Table 11.1 lists the energy density and percentage of mass converted into energy for some common fuels.

specific energy: the amount of energy that can be produced per unit mass of the fuel (i.e. specific energy is measured in J kg^{-1})

Table 11.1 Specific energy and percentage of mass converted to energy for a range of fuels

Fuel	Uses	Specific energy (J kg^{-1})	Percentage of mass converted to energy (%)
Uranium fission	Power generation	8.1×10^{13}	0.1
Deuterium fusion	Stars and experimental fusion reactors	5.8×10^{14}	0.6
LPG (chemical combustion)	Heating and transport	4.6×10^{7}	5×10^{-8}
Diesel (chemical combustion)	Heating and electricity generation	4.8×10^{7}	5.3×10^{-8}
Petrol (chemical combustion)	Transport	4.6×10^{7}	5.1×10^{-8}
Fat (chemical oxidation)	Nutrition	3.7×10^{7}	4.1×10^{-8}
Coal (chemical combustion)	Generating electricity	3.0×10^{7}	3.3×10^{-8}

EXAMPLE 4

Given that the Sun is 150 million kilometres from the Earth and the intensity of the Sun's radiation is 1361 W m^{-2} at the top of the Earth's atmosphere, find:

a the total amount of energy produced by the Sun per second

b the amount of mass converted to energy by the Sun per second.

The sun radiates the same amount of radiant power in every direction

Answer:

a We can find the total energy produced by the Sun per second by multiplying the number of square metres in the surface area of a sphere of radius 150 million kilometres by the energy that passes through each square metre.

The surface area of a sphere is $4\pi r^2$.

Hence to find the energy radiated per second:

Energy radiated per second

$= 4\pi r^2 \times$ energy per m^2

$= 4 \times \pi \times (150 \times 10^9)^2 \times 1361$

$= 3.85 \times 10^{26}$ W

b Applying Einstein's equation:

$E = mc^2$ and hence

$$m = E/c^2 = \frac{3.85 \times 10^{26}}{(3 \times 10^8)^2}$$

$= 4.28 \times 10^9$ kgs^{-1} $= 4.28 \times 10^6$ tonnes per second

EXAMPLE 5

A positron is the anti-particle of the electron. A positron has the same mass as an electron ($m = 9.109 \times 10^{31}$ kg) but the opposite charge. The electrostatic attraction between electrons and positrons bring them together and when they meet their mass is totally converted into energy, in the form of two gamma rays. (You may assume that the kinetic energy of the particles before they collide is negligible.)

a **Find the energy produced in an electron–positron annihilation.**

b **If the emitted gamma rays in an electron–positron annihilation had equal energy, find the wavelength of the emitted gamma rays.**

We can use Einstein's mass–energy equivalence equation whenever mass in converted to energy

Answer:

a Applying Einstein's equation:

$E = mc^2 = (2 \times 9.109 \times 10^{-31})(3 \times 10^8)^2 = 1.64 \times 10^{-13}$ J

b As the energy is shared between the two gamma ray photons, they will have an energy of:

$$E = \frac{1.64 \times 10^{-13}}{2} = 8.2 \times 10^{-14} \text{ J}$$

Now we can apply Planck's equation $E = hf = hc/\lambda$ and hence:

$\lambda = hc/E$

$$\lambda = \frac{6.626 \times 10^{-34} \times 3 \times 10^8}{8.2 \times 10^{-14}} = 2.42 \times 10^{-12} \text{ m}$$

EXAMPLE 6

Consider the fusion reaction:

$H_1^2 + H_1^2 = He_2^4$

Given the mass of H_1^2 (deuterium) is 3.3435×10^{-27} kg and the mass of helium is $6.644\,24 \times 10^{-27}$ kg, find:

a **the amount of mass converted to energy in the reaction**

b **the energy released per fusion**

c **the percentage of the initial mass lost in the reaction.**

Answer:

If mass is lost in a reaction it must have been converted to energy

a Mass lost in reaction:

Δm = mass of products – mass reactants

$= 6.644\,24 \times 10^{-27} - (2 \times 3.3435 \times 10^{-27})$

$\Delta m = -4.276 \times 10^{-29}$ kg

Hence the mass lost is 4.276×10^{-29} kg.

b Applying Einstein's equation:

$E = mc^2 = 4.276 \times 10^{-29} \times (3 \times 10^8)^2 = 3.85 \times 10^{-12}$ J

c The percentage of the initial mass converted to energy is:

$$\Delta m/(\text{mass of reactants}) = \frac{(4.276 \times 10^{-29})}{(2 \times 3.3435 \times 10^{-27})} \times 100$$

$= 0.64\%$

EXAMPLE 7

The chemical combustion of petrol releases 46 MJkg^{-1}.

a **Find the percentage of the initial mass of petrol converted to energy.**

b **Compare the energy released per kg of fuel in combustion, fusion and matter/antimatter annihilation.**

Recall that 0.6% and 100% of the initial mass is converted to energy in fusion and matter–antimatter annihilation, respectively

Answer:

a Applying Einstein's equation to find the mass lost:

$$E = mc^2 \text{ and hence } m = E/c^2 = \frac{46 \times 10^6}{(3 \times 10^8)^2}$$

$= 5.1 \times 10^{-10}$ kg

$$\text{Percentage mass lost} = \frac{5.1 \times 10^{-10}}{1} \times 100 = 5.1 \times 10^{-8}\%$$

b 100% of the mass is converted into energy in matter–antimatter annihilation, approximately 0.6% in nuclear fusion and 5×10^{-8}% in combustion.

Note in terms of energy produced per kilogram, fusion produces about $\frac{0.6}{5.1 \times 10^{-8}} = 1.2 \times 10^7$ times more energy than combustion and matter–antimatter annihilation produces $\frac{100}{0.6} = 166$ times more energy than fusion.

→ KEY QUESTIONS

12 **What is mass–energy equivalence?**

13 **Give three examples of natural phenomena that demonstrate mass–energy equivalence.**

14 **Fuels convert mass to energy. How do chemical reactions, nuclear reactions and matter–antimatter reactions compare in terms of the percentage of mass converted to energy?**

Answers ➲ p. 166

CHAPTER SYLLABUS CHECKLIST

Are you able to answer these questions from the syllabus for this chapter? Tick each question as you go through the checklist if you are able to answer it. If you cannot answer a question, turn to the relevant page in the study guide to find the answer. For NESA key word meanings, go to www.educationstandards.nsw.edu.au and search 'key words'.

	FOR A COMPLETE UNDERSTANDING OF THIS TOPIC:	PAGE NO.	✓
1	Can I state the two postulates upon which Einstein based his theory of special relativity?	153	
2	Can I define an inertial frame of reference?	153	
3	Can I evaluate the evidence for and against Einstein's postulates?	153–155	
4	Can I explain why light travelling in an ether was inconsistent with Einstein's postulates?	154	
5	Can I explain why Einstein used thought experiments?	156	
6	Can I perform calculations using Einstein's equations for time dilation and length contraction?	156–159	
7	Can I explain how time dilation and length contraction occur simultaneously in frames of reference to reconcile the observations made by observers in different frames?	157–158	
8	Can I explain how special relativity accounts for short-lived muons created in the upper atmosphere reaching the ground?	157–158	
9	Can I outline the atomic clock experiments, observations made in particle accelerators and cosmological studies that support the special theory of relativity?	157	
10	Can I perform calculations using the relativistic momentum?	160–161	
11	Can I explain why special relativity imposes a maximum speed for any particle?	160	
12	Can I explain how it is possible for momentum and energy to change without the velocity changing when particles approach the speed of light?	160	
13	Can I perform calculations using Einstein's mass–energy equivalence relationship?	162–163	
14	Can I explain and calculate mass–energy changes in the Sun, in particle–antiparticle annihilation and in the chemical combustion of fuels?	162	

HSC EXAM-TYPE QUESTIONS

Objective-response questions

(1 mark each)

1 **Which of the following statements best defines an inertial frame of reference?**

A An inertial frame is one in which the speed of light is constant.

B An inertial frame is one in which Newton's first law is obeyed.

C An inertial frame is a set of coordinates that can be used to measure the position and velocity of objects in the frame.

D An inertial frame is one in which gravity is always zero.

2 **Why did the Michelson–Morley experiment produce a null result?**

A The experiment was not accurate enough to measure the ether wind.

B The ether moved with the Earth.

C Length contraction of the device made the speed of light appear to be constant in all directions.

D The speed of light was not affected by the velocity of the Earth.

3 **A rocket travelling at 0.9*c* passes the Earth and an observer on the Earth compares his clock with a clock on the rocket ship. What would the Earth observer notice?**

A When one second passed on the Earth, less than one second passed on the rocket.

B When one second passed on the Earth, more than one second passed on the rocket.

C Both clocks agreed on the time that passed because the rocket was not travelling at the speed of light.

D The Earth observer could not compare clocks because at this speed the rocket would look like an infinitely thin plane.

4 **In a particle accelerator, one proton X is moving at 0.45*c* and another proton Y is moving at 0.9*c*. Which of the following best describes how the protons would appear to an observer outside the accelerator?**

A Particle Y would have four times as much kinetic energy as particle X.

B An observer in the lab would measure the time passing for particle X at half the rate as time passed for particle Y.

C Particle Y would have more than twice the momentum of particle X.

D Particle Y would have less than twice the momentum of particle X.

5 **Under what circumstances will a nuclear reaction always release energy?**

A When two nuclei fuse together.

B When a collision results in a nucleus splitting into two smaller nuclei.

C When the combined mass of the products of the reaction is greater than the combined mass of the reactants.

D When the combined mass of the reactants of the reaction is greater than the combined mass of the products.

Extended-response questions

6 **The special theory of relativity is based on two postulates.**

a **State the two postulates.** (2 marks)

b **Explain why the ether model of light transmission conflicted with both postulates.** (2 marks)

7 **The special theory of relativity was initially a theory based on postulates and developed using thought experiments.**

a **Why do scientists use thought experiments?** (3 marks)

b **Describe one of the thought experiments used by Einstein and outline the prediction made by the experiment.** (2 marks)

c **Outline, in terms of the theory of relativity, why it is impossible for an object with a rest mass to be accelerated to the speed of light.** (2 marks)

8 **The half-life of a subatomic particle called a π-meson at rest is 2.6×10^{-8} s. In a high-speed particle accelerator, a π-meson was measured to have a half-life of 7.8×10^{-8} s.**

a **Explain why the two half-lives are different.** (2 marks)

b **Find the velocity of the π-meson in the particle accelerator.** (2 marks)

9 **A space traveller moving at 0.95*c* with respect to the Earth takes six years (space-traveller time) to travel to a nearby star.**

a **What is the distance to the star in light years as measured by the space traveller?** (1 mark)

b **When six years had passed for the space traveller how much time would have passed on the Earth?** (1 mark)

c **What is the distance to the star as measured by an observer on the Earth?** (1 mark)

10 **In a process called *pair production* a gamma ray can spontaneously change into an electron–positron pair. (Mass of an electron = mass of a positron = 9.109×10^{-31} kg.)**

a **Find the minimum gamma energy required for pair production.** (1 mark)

b **Find the minimum gamma ray frequency required for pair production.** (1 mark)

c **Explain why we can only calculate the minimum gamma ray energy.** (1 mark)

ANSWERS

KEY QUESTIONS

Key questions ➲ p. 155

1 The first postulate is that all inertial frames are equivalent for all the laws of physics and the second postulate is that the speed of light is an absolute constant.

2 The Michelson–Morley experiment was designed to measure the speed of the Earth through the ether. The experiment did not produce the expected result and appeared to show that light travelled at the same speed parallel and perpendicular to the Earth's motion.

3 The existence of the ether would have contradicted both of Einstein's postulates but the concept has since proved to be superfluous. Absolute mass, space and time were also inconsistent with Einstein's postulates but were shown by Einstein and later experiments to be erroneous concepts. Einstein's postulates are supported by Michelson–Morley-style experiments, which show that the speed of light is not affected by the speed of the source or the observer. Because the special theory of relativity is based on the postulates, the success of the theory in explaining experimental observations provides strong support for the underlying postulates. Astronomical observations also support the postulates and after 100 years of experimental testing there appears to be no verifiable evidence contradicting either postulate.

Key questions ➲ p. 159

4 Einstein used thought experiments to investigate the implications of his two postulates and make predictions that could later be tested. At the time, the technology was not available to do the thought experiments discussed by Einstein.

5 Time runs slower (time dilation) and lengths are shorter in the direction of motion (length contraction) in a frame that is moving with respect to an observer.

6 Length contraction and time dilation only occur in frames that are moving with respect to the observer.

7 The predictions of the special theory of relativity have been supported by a large array of observations and experimental evidence, which includes: observations of time dilation and length contraction for muons in the atmosphere and other subatomic particles in particle accelerators; atomic clocks in frames moving with respect to one another; and cosmological observations of binary stars and supernovae.

8 Because the universe is moving towards the photon at the speed of light it would contract to a plane in the direction of the photon's movement. Time would have no meaning for a photon as it would be created and destroyed simultaneously.

Key questions ➲ p. 161

9 The Newtonian momentum is a very good approximation of the momentum of a body at low speeds but increasingly underestimates the magnitude of the momentum as a body approaches the speed of light. The Newtonian momentum would have a finite (incorrect) value for a body travelling at the speed of light, while the actual (relativistic) momentum would be infinite if a body with a rest mass could reach the speed of light.

10 As a body is accelerated towards the speed of light its momentum (and effective mass) increases towards an infinite value. As an infinite force would be required to accelerate an infinite mass it is impossible for a body with a rest mass to be accelerated to the speed of light.

11 Collisions in particle accelerators have confirmed the relativistic momentum equation. The high-speed particles have a far greater momentum than would be predicted by Newton's non-relativistic equation.

Key questions ➲ p. 163

12 Einstein showed that a consequence of his two postulates was that energy could be converted into mass and mass into energy and that the energy contained in a mass (m) was given by $E = mc^2$.

13 Examples of mass being converted into energy include nuclear fusion in the core of the Sun and other stars, radioactivity and exothermic (energy releasing) chemical reactions. Energy converted into mass is demonstrated in high-speed collisions in particle accelerators.

14 Only a tiny fraction (10^{-8}%) of the mass of chemical fuels is converted into energy. A much larger fraction (0.5%) can be converted into energy in nuclear reactions and up to 100% of the mass can be converted into energy by matter–antimatter reactions.

HSC EXAM-TYPE QUESTIONS

Objective-response questions

1 **B**. This is the most general definition of an inertial frame. **A** is incorrect because the speed is constant for all observers. **C** is incorrect as this is a definition of any type of reference frame. **D** is incorrect as an inertial frame is not defined by the absence of a gravitational field. (Indeed, a free-falling object in a uniform gravitational field would be an inertial frame.)

2 **D**. The speed of light is an absolute constant and therefore an experiment based on the speed of light travelling at different speeds in different directions would produce a null result. **A** is incorrect because the experiment was designed to be sensitive enough to measure the ether wind caused by the orbital velocity of the Earth (provided such an ether wind existed). **B** is incorrect because there is no ether. **C** is incorrect because lengths do not contract in the observer's frame of reference; they only change in frames moving with respect to the observer.

3 **A**. Einstein's theory tells us that time runs slower in a reference frame moving with respect to the observer. **B** is incorrect because this would imply that time is passing faster in the moving frame than it does in the observer's frame. **C** is incorrect because time dilation occurs at all speeds not just at the speed of light. **D** is incorrect because the length of the rocket would be reduced to a plane only if it was travelling at the speed of light.

4 **C**. Relativistic momentum increases at a faster rate than velocity at high speeds because the effective mass of the particle increases as the particle travels faster. **A** is incorrect because Newtonian energy increases as the square of the velocity, but relativistic velocity increases at an even faster rate due to the effective mass increasing. **B** is incorrect because time would run slower for particle Y when measured from the laboratory frame because Y is moving faster than particle X. **D** is incorrect because this would be true for Newtonian momentum but not true for relativistic momentum.

5 **D**. Mass must be converted to energy as the reactants are more massive than the products. **A** is incorrect because the fusion of

small nuclei release energy but large nuclei require energy to fuse. **B** is incorrect because sometimes energy will be absorbed in this process, such as when a small nucleus is split in half in a high-speed collision. **C** is incorrect because the mass is increased in the reaction and hence energy is absorbed and converted to mass.

Extended-response questions

6 **EM** This question tests students' understanding of the postulates upon which special relativity is based and why they are inconsistent with the ether theory.

- a First postulate: All inertial frames of reference are equivalent. ✓

  Second postulate: The speed of light is an absolute constant independent of the speed of the source or observer. ✓

- b If light travelled in an ether, the ether would act as an absolute reference frame and this would violate the first postulate. ✓ The motion of other inertial reference frames with respect to the ether could be determined, which would mean they were not identical.

  The second postulate is not consistent with the existence of an ether because the speed of light would no longer be an absolute constant. Observers moving at different speeds with respect to the ether would not agree on the speed of light. ✓ The situation would be like moving observers on Earth measuring different velocities for the speed of sound, due to the observers' motion through the atmosphere.

7 **EM** This question tests students' understanding of thought experiments and how Einstein used them to develop his special theory of relativity.

- a Scientists use thought experiments to investigate the implications and consequences of their theories ✓ and to make predictions that can be tested. ✓ Sometimes thought experiments are used because for various reasons actual experiments cannot be conducted or are not required. ✓ For example, the technology to conduct the experiment may not be available at the time.

- b A description of any of the thought experiments proposed by Einstein could be used and the purpose of the experiment outlined. Two sample answers follow:

  Einstein conducted a thought experiment involving clocks that used the passage of light in a moving reference frame perpendicular to the velocity to predict time dilation. The experiment considered one unit of time as the time it took for a light pulse to travel from the bottom to the top of a carriage and return (see Figure 11.3). ✓ By comparing the times measured using this clock when viewed from two frames of reference, a relationship between the times in each frame was derived. ✓

  OR

  To illustrate the difference between Galilean relativity and Einstein's relativity, Einstein used a thought experiment in which a man travelling at the speed of light looked at his reflection in a mirror held in front of him. ✓ Galilean relativity predicts he would not see his reflection because light would not leave his face. But Einstein's relativity predicts he would see his reflection because all inertial frames are equivalent and the speed of light is not affected by the velocity of the observer. ✓

- c The theory of relativity shows that the momentum of an object with a rest mass approaches infinity as the object is accelerated towards the speed of light. ✓ As an infinite force would be required to give the object an infinite momentum, it is impossible for an object with mass to reach the speed of light. ✓ Alternatively, students could explain how the effective mass of an object approaches infinity as the object approaches the speed of light. Because it would take an infinite force to accelerate an infinite mass it is impossible for an object with mass to be accelerated to the speed of light.

8 **EM** This questions students' understanding of special relativity and their ability to solve the time dilation equation.

- a Einstein's special theory of relativity shows that time runs slower in a reference frame moving with respect to an observer. ✓ In this case the π-meson is moving at very high speed with respect to the observer in the laboratory. ✓ Hence time runs slower for the moving muon than for an observer at rest and consequently the observer at rest will see the moving muon's half-life to be much longer than the half-life of a muon at rest.

- b Applying the equation for time dilation:

  $$t = \frac{t_0}{\sqrt{\left(1-\frac{v^2}{c^2}\right)}}$$ ✓

  Rearranging for the velocity:

  $$v = \sqrt{c^2(1-\frac{t_0^2}{t^2})} = \sqrt{(3\times10^8)^2(1-\frac{(2.6\times10^{-8})^2}{(7.8\times10^{-8})^2})}$$

  $= 2.83 \times 10^8\ \text{ms}^{-1}$ ✓ or $0.943c$

9 **EM** This question tests students' understanding of the relationship between time dilation and length contraction in different frames of reference.

- a The distance will be given by:

  Distance = velocity × time

  = 0.95 × 6 = 5.7 light years ✓

- b More time would have passed on Earth than on the spacecraft. The time elapsed on Earth would be given by:

  $$t = \frac{t_0}{\sqrt{\left(1-\frac{v^2}{c^2}\right)}}$$

  $$t = \frac{6}{\sqrt{\left(1-\frac{(0.95c)^2}{c2}\right)}} = 19.215 \text{ years}$$ ✓

- c The distance travelled as measured by an observer on the Earth would be:

  Distance = velocity × time

  = 0.95 × 19.215 = 18.25 light years ✓

  OR

  The distance seen by the space traveller will be reduced as the astronaut sees the stars moving towards them at $0.95c$. We can use the equation for length contraction to find the distance to the star measured from the Earth:

  $$l_0 = \frac{l}{\sqrt{\left(1-\frac{v^2}{c^2}\right)}}$$

  $$= \frac{5.7}{\sqrt{\left(1-\frac{(0.95c)^2}{c2}\right)}} = 18.25 \text{ light years}$$

10 **EM** This question tests students' understanding of Einstein's mass–energy equivalence equation and how energy can be converted into matter.

- a The minimum energy required will be equivalent to the mass of the particles created and hence:

  $E = mc^2 = (2 \times 9.109 \times 10^{-31})(3 \times 10^8)^2 = 1.64 \times 10^{-13}$ J ✓

- b Applying the Planck equation:

  $E = hf$ and hence

  $$f = E/h = \frac{1.64 \times 10^{-13}}{6.626 \times 10^{-34}} = 2.47 \times 10^{20} \text{ Hz}$$ ✓

- c The gamma ray could have more energy than the minimum calculated to create the electron–positron pair because any additional energy could go into the kinetic energy of the particles created. ✓

CHAPTER 12 ORIGINS OF THE ELEMENTS

INQUIRY QUESTION:

What evidence is there for the origins of the elements?

The Big Bang theory is the most widely accepted theory of the evolution of the universe. The theory explains the existing state of the universe and has made predictions that have been verified by observations and experiments. According to the theory, the universe began 13.8 billion years ago as a tiny, extremely hot point that has been expanding and cooling ever since. Hydrogen, helium and a small amount of lithium condensed from the radiation of the Big Bang shortly after the universe began. The heavier elements were formed much later by fusion in the core of stars and in supernova explosions. These theories for the origin of the elements are supported by a wide range of observational evidence.

1 The Big Bang theory

- » Students investigate the processes that led to the transformation of radiation into matter that followed the 'Big Bang'.
- » Students investigate the evidence that led to the discovery of the expansion of the universe by Hubble.
- » Students analyse and apply Einstein's description of the equivalence of energy and mass and relate this to the nuclear reactions that occur in stars.

- ➔ In the beginning of the 20th century scientists believed the universe existed in a steady state, remaining static rather than evolving.
- ➔ When Einstein published his general theory of relativity in 1915, the steady state theory was the dominant model of the universe.
- ➔ The general theory of relativity extended the special theory to accelerating reference frames and gravity. In 1922 Alexander Friedmann applied Einstein's general theory of relativity to the universe and showed that the equations predicted a dynamic, expanding universe rather than a stable, steady state universe. At first Einstein did not believe Friedmann's solution but later conceded that Friedmann was correct even though his solution contradicted the accepted steady state theory of the universe. Einstein tried to prevent his equations from producing a dynamic solution by adding a 'fudge factor' that he called the cosmological constant, but even with this constant the equations still failed to predict a static universe.
- ➔ Friedmann's theoretical prediction was to be spectacularly confirmed by observations only a few years after he published his controversial expanding universe theory.
- ➔ In 1923 Edwin Hubble (1889–1953) was studying nebulae with what was at the time the biggest telescope in the world, the 100-inch Hook telescope at Mount Wilson in the United States.
- ➔ Nebulae were small, fuzzy, bright clouds that dotted the night sky. Some astronomers thought they were embryonic solar systems, some thought they were clouds of dust, and others thought they might be 'island universes' like our own Milky Way.
- ➔ While studying the nebula M21, commonly known as Andromeda, Hubble noticed that it was made up of a huge number of stars. He also saw a number **Cepheid variable stars** in the nebula. Henrietta Leavitt had shown that Cepheid stars had a period of variability that was related to their absolute luminosity. If the absolute luminosity of a light source and the intensity of light reaching an observer are known, the inverse square law of intensity can be used to find the distance to the source. Cepheid variable stars therefore could be used to measure distances.

Cepheid variable star: a variable supergiant star that changes intensity with a period that is related to its luminosity

- ➔ By measuring the period of the Cepheid variable stars in M21, Hubble determined that M21 was 900 000 **light years (ly)** away. Because he could make out individual stars and because the nebula was so far away, Hubble concluded that M21 and many other nebulae that he observed were galaxies like our own but located far outside the Milky Way.
- ➔ Hubble's work was supported by observations made by Vesto Slipher, who showed in 1925 that the **redshift** of nebulae such as the M21 were much greater than for any star in the Milky Way.
- ➔ Hubble followed up the work of Slipher by examining the redshift of the light from 46 galaxies and showed that

there was an approximate linear relationship between the distance to a galaxy and the amount its light is redshifted. This relationship between distance and redshift, illustrated in Figure 12.1, is now called **Hubble's law**.

light year (ly): the distance that light travels in one year (9.4607×10^{12} km)
redshift: the increase in wavelength seen by an observer when the source of the light is moving away from the observer; astronomers measure redshift by comparing the wavelength of atomic absorption lines from the star with the wavelengths of the lines measured in the laboratory on Earth
Hubble's law: the recession velocity of a galaxy is proportional to the distance to the galaxy

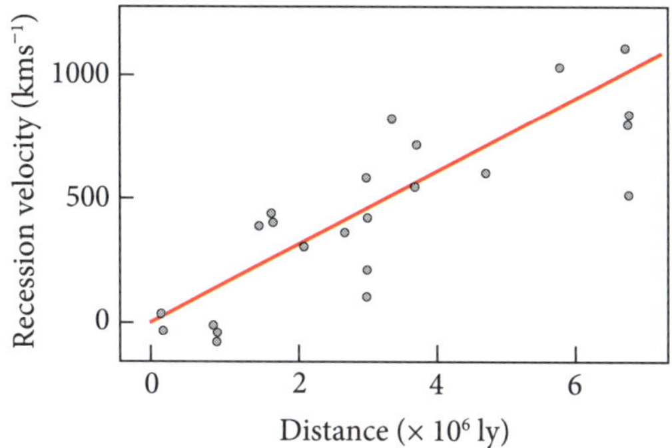

Figure 12.1 Hubble's law

- As redshift is a measure of how fast a light source is moving away from the observer, Hubble's law tells us that the further a galaxy is away from us the faster it is moving away from us. If all the galaxies are moving away from one another, the universe must be expending, just as Friedmann had predicted. The gradient of the graph is now called **Hubble's constant** (H_0) and it gives us an estimate of how fast the universe is expanding.

Hubble's constant: a constant of proportionality between the recession velocity of a galaxy and the distance to the galaxy, which gives a measure of the rate at which the universe is expanding

- Hubble initially measured the Hubble constant to be 160 kms^{-1} per million light years. Modern measurements have the constant at 22 kms^{-1} per million light years. We can use the Hubble constant to estimate the velocity of a galaxy if its distance from the Earth (D) is known or to estimate its distance if the velocity (i.e. redshift) is known, as follows:
  $v = H_0D$ and $D = v/H_0$
  where v is the velocity of the galaxy
  D is the distance to the galaxy and
  H_0 is Hubble's constant
- Einstein's general theory of relativity shows that the galaxies are not moving apart from one another because they are travelling through space but because the space between them is expanding. Recent measurements have shown the rate of the expansion of space is accelerating. To explain this accelerated expansion, astronomers are investigating the possibility that the expansion is being driven by a substance they have called dark energy.
- In 1927 Georges Lemaître proposed that an expanding universe must have had a beginning at some point and time. The Big Bang theory of cosmology was developed from Lemaître's proposal and today the theory is widely accepted by the scientific community as the best explanation of the evolution of the universe. The theory not only accounts for the expansion of the universe but can explain the proportions of the light elements in the universe and the existence and form of the cosmic background radiation that permeates the universe.
- The Big Bang theory asserts that time and space in our universe began from an incredibly hot point called a **singularity** about 13.8 billion years ago. Our current physics cannot be used to describe a singularity because a singularity would have an infinite temperature and density, and the curvature of space within the singularity would be infinite. The Big Bang theory therefore cannot explain the actual beginning of time and just assumes that some sort of fluctuation caused the singularity to expand suddenly. The Big Bang is an expansion of space rather than an explosion and the theory asserts that the universe has been expanding and cooling for the past 13.8 billion years.

singularity: an infinitely small point of infinite density and infinite space curvature

- The Big Bang theory explains how the elements have been formed as the universe has evolved. We can see how this occurred by following the Big Bang explanation of the evolution from when time began in our universe:
  - 10^{-43} – 10^{-12} seconds
    The theory is highly speculative here because the energies involved cannot (as yet) be reached in particle accelerators. But the most widely accepted theories suggest that during this time the universe experienced an exponential growth called *cosmic inflation*, doubling in size 90 times and growing to a diameter of about 10 cm. The four forces of nature (strong and weak nuclear, electrostatic and gravity) decoupled from one another in this time and matter–antimatter particle pairs began to be created from photons and other photon-like (boson) radiation.
  - 10^{-12} – 10^{-6} seconds
    Huge numbers of subatomic matter–antimatter pairs (e.g. quarks, neutrinos and electrons) were produced by photons as the universe continued to expand and cool. For some unknown reason the radiation

produced about one extra matter particle per billion matter–antimatter pairs. This small excess of matter accounts for all the matter we see in the universe today.

- 10^{-6} – 1 second
  The universe has now cooled enough for fundamental particles called *quarks* to collide and stick together to form *hadrons* (e.g. protons and neutrons and their antiparticles). After 1 second almost all the matter and antimatter hadrons annihilated one another to produce huge numbers of *leptons* (e.g. electron and positrons), leaving a small excess of matter particles.
- 1 second – 3 minutes
  Electrons and positrons annihilated one another producing gamma rays, again leaving a slight excess of matter particles.
- 3 minutes – 20 minutes
  The universe has now cooled enough for protons and neutrons to fuse together to form nuclei in a process called **nuclear synthesis**. At the end of this time hydrogen nuclei, helium nuclei and some lithium nuclei have been produced. No heavier elements were formed because nuclei with seven and eight nucleons are unstable. After 20 minutes the temperature of the universe had fallen too far for fusion to continue. The theory predicts the production of 75% hydrogen, 25% helium and a trace quantity of lithium. We find the same ratio of light elements in the universe today, which provides strong evidence for the veracity of the theory.

nuclear synthesis: the cosmic formation of nuclei more complex then hydrogen

- 240 000 – 300 000 years
  The universe is now cool enough for electrons to combine with nuclei to form atoms. This process flooded the universe with photons, released as electrons shed energy as they combined with positive nuclei in a process called *recombination*. The theory predicted that these photons would be stretched over time to microwaves and should still permeate the universe. This prediction of cosmic background was later confirmed by observations and provides strong evidence in support of the Big Bang theory.
- 300 – 500 million years
  The universe has now cooled enough for gravity to collect the primordial gas together in regions where the gas is slightly denser, to form the first stars. When the stars gather enough matter, the core of the star reaches such a high temperature and pressure that hydrogen is fused into helium. When small nuclei fuse together, mass is converted into energy in accordance with Einstein's equation $E = mc^2$.

  Stars exist in a state of equilibrium, with gravity pushing inwards and radiation and gas pressure pushing out. If the star is large enough it will fuse heavier elements up to iron before collapsing and exploding in a **supernova**. Fusion in stars stops at iron because fusing iron to heavier elements consumes energy rather than releases it. When the iron core of the star becomes large enough, fusion in the star will stop and gravity will crush the star inwards until it rebounds in a huge fusion explosion called a *type II supernova* (type I supernova occurs when binary stars coalesce). Elements from iron to bismuth are formed by neutron absorption in shells outside the core and elements heavier than bismuth are formed in supernova explosions.

supernova: a large explosion caused by the gravitational collapse of a supergiant star at the end of the star's life

➔ To summarise our current understanding of the origin of the elements, subatomic particles were produced via matter–antimatter pair production from radiation shortly after the universe was born. For some reason slightly more matter than antimatter was produced, resulting in the matter universe we see in the universe today. Shortly after the subatomic particles were created from radiation, nuclear synthesis produced hydrogen nuclei, helium nuclei and some lithium. Some 300 000 years after the Big Bang, electrons combined with these small nuclei to form atoms. Then over the past 13.5 billion years, elements from helium to iron have been produced by fusion in the core of stars. Elements from iron to bismuth were produced by neutron absorption in shells outside the core and elements heavier than bismuth were produced by fusion in the supernova explosions that occur when huge stars run out of fuel.

➔ KEY QUESTIONS

1. **What evidence led Hubble to conclude that the universe was expanding?**
2. **What is the Big Bang theory?**
3. **Outline how the Big Bang theory explains the transformation of radiation to matter in the early universe.**
4. **Explain how Einstein's mass–energy equivalence explains the energy produced in stars?**
5. **What produced the elements in the universe?**

Answers ➲ p. 180

2 Stellar spectra

» Students account for the production of emission and absorption spectra and compare these with a continuous black-body spectrum.

» Students investigate the key features of stellar spectra and describe how these are used to classify stars.

- You will recall from our studies of light that black-body radiation produces a continuous spectrum containing all wavelengths. As a black body is heated, the wavelength at which most radiation is emitted decreases and the power radiated increases.
- Continuous spectra are quite different to atomic emission spectra, which consist of specific wavelengths of light. *Atomic spectra* can be studied by exciting elemental gases in flames or in electric discharges.
- *Absorption spectra* refer to specific wavelengths of light missing from continuous spectra (i.e. dark lines in a continuous spectrum). Absorption spectra are produced by passing continuous spectra through an unexcited (cool) elemental gas. Figure 12.2 shows the three types of spectra.

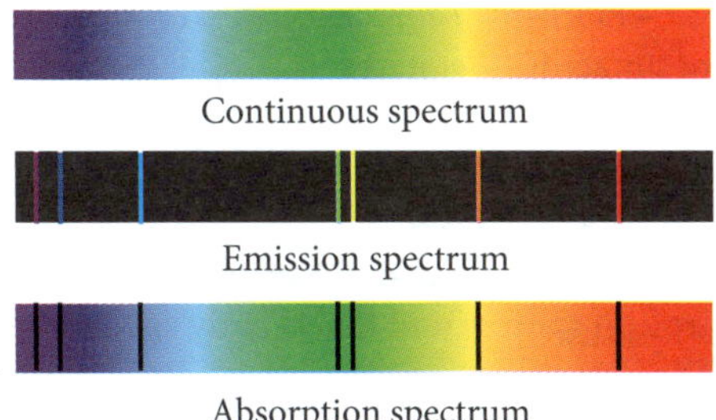

Figure 12.2 Different types of spectra

- Atoms emit and absorb photons with specific energies characteristic of the element involved. Emission spectra are produced by atoms being excited (given energy) and then relaxing back to their unexcited (ground state) by emitting one or more photons with specific energies.
- Absorption spectra are produced by atoms absorbing photons of specific energies. The atom absorbing the photon could be initially in the neutral ground state or it could be in an excited or ionised state.
- The electromagnetic radiation emitted from a star consists of a continuous black-body spectrum with a spectral distribution that corresponds with the temperature of the surface layer (the photosphere) of the star. The stellar spectrum also contains dark absorption lines produced by the atoms, molecules and ions in the star's atmosphere that are absorbing radiation at specific wavelengths. A typical stellar spectrum is illustrated in Figure 12.3.

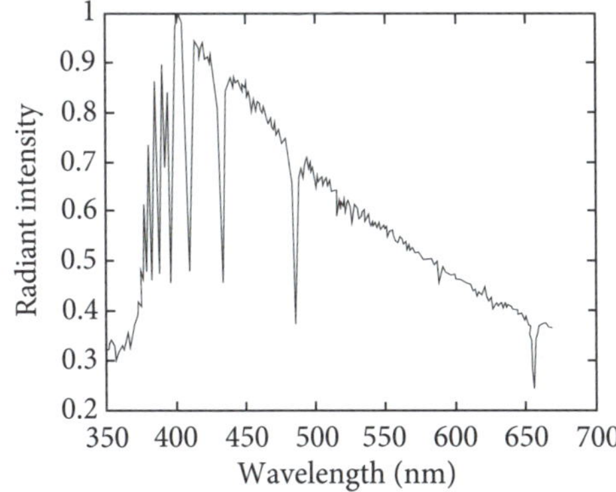

Figure 12.3 Typical stellar spectrum

- When astronomers began studying stellar spectra they noticed that the absorption lines in the spectra of different stars varied, indicating that their atmospheres were different. Astronomers used an alphabetic classification system for stars based on the absorption lines in their spectra. But it was later shown that the main reason for the different absorption lines was the temperature of the star; the original classification scheme was alphabetical but not arranged in terms of temperature. The updated stellar classification scheme based on the star's temperature is called the **Morgan–Kennan (MK) system.** The MK scheme uses the same letters as the initial scheme but because the new scheme is based on temperature the spectral classes are no longer in alphabetical order.

Morgan–Kennan (MK) system: a system of stellar classification based on the surface temperature and certain spectral characteristics of stars

- The absorption lines in stellar spectra can be used to estimate the amount of ionisation in the atmosphere of a star. The degree of ionisation can then be used to classify the star because the temperature of the photosphere of the star determines the density of ions in the star's atmosphere.
- The MK system, shown in Table 12.1, classifies stars into seven **spectral classes** labelled OBAFGKM, where O stars are the hottest and M stars the coolest. The system is subdivided further using numbers from 0 to 9, where 9 represents the coolest star. For example, each of the stars in the sequence F7, F8, F9, G0, G1, G2 has a slightly lower temperature than the preceding star. Historically the mnemonic **O**h **b**e **a** **f**ine **g**irl/guy **k**iss **m**e has been used to recall the MK spectral classification. Our Sun is a G2 main sequence star about halfway through its main sequence lifespan of about 10 billion years. Astronomers call the mass of the Sun a **solar mass** ($M_\odot$) and use it to specify the mass of other stars. Note that $1\ M_\odot = 2 \times 10^{30}$ kg.

spectral class: categories used by the MK system of stellar classification; stars are labelled with a letter from OBAFGKM, with O being the hottest star and M the coolest

solar mass: mass of the Sun

- The lines present in the stellar absorption spectra reflect the density of the absorbing species in the atmosphere of the star. The coolest stars are not hot enough for collisions to break apart molecules and hence show molecular lines in their absorption spectra. Hotter stars do not show molecular lines but as stars get hotter, the collisions in the atmosphere are energetic enough to singly ionise metals. In even hotter stars, metals can be multiply ionised.

Table 12.1 Spectral classification of stars

Spectral class	Temperature range (K)	Colour	Features of absorption spectra
O	> 25 000	blue	Ionised helium strong, multiply ionised heavy elements, hydrogen faint
B	11 000 – 25 000	light blue	Neutral helium and hydrogen moderate, singly ionised heavy elements
A	7500 – 11 000	white	Neutral helium very faint, hydrogen strong, singly ionised heavy elements
F	6000 – 7500	yellow-white	Hydrogen moderate, singly ionised heavy elements, neutral metals
G	5000 – 6000	yellow	Hydrogen faint, singly ionised elements, neutral metals
K	3500 – 5000	orange	Hydrogen faint, singly ionised elements, neutral metals strong
M	< 3500	red	Neutral atoms strong, molecules moderate, hydrogen very faint

Helium has a very high ionisation energy and hence the absorption lines of ionised helium are present only in very hot stars. The hydrogen lines are weak in cooler stars because the first energy level is very high and very few hydrogen atoms will reach this level in collisions at low temperature. Neutral hydrogen lines are also faint in very hot stars because most of the hydrogen would have been ionised by the energetic collisions that occur at these temperatures.

➔ We can explain the colour of each spectral class by recalling that the peak emission wavelength of the black-body radiation emitted by the star is inversely proportional to the surface temperature of the star. Wien showed that the wavelength of maximum emission is related to the surface temperature of a black body by:

$\lambda_{max} = b/T$

where λ_{max} is the wavelength at which the emitted radiation intensity is greatest

$b = 2.898 \times 10^{-3}$ mK is a constant of proportionality called *Wien's displacement constant* and

T is the absolute temperature in kelvins

➔ Figure 12.4 shows the black-body emission for three temperatures and the related star type. Note how the shift towards shorter wavelengths, as the temperature of the star increases, changes the colour of the star.

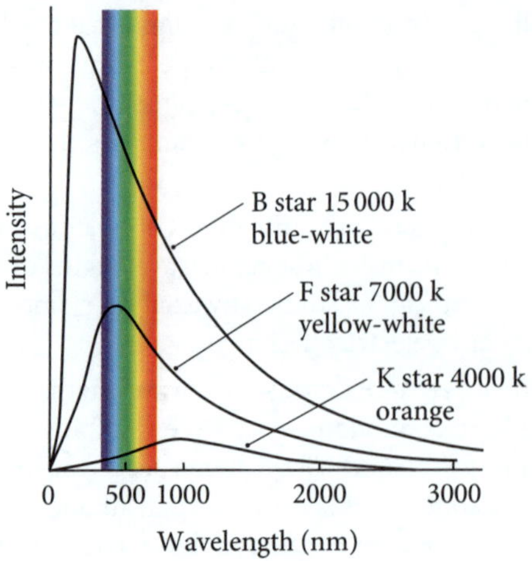

Figure 12.4 Black-body emission curves for stars with different surface temperatures

➔ KEY QUESTIONS

6 **What is the difference between emission spectra, absorption spectra and black-body spectra?**

7 **How can the black-body spectrum of a star be used to determine the temperature of a star?**

8 **Explain how the temperature of a star affects the absorption spectrum of the star?**

9 **Outline the MK stellar classification system and list the spectral classes specified by the MK system.**

Answers ➲ p. 180

3 The Hertzsprung–Russell diagram and stellar evolution

» Students investigate the Hertzsprung–Russell diagram and how it can be used to determine the following about a star: characteristics and evolutionary stage, surface temperature, colour and luminosity.

The Hertzsprung–Russell diagram

➔ In 1911 Ejnar Hertzsprung and Henry Russell independently plotted the spectral class (temperature) against the **luminosity** for the known stars. They discovered that the stars were grouped together in different regions on the graph. This graph, now called the **Hertzsprung–Russell (HR) diagram**, is one of the most important diagrams in astrophysics because it shows us that there are different types of stars and that stars evolve in different ways depending on their initial mass.

luminosity: the total energy radiated by a star per second
Hertzsprung–Russell (HR) diagram: a plot of luminosity against spectral class for stars

- Consider the HR diagram shown in Figure 12.5. Note that the luminosity (total radiated power) is given as a multiple of the Sun's luminosity and is also given in a quantity called the *absolute magnitude*. The **magnitude scale** is used by astronomers to specify the brightness of a star. A large magnitude is a dull star, and a negative magnitude is a very bright star. The difference in intensity between each integer on the magnitude scale is a factor of $\sqrt[5]{100} = 2.512$. Thus a magnitude 3 star is 2.512 times brighter than a magnitude 4 star, and a magnitude 1 star is $2.512 \times 2.512 \times 2.512 = 12.6$ times brighter than a magnitude 4 star. Relative magnitude is the brightness as seen from Earth and absolute magnitude is the brightness the star would have if it was 10 **parcecs (pc)** (32.6 ly) from the Earth. Note also that the temperature decreases from left to right across the graph.

> **magnitude scale:** a logarithmic measure of the brightness of a star; absolute magnitude is a measure of luminosity, while relative magnitude measures how bright the star appears from the Earth
>
> **parsec (pc):** a unit of distance equal to 3.26 light years; one parsec is the distance at which the mean radius of the Earth's orbit subtends an angle of one second of arc

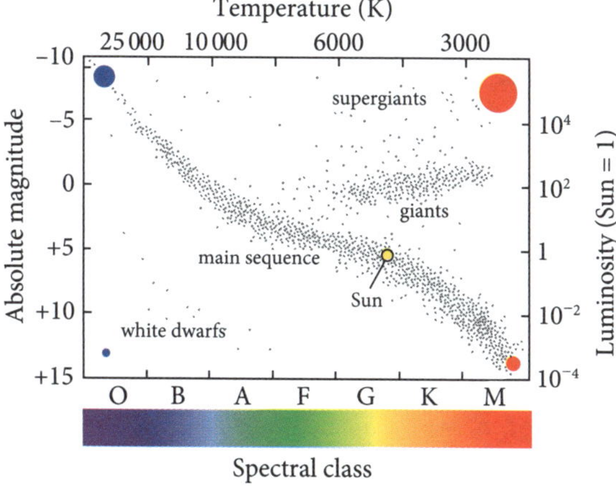

Figure 12.5 Hertzsprung–Russell diagram

- Figure 12.5 shows four types of stars, which we will examine independently. Before we can do this, however, we must refresh our memory regarding the intensity of the radiation emitted from a black body.
- The luminosity of a black body, like a star, is equal to the product of the power radiated per unit area and the area of the body. For a spherical body, the luminosity would be given by:

  $L = \text{intensity} \times (4/3)\pi r^3$

  Now we have previously seen that the power radiated from a black body is proportional to T^4. Combining these two results we see that the luminosity of a star increases with the surface temperature and with the radius of the star (i.e. $L \alpha T^4 r^3$).
- The HR diagram shown in Figure 12.5 has a coloured star drawn in each corner of the diagram. These stars are not drawn to scale but represent the difference between stars in each corner of the diagram. The reason why the colour and size of these stars varies is explained in Table 12.2.

Table 12.2 Features of stars located in each corner of a Hertzsprung–Russell diagram

Star at the top left	**Star at the top right**
This star would have the same high surface temperature as a star on the lower left, making both stars blue-white, but a much greater luminosity. As the stars have the same temperature, the upper star must be much larger as its luminosity is much greater.	This star would have the same low surface temperature as a star on the lower right, making both stars red, but with a much greater luminosity. This could only occur if the star was considerably larger. These red stars would be larger than the blue-white stars at the top left because they are cooler but have the same luminosity.
Star at the bottom left	**Star at the bottom right**
This star would have very a high surface temperature but very low luminosity. This would only occur if the star was very small. Such a star would be blue-white and have a very small radius.	This star would have a low surface temperature and not be very luminous. Such a star would be red and small. These red stars would be larger than the stars at the lower left because they have the same luminosity but are much cooler.

- We can now examine the four types of stars shown on the diagram: main sequence stars, giant stars, white dwarf stars and supergiant stars.

Main sequence stars

- The **main sequence** refers to the group of stars that run diagonally across the diagram from the top left to the lower right. Over 90% of stars in the galaxy, including our Sun, fall into this group. Main sequence stars vary from large, hot, blue stars on the top left to small, red dwarf stars on the lower right. These stars derive their energy from fusing hydrogen to helium in the core of the star. Stars form when large clouds of gas are pulled together by gravity until the core temperature becomes large enough (10^7 K) for hydrogen fusion to begin. The star then comes into hydrostatic equilibrium with the inward force of gravity balanced by the outward thermal pressure from the fusion reactions occurring in the core. The initial mass of a star can vary from 0.1 $M_\odot$ to 200 $M_\odot$ and it is this initial mass that determines the position of a star on the main sequence. Large, massive stars must fuse hydrogen at a prodigious rate to balance the enormous inward pressure of gravity. For this reason, large main sequence stars are hotter (blue) and have comparatively short lifespans (tens of millions of years). Smaller main sequence stars do not have as much inward gravitational pressure and fusion occurs at a much slower rate. That is why very small main sequence stars (red dwarfs) are found on the bottom right of the HR diagram and have comparatively long lifespans (hundreds to thousands of billions of years). Intermediate mass main sequence stars, including our G2 Sun, which has a main sequence lifespan of about 10 billion years, fall between these two extremes.

main sequence: region on the HR diagram that contains stars that fuse hydrogen to helium in the core; main sequence stars fall on the diagonal that runs from the upper left to the lower right on an HR diagram

Giant stars

→ The most common **giant stars** are red giants. These stars have much higher luminosities than main sequence stars that have the same surface temperature and hence are much larger than main sequence red stars. Red giant stars typically have masses of 0.3 $M_\odot$ to 8 $M_\odot$ but are ten to hundreds of times larger than the Sun. All red giant stars are in a late stage of their evolution, having fused all the hydrogen in their core to helium. Some red giants have an inert helium core while others fuse helium to carbon in their core. All red giant stars fuse hydrogen to helium in a shell around the core.

giant star: a star that has finished fusing hydrogen in the core and is now fusing hydrogen in a shell around the core; a giant star can also fuse helium in the core

White dwarf stars

→ **White dwarf** stars represent the final phase in the evolution of main sequence stars with masses of less than 8 $M_\odot$. When main sequence stars in this size range run out of fissionable fuel they blow away their outer layers to form a planetary nebula and the core that remains collapses to form a white dwarf. We can think of a white dwarf as a hot ember in which fusion has ceased. White dwarf stars typically have a similar mass to the Sun but are about the size of the Earth. They consist of degenerate matter, rather than atoms, and have a density about 10^9 kgm^{-3}. A teaspoon of white dwarf matter would weigh several tonnes. White dwarfs slowly radiate their remaining thermal energy into space over a period of about a billion years, becoming duller as they do so.

white dwarf: the hot leftover core of a star that has ceased to fuse elements

Supergiant stars

→ **Supergiant stars** are very massive stars in the latter phase of their evolution. They are formed from very large main sequence stars that have ceased fusing hydrogen and helium in their core and are now fusing carbon or heavier elements, while lighter elements fuse in shells around the core. Supergiants are the largest stars in the universe. They have masses from about 8 $M_\odot$ to over 100 $M_\odot$ and their radii can be 1000 times larger than the radius of the Sun. Many supergiants are variable stars that show periodic changes in their radiant output. For example, the Cepheid variable stars used by astronomers to measure distances are yellow supergiants.

supergiant star: a very large star that has finished fusing hydrogen in the core; the largest of these can fuse heavier elements in shells around the core until they produce iron in the core

EXAMPLE 1

Consider the following stars: O2 main sequence, K1 main sequence, F6 main sequence and K4 supergiant.

a List the stars from hottest to coolest.

b List the stars from largest to smallest.

Recall how temperature and the size of a star change on the HR diagram

Answer:

a Using the MK scale: hottest O2, F6, K1, K4 coolest

b Remembering that the size increases vertically and from left to right: largest K4 supergiant, O2 main sequence, F6 main sequence, K1 main sequence

Stellar evolution

→ The lifecycle of a star is largely determined by a star's initial mass. Stars form from inhomogeneities in gas and dust clouds that cause gravity to accumulate matter. We have seen that such gas clouds would largely consist of hydrogen, with some helium and trace amounts of heavy elements scattered by the death of old stars. If the mass accumulated is between 0.1 $M_\odot$ and 200 $M_\odot$, the temperature in the core will become high enough for hydrogen fusion to begin. A temperature of about 10^7 K is required because the hydrogen nuclei must be moving fast enough to overcome the electrostatic repulsion between the positively charged nuclei for fusion to occur. Instabilities in the collapse of gas clouds are thought to prevent masses over 200 $M_\odot$ from forming stars, and masses below 0.1 $M_\odot$ do not have enough gravitational force to heat the core to a high enough temperature for fusion to begin. Hydrogen is the first element to fuse because it only has one charge in the nucleus and hence has the least electrostatic repulsion and the lowest fusion ignition temperature. Fusing heavier nuclei requires much higher ignition temperatures.

- Once fusion begins the star settles into a stable state of hydrostatic equilibrium in which the inwards force of gravity is balanced by the outward thermal pressure from fusion. This is when the star moves onto the main sequence where it will spend the greatest part of its active life.
- When a main sequence star builds up a large-enough inert helium core, hydrogen fusion will cease and gravity will collapse the star. The fate of the star from this point depends on its mass. Theory suggests that main sequence stars with masses of less than 0.23 $M_\odot$ will simply collapse into white dwarf stars. although none have yet done so as they have much longer lifetimes than the current age of universe. Stars with masses from 0.23 to 10 solar masses are large enough to fuse hydrogen around the inert helium core. This causes the star to expand and cool as it moves from the main sequence to become a red giant star. Because helium produces only one-tenth of the energy that hydrogen fusion produces, the lifetimes of red giants are much shorter than the lifetimes of main sequence stars. Stars with masses greater than 0.5 $M_\odot$ will fuse helium in their cores in the red giant stage, and stars with masses greater than 5 $M_\odot$ will fuse heavier masses. Stars with masses above 10 $M_\odot$ will move from the main sequence to become supergiants.
- Giant and supergiant stars fuse elements of increasing mass until a limit is reached, depending on the initial mass of the star. Very large stars fuse elements up to iron but at this point fusion ceases as the fusion of iron to heavier elements does not release energy. Stars at this point take three mass dependent paths. Stars with masses less than 8 $M_\odot$ will puff away their outer layers (this process is called a *nova*) to form a **planetary nova**, and the core will shrink to become a white dwarf. Stars larger than this will collapse and explode in huge supernova explosions. If the remaining core has a mass between 1.4 $M_\odot$ and 3 $M_\odot$, it will collapse to become a **neutron star**. If the core has a mass greater than 3 $M_\odot$ it will collapse to become a **black hole**. Neutron stars have radii of about 10 km and the density of nuclear matter. A teaspoon of the material that makes up a neutron star would weigh about 10 million tonnes. A black hole is a singularity surrounded by a sphere called the *event horizon*, which marks the radius where the escape velocity is equal to the speed of light. Light emitted below this radius cannot escape from a black hole.

planetary nova: a very hot, expanding shell of ionised gas formed when a red giant puffs off its outer layer late in the life of the star
neutron star: remnant of a massive star
black hole: a remnant of a super-massive star that consists of a region of space from which light cannot escape

- Figure 12.6 shows how the evolutionary path of a star depends on its mass. Note that the cut-off mass values stated on the flowchart are approximate as they are still being actively researched.

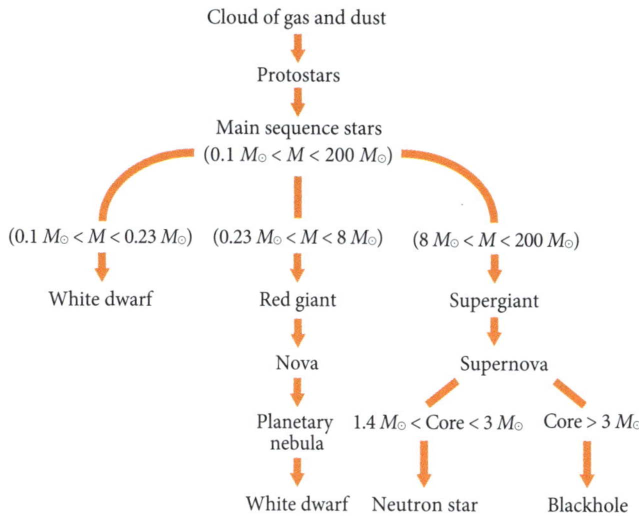

Figure 12.6 The evolutionary path taken by a star depends on its initial mass

- We can also use an HR diagram to draw the lifecycle of a star. For example, the lifecycle of a high-mass star and of the Sun are shown in Figure 12.7. Note that a medium-mass yellow star, like our Sun, will move off the main sequence as it starts to fuse a hydrogen shell around an inert helium core. When the helium core gets big enough, hydrogen fusion in the shell around it ceases and the star collapses, heating the core to a temperature that enables helium fusion to begin. This causes the star to move to the left in the giant phase. It then moves to the right again as a helium shell fuses around a carbon core. When the carbon core gets big enough, helium fusion stops. The star is not massive enough to reach the temperature required to fuse heavier elements and instead it blows away its outer layer in a nova, forming a planetary nebula with a white dwarf core.

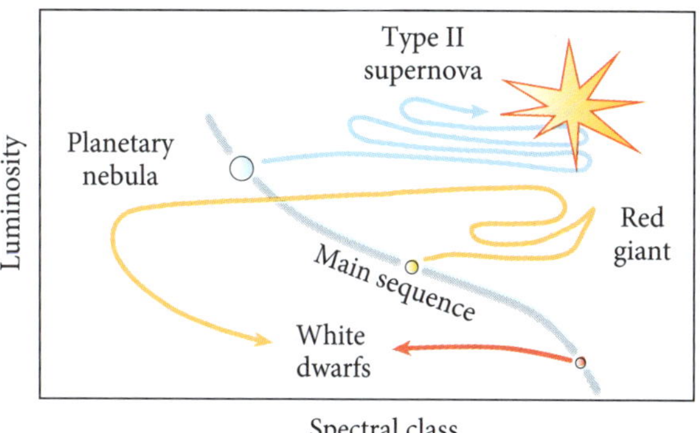

Figure 12.7 HR diagram showing evolutionary paths of stars with different mass

- The heavy mass star moves right and left (heating and cooling) in the supergiant phase as successively heavier elements begin to fuse, until an iron core is produced and the star explodes in a type II supernova. The core of the supernova may collapse to form a neutron star or black hole.

EXAMPLE 2

Describe qualitatively how an HR diagram can be used to estimate the distance to a main sequence star.

Remember that if a star's luminosity and brightness as viewed from the Earth are known, the distance to the star can be calculated

Answer:

The spectrum of the star could be examined to determine the spectral class of the star and, because it is a main sequence star, plotting it on an HR would enable its luminosity to be read off the graph. By measuring the brightness of the star as seen from the Earth and combining this with the luminosity and the inverse square law of intensity, the distance to the star could be calculated.

→ KEY QUESTIONS

10 **Explain how the position of a star on the HR diagram can be used to determine the temperature, luminosity and size of a star.**

11 **Explain why stars are grouped into specific regions of the HR diagram.**

12 **Outline how the lifecycle of a star is related to the initial mass of the star.**

13 **Describe how the evolutionary path of a star can be drawn on an HR diagram.**

Answers ⊃ p. 180

4 Nuclear synthesis in stars

» Students investigate the types of nucleosynthesis reactions involved in main sequence and post-main sequence stars, including but not limited to the proton–proton chain and the CNO (carbon–nitrogen–oxygen) cycle.

➔ We have seen that hydrogen fusion occurs in the core of all main sequence stars. There are two principle reaction paths that enable stars to produce energy by fusing hydrogen to helium: the **CNO cycle** and the **p–p chain reaction**.

CNO cycle: a cyclic process involving carbon isotopes being changed to nitrogen and oxygen and back to carbon in a catalytical process that enables four protons to fuse to helium in the core of stars with mass greater than 1.3 $M_\odot$

p–p chain reaction: the proton–proton chain reaction involves three reactions that enable four protons to fuse to helium in the core of stars with masses less than 1.3 $M_\odot$

➔ The CNO cycle is the predominant hydrogen fusion process in stars with a mass greater than 1.3 $M_\odot$. The CNO cycle uses carbon, nitrogen and oxygen nuclei to catalyse (assist), a reaction that converts four protons to a helium nucleus. The CNO cycle involves a carbon nucleus being changed to nitrogen and then to oxygen before changing back to carbon in an endless loop. Each time the cycle is repeated four protons (^1H) are converted to a helium nucleus (^4He) plus three gamma rays (γ), two positrons (e^+) and two electron neutrinos (ν_e). Electron neutrinos are tiny, neutral, almost massless particles that interact with matter so rarely that they can pass straight through the Earth or a star. We can represent each step in the CNO cycle as a reaction:

$$^{12}C + {}^1H \rightarrow {}^{13}N + \gamma$$
$$^{13}N \rightarrow {}^{13}C + e^+ + \nu_e$$
$$^{13}C + {}^1H \rightarrow {}^{14}N + \gamma$$
$$^{14}N + {}^1H \rightarrow {}^{15}O + \gamma$$
$$^{15}O \rightarrow {}^{15}N + e^+ + \nu_e$$
$$^{15}N + {}^1H \rightarrow {}^{12}C + {}^4He$$

The net results of the CNO process is that each cycle mass is converted to energy by the fusion of hydrogen to helium:

$$4{}^1H \rightarrow {}^4He + 2e^+ + 2\nu_e + \gamma + 26.72 \text{ MeV}$$

Note that an **electron volt (eV)** is a measure of energy used in nuclear physics. It is the energy gained by an electron when it is accelerated across a potential difference of 1 volt: 1 eV = 1.6×10^{-19} J.

electron volt (eV): energy gained by an electron when it is accelerated across a potential difference of 1 volt: 1 eV = 1.6×10^{-19} J

The Sun only produces 1 to 2% of its energy by the CNO cycle but a main sequence star twice the mass of the Sun would produce almost all its energy by the CNO process.

➔ The p–p reaction is the main fusion reaction in main sequence stars with masses of less than 1.3 $M_\odot$. This is because the CNO cycle requires a higher ignition temperature (1.5×10^7 K) than the p–p chain (10^7 K). The p–p chain can follow several possible paths but the most common path is:

$$^1H + {}^1H \rightarrow {}^2H + e^- + \nu_e$$
$$^2H + {}^1H \rightarrow {}^3He + \gamma$$
$$^3He + {}^3He \rightarrow {}^4He + 2{}^1H$$

The p–p chain reaction releases a total energy of 26.72 MeV per ^4He. The Sun produces 98 to 99% of its energy by the p–p chain reaction, with 83% of this being produced by the particular p–p path shown above.

➔ When a star leaves the main sequence, it can fuse helium and may fuse heavier elements. When hydrogen fusion ceases in stars with a mass greater than 0.5 $M_{\odot}$, gravity causes the core to contract until it reaches 100 million kelvins. At this temperature the helium nuclei (alpha particles) in the core can collide at speeds high enough to overcome electrostatic repulsion and fuse together. This initiates the triple-alpha reaction that leads to the release of energy that halts further contraction of the core. The triple-alpha reaction consists of two sequential reactions:

$$^{4}He + ^{4}He \rightarrow ^{8}Be$$

$$^{8}Be + ^{4}He \rightarrow ^{12}C + 2\gamma$$

Some of the carbon produced forms oxygen in the following reaction:

$$^{12}C + ^{4}He \rightarrow ^{16}O + \gamma$$

It is these reactions that have produced the carbon and oxygen atoms that are so essential for life. The Sun is not massive enough to fuse elements heavier than helium and late in its life it will be the inert core of carbon and oxygen that will spell the end of energy production in our star.

➔ More massive stars will fuse heavier and heavier elements. Progressively heavier elements require higher and higher fusion ignition temperatures due to the higher electrostatic forces that must be overcome. Supermassive stars will fuse elements in shells, as shown in Figure 12.8. For the largest stars, fusion will only cease when a core of iron is formed because producing elements heavier than iron consumes energy and hence cannot be used to balance the huge inward force of gravity on the star. Elements from helium to iron are formed via fusion in the core of the star and elements from iron to bismuth are formed by neutron absorption in the shells outside the core. Note that while most elements are formed in stars, some light elements (lithium, beryllium and boron) are also formed in interstellar space by cosmic rays hitting and splitting up larger nuclei in a process called spallation.

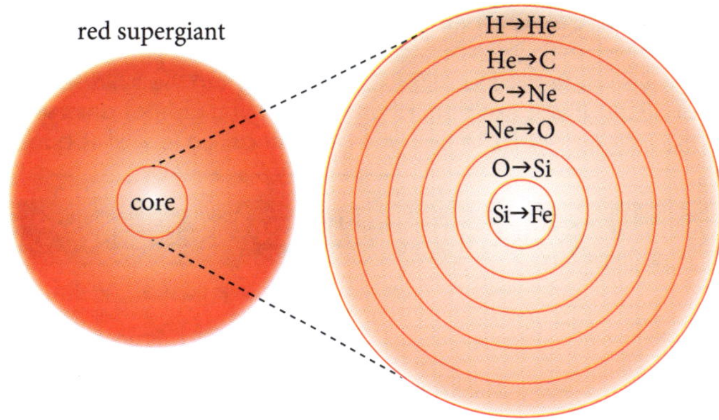

Figure 12.8 Fusion of different elements in shells in a supergiant star near the end of its life

➔ When an inert iron core is formed, gravity collapses the star and causes it to explode in a supernova explosion. Supernova explosions provide the energy to create elements heavier than bismuth and to spread the elements produced in the star and in the supernova throughout space.

➔ KEY QUESTIONS

14 **Compare the CNO cycle and p–p reactions in stars.**

15 **Explain how elements heavier than helium are produced in stars and supernova explosions.**

Answers ➲ p. 180

CHAPTER SYLLABUS CHECKLIST

Are you able to answer these questions from the syllabus for this chapter? Tick each question as you go through the checklist if you are able to answer it. If you cannot answer a question, turn to the relevant page in the study guide to find the answer. For NESA key word meanings, go to www.educationstandards.nsw.edu.au and search 'key words'.

	FOR A COMPLETE UNDERSTANDING OF THIS TOPIC:	PAGE NO.	✓
1	Can I describe how Hubble discovered that the universe was expanding?	168–169	
2	Can I outline the key features of and evidence for the Big Bang theory?	169–170	
3	Can I outline how radiation condensed into matter and antimatter as the universe expanded and cooled?	169–170	
4	Can I recall that more matter than antimatter was formed after the Big Bang?	169–170	
5	Can I describe how mass is converted into energy by fusion in stars and how to apply Einstein's mass–energy relationship to determine the energy produced in a fusion reaction?	170 & 176	
6	Can I describe the appearance and origin of emission spectra, absorption spectra and continuous black-body spectra?	171	
7	Can I describe and explain the general characteristics of stellar spectra?	171–172	
8	Can I outline how stellar spectra are used to classify stars into spectral classes that are related to the surface temperature of the star?	171–172	
9	Can I draw a Hertzsprung–Russell diagram and explain why stars fall into groups on the diagram?	173–174	
10	Can I use an HR diagram to determine the luminosity, colour, surface temperature and energy source of a star?	173–174	
11	Can I describe the production of the elements in the Big Bang, in stars and in supernova explosions?	173–175	
12	Can I describe the principle evolutionary paths of stars of different mass and draw these paths on an HR diagram?	175	
13	Can I outline how the p–p chain reaction and CNO cycle fuse hydrogen to helium in stars?	176–177	

HSC EXAM-TYPE QUESTIONS

Objective-response questions (1 mark each)

1 The spectrum of a main sequence star is found to contain absorption lines of helium and of multiply charged heavy metal ions. All the absorption lines were at a higher frequency than the frequencies observed in the laboratory. What can we conclude about the star?

A It is a small star moving away from the Earth.
B It is a large star moving away from the Earth.
C It is a small star moving towards the Earth.
D It is a large star moving towards the Earth.

2 What observation did Edwin Hubble make that led to the suggestion that the universe was not in a static state?

A Many gas clouds that were thought to be planetary nebula were in fact island galaxies.
B Many galaxies contained Cepheid variable stars that could be used to estimate their distance from the Earth.
C The amount of redshift in the light from a galaxy decreased for galaxies further from the Earth.
D The amount of redshift in the light from a galaxy increased for galaxies further from the Earth.

3 Which of the following correctly shows the stages a star of one solar mass would pass through?

A protostar → main sequence → nova → red giant → white dwarf
B protostar → nova → main sequence → red giant → white dwarf
C protostar → main sequence → red giant → nova → white dwarf
D protostar → red giant → main sequence → nova → white dwarf

4 A Hertzsprung–Russell (HR) diagram plots a star's luminosity against its spectral class. Where on an HR diagram would you find a star that was fusing carbon and heavier elements?

A in the upper centre to the right of the diagram
B in the upper left of the diagram
C in the lower left of the diagram
D in the lower centre to right of the diagram

5 Where were the elements that make up our world produced?

A in the Big Bang, in stars and in supernova explosions
B in stars and in supernova explosions
C in Big Bang and in supernova explosions
D in stars and in nova explosions

Extended-response questions

6 Outline two pieces of evidence that support the Big Bang theory. (4 marks)

7 Globular clusters are collections of about a million stars grouped tightly together that orbit in the halo of our galaxy. Figure 12.9 shows an HR diagram for typical stars in a specific globular cluster.

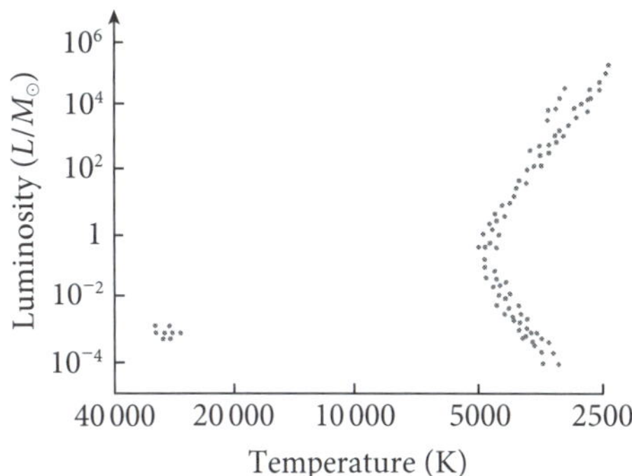

Figure 12.9 HR diagram of stars in a specific globular cluster

a Compare the size and colour of the stars at the bottom left with those at the bottom right. (2 marks)
b Explain why much of the main sequence appears to be missing. (2 marks)

8 Stellar spectra have proved to be a valuable resource for astronomers.

a Explain how an astronomer could obtain and record the spectra of a star. (3 marks)
b Compare the spectra produced by our Sun to the spectra from a red dwarf main sequence star. (3 marks)
c Compare the energy source and lifespan of a red dwarf star to the energy source and lifespan of a red giant star. (4 marks)

9 The reaction in the p–p chain reaction involves two protons combining as follows:

$$^1H + {}^1H \rightarrow {}^2H + e^- + \nu_e$$

If we assume the neutrino (ν_e) is massless and, given the that the mass of a proton (^1H) is 1.6726×10^{-27} kg and the mass of deuterium (^2H) is 3.3435×10^{-27} kg, find:

a the mass change involved in this reaction. (1 mark)
b the energy released or absorbed. (1 mark)
c Explain why this reaction will only occur at extremely high temperatures. (2 marks)

10 During their active lives, stars convert mass to energy.

a Explain why a minimum mass is required for a star to shine. (2 marks)
b Explain why a star like our Sun and even stars with very large masses eventually stop converting mass to energy. (3 marks)
c Outline the three possible paths a star can take when it stops converting mass to energy. (3 marks)

ANSWERS

KEY QUESTIONS

Key questions ➲ p. 170

1 In 1929 Edwin Hubble measured the redshift of the light from 46 galaxies and by observing the period of Cepheid variable stars in each of these galaxies he estimated the distance to each of the galaxies. He then combined the distance data and redshift data from the 46 galaxies to show that the further a galaxy was away from the Earth, the faster it was moving away. These observations showed that the universe was expanding.

2 The Big Bang theory is the accepted theory of the birth and evolution of the universe. It states that the universe began 13.8 billion years ago when an infinitely dense, incredibly energetic singularity began to expand and cool. Energy condensed into matter as the universe expanded; the universe has been expanding and cooling ever since.

3 Subatomic particles condensed from radiation shortly after the universe began in accordance with Einstein's mass–energy relationship. This was followed by a period of nuclear synthesis when hydrogen, helium and a small amount of lithium nuclei were formed. For reasons unknown, slightly more matter than antimatter was formed and this slight imbalance led to the matter we find in the universe today. As the universe expanded and cooled further, electrons and nuclei combined to form atoms and later these atoms came together to form the first stars. The universe now consists of a vast number of galaxies and the space between the galaxies continues to expand. Recent observations show that the rate of expansion is increasing.

4 The enormous temperature and pressure produced by gravity in the core of stars enables them to convert matter to energy through fusion reactions. These reactions result in the conversion of mass to energy in accordance with Einstein's mass–energy equivalence relationship $E = mc^2$.

5 Hydrogen, helium and some lithium were produced by nuclear synthesis in the Big Bang. Elements from helium to iron have been produced by fusion in stars and heavier elements have been formed in supernova explosions.

Key questions ➲ p. 172

6 Emission spectra are specific wavelengths of light emitted from excited elemental gases. Absorption spectra are specific wavelengths absorbed by an elemental gas. A black-body spectrum is a continuous spectrum emitted from a hot body.

7 The overall spectral emission from a star is like that from a black body. Because the wavelength of maximum emission from a black body is related to the temperature of the body, we can determine the temperature of a star by measuring the wavelength for which the intensity of light emitted from the star is greatest. Once the wavelength of maximum emission has been determined, we can apply Wien's law ($\lambda_{max} = b/T$) to find the temperature of the star. Alternatively we can use the relative intensity of the absorption spectra of ions and neutral atoms to estimate the temperature of the star.

8 As the temperature of a star increases, atoms in the gaseous layers around the star are ionised. If the star is very hot, multiple ionisation of heavy metals will begin and helium will be ionised. The strength of the ionic metal lines will increase and the strength of the neutral metal atomic lines will decrease as the temperature of the star increases.

9 The MK system classifies stars in order of temperature and uses the spectral class labels of OBAFGKM, where O is the hottest star and M the coolest. In addition, subcategories from 0 to 9 for each spectral class further differentiate the scale.

Key questions ➲ p. 176

10 An HR diagram plots luminosity against spectral class (temperature from high to low) and hence we can read the luminosity and temperature from the axes. The size of a star increases from the bottom to the top of the diagram and from the left to the right of the diagram. Hence the largest stars are at the top right and the smallest are at the bottom left of the diagram.

11 Stars fuse different elements throughout their lives and the different groups represent different stages in the life cycle of a star. The main sequence stars are fusing hydrogen to helium. The red giants are fusing a shell of hydrogen and possibly a core of helium. The supergiants are fusing elements heavier than helium and the white dwarf stars are the hot embers of stars in which fusion has ceased.

12 Stars with less than 0.23 $M_{\odot}$ fuse hydrogen to helium and then collapse to become white dwarf stars. Intermediate mass stars fuse hydrogen and may fuse helium before ejecting their outer layers in a nova and collapsing to become white dwarfs. Large stars may fuse heavier elements up to iron. If an iron core is formed, the star will collapse and explode in a supernova. If the remaining core is between 1.4 and 3 $M_{\odot}$ it will collapse to form a neutron star, and if the core is bigger it will form a black hole.

13 The lifecycle of a star can be traced out on an HR diagram to show how the star fuses different elements throughout its active life. For example, a star like the Sun would start on the main sequence, move to become a red giant and then collapse to become a white dwarf.

Key questions ➲ p. 177

14 The CNO cycle is the dominant hydrogen fusion process in stars greater than 1.3 $M_{\odot}$. This reaction cycles carbon to nitrogen and oxygen and back to carbon to catalyse the fusion of four protons to helium. In cooler stars, the p–p chain reaction is the dominant reaction. This reaction involves two protons fusing to form deuterium (^2H), then a proton fusing with deuterium to form helium-3 and finally, two helium-3 atoms combing to form helium-4 and two protons. Both the CNO and p–p chain reactions convert four protons to helium and release energy in accordance with Einstein's mass–energy relationship $E = mc^2$.

15 Elements from helium to iron can be produced by fusion reactions in the core of stars. Small stars are only massive enough to provide the temperature required to fuse hydrogen to helium, but super massive stars can fuse elements up to iron. The production of elements heavier than iron consumes energy and hence cannot be produced in the core of stars. These heavy elements are produced when core fusion ceases in super massive stars and they explode in a supernova. Supernovas create heavy elements and disperse them, together with those formed earlier in the core of the star, through the cosmos. This is why we see so many elements in the atmosphere of second-generation stars like our Sun and why our planet is made up of so many elements. Every atom in our bodies heavier than lithium was formed in the distant past in a star.

HSC EXAM-TYPE QUESTIONS

Objective-response questions

1 **D.** The increase in frequency indicates the lines are blue-shifted and hence the star is moving towards the Earth. Only a very large main sequence star would be hot enough to exhibit helium and multiply charged heavy metal ion lines. **A** is incorrect because if the star was moving away, the lines would be shifted to a lower frequency. In

addition, a small-star main sequence would not be hot enough to ionise helium in its atmosphere. **B** is incorrect because if the star was moving away the lines would be shifted to lower frequencies. **C** is incorrect because a small star would not be hot enough to ionise helium or multiply ionised heavy atoms in its atmosphere.

2 **D**. Hubble noted that the light from galaxies further away was more redshifted, indicating they were moving faster from the Earth. **A** and **B** are incorrect as both were observed by Hubble but neither proves the universe is expanding. **C** is incorrect as this would imply the universe was contracting.

3 **C**. This is the correct sequence. **A**, **B** and **D** are incorrect as they do not have the evolutionary stages in the correct order.

4 **A**. Only supergiant stars can fuse heavier elements. **B** is incorrect as the upper left contains main sequence stars that fuse hydrogen. **C** is incorrect as this is where the inert white dwarf stars are found. **D** in incorrect as this region contains main sequence red dwarf stars.

5 **A**. All three are required: the Big Bang produced the hydrogen (and some helium and a trace amount of lithium), stars fused the hydrogen to form atoms up to iron and supernova are required to produce elements heavier than iron. **B**, **C** and **D** are incorrect because all three processes are required to make all the elements.

Extended-response questions

6 EM This question tests students' knowledge of the evidence that supports the Big Bang theory.

Any two pieces of experimental or theoretical evidence are accepted, such as:

- The Big Bang theory predicts that nuclear synthesis would produce about 75% hydrogen and 25% helium ✓ and a small amount of lithium; these are the ratios we find in today's universe. ✓
- The Big Bang theory also predicted that the universe would be bathed in microwave radiation with a specific spectral distribution. ✓ Observations from Earth and space telescopes have confirmed the Big Bang prediction of the cosmic background radiation. ✓

7 EM This question tests students' understanding of the HR diagram and their ability to interpret new data using the diagram and their understanding of stellar evolution.

a Stars at the bottom left would be blue-white because they are very hot (i.e. white dwarf stars), while stars at the bottom left would be red because they are comparatively cool (red dwarf main sequence stars). ✓ Because both stars have the same luminosity but one is cooler, the cooler star must be larger as luminosity increases with the radius (surface area) and temperature of the star. ✓

b The lifetime of stars on the main sequence decreases as we move from the lower right to the upper left. ✓ The stars in the globular cluster must be old enough so that all the larger main sequence stars have stopped fusing hydrogen in their cores and left the main sequence. ✓ The turnoff point is about one solar mass.

8 EM This question tests students' understanding of stellar spectra and their knowledge of the fusion processes that occur in stars as they evolve.

a The astronomer would collect light from the star using a telescope, ✓ pass the light through a prism or grating spectrometer ✓ and record the spectrum photographically or electronically. ✓

b The Sun is hotter than a red dwarf star and hence the peak emission of its black body spectrum would occur at a shorter wavelength (i.e. in the yellow instead of the red). ✓ The red dwarf would also show strong neutral metal absorption lines as it is cooler than the Sun. ✓ The Sun would show some first ionisation lines for metals as it is hot enough to ionise these elements in its atmosphere. ✓

c Red dwarf stars are main sequence stars that fuse hydrogen to helium in their core. ✓ Because gravity does not compress the star strongly, fusion only needs to run at a low rate to provide the radiation pressure to stop the star from collapsing. For this reason these stars stay on the main sequence for hundreds or thousands of billions of years. ✓ In contrast a red giant star is fusing a shell of hydrogen and often also fusing helium in the core of the star. ✓ Because the star is much larger than a red dwarf the fusion reactions must run much faster to balance gravity and hence the lifetime of a red giant is comparatively short; perhaps only a few million years. ✓

9 EM This question tests students' understanding of fusion and their ability to apply Einstein's mass–energy equation to fusion.

a The change in mass is given by:

Δm = mass of products – mass of reactants

$= 3.3435 \times 10^{-27} - 2 \times 1.6726 \times 10^{-27}$

$= -1.7 \times 10^{-30}$ kg (the negative sign tells us that mass is lost)

$\Delta m = 1.7 \times 10^{-30}$ kg mass lost ✓

b Applying Einstein's mass–energy equation:

$E = mc^2 = (1.7 \times 10^{-30})(3 \times 10^8)^2 = 1.53 \times 10^{-13}$ J ✓

c Because both protons are positively charged they repel one another ✓ and hence to induce a fusion reaction the particles must have enough kinetic energy (a high-enough temperature) ✓ to overcome the electrostatic repulsion between the nuclei.

10 EM This question tests students' understanding of how fusion is initiated and maintained in stars.

a A protostar is formed when gravity contracts a large gas and dust cloud. As the gas comes together it loses potential energy, which is converted to heat. ✓ The core temperature therefore rises. If there is enough material (about 0.1 $M_\odot$), the core temperature will reach the ignition temperature (10^7 K) for the p–p chain reaction to begin to fuse hydrogen nuclei (protons) to helium nuclei. ✓ If the gas and dust cloud is not large enough, the core temperature will not become high enough for fusion to begin.

b As an element fuses, the inert product forms. This is heavier and hence pulled by gravity to the centre of the star where it forms an inert core. ✓ Eventually the inert core becomes large enough to move the fusing species far enough from the centre of the star so that the temperature is not high enough to maintain the reaction. ✓ When this occurs the outward forces that balance gravity disappear and gravity contracts the star and heats the core. If there is enough mass gravity it may cause the fusion to begin in the previously inert core. But if there is not enough mass to raise the temperature sufficiently, fusion will stop in the star. Very large stars can fuse elements up to iron but when an iron core forms and becomes large enough, fusion will cease because the fusion of iron consumes energy. ✓ When the fusion stops will depend on the mass of the star. Very small stars will fuse only hydrogen, stars like our Sun can fuse hydrogen and helium, and more massive stars can fuse heavier elements—but even the biggest stars cannot fuse elements heavier than iron.

c Very small stars collapse to become white dwarf stars after they fuse hydrogen. Larger stars may fuse helium before puffing away their outer layers and collapsing to form white dwarf stars. ✓ Very large stars may fuse elements heavier than helium and when fusion ceases in these stars they will collapse and explode in a supernova. If the remaining core has a mass between 1.4 $M_\odot$ and 3 $M_\odot$ it will collapse to become a neutron star. ✓ If the remaining core has a mass greater than 3 $M_\odot$, the core will collapse to become a black hole. ✓

CHAPTER 13 STRUCTURE OF THE ATOM

INQUIRY QUESTION:

How is it known that atoms are made up of protons, neutrons and electrons?

The word 'atom' comes from the Greek word *tomos*, meaning 'uncuttable', because initially atoms were considered to be uncuttable. However, experiments in the late 19th century with cathode rays led to the discovery of a negatively charged subatomic particle, the electron. Investigations conducted in 1917 with radioactive particles led to the discovery of a positively charged subatomic particle, the proton, and in 1932 a neutral subatomic particle, the neutron, was discovered. The modern nuclear model of the atom is built upon these three subatomic particles.

1 The electron

» Students investigate, assess and model the experimental evidence supporting the existence and properties of the electron, including:
- early experiments examining the nature of cathode rays
- Thomson's charge-to-mass experiment
- Millikan's oil drop experiment.

Cathode ray early experiments

→ In early experiments on electricity, physicists passed current through low-pressure gases. When the current flowed through the low-pressure gas, the gas glowed. In 1870 William Crookes found that when the gas pressure was reduced further, a dark space near the cathode (negative electrode) grew until it reached the anode. At this point no light was emitted from the body of the tube. Crookes also found that at extremely low gas pressures (10^{-6} of atmospheric pressure), current continued to flow in the tube and the glass near the anode glowed green. Further investigation showed that rays were being emitted from the cathode towards the anode. These **cathode rays** were a mystery at the time and became an area of active scientific research.

→ The evacuated, current-carrying tubes developed by Crookes became known as **Crooke's tubes** and later as **cathode ray tubes** (CRT). Figure 13.1 shows how cathode rays are produced in a Crooke's tube by placing a high voltage across the electrodes in the tube.

cathode rays: rays emitted from the negative electrode (cathode) in a Crooke's tube
Crooke's tube or **cathode ray tube (CRT):** an evacuated glass tube with inserted metal electrodes used to produce cathode rays

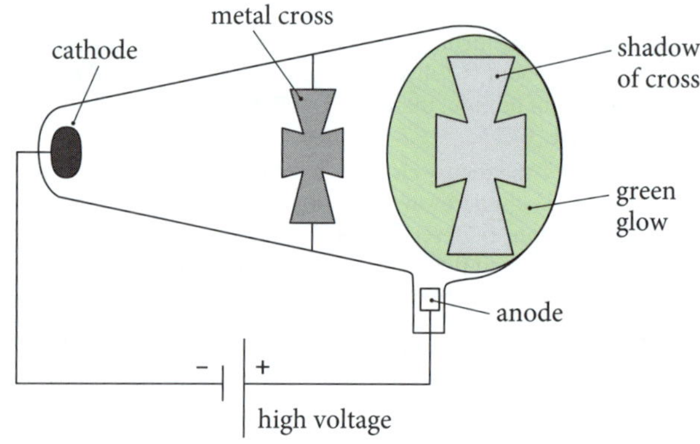

Figure 13.1 Crooke's tube experiment showing how cathode rays travel in straight lines and cast a shadow

→ Experiments with that cathode rays showed they travelled in straight lines, cast a shadow, exposed photographic plate, passed through very thin metal foils, could turn a paddle wheel and were deflected by magnetic fields. Early experiments conducted by Heinrich Hertz indicated the rays were not deflected by electric fields and led him and some other physicists in Germany to suggest the rays were a 'disturbance' in the 'ether' like electromagnetic waves. In contrast the British scientists William Crookes and JJ Thomson thought the waves were made up of charged corpuscles (particles). Some of the evidence used by each side of the debate to support their points of view is listed in Table 13.1.

→ Because the early observations supported both points of view to some degree, the nature of cathode rays remained an area of debate for several decades.

Table 13.1 Early evidence used to support wave and particle theory of cathode rays

Evidence used to support the wave model	Evidence used to support the particle model
Rays expose photographic plate, like light.	Rays turn a paddle wheel and hence must have momentum.
Rays pass through thin metal foils.	Rays are deflected by magnetic fields.
Rays are not deflected by electric fields (this experimental result was later shown to be incorrect).	Rays are emitted at a right-angle to the cathode surface (rather than in all directions like a wave).
Rays travel in straight lines and cast a shadow.	

Thomson's charge-to-mass experiment

- The cathode ray debate was settled in favour of the particle model by a series of experiments conducted by JJ Thomson in 1897.
- Thomson first confirmed that cathode rays were deflected by magnetic fields. He then showed, contrary to early results obtained by Hertz, that the rays were also deflected by electric fields. It appears the vacuum used by Hertz was not low enough to demonstrate a deflection because ions in the tube were attracted to the charged plates and reduced the electric field strength. Thomson's observation that the rays were deflected by electric fields negated one of the main arguments used against the particle model.
- With the knowledge that the rays were deflected by electric and magnetic fields, Thomson constructed a vacuum tube with electric and magnetic fields at right-angles to each other, as shown in Figure 13.2.

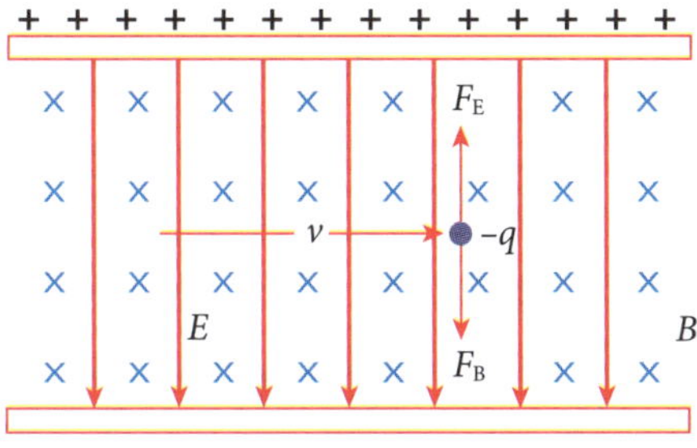

Figure 13.2 Forces on a charge moving perpendicularly to crossed electric and magnetic fields

- The crossed fields produced an opposing electric force and magnetic force on the cathode rays passing through the fields. Thomson then adjusted the voltage on the parallel plates until the cathode ray passed between the crossed fields without being deflected.

When this occurred the upwards electric force and downwards magnetic force would be balanced and Thomson could write:

$F_E = F_B$

$qE = qvB$

And hence:

$$v = E/B$$

where v is the velocity of the charged particle
E is the electric filed and
B is the magnetic field

- Thomson used parallel charged plates separated by a distance d to produce the electric field and hence could calculate the strength of the electric field using $E = V/d$. He used parallel coils (Helmholtz coils) to produce the magnetic field because the field between parallel coils is uniform and can be calculated from the current in and the dimensions of the coils.
- When he conducted the experiment Thomson found the velocity of the cathode rays was only a small fraction of the speed of light. This result strongly favoured the particle model as waves in the ether were expected to travel at the speed of light.
- Thomson's final experiment (now known as *Thomson's experiment*) put the case for the particle model beyond doubt. In this experiment he found the charge-to-mass (q/m) ratio of the particles by measuring the radius of curvature of cathode rays of known velocity in a magnetic field directed perpendicularly to the velocity of the rays. Thomson's experiment is shown in Figure 13.3.

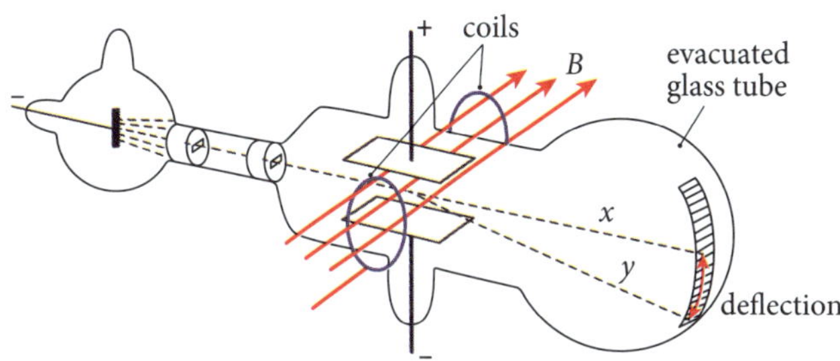

Figure 13.3 Thomson's q/m experiment

- Thomson first adjusted the electric and magnetic fields to ensure the beam passed through the crossed fields without being deflected (path x) and used the values to calculate the velocity of the particles. He then turned off the electric field and measured the deflection of the beam caused by the magnetic field (path y). Thomson then used the deflection to calculate the radius of curvature of the beam when it was in the magnetic field. As the centripetal force on the particles is provided by the magnetic field:

$qvB = mv^2/r$

And hence $q/m = v/Br$

- The q/m value obtained by Thomson was 1836 times greater than the q/m ratio of the smallest known ion (hydrogen). Thomson and others had tried to find the charge on an electron by measuring how fast charged water droplets fell in electric fields. The experiments were not conclusive but implied that the magnitude of the quantum of negative charge would be similar to the charge on a hydrogen ion. Combining this result with the charge-to-mass ratio he had measured, Thomson concluded that the mass of the cathode ray particles must be at least 1000 times smaller than the mass of the smallest atom.
- Thomson then measured the q/m ratio of cathode rays produced using cathodes made of different metals, and always got the same value. Even though the evidence was not overwhelming, Thomson was convinced that the negatively charged corpuscles he had discovered were subatomic particles that would be found in all atoms. His assumption turned out to be correct; he had shown the atom was not indivisible and he had discovered one of its fundamental constituents.
- The negative corpuscles discovered by Thomson were later called **electrons** because the term had been used to describe an 'atom of electricity' in earlier work by George Stoney.
- The method used by Thomson to find the charge-to-mass ratio of electrons was later applied to ions and enabled physicists to identify and separate **isotopes** of elements for the first time. An isotope is an element with a different number of neutrons. The device used is now called a *mass spectrometer* as it separates a sample into its constituent elements and isotopes. Mass spectroscopes utilise the principle that the radius of curvature of an ion moving perpendicular to a magnetic field is proportional to the mass of the ion. Hence if two singularly charged isotopes of an element were sent into the magnetic field with the same velocity, the heavier isotope would have a bigger radius of curvature in the field, enabling it to be identified.

electron: fundamental subatomic particle that carried a negative charge of 1.602×10^{-19} C
isotope: an element with a different number of neutrons in the nucleus

EXAMPLE 1

In a mass spectrometer, shown in Figure 13.4, two isotopes of carbon are projected into a magnetic field aligned at 90° to the velocity of the isotopes. The isotopes were both singularly ionised and had a velocity of $v = 10^4$ ms^{-1} when they entered the magnetic field, which had a strength of 0.05 T. The isotopes traced out circular paths in the field with diameters of 4.18 cm and 5.0 cm.

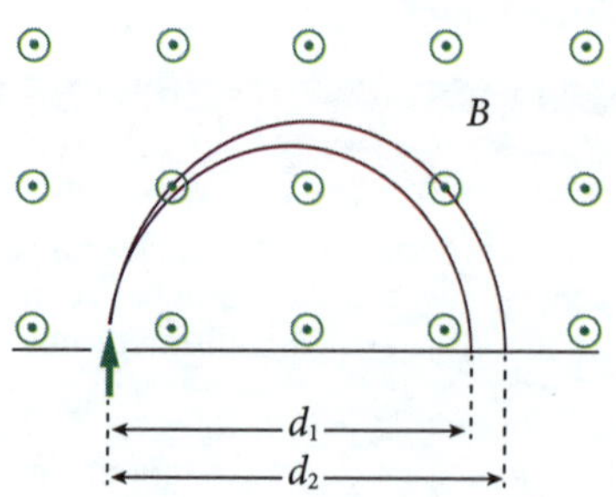

Figure 13.4 Charged isotopes being deflected by a magnetic field in a mass spectrometer

a **Will the energy of the isotopes change as they move through the magnetic field?**
b **Are the isotopes positively or negatively charged?**
c **Find the mass of the isotopes.**

Remember to change to SI units before performing calculations

Answer:

a Because the force applied by the magnetic field is perpendicular to the velocity it changes the direction of motion but not the velocity of the particle and hence the kinetic energy of the particle remains constant.
b Applying the right-hand palm rule to the charged isotopes in Figure 13.4 shows the isotopes are positively charged.
c As the centripetal force is provided by the magnetic force we can write:

$mv^2/r = qvB$

And hence $m = qBr/v$

For the smaller isotope:

$$m_1 = qBr/v = \frac{1.602 \times 10^{-19} \times 0.05 \times 0.0209}{10^4}$$
$$= 1.674 \times 10^{-26} \text{ kg}$$

For the larger isotope:

$$m_2 = qBr/v = \frac{1.602 \times 10^{-19} \times 0.05 \times 0.025}{10^4}$$
$$= 2.00 \times 10^{-26} \text{ kg}$$

FIRSTHAND INVESTIGATION

Cathode rays

Safety: Cathode ray tubes require high voltages and emit low-intensity x-rays. Students should be careful not to touch any high-voltage wires, stand well away from operating tubes and only operate the tubes for a short time.

Students should use cathode ray tubes to investigate some or all of the properties of cathode rays shown below:

1 Cathode rays make the glass at the end of the tube near the anode glow green.
2 Cathode rays travel in straight lines and cast a shadow.

3 Cathode rays are deflected by magnetic fields.
4 Cathode rays are deflected by electric fields.
5 Cathode rays can turn a paddle wheel (this experiment is illustrated in Figure 13.5).
6 Cathode rays have a q/m ratio (i.e. JJ Thomson's experiment—if the equipment is available).

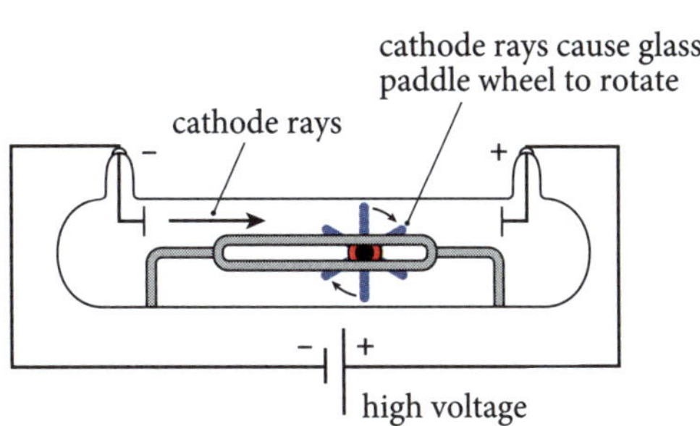

Figure 13.5 Experimental setup required to demonstrate cathode rays turning a paddle wheel

➔ KEY QUESTIONS

1 **What evidence supported the wave and particle models for cathode rays?**

2 **How did Thomson measure the velocity and charge-to-mass ratio for cathode ray particles?**

Answers ➲ p. 193

Millikan's oil drop experiment

- After Thomson found the charge-to-mass ratio of the electron, physicists realised that both the mass and charge could be found if one of the quantities could be determined.
- The charge on the electron was first accurately measured by Robert Millikan and his research student Harvey Fletcher in a series of experiments that were conducted between 1909 and 1913.
- If the electron was the quantum of charge, Millikan reasoned that an ionised object would have a charge that would be an integer multiple of the charge on the electron. To find the charge he designed an experiment that would use an electric field to levitate a small sphere of known mass, as shown in Figure 13.6.

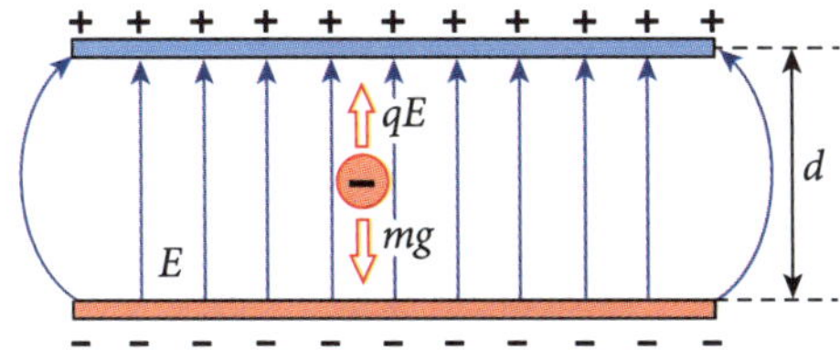

Figure 13.6 Levitating a charged sphere in an electric field

- When the charge is at rest, the forces on the charge must sum to zero and hence:
  $mg = qE = qV/d$
  And therefore:
  $q = mgd/V$
  where q is the charge on the sphere
  m is the mass of the sphere
  d is the separation between the charged plates and
  V is the potential difference that must be placed across the plates to levitate the charge
- Note the charge on the sphere is not the charge on the electron but an integer multiple of the electron charge. To find the quantum of charge the experiment would have to be repeated many times and a common denominator sought that could be divided into the charge on each sphere.
- The apparatus used by Millikan is shown in Figure 13.7. It consisted of two cylindrical chambers with a pin-hole linking the chambers. An **atomiser** was used to spray a mist of tiny oil droplets into the top chamber that drifted downwards. A type of oil was used that had a very high vapour pressure to stop the droplets evaporating in the chamber. Some of the droplets would have been charged by leaving the atomiser and an x-ray source was used to further charge the droplets. Droplets that drifted down into the lower chamber were illuminated with a light and could be observed through a telescopic eyepiece. The top and bottom plates on the lower chamber were connected to a variable voltage supply. By applying a voltage to the plates, it was possible to apply a force to counteract the gravitational force on a droplet and levitate the droplet between the plates.

atomiser: a device used to produce a fine mist of droplets

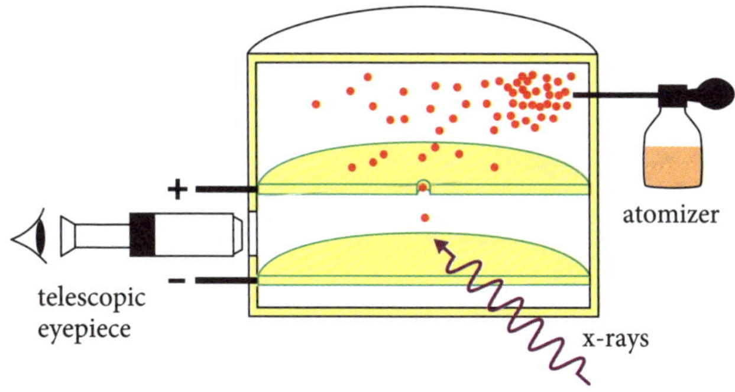

Figure 13.7 Millikan's oil drop experiment apparatus

- By the time the droplets entered the lower chamber they had reached terminal velocity. Millikan would initially let a droplet drift past the eyepiece and estimate its velocity. Terminal velocity of a sphere is related to the weight of the sphere and Millikan could use the measured terminal velocity of the droplet to determine the droplet's weight. Millikan would then increase the positive voltage on the top plate until the droplet would levitate between the plates. The levitating voltage and the weight of the droplet could then be used to calculate the charge on the droplet using the above equation.

- Millikan measured the charge on thousands of droplets and found the common denominator for these charges was:
  $e = 1.592 \times 10^{-19} \mp 0.2\%$ C
- The accepted value of the charge on an electron today is:
  $e = 1.602\ 176 \times 10^{-19}$ C
- Millikan's value was low because the value for the viscosity of air he used was not correct.
- Once the charge on the electron was known, Thomson's charge-to-mass ratio was used to show that the mass of the electron was:
  $m_e = 9.109\ 38 \times 10^{-31}$ kg

EXAMPLE 2

A student repeats Millikan's experiment and finds a potential difference of 6000 V is required to levitate a 0.04 μg oil droplet between parallel plates separated by 2 cm.

a Find the size of the charge on the droplet.

b Explain why this is not the charge of the electron.

Remember that the SI unit of mass is the kilogram (not the gram)

Answer:

a As the charge is at rest, the forces on the charge must sum to zero and hence: $mg = qE = q(V/d)$
Note 0.04 μg = 0.04×10^{-9} kg.

$$\text{Hence } q = mgd/V = \frac{(0.04 \times 10^{-9} \times 9.8 \times 0.02)}{6000}$$

$$= 1.3067 \times 10^{-15} \text{ C}$$

b The charge on the droplet will be much greater than the charge on an electron because a large number of electrons have been removed from the droplet (i.e. the droplet contains a large number of ions).

➔ KEY QUESTION

3 **How did Millikan find the charge on the electron?**

Answers ➲ p. 193

2 The nuclear model of the atom

» Students investigate, assess and model the experimental evidence supporting the nuclear model of the atom, including:
- the Geiger–Marsden experiment
- Rutherford's atomic model
- Chadwick's discovery of the neutron.

Rutherford's model of the atom

- Between 1908 and 1911, under the direction of Ernest Rutherford, Hans Geiger and Ernest Marsden carried out a series of scattering experiments that would lead Rutherford to propose a nuclear model of the atom.
- After Thomson discovered the electron in 1887, he proposed a model of the atom in which negative electrons were arranged in rings in a uniform, positively charged substrate. Thomson's model is sometimes called the **plum pudding model** as the electrons are like the plums and the positive substrate is analogous to the pudding.

plum pudding model: an atomic model suggested by Thomson that had electrons embedded in a positive substrate

- To investigate Thomson's model, Rutherford proposed a scattering experiment using alpha rays and assigned the task of conducting the experiment to Geiger and Marsden. Rutherford had shown alpha particles were doubly charged helium nuclei emitted at high speed from radium and other radioactive atoms. He thought they would be scattered by small angles when they interacted with the electrons in Thomson's atom.
- The experimental setup used by Geiger and Marsden to conduct their investigation is illustrated in Figure 13.8. Lead slits were used to **collimate** a beam of alpha particles that were incident upon a very thin piece of gold foil. Gold was used because it is so **ductile** it can be beaten to a foil that is only about 500 atoms thick. To detect the alpha particles they used a zinc sulphide crystal. When an alpha particle hits a crystal of zinc sulphide it produces a tiny flash of light that can be observed with a microscope. Geiger and Marsden used this method to detect and record the angle of deflection of each scattered alpha particle. They observed and recorded the angle of scatter of over a million individual alpha particles during their investigations.

collimate: make rays of light parallel

ductile: the ability to be formed, beaten thinly, drawn into a wire or to undergo significant plastic deformation without rupturing

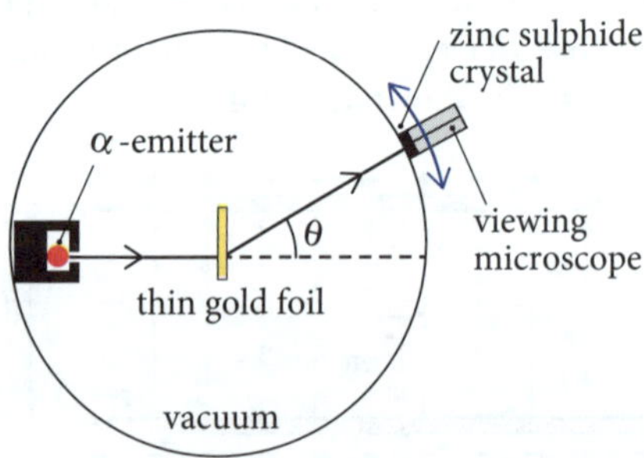

Figure 13.8 Geiger and Marsden experiment

- As Rutherford expected that the energetic, doubly charged alpha particles would be scattered by only a small amount he assigned his senior researcher Geiger to investigate this area. Marsden was an undergraduate student at the time and was given the less important task of looking to see if any alpha particles scattered by large angles.
- As expected most alpha particles passed straight though the foil with no or little deflection but, to Rutherford's surprise, Marsden found that some alpha particles (1 in 8000) were scattered by more than 90°. This observation could not be explained by Thomson's model of the atom. Rutherford reasoned that the energetic alpha particles could only be reflected from the foil by encountering an extremely intense electric field.
- To explain the experimental results, Rutherford suggested that almost all the mass and all the positive charge of the atom was concentrated in a tiny fraction of the total volume of the atom. Rutherford called this tiny, positively charged region the *nucleus* ('little nut' in Latin) and assumed it would reside in the centre of the atom.
- Most alpha particles would not come close to the nucleus and pass straight through the atom with little deflection, but occasionally an alpha particle would approach the nucleus head on and be scattered at an angle greater than 90°. Calculations based on the results of the experiment indicated that the nucleus was smaller than 10 000th of the diameter of the atom.

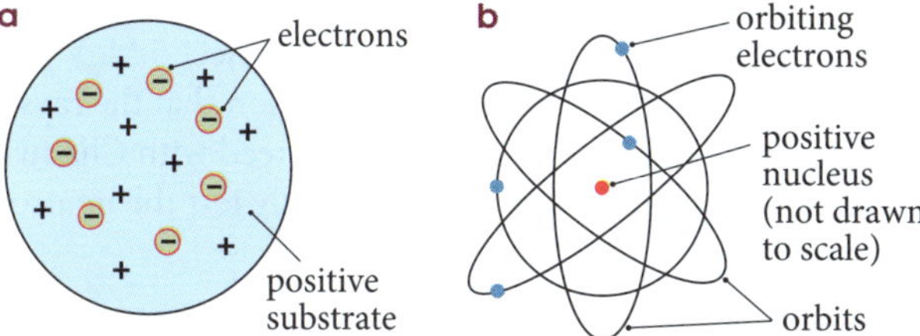

Figure 13.9 a Thomson's atomic model and **b** Rutherford's atomic model

- In 1911 Rutherford published a **planetary model** of the atom with electrons orbiting a tiny, dense, positively charged nucleus (Thomson's atomic model and Rutherford's atomic model are shown in Figure 13.9). Rutherford proposed the electrostatic attraction between the negative electrons and positive nucleus would provide the centripetal force to hold the electrons in their orbit. The expected scatter patterns for Thomson's uniform density atom and for Rutherford's nuclear atom are illustrated in Figure 13.10.

planetary model: an atomic model suggested by Rutherford that consisted of a tiny, positively charged nucleus that contained most of the mass of the atom and orbiting electrons

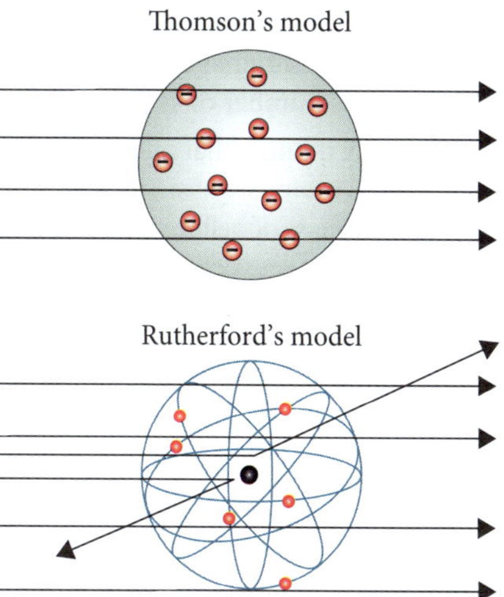

Figure 13.10 Expected alpha particle scattering from Thomson's atom and Rutherford's atom

- Rutherford's nuclear model explained the results of the Geiger–Marsden scattering experiment, but his planetary model was not taken seriously by most physicists at the time because it would have been unstable. This is because an orbiting electron would be accelerating and would therefore emit electromagnetic radiation. This radiation would decrease the energy of the electron, causing the electron to slow down and spiral into the nucleus, as shown in Figure 13.11.

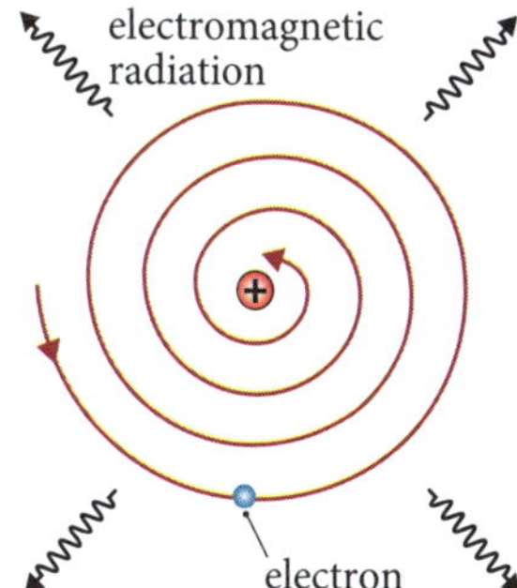

Figure 13.11 Instability of Rutherford's atomic model

- Another shortcoming of Rutherford's atomic model was that it did not explain atomic emission or absorption spectra.
- In 1919 Rutherford reported that when he bombarded nitrogen with alpha particles, a hydrogen ion was emitted. Although it had been suggested earlier, this was the first time an experiment had demonstrated that other elements contained hydrogen nuclei. Rutherford knew hydrogen was the smallest atom and concluded that the nucleus of each of the other elements was made up from hydrogen nuclei. He named this positive particle a *proton*.
- Nuclei made of only protons raised two immediate questions. What held the positive protons together in such a small space? Why, if the periodic table lists elements in term of the number of protons, is the atomic

number so different to the mass number? For example, helium has an atomic number of 2 (two charges in the nucleus) but a mass number of 4.

- Rutherford initially thought that nuclei must contain electrons as well as protons. Physicists at the time would explain the atomic mass of helium with this scheme by saying that the nucleus must contain four protons and two nuclear electrons. At that time there was no strong evidence against having electrons in the nucleus, but Rutherford thought it was improbable so in 1920 he suggested that a proton and an electron might combine in the nucleus to form a neutral particle. In 1921 Rutherford named this hypothetical particle a *neutron*. However, he predicted that because they carried no charge, neutrons would pass through matter easily and be very difficult to detect.

→ KEY QUESTIONS

4 **What was Geiger and Marsden's scattering experiment?**

5 **What did the Geiger–Marsden experiment find?**

6 **Why were the Geiger–Marsden results unexpected?**

7 **What atomic model did Rutherford propose to explain the results of the Geiger–Marsden experiment?**

8 **Why was Rutherford's planetary atomic model incompatible with the laws of electromagnetism?**

Answers ➲ p. 193

The discovery of the neutron by Chadwick

- From 1921 Rutherford conducted a series of experiments with his colleague James Chadwick to try to prove that neutrons existed. However, despite multiple attempts over the next decade, all their experiments failed to detect the illusive particle.
- A breakthrough came in 1930 when Bothe and Becker reported that, when they bombarded beryllium with alpha rays, an intense neutral radiation was produced. This radiation was very penetrating but not very ionising and they assumed that it was gamma rays.
- Marie Curie's daughter Irene and her husband Jean Joliot conducted a series of experiments with the radiation emitted from beryllium in 1931 (see Figure 13.12). They found to their surprise that when paraffin was irradiated with the neutral rays, protons were ejected from the paraffin with energies in the order of 5 MeV. They found the neutral radiation emitted from the beryllium had three times as much energy as the alpha particles that produced the radiation. This led Curie and Joliot to think that some sort of nuclear disintegration was occurring in the beryllium even though no protons were ejected. However, they did not make the link to Rutherford's suggestion of a neutral particle in the nucleus (perhaps they were unaware if it) and continued to assume the rays were intense gamma rays.

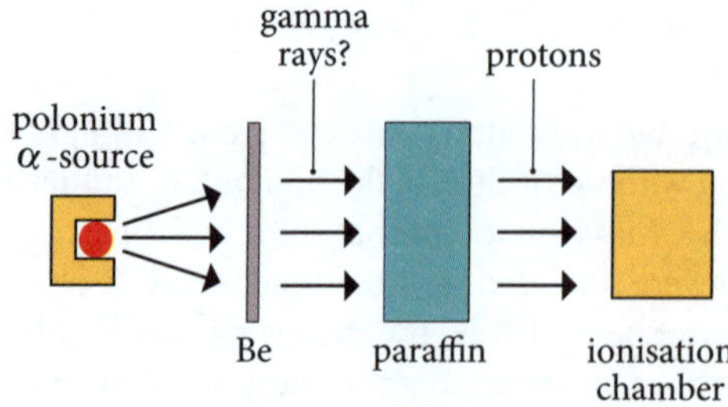

Figure 13.12 Experiment conducted by Curie and Joliot

- Chadwick was surprised by the result obtained by Bothe and Becker, and assigned the task of investigating the result to one of his research students, a young Australian named Hugh Webster. In 1931 Webster reported that the radiation emitted from beryllium in the direction of the alpha particles was more energetic than the radiation emitted in the opposite direction. Chadwick thought this could not be explained by the gamma ray hypothesis because the energy of the gamma rays should have been the same for rays emitted in all directions. He began to wonder if this neutral radiation might be neutrons emitted when beryllium nuclei reacted with alpha rays.
- When Chadwick read a report of the Curie–Joliot experiments, he immediately realised that the gamma rays would not have enough energy to eject 5 MeV protons and strengthened his conviction that the rays consisted of neutrons. Rutherford agreed with Chadwick and encouraged him to experimentally test the neutron hypothesis.
- Because alpha rays travel only a few centimetres in air, Chadwick used an evacuated chamber with a polonium alpha source and a piece of beryllium to produce the neutral radiation. He then passed the radiation through a target material in front of an ionisation chamber (see Figure 13.13). When he irradiated gases, he simply replaced the gas in the ionisation chamber with the target gas.

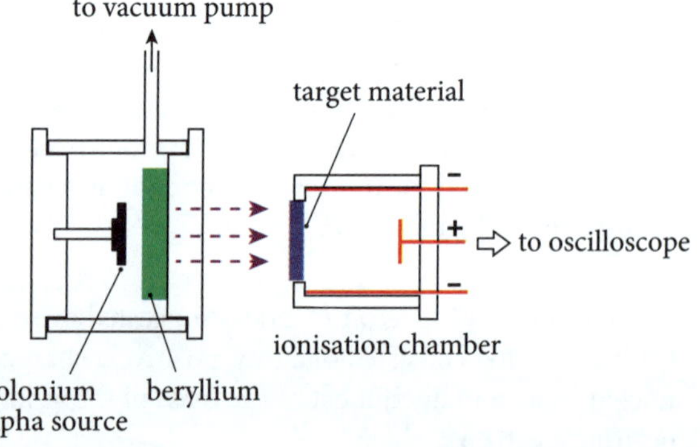

Figure 13.13 Apparatus used by Chadwick

- Fast-moving, charged particles entering the ionisation chamber were slowed as they ionised the gas molecules in the chamber. This produced a pulse of current that was amplified and displayed on an oscilloscope. Because the number of electron–ion pairs produced by a fast-moving proton or ion is related to the initial kinetic energy of the particle, the current pulse detected in the ionisation chamber could be used to calculate the kinetic energy of the charged particle.
- Chadwick found the radiation not only knocked protons out of paraffin but it also knocked particles out of a range of elements, including hydrogen, lithium, beryllium, boron, carbon, helium, oxygen and nitrogen. He assumed the particles were the recoiling atomic nuclei of these atoms. Chadwick subsequently measured the recoil energy of hydrogen and nitrogen. He then used conservation of momentum and kinetic energy to show that if the recoil velocity was caused by a gamma ray it would need to have increasing energy as the mass of the target atom increased. As the gamma rays in the radiation would have the same energy, this would seem to violate the laws of conservation of momentum and energy. Chadwick then showed that the results could be easily accounted for if the recoil was caused by a collision with a neutral particle with a similar mass to the proton.
- By combining his results for hydrogen and nitrogen, and applying the laws of conservation of momentum and kinetic energy, Chadwick showed that the neutrons would have a mass slightly greater than a proton. (The accepted mass of the neutron today is 1.008 664 atomic mass units.)
- Chadwick correctly assumed that the alpha bombardment of beryllium produced neutrons by **transmuting** the beryllium nucleus to a carbon nucleus as follows:

  $\mathrm{Be}^9 + \mathrm{He}^4 + \alpha = \mathrm{C}^{12} + n^1$

transmutation: a nuclear reaction that changes one element into another element

- In a few weeks of intense work, after 12 years of searching, Chadwick had finally showed that Rutherford's farsighted speculation about the existence of a neutral nuclear particle was correct. The third principle constituent of matter had at last been found.

→ KEY QUESTIONS

9 **What evidence led Rutherford to propose that nuclei may contain neutrons?**

10 **What were the key experiments that led Chadwick to investigate the radiation emitted from beryllium?**

11 **How did Chadwick prove that the radiation emitted from alpha-irradiated beryllium was made up of neutrons?**

Answers ➲ p. 193

CHAPTER SYLLABUS CHECKLIST

Are you able to answer these questions from the syllabus for this chapter? Tick each question as you go through the checklist if you are able to answer it. If you cannot answer a question, turn to the relevant page in the study guide to find the answer. For NESA key word meanings, go to www.educationstandards.nsw.edu.au and search 'key words'.

FOR A COMPLETE UNDERSTANDING OF THIS TOPIC:		PAGE NO.	✓
1	Can I outline the contradictory evidence that led to the cathode ray debate?	182–183	
2	Can I explain how Thomson measured the velocity of cathode rays?	183	
3	Can I explain how Thomson measured the charge-to-mass ratio of the particles that make up cathode rays?	183–184	
4	Can I explain how Millikan measured the charge on the electron?	185–186	
5	Can I perform calculations related to Thomson's experiment and Millikan's experiment?	186	
6	Can I describe the Geiger–Marsden scattering experiment and outline the results they obtained?	186–187	
7	Can I outline the nuclear atomic model proposed by Rutherford to explain the results of the Geiger–Marsden experiment?	187	
8	Can I describe the shortcomings of Rutherford's nuclear atomic model?	187	
9	Can I outline the reasoning that led Rutherford to predict that a neutral atomic particle might exist?	187–188	
10	Can I outline the experimental results that led Chadwick to think that neutrons were being produced when beryllium was bombarded with alpha rays?	188	
11	Can I outline how Chadwick used conservation of kinetic energy and momentum to prove the radiation emitted from beryllium was a stream of neutrons?	188–189	

HSC EXAM-TYPE QUESTIONS

Objective-response questions (1 mark each)

1 Millikan measured the charge on the electron by balancing the forces on a charged oil drop. What forces were in balance in Millikan's experiment?

A magnetic force and the gravitational force

B electric force and the gravitational force

C electric force and the magnetic force

D electric, magnetic and gravitational forces

2 An electron is projected at velocity v between two charged parallel plates with an electric field E between the plates, as shown in Figure 13.14.

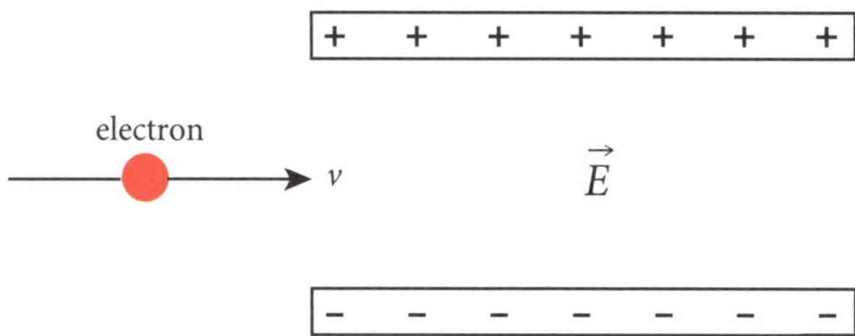

Figure 13.14 Electron projected between charged parallel plates

What needs to be added for the charge to pass through the plates without being deflected?

A a magnetic field directed into the page with a strength $B = E/v$

B a magnetic field directed out of the page with a strength $B = vE$

C a magnetic field directed out of the page with a strength $B = E/v$

D a magnetic field directed into the page with a strength $B = vE$

3 Which of the following best describes the result of the Geiger–Marsden scattering experiment?

A Most alpha particles were deflected.

B Some alpha particles were slightly deflected but most passed straight through.

C About half of the particles passed straight through with minimal or no deflection and half were deflected by a large angle.

D Almost all of the alpha particles passed straight through with minimal or no deflection and a small number were scattered by a large angle.

4 When beryllium is bombarded with alpha particles, neutrons are ejected from the beryllium with three times more energy than the incident alpha particles. Where does the neutron's energy come from?

A The alpha particle causes a nuclear reaction in which mass is lost.

B The electrons in beryllium move to a lower energy level and release energy.

C The alpha particle causes the beryllium atom to split and release energy.

D The alpha particle causes a nuclear reaction in which mass is gained.

5 JJ Thomson measured the radius of curvature of cathode ray beams (electrons) in a magnetic field to find the charge-to-mass ratio of the electron. For the electron shown in Figure 13.15, what would the radius of the path be if the electron entered the field travelling three times faster and the strength of the magnetic field was halved?

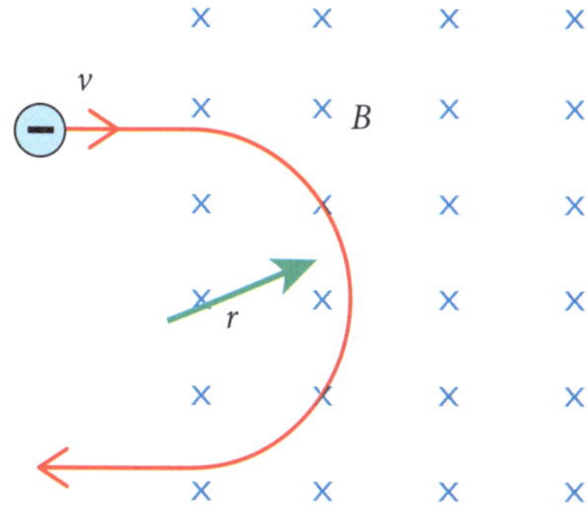

Figure 13.15 Path of a charge moving through a magnetic field

A $6r$

B $2r/3$

C $3r/2$

D $r/6$

Extended-response questions

6 Cathode rays were first observed in 1855 but it was not until 1897 that Thomson proved they were made up of negatively charged subatomic particles.

a Describe a laboratory experiment that could be used to produce and observe cathode rays. (3 marks)

b Outline three properties of cathode rays and explain how they could be demonstrated in the school laboratory. (3 marks)

7 Millikan measured the charge on the electron in his famous oil drop experiment.

a Explain how Millikan found the charge on a charged droplet of oil. (3 marks)

b How did Millikan find the charge on the electron from his measurements of the charge on oil drops? (2 marks)

8 JJ Thompson ended the cathode ray debate by finding the charge-to-mass ratio of the charged particles (electrons) in the beam.

a Explain how Thomson measured the velocity of a cathode ray beam. (3 marks)

b Explain the relevance of the velocity Thomson measured to the cathode ray debate. (1 mark)

9 Ernest Rutherford proposed a new atomic model in 1911.

a Describe Rutherford's atomic model. (2 marks)

b Outline the evidence that led Rutherford to propose this atomic model. (3 marks)

c Explain why the model was incompatible with the laws of electromagnetism. (2 marks)

10 Of his own work Isaac Newton said, 'If I have seen further than other men, it is because I have stood on the shoulders of giants.' Evaluate the relevance of this statement with reference to Chadwick's discovery of the neutron. (4 marks)

ANSWERS

KEY QUESTIONS

Key questions ➲ p. 185

1 The wave theory evidence included: the rays travel in straight lines and cast a shadow like light rays; the rays were not deflected by electric fields (though this evidence was later shown to be erroneous); and the rays exposed photographic plate and passed through thin metal foils. The particle theory was supported by evidence that included: the rays turned a paddle wheel (indicating they had momentum); they were deflected by magnetic fields; and they left the cathode at 90° to the surface (not in all directions like a wave).

2 Thomson first passed the rays through electric and magnetic fields that were oriented at 90° to one another and to the velocity of the rays. He adjusted the strength of the fields until the ray passed straight through the crossed fields without being deflected. Thomson then equated the electric and magnetic forces on the charge to obtain the equation $v = E/B$ and used it to determine the velocity of the particles. Thomson then passed the beam into a magnetic field at 90° to the particle velocity and measured the radius of curvature of the path taken by the particle in the field. As the centripetal force on the charge was provided by the magnetic field he was able to write $qvB = mv^2/r$ and rearranged it to obtain $q/m = v/Br$, which he used to find the q/m ratio.

Key question ➲ p. 186

3 Millikan used x-rays to charge oil droplets and then levitated them in an electric field. By equating the force of gravity and the electric force on the charge he could write $mg = qE = qV/d$, which can be rearranged to give the charge on the droplet $q = mgd/V$. As the droplets carried multiple charges, Millikan measured the charge on a large number of oil droplets and then looked for a common denominator that would divide into all the charge values he had obtained. This common denominator was the quantum of electric charge, the charge on an electron.

Key questions ➲ p. 188

4 Geiger and Marsden fired high-speed alpha particles (He^{++}) through thin gold foil and measured the degree to which each alpha particle was scattered. They used a piece of polonium as an alpha particle source and detected the alpha particles by observing the flash of light made by the particles when they were incident upon a zinc sulphide crystal.

5 Geiger and Marsden found that most alpha particles passed straight through the gold foil with little or no deflection, but 1 in 8000 were deflected by an angle greater than 90° (i.e. they were reflected).

6 Rutherford and Geiger expected the alpha particles to pass through the foil with little deflection as the density of the Thomson atomic model would have been very low. It would be like firing bullets through jelly with small glass marbles (electrons) in it.

7 Rutherford suggested a nuclear model in which all the positive charge and almost all the mass in the atom was concentrated in a tiny nucleus near the centre of the atom. The electrons orbited some distance from the nucleus and were held in their orbit by electrostatic attraction between the electrons and the positively charged nucleus. Because most of the atom was empty space, most alpha particles would pass straight through the atom. However, occasionally an alpha particle would come near the nucleus and the strong electrostatic repulsion would deflect the alpha particle by a large angle (the nucleus would not move significantly because gold is a very heavy atom).

8 Rutherford's model was based on orbiting electrons. These electrons would be accelerating and would consequently radiate electromagnetic radiation. The energy going into the radiation would come from the kinetic energy of the electron, slowing it down. If the electron slowed down it would spiral into the nucleus.

Key questions ➲ p. 189

9 Rutherford had discovered that nuclei contained positive particles called protons (hydrogen nuclei) and that the periodic table was a list of the elements in terms of the number of protons in the nucleus (the atomic number). Rutherford also knew that essentially all the mass of the atom was in the nucleus but the mass of the elements was about twice as great as the number of protons in the nucleus. That is, the mass number was about twice as great as the atomic number for all elements heavier than hydrogen. Rutherford at first thought electrons and protons must combine in the nucleus to form neutral 'doublets' but could not understand how this could occur for only some protons in the nucleus. This led Rutherford to suggest in 1920 that a neutral particle might exist in the nucleus.

10 The key experiments that preceded Chadwick's work were:
- Bothe and Becker found an intense neutral radiation was emitted from beryllium when it was irradiated with alpha particles and assumed it was an energetic gamma ray.
- Curie and Joliot found the radiation emitted from beryllium was three times more energetic than the alpha rays that initiated it, and that the radiation could knock protons out of paraffin. Curie and Joliot also assumed the particles were gamma rays.
- Webster found that the radiation emitted from beryllium was more energetic in the direction of the alpha bombardment than against the direction of the bombardment.

11 Chadwick measured the recoil velocity of hydrogen and nitrogen nuclei after they had been hit by the radiation from beryllium and then used the conservation of kinetic energy and momentum to determine the mass of the particle in the neutral radiation. His calculations showed that the neutral radiation consisted of particles that had a mass slightly greater than the proton.

HSC EXAM-TYPE QUESTIONS

Objective-response questions

1 **B**. Millikan used the electric field between two parallel charged plates to balance the weight force on the charged oil droplet. **A**, **C** and **D** are all incorrect as they all include a magnetic force, which is not related to Millikan's experiment.

2 **A**. The right-hand palm rule shows that the magnetic force on the charge will be down the page if the magnetic field is directed into the page (remember the electron is a negative charge). For the charge to go straight through, $qE = qvB$ and hence $B = E/v$. **B** and **C** have the field in the wrong direction. **D** is incorrect as it has the wrong equation for the magnetic field.

3 **D**. This is the result reported by Geiger and Marsden. **A** is incorrect as the atom is mostly empty space and most alpha particles would pass straight through. **B** is incorrect because the particles that came close to the nucleus were deflected by a large amount. **C** is incorrect because this would require the nucleus to take up most of the atom's interior. The nucleus is less than one-10 000th the size of the atom.

4 **A**. For energy to be conserved some mass must be converted into energy. **B** is incorrect because the beryllium atoms were not initially excited and because the energy involved in changing electrons'

energy levels is tiny compared with the energy involved in nuclear reactions. **C** is incorrect because fusing small atoms releases energy and splitting them therefore must consume energy. **D** is incorrect as increasing the mass would consume energy rather than release it.

5 **A**. The radius is given by $qvB = mv^2/r$ and hence $r = mv/qB$ or $r \alpha v/B$. **B**, **C** and **D** are all incorrect as they use the incorrect relationship for the radius.

Extended-response questions

6 EM This question tests students' understanding of the properties of cathode rays and how they are produced in firsthand experiments.

a The electrodes in an evacuated Crooke's tube ✓ are attached to a high-voltage power supply. ✓ Care should be taken not to touch any high-voltage wires. When the current flows in the tube the glass near the anode will glow green, indicating the presence of cathode rays. ✓ Care should be taken to switch the device on for only a short time and to stand back because the device releases some x-rays.

b Placing a metal Maltese cross in the path of the rays in a Crooke's tube can be used to show that the rays cast a shadow ✓ and that the rays travel in straight lines ✓ as a shadow of the cross will form on the glass at the anode end of the tube (see Figure 13.1). Bringing a magnet near the tube will cause the glowing region on the glass to deflect, showing that the beam is deflected by magnetic fields. ✓

7 EM This question tests students' understanding of Millikan's oil drop experiment.

a Millikan used the electric field produced by a pair of charged parallel plates to balance the force of gravity on the oil droplet. ✓ The net force on the levitating oil droplet would be zero and hence the forces on the drop must sum to zero. Therefore $mg = qE = q(V/d)$ and rearranging this gives:

$q = mgd/V$ ✓

Millikan found the mass (m) of the droplet by measuring its terminal velocity. ✓ The separation between the plates (d) was known and the potential difference (V) across the plates was the voltage required to levitate the charge between the plates.

b Each oil droplet had an integer number of electrons missing from the droplet. ✓ Millikan measured the charge on a large number of droplets and then looked for a common denominator ✓ for all the values of charge he measured. This common denominator was the quantum of charge; the charge on the electron.

8 EM This question tests students' understanding of Thomson's experiment and electromagnetism.

a Thomson directed the cathode rays into crossed electric and magnetic fields that were oriented at 90° to each other and at 90° to the velocity of the charge. ✓ He chose the direction of the magnetic field so that the magnetic force and electric force on the charge would be in opposite directions and perpendicular to the particle velocity, as shown in Figure A13.1.

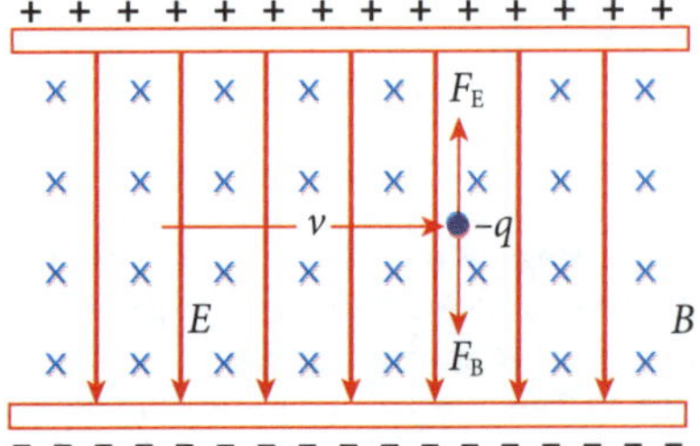

Figure A13.1 Experiment to determine the velocity of the charged particles in a cathode ray

Thomson adjusted the voltage across the plates until the charged particle moved through the fields without being deflected. ✓ When this occurred the net force on the charge was zero and hence:

$qE = qvB$ or $v = E/B$ ✓

Thus by calculating the electric field using $E = V/d$ and calculating the magnetic field, Thomson could substitute values into the above equation to find the velocity of the charged particle.

b The cathode ray debate was about the nature of cathode rays. One group of scientists (Crooke and Thomson) thought they were charged particles and others (Hertz and Goldstein) thought they were waves in the ether. Waves in the ether would travel at the speed of light but Thomson measured the speed of cathode rays to be much less than the speed of light. Hence his measurement provided strong evidence against the cathode ray wave hypothesis. ✓

9 EM This question tests students' understanding of Rutherford's model of atom.

a Rutherford proposed a planetary model of the atom with almost all the mass of the atom concentrated in a tiny, very dense, positively charged nucleus in the centre of the atom. ✓ His model had electrons orbiting the positive nucleus, held in their orbit by electrostatic attraction between the positive nucleus and negative electrons. ✓ Rutherford's atom, like the solar system, was mostly empty space.

b Geiger and Marsden had bombarded thin gold foil (500 atoms thick) with alpha rays ✓ and found that most alpha rays went through the film with little or no deflection, ✓ but about 1 in 8000 alpha particles were scattered at more than 90°. ✓ This led Rutherford to suggest the atom was largely empty space with a small, extremely dense, positively charged nucleus. Most alpha particles went straight through the empty space but occasionally one would approach the nucleus and be deflected by the intense electric field near the nucleus.

c Because the orbiting electrons in Rutherford's model were accelerating they would emit electromagnetic radiation. ✓ The energy carried away in the radiation would be provided by the kinetic energy of the electron, which would cause it to slow down and spiral into the nucleus. ✓

10 EM his question tests students' understanding of the experiments that led to the discovery of the neutron.

This statement was an acknowledgement by Newton that science progresses sequentially, with each scientist building on the foundations laid by previous scientists. ✓ The statement is certainly correct when applied to Chadwick's discovery of the neutron. The observations and results of other scientists were essential to Chadwick's discovery. They were, for example:

1 Bothe and Becker, who found a penetrating, neutral radiation was emitted when beryllium was irradiated with alpha particles. ✓

2 Curie and Joliot, who showed the radiation was three times more energetic than the alpha particles that initiated it and that it could knock protons out of paraffin. ✓

3 Webster, who found that the neutral radiation was more intense in the direction of the alpha particle bombardment than against it. ✓

The work of these scientists led Chadwick to believe that the neutral radiation might be neutrons. Chadwick also relied on the previous work of Newton, who formulated the laws of conservation of momentum and kinetic energy, because Chadwick used these laws to show the neutral radiation was a stream of neutrons. There is no doubt Chadwick could never have made his discovery without 'standing on the shoulders of giants'. (Note that full marks would be awarded for an evaluation statement supported by any three relevant pieces of evidence.)

CHAPTER 14 QUANTUM MECHANICAL NATURE OF THE ATOM

MODULE 8 FROM THE UNIVERSE TO THE ATOM

INQUIRY QUESTION:

How is it known that classical physics cannot explain the properties of the atom?

Many observations of atomic phenomena could not be explained by classical physics. For example, classical physics could not explain black-body radiation, the photoelectric effect, the stability of the hydrogen atom, the periodicity of the elements or atomic spectra. In this chapter we see that classical physics predicted that the planetary atomic model proposed by Rutherford would not have been stable. We also see that classical physics could not explain why Balmer's equation was able to predict the visible emission lines of the hydrogen spectrum.

1 The spectrum of hydrogen

» Students investigate the line emission spectra to examine the Balmer series in hydrogen.

- In previous chapters we saw that excited elemental gases produce light at specific wavelengths that are characteristic of the gaseous element being excited.
- The lightest element, hydrogen, produces a simple spectrum with four visible wavelengths. The visible spectrum of hydrogen is made up of a red line at 656 nm, a blue-green line at 486 nm and two violet lines at 434 nm and 410 nm. Only excited hydrogen emits these four specific wavelengths of light.
- In 1885 Johann Balmer, a Swiss high school maths teacher, noticed that the wavelengths in the visible hydrogen spectrum came closer together as the wavelength decreased, like a limiting mathematical series. Balmer then tried to find a mathematical series that would replicate the observed wavelengths. He found that the following relationship, now known as the **Balmer equation**, could reproduce the series of wavelengths:

$$1/\lambda = R(1/4 - 1/n^2)$$

where λ is the wavelength
R is a constant and
n is an integer that can take the values of $n = 3, 4, 5, 6$, etc.

The constant $R = 1.097 \times 10^7\ \text{m}^{-1}$ is called **Rydberg's constant** after a Swedish physicist who devised a more general expression of Balmer's equation that could predict the spectra of some elements other than hydrogen. We call an equation like this, which is fitted to the data rather than derived from fundamental theory, an **empirical equation**.

Balmer's equation: an empirical equation that could predict the visible lines emitted from excited hydrogen gas
Rydberg's constant: an empirical constant used in the Balmer equation and later shown by Bohr to be made up from fundamental physics constants
empirical equation: an equation that is fitted to the data rather than derived from fundamental principles

- Balmer's equation predicted ultraviolet wavelengths that were later observed in the spectra. It also predicted that the series would have a limiting wavelength in the ultraviolet. He showed that as n approached infinity (i.e. $n \to \infty$), the wavelength would approach $\lambda = 364$ nm. The Balmer series of wavelengths is illustrated in Figure 14.1.

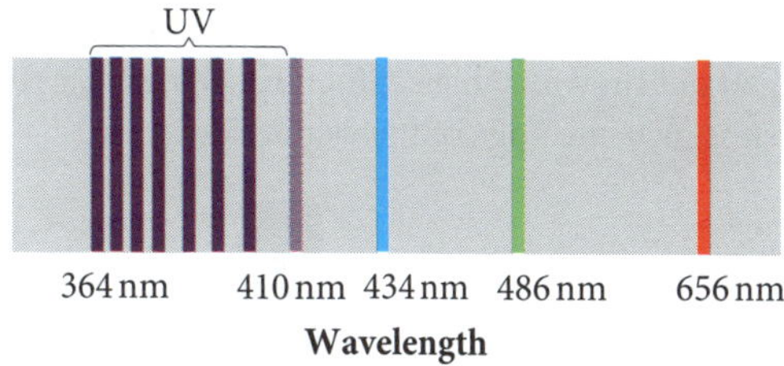

Figure 14.1 Balmer series showing the four visible wavelengths, some of the UV wavelengths and the UV-limiting wavelength

- The success of Balmer's empirical equation in predicting the visible spectrum of hydrogen strongly implied there must be some mechanism inside the atom producing these specific wavelengths. Clearly an atomic model was needed that could explain why Balmer's equation gave the wavelengths of light emitted by excited hydrogen atoms.

EXAMPLE 1

a Find the frequency of the longest wavelength line in the Balmer series.

b Find the shortest limiting wavelength of the Balmer series.

Remember that the longest wavelength (i.e. the least energetic photon) corresponds to a transition from the first excited state above the $n = 2$ level, and the shortest wavelength (most energetic photon) corresponds to a transition from the most energetic ($n = \infty$) orbital

Answer:

a The longest wavelength line in the Balmer series corresponds to $n = 3$:

$$1/\lambda = R(1/4 - 1/n^2) = 1.097 \times 10^7\left(\frac{1}{4} - \frac{1}{3^2}\right)$$

$\lambda = 656$ nm

$c = f\lambda$ and hence

$$f = c/\lambda = \frac{3 \times 10^8}{656 \times 10^{-9}}$$

$= 4.573 \times 10^{14}$ Hz

b The shortest limiting wavelength corresponds to $n \to \infty$. Applying the Balmer equation:

$$1/\lambda = R(1/4 - 1/n^2)$$

$$= 1.097 \times 10^7\left(\frac{1}{4} - 0\right) = 364 \text{ nm}$$

FIRSTHAND INVESTIGATION

Observing the visible lines of the hydrogen spectrum

Danger: This experiment involves the use of high voltages and students should take care to avoid being near any high-voltage wires or connections.

Students should connect a hydrogen discharge tube to a high-voltage supply or induction coil to examine the visible lines of hydrogen. Figure 14.2 shows how a spectrometer or hand spectroscope can be used to investigate the visible spectra. Careful observation should enable students to witness the four visible lines of hydrogen but they may have difficulty seeing the 410 nm line as it is near the edge of the visible spectrum.

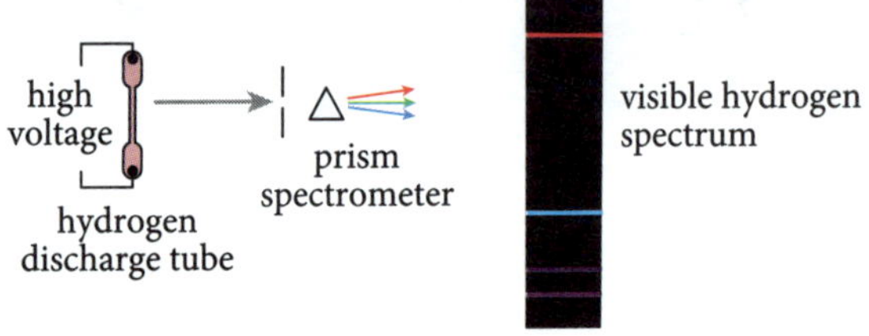

Figure 14.2 Schematic diagram of an experiment to observe visible hydrogen spectrum

→ KEY QUESTIONS

1 How many visible emission lines does excited hydrogen emit?

2 How was the Balmer equation derived?

3 Does the Balmer series have a long and short wavelength limit?

Answers ➲ p. 206

2 Bohr's model of the atom

» Students assess the limitations of the Rutherford and Bohr atomic models.

» Students relate qualitatively and quantitatively the quantised energy levels of the hydrogen atom and the law of conservation of energy to the line emission spectrum of hydrogen using:

- $E = hf$
- $E = \frac{hc}{\lambda}$
- $\frac{1}{\lambda} = R\left[\frac{1}{n_f^2} - \frac{1}{n_i^2}\right]$

➔ We saw in the previous chapter that Rutherford's planetary model of the atom had two major limitations. The first was that the model was unstable because the orbiting electrons would emit electromagnetic radiation, slow down and spiral into the nucleus. The second limitation of the model was that it said nothing about atomic emission or absorption spectra.

➔ Niels Bohr, a young Danish physicist who joined Rutherford's research group after Geiger and Marsden had completed their scattering experiments, sought ways to overcome the limitations of Rutherford's model.

➔ Bohr was familiar with the work of Planck and Einstein on light quanta and, when a colleague asked him how an atomic model might be used to explain the Balmer equation, he began to think about how energy quanta could be applied to the atom.

➔ In 1913 Bohr proposed a new model of the atom that used quantised electron orbits to explain the specific wavelengths of light emitted and absorbed by atoms.

➔ Bohr's model was based on three postulates:

1 Rutherford's atomic model is correct but electrons can only orbit in specific (quantised) **orbitals**. In these specific orbitals, which Bohr called **stationary states**, the electrons do not emit electromagnetic radiation.

2 An electron can move between orbitals by emitting or absorbing a photon of radiation with an energy equal to the energy difference between the orbitals. That is:

$$\Delta E = E_f - E_i = hf = hc/\lambda$$

where ΔE is the photon energy
h is Planck's constant
f is the frequency of the photon
E_i is the energy of the electron in the initial orbital and
E_f is the energy of the electron in the final orbital

orbital: stationary state orbit

stationary state: an orbit from which an electron does not emit electromagnetic radiation; stationary states are the only orbits an electron can have in the Bohr atom

3 The stationary states (orbitals) can only have an angular momentum given by:

$$mvr = nh/2\pi$$

where m is the mass of the electron
v is the velocity of the electron
r is the radius of the orbit
h is Planck's constant and
n is an integer called the principal quantum number

principle quantum number: the number n of the Bohr orbital

- The first postulate overcomes the Rutherford atom-instability problem by saying that for some orbitals the laws of electromagnetism do not apply. At the time Bohr could not justify this postulate other than by saying it gave the correct answer. He also suggested that at the atomic level the laws of mechanics and electromagnetism may need to be altered.
- The second postulate explains emission and absorption spectra by saying that each line in the spectra corresponds to a specific photon energy related to an electron moving between two specific orbitals in the atom. Photons would be absorbed when the electron moved to a higher energy orbital and emitted when an electron moved from a higher to a lower energy orbital.
- The third postulate specifies the allowed orbitals. It can also be written in terms of energy, as we will see later. Bohr again could not justify why this condition was chosen other than that it produces the correct spectrum for hydrogen.

SECONDARY-SOURCED INVESTIGATION

Using Bohr's postulates

Students should investigate how Bohr's postulates enable the radius, ionisation energy and emission spectra of hydrogen to be determined. A brief derivation of each quantity follows.

This investigation is not tested in the HSC but is included here to assist students with gaining a deeper understanding of Bohr's atomic model.

ionisation energy: the energy required to remove an electron from an atom

1 The Bohr radius

As the electrostatic attraction to the nucleus provides the centripetal force that holds the electron in its orbit, we can write:

$$\frac{1}{4\pi\varepsilon_0}\frac{e^2}{r_2} = \frac{mv^2}{r}$$

where e is the magnitude of the charge on the electron and proton

We can remove the velocity (v) by substituting its value from Bohr's postulate:

$mvr = nh/2\pi$ and hence $v = nh/2\pi mr$

Substituting this value of the velocity into the equation above and rearranging yields:

$$r_n = \frac{n^2h^2\varepsilon_0}{\pi me^2} = n^2 r_1$$

Substituting the values into this equation gives the radius of the $n = 1$ ground state hydrogen atom:

$r_1 = 5.3 \times 10^{-11}$ m

2 The energy of an orbiting electron

Equating the electrostatic force and the centripetal force gives:

$$v^2 = \frac{e^2}{4\pi\varepsilon_0 mr}$$

We can now use this value of v^2 to derive an expression for the kinetic energy (K) of the orbiting electron:

$$K = \tfrac{1}{2}mv^2 = \frac{e^2}{8\pi\varepsilon_0 r}$$

If we assume a free electron has zero potential energy (in an analogous manner to the gravitational potential energy of an orbiting planet), the potential energy of an electron a distance r from the hydrogen nucleus will be given by:

$$U = \frac{-e^2}{4\pi\varepsilon_0 r}$$

The total energy of the electron is therefore:

$$E = K + U = \frac{e^2}{8\pi\varepsilon_0 r} - \frac{e^2}{4\pi\varepsilon_0 r} = \frac{-e^2}{8\pi\varepsilon_0 r}$$

By substituting the value of r_n from above we can obtain the total energy of the electron in the nth orbital:

$$E_n = \frac{-e^4 m}{8{\varepsilon_0}^2 h^2 n^2} = \frac{E_1}{n^2}$$

3 Ionisation energy of hydrogen

The magnitude of the energy of the electron in the $n = 1$ orbit will be the energy required to free the electron from the atom. Hence the ionisation energy of a ground state ($n = 1$) hydrogen atom is:

$$E_1 = \frac{-e^4 m}{8{\varepsilon_0}^2 h^2} = 2.184 \times 10^{-18}\text{ J} = 13.6\text{ eV}$$

4 Hydrogen line spectra

A photon of light will be emitted or absorbed when an electron moves to a lower energy or higher energy orbital, respectively. The difference in energy between the initial stationary state (n_i) and the final stationary state (n_f) will be given by:

$$\Delta E = E_f - E_i = -E_i\left(\frac{1}{n_f^2} - \frac{1}{n_i^2}\right)$$

The negative sign tells us that the electron loses energy when $n_i > n_f$. If we use this equation to give us the energy of the released photon, we write the equation without the negative sign. A positive change in energy then tells us that a photon is emitted and a negative change in energy tells us that a photon is absorbed. Now we can apply Einstein's photon equation.

As $\Delta E = hc/\lambda$, we can write:

$$\Delta E = \frac{hc}{\lambda} = E_i\left(\frac{1}{n_f} - \frac{1}{n_i}\right)$$

Or $\frac{1}{\lambda} = \frac{E_1}{hc}\left(\frac{1}{n_f} - \frac{1}{n_i}\right)$

This equation gives the hydrogen emission series.

For example, provided E_1/hc is equal to the Rydberg constant, setting $n_f = 2$ and $n_i = 3, 4, 5$, etc., will give the Balmer series.

To check the derivation of Rydberg's constant we substitute the value of E_1 from above into the equation:

$$R = \frac{E_1}{hc} = \frac{\frac{-e^4m}{8\varepsilon_0^2h^2}}{hc} = \frac{-e^4m}{8\varepsilon_0^2h^3c}$$

$= 1.097 \times 10^7\ \text{m}^{-1}$ = Rydberg's constant, as expected

➔ By assuming the electron was held in its orbit by electrostatic attraction to the positive nucleus, Bohr could derive the following relationships for the hydrogen atom.

a The radius of the nth orbital is given by:

$$r_n = n^2r_1 = \frac{n^2h^2\varepsilon_0}{\pi me^2}$$

Substituting $n = 1$ gives the radius of the lowest energy, unexcited, ground-state hydrogen atom. We call this the **Bohr radius**:

$r_1 = 5.3 \times 10^{-11}$ m

This radius agreed with experimental measurements at the time.

Bohr radius: the radius of the $n = 1$ ground-state orbital of the Bohr hydrogen atom

b The total energy of an electron in nth orbital is given by:

$$E_n = \frac{-e^4m}{8\varepsilon_0^2h^2n^2} = \frac{E_1}{n^2} = \frac{13.6\ \text{eV}}{n^2}$$

The negative sign tells us that the electron is trapped in its atomic orbit, like a satellite in orbit. The energy of the $n = 1$ level is the energy required for the electron to escape from the atom. This energy (i.e. 13.6 eV) was in good agreement with the experimental value for the ionisation energy of hydrogen.

c The difference between two energy levels is therefore:

$$\Delta E = -E_1\left[\frac{1}{n_f^2} - \frac{1}{n_i^2}\right]$$

And as $\Delta E = hf = \frac{hc}{\lambda}$:

$$\frac{1}{\lambda} = \frac{E_1}{hc}\left[\frac{1}{n_f^2} - \frac{1}{n_i^2}\right]$$

We removed the negative sign so that a positive energy difference would represent an emitted photon.

Note that if we set $n_f = 2$, this is Balmer's equation and Rydberg's constant is therefore:

$$R = \frac{E^1}{hc} = \frac{me^4}{8ch^3\varepsilon_0^2}$$

➔ Balmer's equation represents one series of wavelengths produced by Bohr's more general equation:

$$\frac{1}{\lambda} = R\left[\frac{1}{n_f^2} - \frac{1}{n_i^2}\right]$$

We get the Balmer series by setting $n_f = 2$ and $n_i = n_f + 1$, $n_f + 2$, $n_f + 3$, etc.

➔ Bohr's equation predicts another series of lines will be found in the ultraviolet if we set $n_f = 1$ and $n_i = n_f + 1$, $n_f + 2$, $n_f + 3$, etc. This ultraviolet series, now called the *Lyman series*, contained the wavelengths predicted by the Bohr model.

➔ By setting $n_f = 3$ and $n_i = n_f + 1$, $n_f + 2$, $n_f + 3$, etc., we get the first infrared series of lines, which is called the *Paschen series*.

➔ Note that the longest wavelength emitted from each series is produced by the transition between the $n_i = n_f + 1$ level and the n_f level, and the shortest wavelength is emitted from the $n_i \rightarrow \infty$ orbital and the n_f. The shortest wavelength in a series represents the minimum energy photon that would ionise the atom if the electron was in the lowest energy state of that series.

➔ The first three series of hydrogen spectra are shown in Figure 14.3.

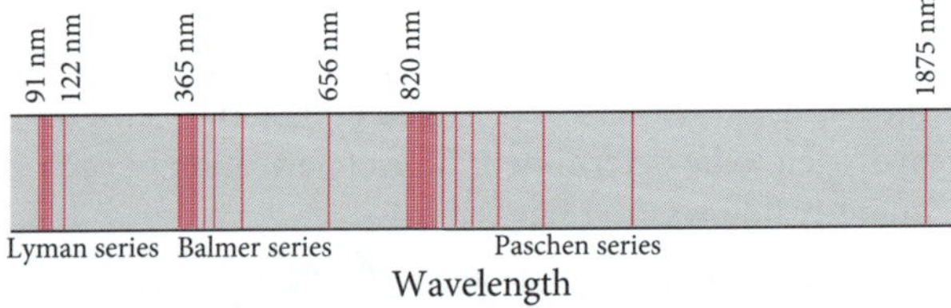

Figure 14.3 Hydrogen emission spectra showing UV, visible and first IR line series and their limiting wavelengths

➔ The relationship between the stationary state orbitals and the emissions series is shown in Figure 14.4. Note that a series of lines is produced by all the transitions that end on a specific orbital. For example, the UV Lyman series is produced by transitions from other levels to the ground state ($n = 1$) level of hydrogen. The Balmer series is produced by all those transitions that end on the $n = 2$ orbital, etc.

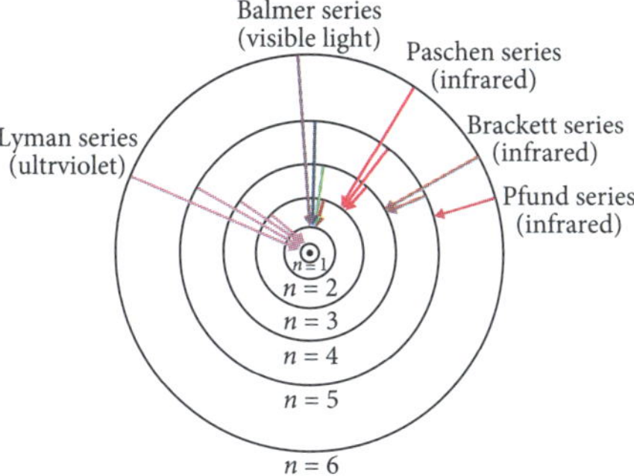

Figure 14.4 First four spectral series of hydrogen (note the orbital radii are not drawn to scale)

➔ The orbitals and transitions can also be represented in energy-level diagrams. The energy level diagram shown in Figure 14.5 fixes the energy for an electron in the $n \to \infty$ state at zero and hence shows the energy of the electron in each orbital as a negative number. Sometimes energy level diagrams set the ground state $n = 1$ energy at zero and show the energy of the orbitals as a positive number, representing the energy above the ground-state energy. The energy difference between levels can be read directly from the diagram and can be substituted into $E = hf = hc/\lambda$ to determine the frequency or wavelength of the photon that would be released or absorbed when an electron moved between the levels.

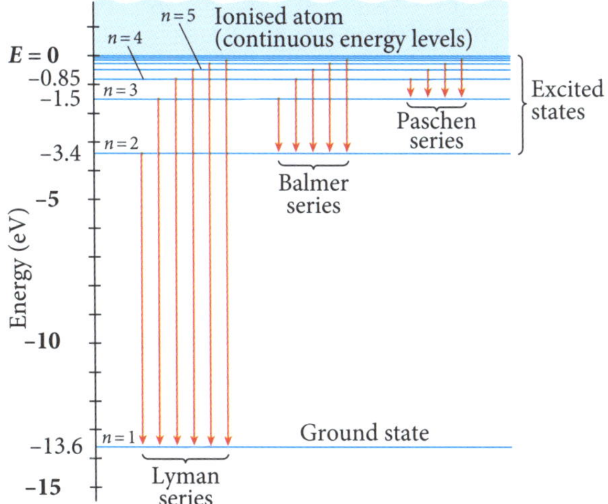

Figure 14.5 Energy level diagram for the hydrogen atom

➔ The Bohr model was the first attempt to quantise the atom. It could predict the size and ionisation energy of the hydrogen atom. The model also provided a plausible mechanism to explain atomic emission and absorption spectra and could be used to derive the Balmer equation and to show that Rydberg's constant was a combination of more fundamental constants.

➔ However, the Bohr model had some serious limitations. It could not predict the spectra of atoms or ions with more than one electron. It said nothing about the fine splitting of lines that had been observed in the hydrogen spectra or the intensity of the emission lines. It also failed to account for the splitting of spectral lines when excited atoms were placed in a magnetic field (**Zeeman effect**) or an electric field (**Stark effect**). In addition Bohr's quantisation postulate seemed to be engineered to produce the correct answer. While Bohr was happy to use classical electrostatics and Newtonian mechanics, he ignored the laws of electromagnetism.

➔ A good model should unify and explain existing theories and observations, and should be able to make predictions that can be tested. Bohr's model was undoubtedly a step in the right direction but it was far from the complete solution.

Zeeman effect: the splitting of spectral lines that occurs when the excited atoms are placed in a magnetic field

Stark effect: the splitting of spectral lines that occurs when the excited atoms are placed in an electric field

EXAMPLE 2

A photon was absorbed by a ground-state hydrogen atom that excited the atom in the $n = 4$ state. The atom then relaxed by releasing three photons as it transitioned through each successive orbital until it returned to the ground state.

a **Find the wavelength of the absorbed photon.**

b **Find the wavelength of the three photons emitted from the atom.**

c **Explain the relationship between the energy of the absorbed photon and the three emitted photons.**

Recall that the ground state corresponds to the $n = 1$ level and that energy must be conserved when photons are absorbed and emitted

Answer:

a The photon caused the atom to move from the ground state ($n = 1$) to the $n = 4$ state. Applying Bohr's equation:

$$\frac{1}{\lambda} = R\left[\frac{1}{n_f^2} - \frac{1}{n_i^2}\right] = 1.097 \times 10^7\left(\frac{1}{1^2} - \frac{1}{4^2}\right)$$

$\lambda = 97$ nm

b The first photon is released when the electron moves from the $n = 4$ to the $n = 3$ orbital and hence:

$$\frac{1}{\lambda} = R\left[\frac{1}{n_f^2} - \frac{1}{n_i^2}\right] = 1.097 \times 10^7\left(\frac{1}{3^2} - \frac{1}{4^2}\right)$$

$\lambda = 1875$ nm

The second photon would be released when the electron moved from the $n = 3$ to the $n = 2$ state and hence:

$$\frac{1}{\lambda} = R\left[\frac{1}{n_f^2} - \frac{1}{n_i^2}\right] = 1.097 \times 10^7\left(\frac{1}{2^2} - \frac{1}{3^2}\right)$$

$\lambda = 656$ nm

The final photon is released when the electron moves from $n = 2$ to $n = 1$ and hence:

$$\frac{1}{\lambda} = R\left[\frac{1}{n_f^2} - \frac{1}{n_i^2}\right] = 1.097 \times 10^7\left(\frac{1}{1^2} - \frac{1}{2^2}\right)$$

$\lambda = 121$ nm

c Because energy is conserved, the energy of the incident photon must be equal to the sum of the energies of the emitted photons.

EXAMPLE 3

Find the minimum energy to ionise a hydrogen atom from:

a the ground state and

b the $n = 3$ state.

Ionising a hydrogen atom requires the electron to be excited to the $n = \infty$ level

Answer:

a To ionise an atom we must take it to the $n = \infty$ level and hence, by applying Bohr's equations to the excitation from the ground state ($n = 1$):

$$\frac{1}{\lambda} = R\left[\frac{1}{n_f^2} - \frac{1}{n_i^2}\right] = 1.097 \times 10^7\left(0 - \frac{1}{1^2}\right)$$

$\lambda = -91$ nm

The negative sign reminds us that energy was put into the atom.

$$\text{Energy} = hc/\lambda = \frac{6.626 \times 10^{-34} \times 3 \times 10^8}{91 \times 10^{-9}} = 2.184 \times 10^{-18}\ \text{J} = \frac{2.184 \times 10^{-18}}{1.602 \times 10^{-19}} = 13.6\ \text{eV}$$

b Applying Bohr's equations to the $n = 3$ orbital:

$$\frac{1}{\lambda} = R\left[\frac{1}{n_f^2} - \frac{1}{n_i^2}\right] = 1.097 \times 10^7\left(0 - \frac{1}{3^2}\right)$$

$\lambda = -820$ nm

Again, the negative sign simply reminds us that energy has been added to the atom.

$$\text{Energy} = hc/\lambda = \frac{6.626 \times 10^{-34} \times 3 \times 10^8}{820 \times 10^{-9}} = 2.424 \times 10^{-19}\ \text{J} = \frac{2.424 \times 10^{-19}}{1.602 \times 10^{-19}} = 1.51\ \text{eV}$$

➔ KEY QUESTIONS

4 **What are the limitations of the Rutherford atomic model?**

5 **What three postulates formed the foundation of Bohr's atomic model?**

6 **How did Bohr use conservation of energy in his model?**

7 **How did Bohr explain the Balmer series and other line series in the hydrogen spectra?**

8 **What are the limitations of the Bohr atomic model?**

Answers ➲ p. 206

3 De Broglie's matter-waves

» Students investigate de Broglie's matter waves, and the experimental evidence that developed the following formula:

$\lambda = \frac{h}{mv}$

➔ The next step forward in our understanding of the atom came from an unlikely source. In 1924 a French student, Louis de Broglie, wrote a radical doctoral thesis. De Broglie combined the ideas of special relativity with light quanta theory to suggest that if electromagnetic waves act like particles, perhaps other particles could act like waves.

➔ De Broglie equated Einstein's mass–energy equivalence equation with energy of the photon:

$E = hf = hc/\lambda = mc^2$

Because particles with mass cannot travel at the speed of light, he replaced the expression let $c = v$ and obtained an expression for the wavelength of a particle:

$mv^2 = hv/\lambda$, or

$$\lambda = h/mv = h/p$$

where λ is the wavelength associated with the particle
h is Planck's constant and
$p = mv$ is the momentum of the particle

➔ De Broglie proposed that all objects had a wave–particle nature. The wavelength associated with everyday objects would be too small to notice any wave effects but tiny particles like electrons would have wavelengths that were large enough to produce wave effects.

➔ The first success of de Broglie's **matter-wave theory** was that it could be used to explain quantisation of the orbitals in Bohr's model. De Broglie suggested that the stationary states in Bohr's model correspond to electron standing waves around the nucleus. For a standing wave to remain in phase around an orbit, the circumference of the orbit would have to be equal to a whole number of electron wavelengths. That is:

$2\pi r = n\lambda$

Now as $\lambda = h/mv$, we can write $2\pi r = n(h/mv)$ or:

$mvr = nh/2\pi$

This is Bohr's third postulate. The $n = 1$ state in Bohr's model would correspond to one complete wavelength, the $n = 2$ to two complete wavelengths, etc. Figure 14.6 shows how standing waves explain Bohr's $n = 4$ stationary-state orbital.

matter-wave theory: a theory that suggests particles have wave properties and an associated wavelength

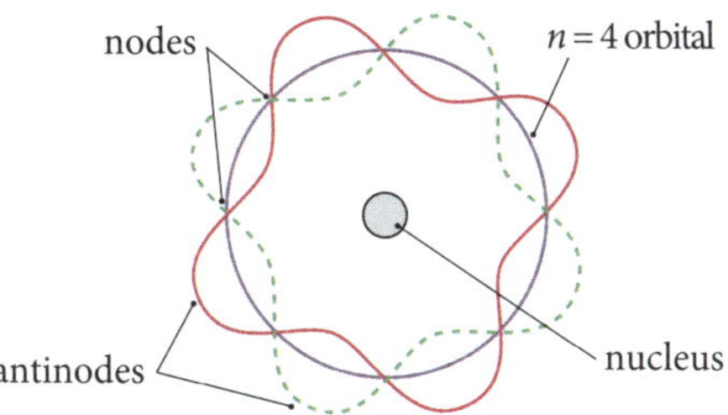

Figure 14.6 Standing waves in an $n = 4$ orbital

- The idea that particles have wave properties was first confirmed experimentally in 1927 when Davisson and Germer reported electron interference.
- Davisson and Germer were investigating electron scattering from nickel when they noticed that the electrons preferentially scattered in a specific direction. Their experiment, illustrated in Figure 14.7**a**, involved accelerating electrons over a potential of 50 to 65 V in an evacuated chamber and observing the number of electrons scattered at different angles.
- They expected an even distribution of scattered electrons but found that most electrons were scattered at a specific angle (see Figure 14.7**b**). This implied that the electrons were acting like waves and constructively interfering to give a peak of intensity. Physicists were familiar with this type of behaviour from x-ray diffraction, which at the time was a well-established experimental technique for studying the structure of crystal lattices. The atoms in the crystal act as individual scattering centres, like a diffraction grating, and produce interference maxima in the diffracted rays.

a

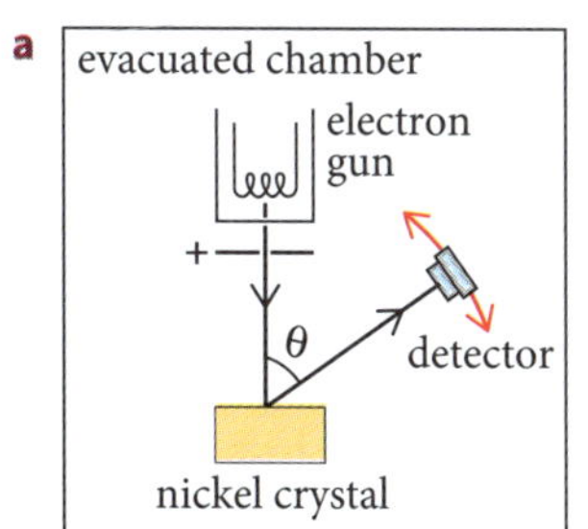

b

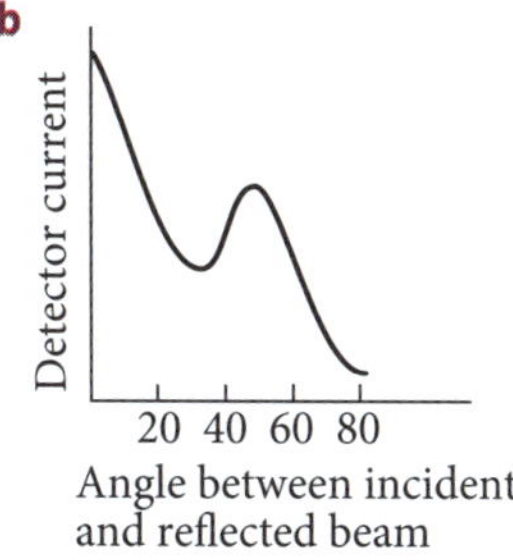

Figure 14.7 **a** The Davisson–Germer experiment and **b** the results obtained

- It was known from x-ray diffraction studies that changing the wavelength of the x-rays changed the angle at which the diffraction maxima occur. **Bragg's law** for x-rays diffracted from a crystal states that the maxima of interference occur when:

  $n\lambda = 2d\sin\theta$

  where λ is the x-ray wavelength
  d is the spacing between the planes of atoms in the crystal
  θ is the angle at which the maxima of interference occur and
  n is an integer

Bragg's law: a relationship between the angle at which diffraction maxima occur for incident waves and the spacing between the planes of atoms in the crystal

- When Davisson and Germer changed the accelerating voltage they found that the angle at which the peak in emission occurred also changed in accordance with Bragg's law of diffraction. That is, by changing the accelerating voltage, they changed the momentum and hence the wavelength associated with the electron. This change in wavelength resulted in a change in the angle at which most electrons were scattered in the same way as changing an x-ray wavelength would change the angle at which the diffraction maxima occurred.
- Interference experiments similar to one carried out by Davisson and Germer have now been conducted with protons, neutrons, neutral atoms and molecules.
- The wave nature of matter is used today in electron microscopes, which can image objects on the atomic scale. Electron microscopes are powerful because the resolving power of a microscope is related to the wavelength used and electrons can be produced with wavelengths that are orders of magnitude shorter than visible light.
- Because the influence of a small particle is smeared out like a wave, de Broglie's matter-wave hypothesis led physicists to re-examine how accurately physical quantities could be measured. This re-examination led to the famous **Heisenberg uncertainty principle**. In 1926

Heisenberg's uncertainty principle: a principle that states that there is a fundamental limit to the accuracy to which certain pairs of variables can be simultaneously measured

Werner Heisenberg showed that there was a fundamental limit to the accuracy with which some variables could be simultaneously measured. He showed that:

$\Delta x\Delta p > h/4\pi$

where Δx is the uncertainty in the position of a particle
Δp is the uncertainty in the momentum and
h is Planck's constant

What this tells us is that if we know the exact position of a particle (i.e. $\Delta x = 0$), then we do not know anything about the particle's momentum. Or if we know exactly how fast a particle is travelling and in what direction (i.e. Δp), then we have no idea where the particle is.

- Heisenberg also showed the energy of a particle must vary over very small time-intervals and that the law of conservation of energy was actually an average over time because:

  $\Delta E \Delta t > h/4\pi$

  where ΔE is the uncertainty in the energy of a particle
  Δt is the uncertainty in the time and
  h is Planck's constant

- Heisenberg's uncertainty principle is one of the principle tenets of modern quantum physics and has been verified by a huge variety of experimental observations. Notice that Heisenberg is not saying you cannot measure a specific quantity accurately; he is saying that you cannot measure certain pairs of quantities accurately at the same time.
- Modern quantum physics is also based on the particle-wave nature of matter. Interestingly, experiments on the atomic scale have always shown that matter behaves like a wave or a particle but never as both at the same time. It is a bit like the head and the tail of a coin. You know the coin has a head and a tail but you can only see the head or the tail at the one time. You can never see both sides of the coin at once.

EXAMPLE 4

Find the momentum and velocity of an electron with a de Broglie wavelength of 6.400×10^{-9} m.

De Broglie's equation relates momentum and wavelength for a particle

Answer:

Applying de Broglie's equation:

$\lambda = h/p$ and hence

$$p = h/\lambda = \frac{6.626 \times 10^{-34}}{6.4 \times 10^{-9}} = 1.035 \times 10^{-25}\ \text{kgms}^{-1}$$

$$\text{As } p = mv,\ v = p/m = \frac{1.035 \times 10^{-25}}{9.109 \times 10^{-31}} = 1.14 \times 10^{5}\ \text{ms}^{-1}$$

EXAMPLE 5

The Compton effect involves a photon scattering off an electron and imparting some momentum to the electron. In a Compton effect experiment an incident x-ray photon with a wavelength of 0.8 nm scatters off an electron that was initially at rest and transfers half its momentum to the electron.

a Find the initial momentum of the photon.

b Find the velocity of the photon after it has scattered the electron.

c Find the wavelength of the scattered photon.

De Broglie's equation shows that a photon has momentum

Answer:

a Applying de Broglie's equation:

$$p = h/\lambda = \frac{6.626 \times 10^{-34}}{0.8 \times 10^{-9}} = 8.28 \times 10^{-25}\ \text{kgms}^{-1}$$

b Half the momentum is transferred to the electron. Applying the equation for momentum:

$$p = mv \text{ or } v = p/m = \frac{4.14 \times 10^{-25}}{9.109 \times 10^{-31}} = 4.55 \times 10^{5}\ \text{ms}^{-1}$$

c For momentum to be conserved the scattered photon must have half the momentum of the initial photon and hence:

$$\lambda = h/p = \frac{6.626 \times 10^{-34}}{4.14 \times 10^{-25}} = 1.6 \times 10^{-9}\ \text{m} = 1.6\ \text{nm}$$

Or, as the wavelength is inversely proportional to the momentum (i.e. $\lambda \alpha 1/p$), halving the momentum would double the wavelength and hence $\lambda = 1.6$ nm.

→ KEY QUESTIONS

9 **What was de Broglie's hypothesis?**

10 **How did de Broglie's hypothesis support the Bohr model?**

11 **How was de Broglie's hypothesis experimentally verified?**

12 **How did de Broglie's hypothesis change our view of what can be measured?**

Answers ⊃ p. 206

4 Schrödinger's model of the atom

» Students analyse the contribution of Schrödinger to the current model of the atom.

- A full quantum model of the atom was independently developed in the 1920s by Werner Heisenberg and Erwin Schrödinger. The two models used a different mathematical approach but both theories were eventually shown to be equivalent. Heisenberg's quantum mechanics used matrix algebra to explain the behaviour of matter on the atomic scale, while Schrödinger's wave mechanics used de Broglie's matter-wave hypothesis and wave theory to solve the problem.
- Physicists initially preferred Schrödinger's theory because they found it less complex and easier to use. Our current

model of the atom is based on Schrödinger's theory but contains elements of both theories.

- Because you have not as yet studied the mathematics used in either theory, it is difficult to explain the theory in detail but we can get a feel for how Schrödinger approached the problem.
- Schrödinger used de Broglie's equation to replace the wavelength with the momentum in a standard wave equation. He then wrote an equation for the electron wave in terms of energy. Next he added the potential of the nucleus and solved the equation to find the modes of the three-dimensional electron standing waves that could exist around the nucleus. This is analogous to solving for the one-dimensional standing waves that can exist on a stretched string.
- Each standing wave solution was expressed by a **wave function** (ψ). The energy of each standing-electron wave mode represented the energy of the excited states of the electron in the atom (i.e. like the Bohr orbitals), and the square of the wave amplitude at a point gave the probability of finding the electron at that point.

wave function: a complex-valued function that describes the wave properties of a particle; the wave function is sometimes called the *probability amplitude* because squaring the amplitude of the function at a point gives the probability of finding the particle at the point

- The wave functions are complex functions (i.e. involving $i = \sqrt{-1}$) and hence they are imaginary rather than real waves. The generally accepted interpretation of this is that the particle exists as a wave (i.e. a complex wave function) until the particle is observed. When we observe the particle, the wave function collapses into the reality (i.e. the particle) that we observe.
- We have seen that de Broglie showed how the principle quantum number (n) used by Bohr was related to number of wavelengths in the circular standing wave orbitals. Schrödinger's standing waves were in three dimensions and he needed three quantum numbers to describe them. Later a fourth quantum number called **spin** was introduced to explain the Zeeman effect and some other observations. The four quantum numbers are:

  The principle quantum number (n): $n = 1, 2, 3, \ldots$

  The azimuthal quantum number (l): $l = 0, 1, 2, \ldots n - 1$

  The magnetic quantum number (m_l): $m_l = -l, 0, +l$

  Spin (m_s): $m_s = +1/2, -1/2$

spin: an intrinsic form of angular momentum carried by a particle; it is a quantum mechanical property and should not be confused with mechanical spin

The fine structure of the spectrum is explained by the sublevels in Schrödinger's solution. For example, Bohr had one $n = 2$ energy level but Schrödinger's rules give eight sublevels at $n = 2$:

$n = 2, l = 0, m_l = 0, m_s = +1/2$

$n = 2, l = 0, m_l = 0, m_s = -1/2$

$n = 2, l = 1, m_l = -1, m_s = +1/2$

$n = 2, l = 1, m_l = -1, m_s = -1/2$

$n = 2, l = 1, m_l = 0, m_s = +1/2$

$n = 2, l = 1, m_l = 0, m_s = -1/2$

$n = 2, l = 1, m_l = +1, m_s = +1/2$

$n = 2, l = 1, m_l = +1, m_s = -1/2$

Electrons moving from and to these sublevels explained the observed fine structure of the hydrogen spectrum.

- The following letters are often used to describe the azimuthal quantum numbers:

  $l = 0$ is called the s level

  $l = 1$ is called the p level

  $l = 2$ is called the d level

  $l = 4$ is called the f level

  We can then specify a level using the principle quantum number and the azimuthal quantum number as a letter. For example, a 2d orbital has the quantum numbers $n = 2, l = 1$.
- The problem of what happens with atoms with more than one electron was solved by Pauli in 1925. Pauli's exclusion principle states that no two electrons in an atom can have the same quantum number. By abiding by this principle, each extra electron falls into the next available energy level. This explains the periodicity of the periodic table. We see in Figure 12.8 that the horizontal rows of the periodic table refer to the principle quantum number and the vertical columns are related to the filling of the subshells.

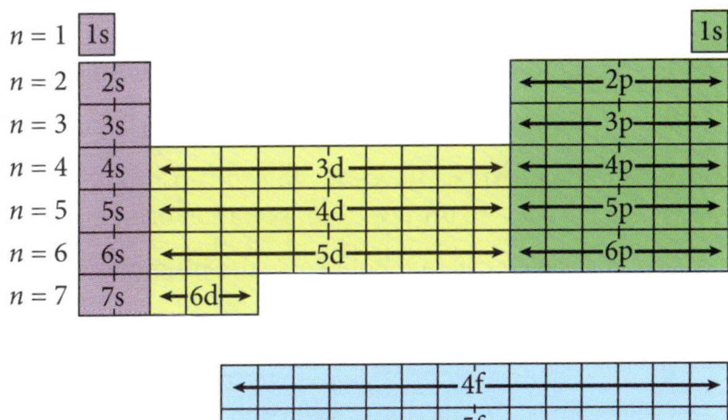

Figure 14.8 Quantum numbers and the periodic table

- Today we still think of the atom as having a tiny nucleus containing protons and neutrons. We also still picture the electron existing in stationary states with specific energies and being able to move between states by emitting or absorbing a photon with the energy difference between the states. We no longer believe, however, that electrons orbit the nucleus. The best we can say is that there is a certain probability of finding an electron in some position around the atom. If we do not look, then the electron is 'spread' around the nucleus more like a wave than a particle (but in an imaginary space). Squaring of the amplitude of the wave function gives a probability distribution that shows where the electron might be found. We can think of this probability as the intensity of the standing wave at that point. Figure 14.9 shows the non-zero probability distribution for three orbitals. Note the s-orbital is spherically symmetric but the other orbitals are not.

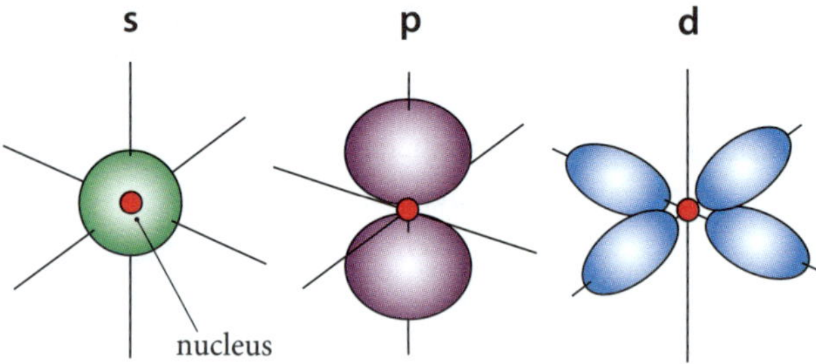

Figure 14.9 Probability distributions for three orbitals

- Schrödinger made a significant contribution to the current model of the atom by using de Broglie's matter-wave hypothesis to produce the first full quantum model of the atom. Schrödinger's model overcame most of the limitations of Bohr's model and still forms the basis of our current model of the atom.
- Our current model of the atom incorporates special relativity, Heisenberg's uncertainty principle, electron spin and Pauli's exclusion principle into Schrödinger's model; it has none of the limitations of the earlier Bohr model.

KEY QUESTIONS

13 **How did Schrödinger use de Broglie's matter-wave hypothesis in his atomic model?**

14 **How did Schrödinger's model differ from Bohr's model?**

15 **How important was Schrödinger's contribution to our current atomic model?**

Answers p. 206

CHAPTER SYLLABUS CHECKLIST

Are you able to answer these questions from the syllabus for this chapter? Tick each question as you go through the checklist if you are able to answer it. If you cannot answer a question, turn to the relevant page in the study guide to find the answer. For NESA key word meanings, go to www.educationstandards.nsw.edu.au and search 'key words'.

FOR A COMPLETE UNDERSTANDING OF THIS TOPIC:		PAGE NO.	✓
1	Can I describe the Balmer series in hydrogen and explain how its visible lines could be observed?	195	
2	Can I use the Balmer equation to calculate the wavelengths emitted in the Balmer series?	196	
3	Can I describe the limitations of the Rutherford model?	196	
4	Can I outline the postulates upon which Bohr's model of the atom was based?	196–197	
5	Can I explain how emission and absorption spectra can be explained in term of quantised energy levels and the law of conservation of energy?	196–197	
6	Can I recall that Bohr could use his model to derive the Balmer equation and the Rydberg constant?	197–198	
7	Can I outline the limitations of Bohr's model?	199	
8	Can I use Bohr's equation to calculate wavelengths in the line spectra of hydrogen and the ionisation energy of hydrogen?	199–200	
9	Can I outline de Broglie's hypothesis and how it supported Bohr's stationary state postulate?	200–201	
10	Can I describe the experimental evidence that supports de Broglie's hypothesis?	201	
11	Can I use de Broglie's equation to calculate the wavelength of a particle?	202	
12	Can I compare the current model of the atom with Bohr's model?	202–204	
13	Can I analyse the contribution that Schrödinger made to the current model of the atom?	202–204	

HSC EXAM-TYPE QUESTIONS

Objective-response questions (1 mark each)

1 **What produces the Balmer emission spectrum in hydrogen?**

A Electrons moving from the $n = 2$ level to higher levels.

B Excited electrons moving to the ground state.

C Excited electrons moving to the $n = 2$ state.

D Electrons moving from the ground state to higher level.

2 **How did Rutherford explain the results of the Geiger–Marsden experiment?**

A The atom must be almost entirely made up of empty space.

B The Thomson plum pudding model was correct.

C The nucleus of the atom must contain protons and neutrons.

D The energy of the electron's orbit is quantised.

3 **What was the main difference between Bohr's model and Rutherford's model?**

A Rutherford included a positive nucleus in his model.

B Bohr's model did not have orbiting electrons.

C Bohr's model could not be used to explain the Geiger–Marsden experiment.

D Bohr's model was stable.

4 **If an ultraviolet photon of wavelength λ is absorbed by an atom in its ground state, what cannot occur?**

A A photon with a wavelength longer than λ could be released by the atom.

B A photon with a higher frequency than the incident photon could be released.

C Two photons with wavelengths less than in the incident photon could be released.

D The atom could be ionised by the incident photon.

5 **When an electron is accelerated over a potential difference of V volts the electron has a de Broglie wavelength of λ. What would the electron's de Broglie wavelength be if it was accelerated over a potential difference of 4 volts?**

A $(\sqrt{2})\lambda$

B $\lambda/4$

C 2λ

D $\lambda/2$

Extended-response questions

6 **Bohr proposed an atomic model to overcome some of the limitations of Rutherford's atomic model.**

a **Describe Rutherford's atomic model.** (2 marks)

b **Outline how Bohr's model overcame two shortcomings of Rutherford's model.** (4 marks)

7 **A hydrogen atom absorbs a photon and goes from the $n = 2$ level to the $n = 5$ level.**

a **Find the wavelength and energy of the absorbed photon.** (2 marks)

b **What is the minimum wavelength photon that could ionise the atom from the $n = 5$ level?** (2 marks)

8 **De Broglie proposed that particles had an associated wavelength.**

a **Explain how de Broglie's hypothesis supported Bohr's atomic model.** (2 marks)

b **Outline an experiment that supports de Broglie's matter-wave hypothesis.** (3 marks)

9 **In the Bohr model electrons in the ground-state orbit with a radius of 5.3×10^{-11} m.**

a **Find the velocity of an electron in the ground-state orbit.** (1 mark)

b **Find the de Broglie wavelength of an electron in the ground state.** (1 mark)

c **Why is the total energy of an electron in the ground-state negative?** (2 marks)

10 **Compare Bohr's model of the atom with Schrödinger's model of the atom.** (4 marks)

ANSWERS

KEY QUESTIONS

Key questions ➲ p. 196

1 There are four visible lines in the visible spectrum of hydrogen.

2 The Balmer equation is an empirical equation that enables the wavelengths in the hydrogen Balmer emission series to be calculated. It was not derived from theory but simply fitted to give the right answer.

3 The Balmer series has a maximum wavelength determined by setting $n = 3$ and a minimum wavelength determined by setting $n \to \infty$, in Balmer's equations.

Key questions ➲ p. 200

4 Rutherford's model was unstable as orbiting electrons would radiate electromagnetic radiation and spiral into the nucleus. Rutherford's model also said nothing about atomic spectra.

5 i Electrons orbited the nucleus in specific stationary states and when in these states they did not radiate electromagnetic radiation.

ii Electrons could move between stationary states by emitting or absorbing a photon with an energy equal to the energy difference between the states.

iii The angular momentum (and energy) of the stationary states was quantised. Stationary states had to have an angular momentum given by $mvr = nh/2\pi$.

6 To conserve energy, when an electron moved to a lower energy level a photon was emitted with an energy equal to the energy difference between the states. Similarly for an electron to move to a more energetic level it had to absorb a photon of the correct energy.

7 Bohr said the series resulted from all the electron transitions that terminated at a specific stationary state. For example, the Balmer series is produced by all the transitions that terminate at the $n = 2$ stationary state.

8 Bohr's model could only be used for one-electron atoms or ions; it could not explain the fine structure of the hydrogen spectrum, the Zeeman and Stark effects or the intensity of spectral lines.

Key questions ➲ p. 202

9 De Broglie proposed that, because waves behaved like particles, particles might behave like waves. He predicted that the wavelength would be related to the momentum of the particle by $\lambda = h/p$.

10 De Broglie showed Bohr's stationary states represented electron standing waves and, by assuming the circumference of the orbitals was a whole number of wavelengths, he could derive Bohr's quantisation postulate (i.e. $mvr = nh/2\pi$).

11 Experiments by Davisson and Germer showed electrons produced an interference pattern when they were scattered off a nickel crystal that was like the diffraction patterns produced by x-rays.

12 De Broglie's hypothesis led to the Heisenberg uncertainty principle, which states that certain pairs of variables cannot simultaneously be measured with arbitrary accuracy.

Key questions ➲ p. 204

13 Schrödinger used de Broglie's relationship to produce a wave equation for the electron. He then solved the equation for the three-dimensional standing wave modes that the electron wave could assume around a positive nucleus. Hence de Broglie's work was central to Schrödinger's atomic model.

14 Schrödinger's model did not have orbiting electrons like Bohr's model and Schrödinger's model used four quantum numbers rather than one. In Bohr's model the electron was treated as a particle, while in Schrödinger's model the electron was treated as a three-dimensional complex wave function. Bohr's model would enable the electron's position to be found exactly but Schrodinger's model only gave the probability of finding an electron at any point in space.

15 Schrödinger's work was extremely important in the development of the current atomic model, and his equation in a slightly modified form is still used to describe the behaviour of electrons in atoms and molecules. It is possible that Heisenberg's quantum mechanics would have led to the current model but because of the complexity of Heisenberg's approach it may well have taken longer to develop the current model of the atom.

HSC EXAM-TYPE QUESTIONS

Objective-response questions

1 **C**. The Balmer emission series is made up of the hydrogen emission lines that originate from electrons moving to the $n = 2$ orbital from more energetic orbitals (i.e. from orbitals $n = 3, 4, 5, \ldots$ etc). **A** is incorrect as this would absorb photons. **B** refers to one of the lines from the UV (Lyman) emission series. **D** refers to the UV (Lyman) absorption series.

2 **A**. Most alpha particles passed straight through the foil. **B** is incorrect as the Geiger–Marsden experiment disproved Thomson's model. **C** is incorrect as the experiment had nothing to do with what was in the nucleus. **D** is incorrect as the experiment said nothing about quantisation.

3 **D**. In Bohr's model electrons in stationary states did not radiate electromagnetic radiation. **A** is incorrect as both models included a positive nucleus. **B** is incorrect as both models had orbiting electrons. **C** is incorrect as both models are consistent with the results of the Geiger–Marsden experiment.

4 **B**. Energy would not be conserved because the emitted photon would have more energy than the incident photon. **A** and **C** are possible as the atom might relax, passing through several levels emitting photons at each level. **D** is possible as the photon might have energy to ionise the atom.

5 **D**. The velocity gained by the electron is given by equating the work done to the kinetic energy gained, $\frac{1}{2}mv^2 = qV$ and hence $v \propto \sqrt{V}$. Hence $v \propto \sqrt{4} = 2$. As the velocity doubles the momentum also doubles and, as $\lambda = h/p$, the wavelength halves. **A**, **B** and **C** are all incorrect as they give the wrong numerical answer.

Extended-response questions

6 EM This question tests students' understanding of the relationship between the Rutherford and Bohr models of the atom.

a The Rutherford model consisted of a tiny, very dense, positively charged nucleus ✓ in the centre of the atom and orbiting electrons. ✓ The atom was mostly empty space and the electrons were held in their orbits by electrostatic attraction to the nucleus.

b Rutherford's model was unstable as orbiting electrons should emit electromagnetic radiation which would cause them to lose energy and spiral into the nucleus. ✓ Bohr overcame this limitation by introducing stationary-state orbitals from which the electron would not emit electromagnetic radiation. ✓

Rutherford's model could not explain atomic emission spectra. ✓ Bohr overcame this limitation by proposing that photons would be emitted when excited electrons relaxed by moving to a lower energy stationary state. This transition would be accompanied by the emission of a photon with an energy equal to the energy difference between the orbitals. ✓

7 **EM** This question tests students' ability to use Bohr's equation.

a Applying Bohr's equation:

$$\frac{1}{\lambda} = R\left[\frac{1}{n_f^2} - \frac{1}{n_i^2}\right] = 1.097 \times 10^7\left(\frac{1}{5^2} - \frac{1}{2^2}\right) ✓$$

$\lambda = 434$ nm ✓

b To ionise an atom the electron must be excited to the $n \rightarrow \infty$ limit. ✓ Applying Bohr's equation:

$$\frac{1}{\lambda} = R\left[\frac{1}{n_f^2} - \frac{1}{n_i^2}\right] = 1.097 \times 10^7\left(0 - \frac{1}{5^2}\right)$$

$\lambda = 2.28 \times 10^{-6}$ m ✓

8 **EM** This question tests students' understanding of the relationship between de Broglie's matter waves and the Bohr atom and their knowledge of the experimental verification of de Broglie's hypothesis.

a De Broglie showed that the stationary states in Bohr's model corresponded to the standing waves of electrons; that is, the circumference of the stationary state orbitals was a whole number of wavelengths. The $n = 1$ corresponded to one complete electron wave, the $n = 2$ orbital to two complete electron waves, etc. ✓ De Broglie then used the standing wave hypothesis to derive Bohr's postulate as follows:

$n\lambda = 2\pi r$

And as $\lambda = h/p$ we can write $mvr = nh/2\pi$, which was Bohr's quantisation postulate. ✓

b Davisson and Germer showed that electrons had wave properties. They accelerated electrons in an evacuated chamber and examined how the electrons were scattered from a nickel crystal. ✓ They found that most electrons were scattered at a specific angle corresponding to the interference maxima that would occur if the electrons were waves with a wavelength equal to the de Broglie wavelength. ✓ The crystal acts like a diffraction grating for x-ray and particles with similar wavelengths to x-rays (i.e. slow electrons). When Davisson and Germer changed the velocity of the incident electrons they found the angle of maximum diffraction changed in the way that it should if the wavelength changed in accordance with de Broglie's equation. ✓

9 **EM** This question test students' ability to apply de Broglie's equation and their understanding of electron energy levels in atoms.

a As electrostatic attraction provides the centripetal force that holds the electron in its orbit we can write:

$mv^2/r = q_1q_2/4\pi\varepsilon_0 r^2$

$$\text{Or } v = \sqrt{\frac{q_1q_2}{mr4\pi\varepsilon_0}}$$

$$= \sqrt{\frac{(1.602 \times 10^{-19})(1.602 \times 10^{-19}) \times (9 \times 10^9)}{(5.3 \times 10^{-11})(9.109 \times 10^{-31})}}$$

$= 2.2 \times 10^6$ ms^{-1}

Or using Bohr's postulate:

$mvr = nh/2\pi$ with $n = 1$, as we are interested in the ground state:

$v = h/2\pi mr$

$$= \frac{6.626 \times 10^{-34}}{2 \times \pi \times 9.109 \times 10^9 \times 10^{-31} \times 5.3 \times 10^{-11}}$$

$= 2.2 \times 10^6$ ms^{-1} ✓

b The de Broglie wavelength is equal to the circumference:

$\lambda = 2\pi r = 2\pi \times 5.3 \times 10^{-11} = 3.3 \times 10^{-10}$ m

Or we can use the de Broglie equation and the velocity calculated in part **a**:

$$\lambda = h/mv = \frac{6.626 \times 10^{-34}}{9.109 \times 10^{-31} \times 2.2 \times 10^6} = 3.3 \times 10^{-10} \text{ m} ✓$$

c A free electron is considered to have zero potential energy. ✓ Hence an electron trapped in the potential of the nucleus has a negative energy as work would have to be done on the electron to free it from the atom ✓ (i.e. to increase its energy to zero).

10 **EM** This question tests students' understanding of the changes that Schrödinger made to Bohr's model. (two marks for any two similarities and two marks for any two differences between the models)

Both models had a tiny, positively charged nucleus in the centre of the atom. ✓ Both also had electrons outside the nucleus, but while the Bohr atom had electrons as particles orbiting the nucleus, Schrödinger treated the electrons as matter waves. ✓ Both models proposed that photons could be released or absorbed when electrons changed energy states and the photon energy would be equal to the difference in energy between the states. ✓ The Bohr model used one quantum number but the Schrödinger model used three quantum numbers. ✓ (A fourth quantum number was later introduced.)

The Bohr model could not explain the fine spectra of hydrogen, the spectra of atoms with more than one electron, the Zeeman effect or the intensity of spectral lines, but Schrödinger's model could explain all these phenomena.

CHAPTER 15 PROPERTIES OF THE NUCLEUS

INQUIRY QUESTION:

How can the energy of the atomic nucleus be harnessed?

Some nuclear reactions convert a significant amount of the mass of the nucleus to energy. We can harness the energy of the atomic nucleus by initiating nuclear reactions that release energy. Examples of how we already do this include electrical power supplies that use radioactive decay, fission power generators, experimental fusion generators and nuclear weapons that use fission and fusion reactions.

1 Radioactive decay

» Students analyse the spontaneous decay of unstable nuclei, and the properties of the alpha, beta and gamma radiation emitted.

- When a nucleus is unstable it releases particles and energy until it becomes stable. We call an unstable nucleus **radioactive** and the spontaneous emission of particles from the nucleus radioactivity.

radioactive: exhibiting the emission of ionising radiation

- The particles and radiation emitted from the nucleus are millions of times more energetic than the photons emitted from the electrons surrounding the nucleus. Because nuclear radiation is energetic enough to cause ionisation as it moves through materials, it is called **ionising radiation**.
- The mass of the products is always less than the mass of the parent nucleus in radioactive decay. The missing mass is converted into the energy of the emitted particles in accordance with Einstein's mass–energy equivalence equation $E = mc^2$.

ionising radiation: radiation that is energetic enough to ionise the material through which it passes

- A nucleus will be unstable if the proton-to-neutron ratio is not correct or if the nucleus is too large. All elements heavier than lead (atomic number 82) are unstable and hence radioactive. There are also some naturally occurring and many synthetic isotopes lighter than lead that are radioactive.
- You will recall that *isotopes* of an element refer to two or more forms of the element that have the same number of protons but a different number of neutrons. Note also that we call a radioactive isotope a **radioisotope** and that the term **nucleon** is used to describe a neutron or a proton, as they exist in the nucleus.

radioisotope: a radioactive isotope
nucleon: a proton or a neutron

- To specify a specific isotope of an element, we follow the element's name with the number of nucleons; for example, carbon-14 or nitrogen-15.
- We can also specify isotopes using such symbols as:

  $^{A}_{Z}\text{X}$

  where X is the symbol for the element
  Z is the atomic number (number of protons) and
  A is the mass number (number of nucleons)

  Hence we can call an isotope of carbon with eight neutrons carbon-14, or represent it as $^{14}_{6}\text{C}$.
- Radioactive decay can occur in three ways as it involves the emission of three different types of particle. The three decay paths and the associated radioactive particles emitted are named after the first three letters of the Greek lowercase alphabet: alpha (α), beta (β) and gamma (γ). Alpha and beta decay result in the unstable element being transmuted to a different element and, because either process does require initiation, they are called **spontaneous transmutations**.

spontaneous transmutation: conversion of one element into another element via the process of radioactive decay

- Each type of radioactive decay involves the emission of a different particle, outlined as follows.
  1 **Alpha particles** are helium nuclei. That is, a tightly bound, doubly charged nucleus containing two protons and two neutrons. Alpha radiation is highly

ionising but it is not very penetrating. It can be stopped by a few sheets of paper or a few centimetres of air. We can represent alpha decay with an equation of the form:

$$^{A}_{Z}\text{X} \rightarrow {}^{(A-4)}_{(Z-2)}\text{Y} + {}^{4}_{2}\text{He}$$

where X is called the *parent nucleus* and Y is the *daughter nucleus*

alpha particle: helium nucleus

Hence to represent the alpha decay of uranium-238 we would write:

$$^{238}_{92}\text{U} \rightarrow {}^{234}_{90}\text{Th} + {}^{4}_{2}\text{He}$$

Note that the number of nucleons and the total charge are conserved in nuclear reactions. This means that the number of nucleons on each side of the equation and the total charge on each side must balance.

2 **Beta particles** are high-speed electrons. Beta radiation is less ionising than alpha radiation but is more penetrating. Beta particles will pass through a few millimetres of aluminium or about a metre of air. Beta decay is always accompanied by the emission of a tiny, neutral particle, called an *antineutrino* ($\overline{\upsilon}$). Beta emission occurs when one of the neutrons in the parent nucleus decays into a proton and ejects a high-speed electron. We therefore represent beta decay as:

$$^{A}_{Z}\text{X} \rightarrow {}^{A}_{(Z+1)}\text{Y} + {}^{0}_{-1}\text{e} + \overline{\upsilon}$$

Note that in beta decay equations electrons are usually written as e rather than β.

beta particle: high-speed electron

Note that the number of nucleons does not change but the number of protons in the nucleus (atomic number = Z) increases by one when a beta particle is emitted. We can think of beta decay as a method of reducing the neutron-to-proton ratio in the parent nucleus. Note also that the charge on the beta particle is shown as a subscript to balance the charge in the reaction. Hence for the beta decay of carbon-14, we would write the equation:

$$^{14}_{6}\text{C} \rightarrow {}^{14}_{7}\text{N} + {}^{0}_{-1}\text{e} + \overline{\upsilon}$$

There is a similar decay reaction called *beta-plus decay*. In this process a positive electron called a positron (the antiparticle of the electron) is emitted when a proton in the nucleus changes to a neutron. This process increases the neutron-to-proton ratio and is accompanied by the emission of a neutrino rather than an antineutrino. There are no naturally occurring beta-plus emitting radioisotopes but proton rich, beta-plus emitting isotopes can be produced artificially using particle accelerators. Because beta-minus decay is the more common process, beta-minus decay is simply called *beta decay*. When a positive electron is emitted we refer to the process as beta-plus decay.

3 **Gamma ray** particles are very energetic (i.e. very high frequency) photons. They are often emitted after an alpha or beta decay to remove excess energy from the nucleus. Gamma rays are poor ionisers but very penetrating. They will pass through several centimetres of lead. Unlike alpha and beta decay, gamma decay does not change the number of protons and hence does not change the parent atom to a different element. We would write a gamma decay as follows:

$$^{A}_{Z}\text{X} = {}^{A}_{Z}\text{X} + {}^{0}_{0}r$$

gamma rays: high-frequency photon electromagnetic radiation

➔ The properties of the principle radioactive particles are listed in Table 15.1.

Table 15.1 Properties of radioactive particles

Radiation	Symbol	Nature of particle	Charge ($\times 1.6 \times 10^{-19}$ C)	Ionising power	Stopped by
alpha	α $^{4}_{2}\text{He}$	helium nucleus	+2	strong	a few sheets of paper
beta	β $^{0}_{-1}\text{e}$	electron	−1	medium	3 mm of aluminium
gamma	γ	gamma ray photon	0	weak	5 cm of lead

➔ The daughter nucleus often remains radioactive after a nuclear decay and a series of radioactive decays may occur before the nucleus becomes stable. Figure 15.1 shows the radioactive decay series for uranium. Note that each diagonal transition represents an alpha decay and each horizontal transition represents a beta decay.

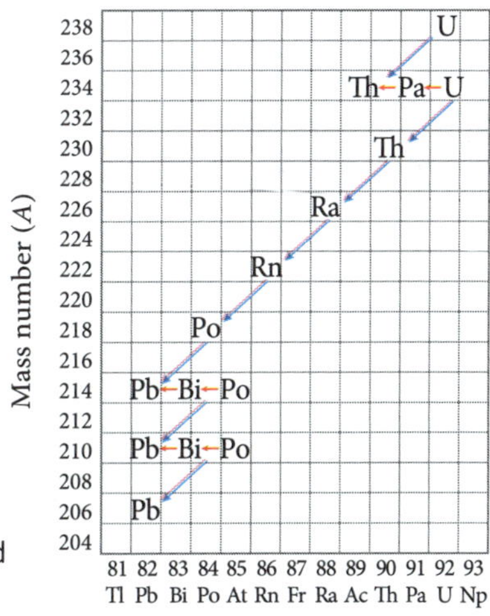

Figure 15.1 Simplified radioactive decay series for uranium

- We can use the different penetrating power or the charge of alpha, beta and gamma radiation to determine what type of radiation is being emitted from a specific radioisotope. For example, Figure 15.2 shows how a magnetic field could be used to separate the different types of ionising radiation.

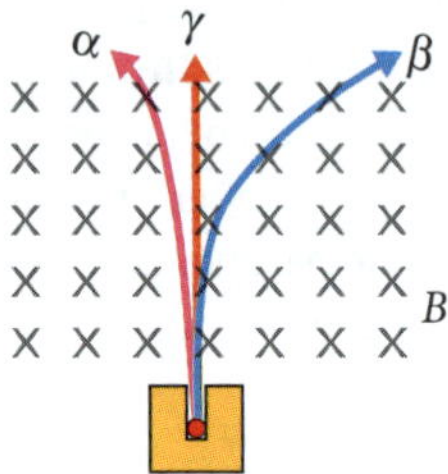

Figure 15.2 Radioactive particles moving in a magnetic field

- Momentum is conserved in radioactive decay and hence when a particle is emitted, the recoil momentum of the nucleus will be equal to the momentum of the emitted particle. Because the nucleus generally has a much greater mass than the emitted particle, this conservation of momentum will mean that the velocity of the emitted particle will be much greater than the recoil velocity of the nucleus. The emitted particle will also have a much higher kinetic energy. We can understand why this occurs by writing the kinetic energy in terms of momentum:

  $K = 1/2mv^2 = p^2/2m$

  This equation tells us that if two particles have the same momentum, the less massive particle will have the higher kinetic energy.

KEY QUESTIONS

1. **What causes an atom to be radioactive?**
2. **What types of radioactive decay are possible?**
3. **How can we differentiate between different types of radioactive emissions?**
4. **Why do some radioisotopes undergo a series of decays?**

Answers p. 219

2 Half-life

» Students examine the model of half-life in radioactive decay and make quantitative predictions about the activity or amount of a radioactive sample using the following relationships:

- $N_t = N_0 e^{-\lambda t}$
- $\lambda = \frac{ln2}{t_{1/2}}$

  where N_t = number of particles at time t
  N_0 = number of particles present at $t = 0$
  λ = decay constant
  $t_{1/2}$ = time for half the radioactive amount to decay

- We saw in the previous chapter that because of the wave-particle nature of matter at the atomic level we can only calculate the probability for specific events to occur. The same limitation applies to radioactive decay. Because radioactive decay is a quantum process, we cannot predict exactly when a specific isotope will decay. The best we can do is give the probability of the isotope decaying in a specific time interval. If we are dealing with a large number of radioactive particles we can, however, use our knowledge of the probability to accurately predict the time for half the isotopes in the sample to undergo radioactive decay. This is analogous to predicting how many times a coin will come up heads or tails in a large number of throws. We cannot predict what a specific throw will produce but if there is a 50% chance of throwing a head, we know that in a large number of throws 50% will come up heads and 50% tails.
- The activity of a radioactive isotope is measured by the isotope's **half-life** ($t_{1/2}$). Half-life is the time it takes for half the nuclei in a radioactive sample to undergo radioactive decay. Thus after one half-life, half of the sample would have decayed; after two half-lives, three-quarters of the sample would have decayed; after three half-lives, seven-eighths would have decayed, etc. A very active sample would have a short half-life, while a less active sample would have a long half-life. If we call the initial number of particles present N_0 and the number that have not decayed N_t we can express this relationship as:

  $N_t = N_0(1/2)^n$

  where n is the number of half-lives that have elapsed

half-life: time for half the atoms in a sample of radioactive material to undergo radioactive decay

- Radioisotopes can have half-lives from fractions of a second to billions of years. Figure 15.3 shows how the number of atoms that have yet to undergo radioactive decay decreases as a function of the number of half-lives that have elapsed.

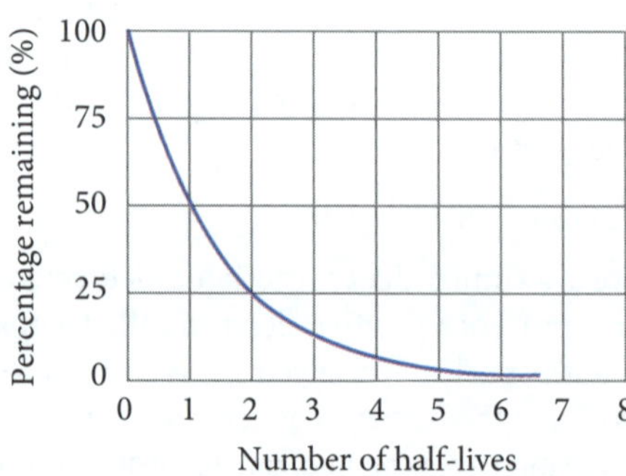

Figure 15.3 Decay of a radioisotope over eight half-lives

- The number of decays per second in a sample depends on the total number of radioactive isotopes in the sample and the half-life of the sample.

➔ Because the number of decays per second ($\Delta N/\Delta t$) depends on the amount of material we can write:

$\Delta N/\Delta t = -\lambda N$

where λ is called the **decay constant** as it is a measure of how fast the isotope will decay

decay constant: a measure of the rate at which radioactive decay occurs

You may recall from your studies in mathematics that this type of equation has a solution of the form:

$$N_t = N_0 e^{-\lambda t}$$

where N_t is the number of undecayed radioisotopes present after a time t
N_0 is the initial number of radioisotopes and
λ is the decay constant

➔ When one half-life ($t_{1/2}$) has elapsed, half the radioisotopes would have undergone radioactive decay. Hence after one half-life $N_t = N_0/2$ and we can write the equation above as:

$\frac{N_0}{2} = N_0 e^{-\lambda t_{1/2}}$, or by cancelling out N_0:

$\frac{1}{2} = e^{-\lambda t_{1/2}}$, which we can rearrange to give $2 = e^{\lambda t_{1/2}}$

Taking the natural log (ln) of both sides of this equation gives us a relationship between the decay constant and the half-life: $ln2 = \lambda t_{1/2}$ or

$$\lambda = \frac{ln2}{t_{1/2}}$$

where λ in the decay constant and
$t_{1/2}$ is the half-life

Note that the decay constant has the SI units of s^{-1} but we use any unit of time in this equation provided the same time unit is used for the decay constant and for the half-life. Note also that the decay constant is inversely proportional to the half-life and hence the decay constant is indeed a measure of the activity of the sample.

EXAMPLE 1

Radon-220 is an alpha emitter with a half-life of 53 seconds.

a Find the decay constant for radon-220.
b Find what percentage of radon-220 atoms would have decayed after 3 minutes.
c Write a nuclear decay equation for radon-220.

Recall that half-life is related to the decay constant and we can work out percentages by setting the initial amount of a sample to 100

Answer:

a Applying the relationship between the decay constant and half-life:

$$\lambda = \frac{ln\,2}{t_{1/2}} = \frac{ln\,2}{53} = 1.31 \times 10^{-2}\ s^{-1}$$

b Taking $N_0 = 100\%$ and applying the decay equation:

$N_t = N_0 e^{-\lambda t} = 100 e^{-(0.0131 \times 180)} = 9.5\%$ remaining

Hence 100 – 9.5 = 90.5% of the atoms have undergone radioactive decay.

c We can use a periodic table to find the atomic number of radon and the element with two less protons than radon and we can then write:

$$^{220}_{86}Rn \rightarrow {}^{216}_{84}Po + {}^{4}_{2}He$$

EXAMPLE 2

Strontium-90 decays by emitting a beta particle. After 10 years, 21% of a sample of strontium-90 is observed to have decayed.

a Find the decay constant and half-life of strontium-90.
b Write a decay equation for strontium-90.

The half-life equations use the amount of sample remaining, not the amount that has decayed

Answer:

a If 21% decays, 79% must remain and hence:

$N_t = N_0 e^{-\lambda t}$ or $N_t/N_0 = e^{-\lambda t}$

Taking the natural log of each side and using the time in years: $ln(N_t/N_0) = -\lambda t$ and hence:

$$\lambda = -(ln(N_t/N_0))/t = -\frac{ln(\frac{79}{100})}{10} = 2.36 \times 10^{-2}\ yr^{-1}$$

The half-life will therefore be given by:

$$t_{1/2} = \frac{ln2}{\lambda} = \frac{ln2}{0.0236} = 29 \text{ years}$$

b After using the periodic table to find the atomic number of strontium and the element with one extra proton than strontium:

$$^{90}_{38}Sr \rightarrow {}^{90}_{39}Y + {}^{0}_{-1}e + \bar{\upsilon}$$

➔ KEY QUESTIONS

5 **How is the half-life and number of atoms in a radioactive sample related to the rate at which particles are emitted from the sample?**

6 **How is half-life related to the stability of the nucleus?**

Answers ➲ p. 219

3 Nuclear fission

» Students model and explain the process of nuclear fission, including the concepts of controlled and uncontrolled chain reactions, and account for the release of energy in the process.

- When a large nucleus splits into two smaller nuclei, the products are less massive than the parent nuclei and hence energy is released in accordance with Einstein's mass–energy equivalence equation $E = mc^2$. The process of splitting a nucleus into two smaller nuclei is called **fission**.

> **fission:** the splitting of a large nucleus into two smaller nuclei

- Fission can be initiated in some large isotopes by the nucleus absorbing a neutron. The additional neutron makes the nucleus unstable and causes it to split apart into two smaller nuclei. Fission is often accompanied by the release of neutrons, which can be used to trigger further fissions in surrounding nuclei and produce a **chain reaction**. Figure 15.4 illustrates a fission chain reaction. Fission triggered by neutron absorption is an example of **artificial transmutation**.

> **chain reaction:** a self-sustaining reaction
> **artificial transmutation:** changing the one chemical element into another by artificial means

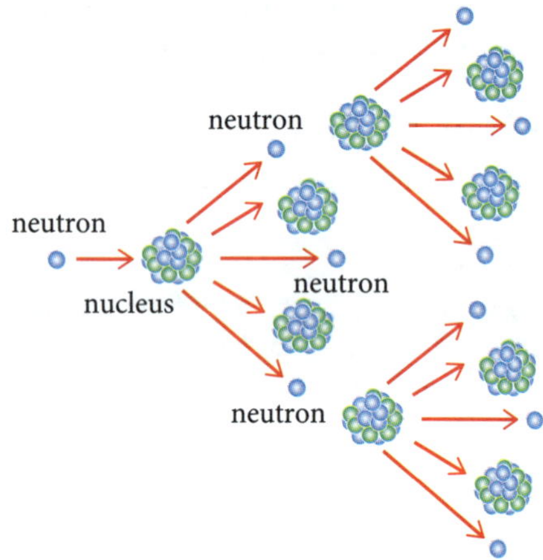

Figure 15.4 Uncontrolled fission chain reaction

- Some isotopes of uranium, including uranium-235 and 233, undergo fission and release neutrons and energy when they absorb a neutron. Note that nuclei do not split into fixed portions and consequently a variety of the daughter products may be produced. For example, uranium-235 may split into a range of smaller nuclei but, on average, 2.3 neutrons and approximately 200 MeV of energy are released per fission. Neutrons are released in fission reactions because the smaller nuclei require a lower neutron-to-proton ratio than the heavy parent nucleus. A typical uranium fission reaction would be:

$${}^{1}_{0}n + {}^{235}_{92}\text{U} \rightarrow {}^{89}_{36}\text{Kr} + {}^{144}_{56}\text{Ba} + 3{}^{1}_{0}n$$

- Note that fission chain reactions do not occur in natural uranium because natural uranium is almost entirely (99.3%) made up of U-238, which absorbs neutrons but does not undergo fission. To initiate a chain reaction the uranium must be **enriched** to increase the percentage of U-235 in the material. Fission weapons use almost pure U-235, while fission power reactors use uranium oxide fuel rods that contain 3.5–4.5% U-235.

> **isotope enrichment:** increasing the percentage of a specific isotope in a sample

- Studies in the 1930s showed that slow neutrons were much more likely to be absorbed by nuclei than fast neutrons. This led to the concepts of **neutron moderation** and **critical mass**. Neutron moderation refers to the process of slowing neutrons to increase the probability of them initiating a fusion reaction, while critical mass is the minimum mass of a sphere of fissionable material that will sustain a chain reaction. Subcritical masses will not sustain a chain reaction because the high-speed neutrons produced in fission escape before being absorbed and initiating another fission. In a mass greater than the critical mass, neutrons have enough collisions to be slowed enough to initiate further fissions and a chain reaction will occur.

> **neutron moderation:** using a material containing small nuclei that does not absorb neutrons, to slow neutrons down (slow neutrons are more likely to be absorbed and initiate fission reactions)
> **critical mass:** the minimum mass required to sustain a fission chain reaction in a solid sphere of fissionable material

- An **uncontrolled fission chain reaction** runs at an increasing rate. That is, each fission initiates more than one more fission and the number of fission reactions per unit time increases. This type of uncontrolled chain reaction is shown in Figure 15.4.

> **uncontrolled fission chain reaction:** a fission chain reaction that occurs at an increasing rate

- The fission bombs used in World War II used uncontrolled fission reactions. To detonate the bombs, a critical mass of material (64 kg of U-235) was assembled in the weapon in the presence of a neutron source. Figure 15.5 shows how conventional explosives were used in one of these weapons to rapidly assemble a critical mass of uranium-235 from two subcritical masses. This weapon converted only 0.7 g of mass to energy but it released an amount of energy that was equivalent to 16 000 tonnes of TNT (a chemical explosive).

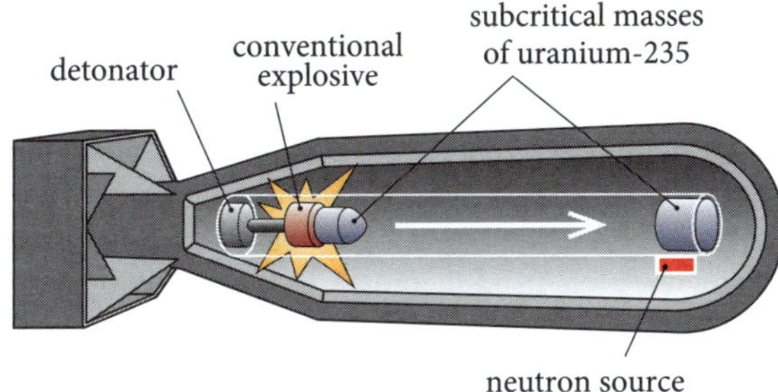

Figure 15.5 Fission bomb showing how a critical mass is assembled from two subcritical masses

➔ Controlled fission reactions run at a constant rate. That is, each fission initiates one more fission. As shown in Figure 15.6 the reaction is controlled by stopping some of the neutrons emitted from fission reactions from initiating further fission reactions.

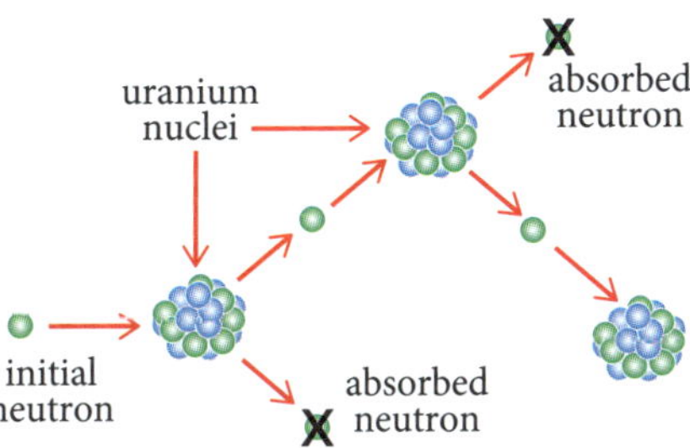

Figure 15.6 Controlled fission reaction requires excess neutrons to be absorbed to ensure a constant rate of reaction

➔ **Controlled fission chain reactions** are used in nuclear power stations.

controlled fission chain reaction: a fission chain reaction that occurs at a constant rate

➔ There are many types of fission power reactors but they all contain the following elements:

1 Fuel: Generally this is slightly enriched uranium (3.5–4.5%) in the form of uranium oxide rods.
2 Moderator: Usually water is used around the fuel to slow or moderate the neutrons in order to increase the probability that they will initiate further fission reactions.
3 Coolant: Water is usually circulated to remove heat from the reactor core.
4 Control rods: These are boron or cadmium rods that absorb neutrons. These rods can be lowered into the reactor to control the rate of reaction.
5 Shielding: Concrete and lead is used to surround the reactor core to protect the operators and the reactor is housed in a larger containment vessel to stop radioactive material escaping in the event of an accident.
6 Waste: Because uranium is a very heavy atom it has a high neutron-to-proton ratio. The neutron-rich daughter products of fission are consequently very radioactive. One-third of the spent fuel rods are removed from the reactor core approximately every year and must be stored for very long periods (thousands of years) before their rate of activity is reduced to safe levels.

➔ A schematic diagram of a nuclear fission power reactor is shown in Figure 15.7. Note that water acts as a neutron moderator and coolant in this reactor and that two independent water loops are used to reduce the risk of accidents.

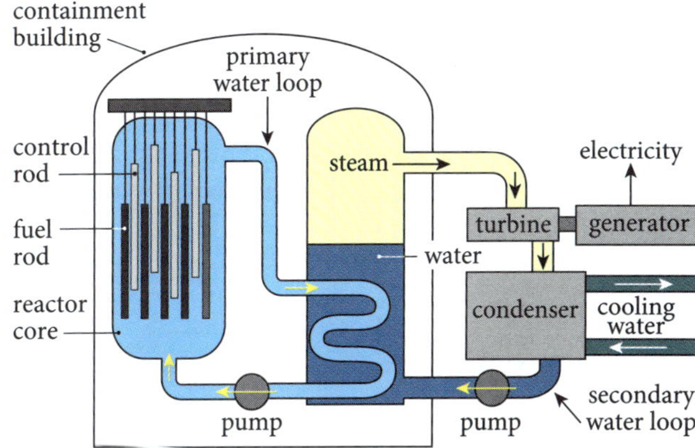

Figure 15.7 Schematic diagram of nuclear fission power reactor that uses water as a coolant and a moderator

➔ KEY QUESTIONS

7 **How are fission reactions initiated and sustained?**

8 **How are uncontrolled fission chain reactions produced?**

9 **How is a controlled fission reaction produced?**

Answers ➲ p. 219

4 Nuclear stability and nuclear reactions

» Students analyse relationships that represent conservation of mass-energy in spontaneous and artificial nuclear transmutations, including alpha decay, beta decay, nuclear fission and nuclear fusion.

» Students account for the release of energy in the process of nuclear fusion.

» Students predict quantitatively the energy released in nuclear decays or transmutations, including nuclear fission and nuclear fusion, by applying:
- the law of conservation of energy
- mass defect
- binding energy
- Einstein's mass–energy equivalence relationship $E = mc^2$.

- Nucleons are held in the nucleus by the **strong nuclear force**. This force acts between all nucleons and is significantly stronger than the electrostatic force, which tries to push the protons apart in the nucleus. As shown in Figure 15.8, the strong nuclear force has a very short range and drops essentially to zero for nucleon separations greater than 2.5×10^{-15} m. This is why nuclei with more than 82 protons are unstable and hence radioactive. Adding another proton increases the force of attraction on the adjacent nucleons but only adds to the electrostatic repulsive force on the protons on the other side of the nucleus (i.e. protons on the far side of the nucleus are too far from the added proton to experience the strong force).

strong nuclear force: a strong, very short-range force of attraction between nucleons in the nucleus of atoms

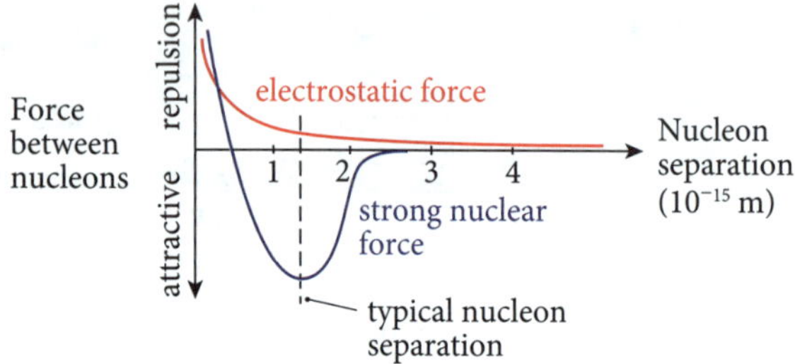

Figure 15.8 Strong nuclear force as a function of the separation between nucleons

- The strong force causes the individual nucleons to combine to form a nucleus. When a stable nucleus is formed, mass is lost and energy is released. Hence, as shown in Figure 15.9, the nucleus is less massive than its constituent nucleons. We call this difference in mass the **mass defect** and the energy released when the nucleus is formed is called the **binding energy** of the nucleus. We can also think of the binding energy as the energy required to separate the nucleus into its constituent nucleons. The energy we put into separating the nucleons is converted into mass and therefore the nucleons are more massive when separated than when they are in a nucleus.

mass defect: the difference in mass between a nucleus and its constituent nucleons
binding energy: the energy required to separate a nucleus into its constituent nucleons; the binding energy is the energy equivalent of the mass defect

Figure 15.9 Stable nuclei are less massive than their constituent nucleons

- The **binding energy per nucleon** is a measure of how tightly the average nucleon is bound to the nucleus and hence it is also a measure of the stability of the nucleus. If we plot the binding energy per nucleon against mass number we obtain a plot, as shown in Figure 15.10, that can be used to compare the nuclear stability of the elements. Note from Figure 15.10 that the stability of the nucleus increases with mass number up to iron and then decreases. This is why the fusion of light elements and fission of very heavy elements both release energy. If two light elements fuse together they produce a more stable nucleus with a greater mass defect, hence mass is lost when fusion occurs and energy is released. Similarly, if a heavy nucleus broke in half the products would be more stable than the parent nucleus. This would mean they had a greater binding energy per nucleon and again mass would be lost and energy released when the reaction occurred. We saw in Chapter 12 that fusion in massive stars stops when they form a core of iron and we now see why this occurs. Adding nucleons to iron consumes energy and hence would not oppose the gravitational compression of the star.

binding energy per nucleon: the average energy to remove a nucleon from a nucleus; the binding energy per nucleon is a measure of the stability of the nucleus

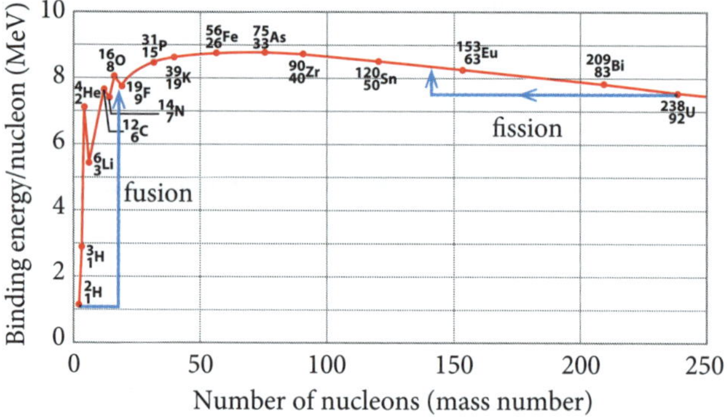

Figure 15.10 Binding energy per nucleon as a function of mass number

- Another important aspect of stability is the neutron-to-proton ratio. Figure 15.11 plots the number of neutrons against the number of protons for the stable isotopes. Note that the number of neutrons equals the number of neutrons in small nuclei but as the size of the nucleus increases, more neutrons are needed to maintain the stability of the nucleus. This is because adding neutrons adds to the strong nuclear attraction but does not add to the electrostatic repulsion. However, note that there is a limit to adding extra neutrons because the **Pauli exclusion principle** (see Chapter 14) applies to protons and neutrons in the nucleus as well as to electrons. This means that each extra neutron that is added must have a higher energy than the last and eventually this will mean that the added neutron has enough energy to escape from the nucleus.

Pauli exclusion principle: no two identical spin 1–2 particles (e.g. electrons, neutrons or protons) can have the same quantum numbers (energy) in an atom

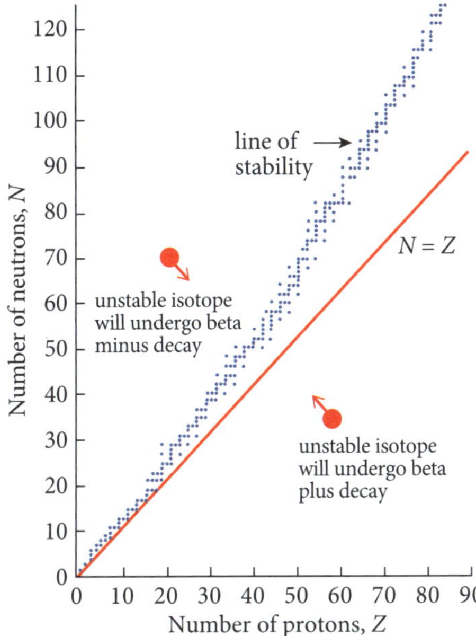

Figure 15.11 Number of neutrons in nucleus against number of protons for stable isotopes

- If an isotope falls off the line of stability it will undergo a series of radioactive decays (i.e. transmutations) until it returns to the line of stability as a stable isotope.
- Note, from Figure 15.11, that if an isotope falls to the left of the line of stability it would have too many neutrons to be stable and would emit beta particles to return to the line of stability. If an isotope fell on the right side of the line it would have too many protons and emit beta plus particles and/or alpha particles to move closer to the line of stability. The further from the line of stability an isotope falls, the more unstable the isotope would be.
- We have examined radioactivity and fission reactions above. You will also recall that we looked at nuclear fusion reactions previously in Chapter 12 and saw that when two small nuclei fuse together to produce a larger nucleus, mass is lost and energy is released. **Nuclear fusion** requires some energy to initiate the process because the electrostatic repulsion between nuclei must be overcome to bring the nuclei close enough together for the strong nuclear force to bind the nucleons together. One way this can be achieved is to use very high temperatures. For example, in the core of the Sun where the temperature is approximately 10^7 K, hydrogen nuclei are moving so fast that some head on collisions result in the nuclei fusing together. You will remember that, in the core of main sequence stars, hydrogen fusion is accomplished via the p–p chain and CNO fusion reaction paths.

nuclear fusion: a nuclear reaction in which two or more small nuclei combine to form a larger nucleus

- Uncontrolled fusion reactions are used in thermonuclear weapons (hydrogen bombs). These devices use an uncontrolled fission reaction to produce the incredibly high temperatures required to start the fusion reaction and can produce thousands of times more energy than a fission bomb.
- Fusion would be a better reaction to use to produce nuclear power on Earth than fission because fusion would have more readily available fuel (hydrogen) and would not produce radioactive waste. However, achieving the ignition temperatures required and containing a plasma at such a high temperature have been difficult technological problems to overcome. Recently there has been significant progress towards achieving controlled fusion on Earth. Extremely high-density currents have been induced in plasmas that have reached the fusion ignition temperature and magnetic fields have been designed to contain these hot plasmas.

If energy is to be released in any spontaneous or artificial nuclear reaction, the products must be less massive than the reactants, because mass must be lost to provide the energy released. When calculating the energy released in a nuclear reaction, such as fission or radioactivity, we find the change in mass (mass defect) and then use Einstein's equation to determine the mass released in joules. Alternatively, if the masses are given in atomic mass units (u), we simply multiply the change in mass by 931.5 MeVc^{-2} to obtain the energy released in millions of electron volts (MeV). The conversion factor has the units MeVc^{-2} because it is a mass. Physicists express the mass of one atomic mass unit as 931.5 MeVc^{-2}. Working through the examples that follow will help you see how the energy released or absorbed in nuclear reactions can be calculated.

EXAMPLE 3

Given that an alpha particle has a mass of 4.001 506 u and a neutron and proton have masses of 1.008 664 u and 1.007 276 u, respectively, find:

a the binding energy of an alpha particle.

b the binding energy per nucleon for an alpha particle (i.e. heium-4).

Binding energy can be calculated from the mass defect involved when the nucleus is formed from its constituent nucleons

Answer:

a We first find the mass defect:

Δm = mass nucleus – mass nucleons

$= 4.001\,506 - (2 \times 1.008\,664 + 2 \times 1.007\,276)$

$\Delta m = -0.030\,374$ u, which means a mass of 0.030 374 u was converted to energy when the nucleus formed.

Therefore the binding energy will be:

Binding energy = $\Delta m \times 931.5\ \text{MeV}c^{-2}$
$= 0.030\,374 \times 931.5 = 28.29\ \text{MeV}$

Note that this is the energy required to separate a helium-4 nucleus into its constituent nucleons and the energy that would be released if a helium nucleus was formed from its constituent nucleons.

b The binding energy per nucleon would be given by:

$\frac{28.29}{4} = 7.07$ MeV per nucleon

EXAMPLE 4

Consider the following reaction:

$^{238}_{92}\text{U} \rightarrow {}^{234}_{90}\text{Th} + {}^{4}_{2}\text{He}$

Note that the mass of uranium 238 is 238.0507 u, the mass of thorium 234 is 234.0436 u and the mass of an alpha particle is 4.001 506 u.

a **Find the amount of energy that would be released or absorbed in this reaction.**

b **Would all the energy go into the kinetic energy of the alpha particle? (Justify your answer.)**

c **Write an equation for the beta minus decay of thorium 234.**

Momentum is conserved in nuclear reactions

Answer:

a We first need to determine if mass is gained or lost in the reaction:

Δm = mass products – mass reactant
$= (234.0436 + 4.001\,506) - 238.0507$

$\Delta m = -0.0056$ u, which means a mass of 0.0056 u was converted to energy.

The energy released will be given by:

$E = \Delta m \times 931.5\ \text{MeV}c^{-2} = 0.0056 \times 931.5 = 5.21\ \text{MeV}$

b Because momentum must be conserved, the particles will share the momentum evenly and recoil from one another. Hence some of the energy will go into the kinetic energy of each particle.

Alternatively we could answer the question quantitatively by calling the momentum of the large particle MV and the momentum of the alpha particle mv. Because momentum is conserved we can write: $MV = mv$ hence the velocities of the particles are related by $v = (M/m)V$.

Now the kinetic energy of the alpha particle would be:

$K_\alpha = \frac{1}{2}mv^2 = \frac{1}{2}m(MV/m)^2 = M/m(\frac{1}{2}MV^2) = (M/m)K_U$

Because for this reaction $M/m = \frac{234.0436}{4.001\,506} = 58.5$, the alpha particle will have 58.5 times more kinetic energy than the thorium nucleus after the alpha particle is emitted.

c In a beta-minus decay a neutron is converted to a proton and hence:

$^{234}_{90}\text{Th} \rightarrow {}^{234}_{91}\text{Pa} + {}^{0}_{-1}\text{e} + \overline{\upsilon}$

EXAMPLE 5

Consider the fusion reaction:

$^{6}_{3}\text{Li} + {}^{2}_{1}\text{H} \rightarrow 2({}^{4}_{2}\text{He})$

Given the mass of lithium-6 is 6.015 121 u, the mass of hydrogen-2 (deuterium) is 2.0140 u and the mass of helium-4 is 4.001 506 u, find:

a **the energy released in joules.**

b **the percentage of initial mass converted to energy in this fusion reaction.**

Remember that 1 MeV = $10^6 \times 1.6 \times 10^{-19}$ J

Answer:

a First we must find the change in mass:

Δm = mass products – mass reactant
$= 2(4.001\,506) - (6.015\,121 + 2.0140)$

$\Delta m = -0.00261$ u, which means a mass of 0.00261 u was converted to energy.

The energy released will be given by:

$E = \Delta m \times 931.5\ \text{MeV}c^{-2} = 0.00261 \times 931.5 = 24.3\ \text{MeV}$

Converting to joules:

$E = 24.3 \times 10^6 \times 1.602 \times 10^{-19} = 3.89 \times 10^{-12}$ J

b The percentage of the mass converted to energy will be given by:

Δm/(mass of reactants) × 100%

$= \frac{0.002\,61}{(6.015\,121 + 2.0140)} \times 100$

= 0.03%

→ KEY QUESTIONS

10 **What is the difference between a spontaneous and artificial transmutation?**

11 **What is binding energy and how is the stability of a nucleus related to the binding energy?**

12 **How is the stability of a nucleus related to the forces operating within the nucleus?**

13 **Why is energy released when light atoms fuse and heavy atoms fission?**

14 **How can we calculate the energy released in a nuclear reaction?**

Answers ➲ p. 219

CHAPTER SYLLABUS CHECKLIST

Are you able to answer these questions from the syllabus for this chapter? Tick each question as you go through the checklist if you are able to answer it. If you cannot answer a question, turn to the relevant page in the study guide to find the answer. For NESA key word meanings, go to www.educationstandards.nsw.edu.au and search 'key words'.

FOR A COMPLETE UNDERSTANDING OF THIS TOPIC:		PAGE NO.	✓
1	Can I describe the properties of alpha, beta and gamma radiation?	208–209	
2	Can I explain why some nuclei are unstable?	208	
3	Can I write equations to represent alpha, beta and gamma decay?	209	
4	Can I explain why unstable isotopes undergo a series of radioactive decays?	209	
5	Can I explain the concept of *half-life* and relate it to the activity of a sample?	210–211	
6	Can I make quantitative predictions about the activity and rate of decay of a radioactive sample?	210–211	
7	Can I write a nuclear fission equation and calculate the energy released from such a reaction?	212	
8	Can I explain the conditions required to produce controlled and uncontrolled nuclear fission?	212–213	
9	Can I explain why only the fission of large nuclei and the fusion of small nuclei can release energy?	214	
10	Can I relate the energy gained or lost in a reaction to an increase or decrease in the binding energy per nucleon?	214	
11	Can I calculate the mass defect and binding energy of a nucleus?	214–215	
12	Can I outline how the binding energy per neutron and neutron-to-proton ratio affect the stability of a nucleus?	214–215	
13	Can I convert mass gained or lost into energy using Einstein's mass–energy relationship $E = mc^2$ and by using the conversion factor 1 u = 931.5 MeVc^{-2}?	215–216	
14	Can I apply the law of conservation of energy to fission, fusion and other nuclear reactions to calculate how much energy is consumed or released in the reaction?	216	

HSC EXAM-TYPE QUESTIONS

Objective-response questions (1 mark each)

1 Consider the following nuclear decay:

$^{90}_{38}\text{Sr} \rightarrow \text{X} + ^{0}_{-1}\text{e} + \overline{\upsilon}$

What element does the symbol X represent?

A rubidium-90
B yttrium-90
C rubidium-89
D yttrium-89

2 Isotope A has a half-life T and isotope B has a half-life of $4T$. A sample of isotope A contained the same number of atoms as a sample of isotope B. If both samples were alpha emitters, which of the following statements about the activity of the two samples is correct?

A Both samples would emit the same number of alpha particles per second.
B Sample B would emit four times more alpha particles per second than sample A.
C Sample A would emit twice as many alpha particles per second than sample B.
D Sample A would emit four times more alpha particles per second than sample B.

3 Which of the following statements best describes what happens in a fusion reaction that releases energy?

A The binding per nucleon increases.
B The binding energy per nucleon decreases.
C The product nucleus is less stable than the reactant nuclei.
D The binding energy per nucleon remains constant but mass is lost.

4 How can we ensure that a fission chain reaction runs at a constant rate?

A by slowing the neutrons with a moderator
B by absorbing some neutrons with control rods
C by using more highly enriched fuel
D by ensuring the mass of fissionable material is exactly twice the critical mass

5 Why are the daughter products of fission highly radioactive?

A The product nuclei will have a neutron-to-proton ratio that is too high.
B The product nuclei will have a lower binding energy per nucleon than the parent nucleus.
C The product nuclei will have a bigger mass defect than the parent nucleus.
D The product nuclei will have a neutron-to-proton ratio that is too low.

Extended-response questions

6 Polonium-218 undergoes an alpha decay followed by two beta decays.

a Write equations to represent each of these decays of polonium-218. (3 marks)
b Compare the nuclear stability of polonium-218 with the stability of the daughter isotopes produced after the three decay reactions. (2 marks)

7 You are given a radioactive isotope but do not know what type of radiation is produced by the isotope. Outline an experiment you could conduct in the school laboratory that would enable you to determine what type of radiation was being emitted from the isotope. (4 marks)

8 Carbon-14 is a beta emitter with a half-life of 5730 years. It decays to nitrogen-14 as follows:

$^{14}_{6}\text{C} \rightarrow ^{14}_{7}\text{N} + ^{0}_{-1}\text{e} + \overline{\upsilon}$

a Explain why carbon-14 spontaneously decays in this way. (2 marks)
b Find what percentage of a sample of carbon-14 isotopes would have decayed after 20 000 years. (2 marks)

9 Consider the following fission reaction and the masses given in Table 15.2.

$^{1}_{0}n + ^{235}_{92}\text{U} \rightarrow ^{92}_{36}\text{Kr} + ^{141}_{56}\text{Ba} + 2^{1}_{0}n$

Table 15.2 Mass of fission reactants and products

Nucleus or nucleon	Mass (u)
neutron	1.008 664
uranium-235	235.043 929
kryton-92	91.926 156
barium-141	140.914 411

a Calculate the mass change in this reaction. (1 mark)
b Calculate the energy released per reaction. (1 mark)
c How many reactions of this type would be required per second to provide 150 MW of power? (2 marks)

10 Main sequence stars fuse hydrogen to helium in their cores via the p–p chain or CNO reaction. The net effect of both these reactions is that four protons fuse to form a helium-4 atom.

a Explain, with reference to the forces involved, why fusion requires temperatures in the order of 10^7 K. (4 marks)
b Given that the binding energy per nucleon of helium-4 is 7.1 MeV, determine how much energy is released in the fusion reaction that produces helium in stars. (1 mark)
c Find the net change in mass for each helium nucleus formed in a star. (1 mark)

ANSWERS

KEY QUESTIONS

Key questions ➲ p. 210

1 If a nucleus has more than 82 protons or the wrong neutron-to-proton ratio, it will be unstable and radioactive. Unstable nuclei emit particles and radiation until they are transmuted into stable nuclei.

2 The three most common types of radioactive decay are alpha (He nuclei), beta (electrons) and gamma (high-energy electromagnetic photons).

3 We can use the penetrating power of the radiation or the charge of the radiation to separate the different types of radiation.

4 After an alpha, beta or gamma ray has been emitted from a radioisotope the daughter product may still be unstable and radioactive. Energy and particles will continue to be emitted from the nucleus until it is transmuted into a stable nucleus.

Key questions ➲ p. 211

5 The number of particles emitted per second from a sample is inversely proportional to the half-life of the radioisotopes in the sample times the number of atoms in the sample.

6 Very unstable isotopes decay very quickly and hence have very short half-lives.

Key questions ➲ p. 213

7 Fission reactions are initiated by bombarding fissionable material with neutrons. When the fissionable isotope absorbs a neutron, it becomes so unstable that it splits into two small nuclei and releases several neutrons. These neutrons can be used to initiate further fission reactions that sustain the reaction.

8 If a sphere of fissionable material (e.g. uranium-235) greater than the critical mass for this material is bombarded with slow neutrons, an uncontrolled fission reaction will be initiated. The rate of reaction in this case increases as time passes because each fission produces more than one additional fission.

9 Controlled nuclear reactions are produced in fission power reactors by using slightly enriched uranium as fuel (3.5–4.5% U-235), a moderator to slow the neutrons to increase the probability that they will initiate fission reactions, and control rods to absorb neutrons and control the rate of reaction. The control rods can be lowered into the reactor core to change the rate at which neutrons are absorbed and hence the rate at which the fission chain reaction proceeds. When running at the required rate, each fission produces exactly one more fission and the rate of reaction remains constant.

Key questions ➲ p. 216

10 A spontaneous transmutation occurs when an unstable nucleus emits an alpha or beta particle. We cannot predict exactly when this will occur but we can predict the probability of it occurring in a specific time interval. In contrast, an artificial transmutation is one that is artificially initiated; for example, by bombarding the nucleus with a high-speed proton or a slow neutron.

11 The binding energy is the energy required to separate a nucleus into its constituent nucleons. The binding energy is the energy equivalent to the mass lost when the nucleus is formed (the mass defect). The binding energy increases with the size of the nucleus and is not related to stability. However, the binding energy per nucleon is a measure of the stability of the nucleus.

12 The stability of the nucleus is directly related to the forces between the nucleons; that is, the strong force of attraction and the electrostatic repulsion between nucleons. (The stability is also affected by the Pauli exclusion principle, which prevents too many neutrons being added to the nucleus.)

13 In both cases the reactions produce products that have a higher binding energy per nucleon and hence are more stable. This means that mass is converted into energy in both the fusion of light elements and the fission of heavy elements.

14 We find the change in mass by subtracting the mass of the reactants from the mass of the products. If mass is lost, energy must be released and if mass is gained, energy must be consumed. If the mass change is known in kilograms, we can use Einstein's mass–energy equivalence equation ($E = mc^2$) to calculate the energy released or consumed in joules. Alternatively, if we know the mass change in atomic mass units (u), we can calculate the energy in MeV by multiplying the change in mass by the conversion factor 931.5 MeVc^{-2}.

HSC EXAM-TYPE QUESTIONS

Objective-response questions

1 **B**. A beta particle is emitted when a neutron in the nucleus changes into a proton and an electron. Hence the daughter nucleus must have the same number of nucleons but one extra proton. **A** is incorrect as rubidium has one less proton than strontium. **C** and **D** are incorrect as they both have one less nucleon than strontium.

2 **D**. The activity is proportional to the inverse of the half-life. Because isotope A has a half-life that is quarter of the half-life of isotope B it will emit four times as many alpha particles per second than isotope B. **A**, **B** and **C** are all incorrect as they predict the wrong rate of emission.

3 **A**. To release energy, the stability of the nuclei must increase. An increased binding energy per nucleon requires a greater mass defect and hence mass is lost and energy emitted in the reaction. **B** and **C** are incorrect as decreasing the binding energy per nucleon or the stability of the nuclei would consume energy. **D** is incorrect as the only way to lose mass is to increase the binding energy per nucleon.

4 **B**. To ensure a controlled rate of reaction we must absorb some neutrons to stop them increasing the rate at which fission occurs. **A** and **C** are incorrect as either would increase the rate of reaction. **D** is incorrect as it would produce an uncontrolled fission chain reaction.

5 **A**. Large nuclei have higher neutron-to-proton ratios than small nuclei. If a large nucleus splits in half, the daughter products will have a neutron-to-proton ratio that is far too high. **B** is incorrect because the products will have a higher binding energy per nucleon. **C** is true but does not cause the products to be radioactive. **D** is incorrect as the products have too many neutrons, rather than too many protons, to be stable.

Extended-response questions

6 **EM** This question tests students' ability to write and balance nuclear decay equations and their understanding of how radioactive decay changes the stability of the nucleus.

a Using the periodic table to find the atomic numbers and symbols:

$$^{218}_{84}\text{Po} \rightarrow {}^{214}_{82}\text{Pb} + {}^{4}_{2}\text{He}$$ ✓

$$^{214}_{84}\text{Pb} \rightarrow {}^{214}_{83}\text{Bi} + {}^{0}_{-1}e + \bar{\upsilon}$$ ✓

$$^{214}_{83}\text{Bi} \rightarrow {}^{214}_{84}\text{Po} + {}^{0}_{-1}e + \bar{\upsilon}$$ ✓

b Because energy is emitted in each decay, mass is lost and the stability of the nucleus must increase. ✓ Hence polonium-214 is more stable than polonium-218. ✓ Students could also talk about the binding energy per nucleon being greater for polonium-214 than for polonium-218 or the neutron-to-proton ratio of polonium-214 being higher.

7 **EM** This question tests students' understanding of the properties of alpha, beta and gamma radiation.

Students could use absorption or pass the radiation into a magnetic field or electric field. They should also explain how they detected the ionising radiation; for example, by using a Geiger counter or cloud chamber. A typical answer would be:

We could use a Geiger counter to measure the proportion of particles that could penetrate various materials. ✓ Before taking each measurement, we would measure and subtract background radiation. We would then measure the intensity of radiation before and after it passes through the absorbing material. If the radiation was stopped by a few sheets of paper it would be alpha radiation. ✓ If it was stopped by a few millimetres of aluminium it would be beta ✓ and if it could pass through a few millimetres of lead it would be gamma radiation. ✓

8 **EM** This question tests students' understanding of the relationship between the neutron-to-proton ratio and the stability of the nucleus. It also examines students' ability to calculate quantities using the half-life equation.

a The neutron-to-proton ratio of carbon-14 is too high and consequently the nucleus is unstable and radioactive. ✓ When it emits a beta particle a neutron changes into a proton, which decreases the neutron-to-proton ratio and increases the stability of the nucleus. ✓

b We first find the decay constant:

$$\lambda = \frac{ln2}{t_{1/2}} = \frac{ln2}{5730} = 1.21 \times 10^{-4}\ \text{yr}^{-1}$$ ✓

Now setting the initial amount to 100%, we can use the decay equation:

$N_t = N_0 e^{-\lambda t} = 100e^{-(0.000\,121 \times 20\,000)}$

$= 8.9\%$

Hence 91.1% would have decayed. ✓

Or students may find the number of half-lives $\left(\frac{20\,000}{5730} = 3.49 \text{ half-lives}\right)$ and then use the alternative decay equation:

$N_t = N_0\left(\frac{1}{2}\right)^n = 100\left(\frac{1}{2}\right)^{3.49} = 8.9\%$

Hence 91.1% would have decayed.

9 **EM** This question tests students' ability to calculate the energy released in a nuclear reaction.

a Δm = mass of products − mass of reactants

$= 235.043\,929 - (91.926\,156 + 140.914\,411 + 2 \times 1.008\,664)$

$= -0.186$ u ✓

b This amount of mass is lost in the reaction. Therefore the energy released will be given by:

$E = 0.186 \times 931.5 = 173.3$ MeV ✓

c We must first convert the energy emitted per reaction to joules:

$E = 173.3$ MeV

$= 173.3 \times 10^6 \times 1.602 \times 10^{-19}$

$= 2.776 \times 10^{-11}$ J ✓

As the power is the energy used per second, the number of reactions per second will be given by:

Energy emitted per second/energy released per reaction

$= \frac{150 \times 10^6}{2.776 \times 10^{-11}}$

$= 5.4 \times 10^{18}$ reactions per second ✓

10 **EM** This question tests students' understanding of nuclear fusion reactions and the relationship between nuclear binding energy and mass defect.

a Because nuclei are positively charged there is an electrostatic, repulsive force between them. ✓ The strong nuclear force attracts nucleons together but it is only stronger than the electrostatic repulsion when the separation between the nuclei is very small (about 2×10^{-15} m). ✓ For two nuclei to fuse together, they must have enough energy to push the nuclei close enough together for the short-range strong nuclear force to overcome the electrostatic repulsion. ✓ The average kinetic energy of particles increases with temperature. Temperatures of 10^7 K are required to ensure that some of the nuclei will have enough kinetic energy in head-on collisions to overcome the electrostatic repulsion and come close enough together for the strong nuclear force to combine them. ✓

b A proton has no binding energy and, because four protons are fused to produce a helium-4 nucleus in stars, the binding energy of helium-4 must be released when it is formed. Hence:

Energy released = (binding energy per nucleon) × (number of nucleons)

$= 7.1 \times 4 = 28.4$ MeV ✓

c The mass change will be the mass defect as the helium nucleus is formed from nucleons. The mass defect is the mass change that produces the binding energy and hence:

Mass defect $= \frac{28.4}{931.5} = 0.03$ u ✓

CHAPTER 16 DEEP INSIDE THE ATOM

INQUIRY QUESTION:

How is it known that human understanding of matter is still incomplete?

Physicists have developed a model to describe the fundamental particles of nature and the forces between them. This model, referred to as the standard model of particle physics, has been extremely successful but is still considered to be incomplete. For example, the model does not include gravity, dark matter or dark energy. The model also assumes (incorrectly) that neutrinos have zero mass and it cannot explain the matter–antimatter asymmetry after the Big Bang that led to our matter-dominated universe. In addition, the model requires 19 experimentally determined parameters to be inserted. Our understanding of matter has progressed enormously over the past century but it is still far from being complete. This chapter examines the fundamental forces and elementary particles encompassed by the standard model and investigates the operation and the role of particle accelerators in the development of the model.

1 Subatomic particles

» Students analyse the evidence that suggests:
- that protons and neutrons are not fundamental particles.
- the existence of subatomic particles other than protons, neutrons and electrons.

➔ With the discovery of the neutron in 1932 physicists had accounted for the three principle constituents of the atom: the electron, the proton and the neutron. They had also developed a full quantum model of the atom that explained atomic spectra, the periodic table and chemical bonding. But experimental studies were soon to show that their understanding of matter was far from complete.

➔ The same year the neutron was discovered, a positive electron (positron) was observed after a cosmic ray collision in a cloud chamber.

➔ **Cosmic rays** are very high-energy, charged particles (mostly protons, alpha particles and electrons) that are produced in the Sun and other stars. Cosmic rays continually bombard the Earth's upper atmosphere. Before particle accelerators were developed physicists worked at high altitudes, on mountain tops and in aeroplanes, where they could use cosmic rays to investigate high-energy collisions. In such energetic collisions, energy is sometimes converted into mass in accordance with Einstein's equation $E = mc^2$ and new particles are formed. By examining these particles using a range of experimental techniques, the mass, energy, charge, lifetime and decay products of the particles could be determined.

cosmic rays: highly energetically charged particles produced by the Sun and other stars that continually bombard the Earth's upper atmosphere

➔ In the 1930s and 1940s cosmic ray experiments let to the discovery of several new subatomic particles. Physicists named these new particles the muon, meson, kaon and lambda. The new subatomic particles were short lived and had large masses comparable with or larger than a proton. These large subatomic particles were called **hadrons** (Greek for 'stout' or 'thick') to distinguish them from electrons and neutrinos that were called **leptons** (Greek for 'small' or 'thin'). The existence of these new particles could not be explained by the atomic or nuclear models of the time. Physicists also began to wonder if these new particles were **elementary** or **composite** particles.

hadrons: heavy composite particles made up of quarks that are affected by the strong nuclear force

leptons: elementary particles that are not affected by the strong nuclear force

elementary particle: a fundamental particle that is not made up of other particles

composite particle: a particle that is composed of two or more elementary particles

➔ In the 1950s and 1960s more and more powerful particle accelerators were developed to study high-energy collisions under more controlled conditions and this led to over 150 new hadrons being discovered. Physicists at the time described this plethora of different particles as the 'particle zoo' and tried to sort the particles into some sort of order based on their properties. Initially physicists thought all these particles were elementary particles but they were later shown to be composite particles made up of elementary particles called **quarks**.

quarks: elementary particles that are the building blocks of hadrons; quarks have fractional charge, interact with the strong (colour) force and, because they have half-integer spin, they obey the Pauli exclusion principle

➔ There was strong evidence to suggest that protons and neutrons were composite rather than fundamental particles. This evidence included:
- Beta-minus decay involved a neutron being transformed into a proton and beta-plus decay involved a proton changing into a neutron. Because electrons or positrons were emitted, these changes inferred an internal structure.
- Measurements of the **magnetic moment** of nuclei showed that the neutron had an associated magnetic field, which strongly implied that some moving charge or charges existed within the neutron.
- In 1969, inelastic scattering experiments between electrons and protons conducted using high-energy particle accelerators indicated there were tiny, scattering centres inside protons. These experiments were in some ways analogous to the early scattering experiments conducted by Rutherford that implied the existence of the nucleus.

magnetic moment: a quantity that represents the magnetic strength and orientation of a magnet; in classical physics, spinning, charged particles have a magnetic moment that is parallel to the spin axis

➔ KEY QUESTIONS

1. **What evidence suggests that protons and neutrons are not elementary particles?**
2. **How were cosmic rays used to produce new subatomic particles?**
3. **What evidence suggests that particles in addition to protons, neutrons and electrons exist?**

Answers ➲ p. 234

2 The standard model of particle physics

» Students investigate the standard model of matter, including quarks and the quark composition:
- hadrons
- leptons
- fundamental forces.

Fundamental forces

➔ All the pushes and pulls in the universe are ultimately due to four fundamental forces. In order from weakest to strongest, these forces are:
1. the force of gravity
2. the electromagnetic force
3. the weak nuclear force
4. the strong (colour) nuclear force.

➔ All the physical interactions and changes we see in the world are due to the electromagnetic force and gravity because the strong and weak forces are confined in the nucleus of atoms. We have seen that gravity holds us to the planet, the planets in their orbits and the stars in their galaxies. All chemical bonds are produced by electromagnetic forces and hence it is electromagnetic forces that hold us together and provide the forces that cause our muscles to contract when we move. Electromagnetic forces also hold the ink to this page and it is electromagnetic forces that we feel when we touch something.

➔ We saw in the last chapter that the strong nuclear force is a very short-range force between nucleons that holds the nucleons together in the nucleus. The weak nuclear force is important in beta decay, where it is instrumental in changing neutrons to protons and protons to neutrons.

➔ We have previously investigated the classical theories of gravity and electromagnetism, which use forces and fields. With the advent of quantum mechanics physicists began to work on producing quantum field theories to describe the fundamental forces of nature.

➔ The electromagnetic field was the first force field to be successfully quantised. The new quantum field theory, developed in the late 1940s, was called **quantum electrodynamics (QED)**. QED is a very successful theory that provides us with an incredibly accurate description of electromagnetic phenomena. Remarkably, some of the predictions made by QED have been experimentally verified to one part in 10^{12}.

quantum electrodynamics (QED): the quantum field theory of electromagnetism

➔ The quantum of a force field is called a **gauge boson**. The photon is therefore the gauge boson of the electromagnetic force.

gauge boson: a force-carrying particle that mediates one of the fundamental forces; bosonic particles have spin 1, which means they do not obey the Pauli exclusion principle

➔ When we studied classical field theory we saw that charged particles were surrounded by electric fields and that a charged particle placed in an electric field experienced a force. The classical theory explains the force between charged particles by saying that one charged particle is in the electric field produced by the other particle.

- QED suggests that the field around a charged particle is made up of a sea of **virtual photons** that are continually being sent out and snapping back to the charged particle. When the virtual photons around two charged objects interact, the objects exchange photons and a force results. In a loose analogy we can think of this as being like two people throwing a heavy ball back and forth, as shown in Figure 16.1. The reaction force on the thrower and the force exerted by the ball on the catcher would cause each person to experience a repulsive force. We say the force-carrying particle (i.e. the photon or ball) mediates the force.

virtual photon: the gauge boson of the electromagnetic force; virtual photons cannot be observed or detected but their exchange causes the electromagnetic force between charges

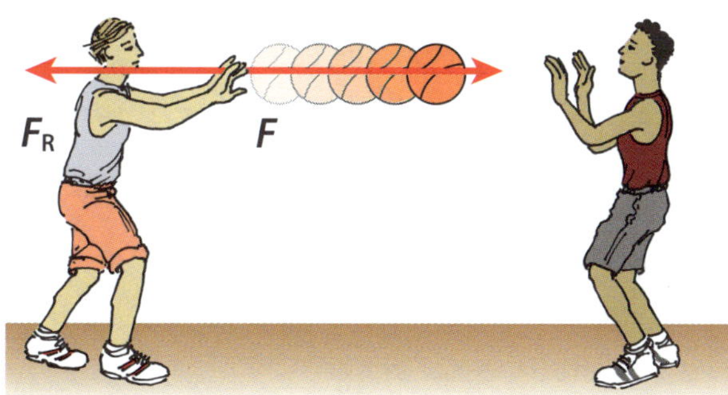

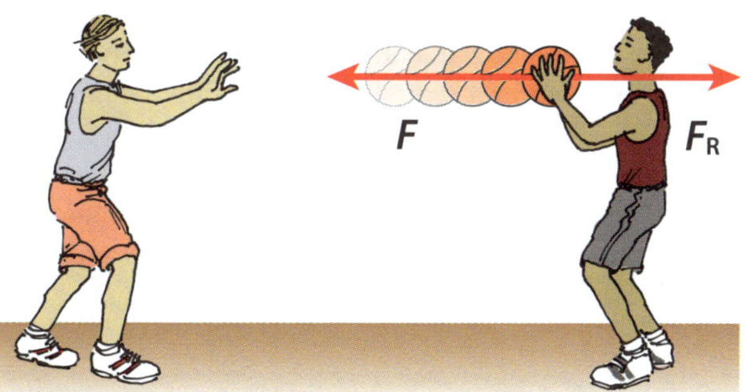

Figure 16.1 Two people throwing a ball to each other experience a force of repulsion

- The theoretical structure of QED was successfully extended to the weak nuclear force to produce a single theory that encompassed both electromagnetism and the weak force. Mediating the weak force required three gauge bosons that were called W^+, W^- and Z^0. Unlike the photon, these bosons had to have mass and this was one of the major difficulties in formulating the theory.
- The weak force has a very short range (about 10^{-16} m) because the virtual bosons have mass and therefore can exist for only a tiny amount of time. This is because they exist due to the uncertainty in energy predicted by Heisenberg's uncertainty principle. Because energy and mass are equivalent, the uncertainty relationship for energy ($\Delta E \Delta t > h/4\pi$) also applies to mass. Because the energy of empty space varies over very short time intervals, a vacuum is not empty but consists of a sea of virtual particles coming in and out of existence for the briefest of instants. The more mass a virtual particle has, the shorter its lifetime would be. Because W and Z bosons are 80 times more massive than protons they can exist for only the briefest of instants and cannot travel far from the parent particle. Photons, in contrast, have no mass and this is why the electromagnetic force extends to infinity.
- Physicists knew that the W and Z bosons had to have mass because the weak force had such a short range but they were initially puzzled about where this mass would come from. The problem was solved by Peter Higgs in the 1960s, who introduced a new field (now called the *Higgs field*) and a new boson (now called the **Higgs boson**). Because of the high energies that were required to produce the Higgs particle, it was not until 2014 that the existence of the Higgs boson was experimentally confirmed.

Higgs boson: the quantum of the Higgs field, which is responsible for giving particles mass

- The properties of the weak-force bosons were theoretically predicted by the **electroweak theory** in 1968. In 1983, when particle accelerators became powerful enough to produce particles of this mass, all three of the weak-force bosons were produced and shown to have the properties that the theory predicted.

electroweak theory: the quantum field theory that encompasses both the weak force and the electromagnetic force

- Electroweak theory predicts that, at very high energies, the weak force and electromagnetic force are identical. Hence for a short time after the Big Bang the weak and electromagnetic forces would have been indistinguishable.
- An extension of the QED formalism to the strong nuclear force proved to be even more difficult to attain but it was eventually accomplished in the 1970s. The theory describing the strong force was called **quantum chromodynamics (QCD)** because it involves colour charge that is related to a force called a **colour force**. We will see in the next section that the colour force holds

quarks together in protons and neutrons. It is the residual of this colour force between quarks that provides the strong force that binds nucleons to one another in nuclei.

quantum chromodynamics (QCD): quantum field theory of strong interactions (colour force and associated strong nuclear force)
colour force: the force that binds quarks together; the residual of the colour force produces the strong nuclear force that binds nucleons together

- **Strong interactions** were found to operate on two scales and are mediated by two types of bosons. The strong force between nucleons is mediated by a gauge boson called a **pion** and the colour force that binds the quarks together within nucleons is mediated by **gluons**.

strong interactions: this term is used to cover the colour force between quarks and the strong nuclear force between nucleons
pion: a hadron made up from two quarks with a mass between an electron and a proton; they are the bosons responsible for the strong force that binds nucleons to one another in the nucleus
gluon: the gauge boson that mediates the colour force that binds quarks together; like quarks, gluons also have colour charge and are affected by strong interactions

- Because pions are made up of two quarks they have mass and hence the strong nuclear force has a very short range (about 3×10^{-15} m). This is why the strong nuclear force does not extend beyond the nucleus.
- Strong interactions are quite different to other forces because the force increases (rather than decreases) when the separation between the interacting bodies is increased.
- QCD predicts that, at the extremely high energies that existed just after the Big Bang, the strong force, weak force and electromagnetic force would have been the one force.
- To summarise, the relative strengths, force-carrying particles (gauge bosons) and range of the four fundamental forces are set out in Table 16.1.

Table 16.1 Properties of the four fundamental forces

Force	Acts on	Relative strength	Quantum field theory	Gauge bosons	Range
gravity	objects with mass	10^{-38}	?	graviton (?)	∞
electromagnetism	objects with charge	10^{-13}	quantum electrodynamics (QED)	photon	∞
weak nuclear	quarks and leptons	10^{-2}	electroweak theory	W^+, W^- and Z	10^{-16} m
strong nuclear (colour force)	between quarks and between nucleons	1	quantum chromodynamics (QCD)	pions (between nucleons) eight gluons (between quarks)	3×10^{-15} m (between nucleons) 0.8×10^{-15} m (between quarks)

- Because gravity is such a weak force compared with the other fundamental forces it has been difficult to carry over the formalism used to quantise the other forces. A boson called a *graviton* has been proposed as the field quanta for gravity, but to date a successful quantum theory of gravity has not been constructed. However, this is an area of active research and theories such as string theory, super symmetry and M theory are being developed in the hope of quantising the gravitational force.

EXAMPLE 1

Consider two protons in a nucleus separated by 10^{-15} m.

a Calculate the electrostatic force between the protons.

b Calculate the gravitational force between the protons.

c Explain why a third force must be operating on the protons.

Protons have a positive charge of the same magnitude as the charge on an electron

Answer:

The mass and charge on the proton is given on the data sheet (see the inside cover of this book).

a The electrostatic force is given by:

$$F = \frac{q_1q_2}{4\pi\varepsilon_0 r^2}$$

$$= 9 \times 10^9 \times \frac{(1.602 \times 10^{-19})^2}{(10^{-15})^2}$$

$$= 231 \text{ N repulsion}$$

b The force of gravitational attraction is given by:

$$F = \frac{Gm_1m_2}{r^2}$$

$$= 6.67 \times 10^{-11} \times \frac{(1.673 \times 10^{-27})^2}{(10^{-15})^2}$$

$$= 1.87 \times 10^{-34} \text{ N attraction}$$

c The electrostatic repulsion is 36 orders of magnitude greater than the gravitational attraction between the protons and yet protons remain within the nucleus. The only way that the protons could remain together is by having another attractive force pulling them together that was stronger than the electrostatic repulsion pushing the protons apart. (This additional force is the strong nuclear force.)

EXAMPLE 2

Compare the strong nuclear force to the electrostatic force.

Recall that the electrostatic force varies as the inverse square of the separation and operates only between charged particles

Answer:

The strong nuclear force is stronger than the electrostatic repulsion between protons for proton separations up to 3×10^{-15} m. The strong nuclear force is an attractive force that operates between all nucleons, while the electrostatic force is a repulsive force that acts only on charged particles (i.e. protons). Electrostatic repulsion decreases as the inverse square of the distance between the protons. In contrast the strong nuclear force increases as the separation is increased up to about 2.5×10^{-15} m and then drops to zero very quickly.

→ KEY QUESTIONS

4 **What are the four fundamental forces of nature and on what type of particles do they act?**

5 **How do the four fundamental forces compare in terms of the strength and range of their interactions?**

6 **What force-carrying particle (gauge boson) mediates each of the fundamental forces?**

Answers ➲ p. 234

Elementary particles

- In the first section of this chapter we saw that over 150 new hadrons were discovered in particle accelerators in the 1950s and 1960s. Physicists hoped they might be able to find some way to order the particles in a similar way to the elements in the periodic table. Organising the elements into the periodic table led to the discovery of atomic numbers and physicists hoped something similar could be achieved with the newly discovered subatomic particles.
- Murry Gell-Mann suggested such an arrangement, called the *eightfold way* in 1961, and he used it to predict the properties of a new particle called the omega minus. A particle with the properties Gell-Mann predicted was discovered in 1964, which provided strong evidence to support the theory.
- The eightfold way led Gell-Mann and George Zweig to suggest independently that hadrons were not elementary particles but were composed of particles that Gell-Mann called quarks.
- It was later shown that six different quarks and six antiquarks could account for all the observed hadrons, including protons and neutrons. Surprisingly quarks had fractional electric charge and a property called **quark colour**.

quark colour: the colour charge is related to the colour force between quarks; quarks can have one of three colour charges—red, blue or green

- We have seen that the force that holds quarks to one another is called the *colour force*. The colour force has three associated colour charges called red, green or blue. The three colour charges are associated with the colour force in a similar way that the electromagnetic force requires two types of electric charge. We can think of the strong nuclear force between nucleons as being due to the residual colour force that extends outside the nucleons.
- Gell-Mann called the different types of quarks *flavours* and placed them into three generations based on the type of particles they formed. The **quark flavours** in each generation of particles are listed in Table 16.2.

quark flavour: refers to the type of quark; there are six quark flavours—up, down, strange, charm, top and bottom

Table 16.2 Quark generation and flavours

Generation	Quark flavour	Quark charge (units of electron charge)	Approximate mass (x mass of proton)
1st	up	+2/3	0.0025
	down	−1/3	0.0053
2nd	charm	+2/3	1.3
	strange	−1/3	0.10
3rd	top	+2/3	182.0
	bottom	−1/3	4.43

- Quarks and the composite particles (hadrons) formed from quarks experience the strong nuclear force. A second group of elementary particles called *leptons* are not influenced by the strong nuclear force. The electron and the electron neutrino, which we introduced previously when we examined beta decay, make up the first generation of leptons. Leptons can also be separated into three generations, as shown in Table 16.3.

Table 16.3 The three generations of leptons

Generation	Lepton
1st	electron and electron neutrino
2nd	muon and muon neutrino
3rd	tau and tau neutrino

- The final group of fundamental particles are the force-mediating bosons. Bosons are quite different to quarks and leptons because, unlike quarks and leptons, bosons have spin 1 and hence do not obey the Pauli exclusion principle. Therefore bosons can have the same quantum state (energy). This is why lasers can produce huge numbers of identical photons with the same energy (quantum state).

- There are specific bosons for each of the fundamental forces and one for the Higgs field that is responsible for giving elementary particles mass. We can think of bosons as field quanta. The bosons associated with the four fundamental forces were listed in Table 16.1. If we add in the Higgs boson we see there are 5 types of bosons and 14 bosons in total.
- In summary, all the observed matter in the universe is made up from a handful of elementary particles: six quarks, six leptons and five types of bosons. All the known elementary particles are listed in Table 16.4.

Table 16.4 Elementary particles of nature

Generation	Quarks (flavour)	Leptons	Bosons
1st	up and down	electron and electron neutrino	photon, W and Z, eight gluons, Higgs, graviton (?)
2nd	charm and strange	muon and muon neutrino	
3rd	top and bottom	tau and tau neutrino	

- The flowchart shown in Figure 16.2 helps to distinguish between the elementary particles.

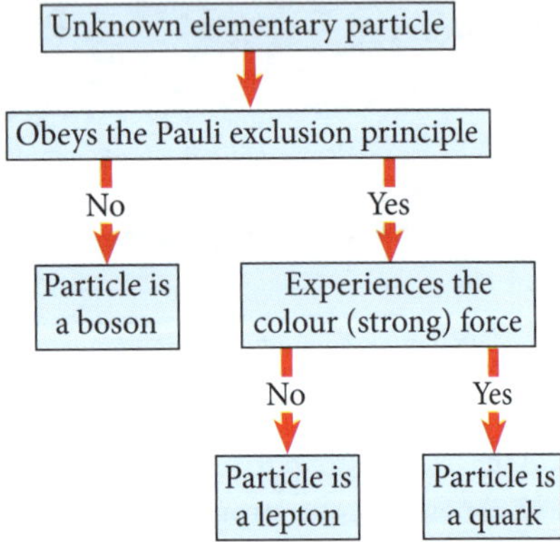

Figure 16.2 Distinguishing between the three types of elementary particles

> **KEY QUESTION**
>
> 7 **What are three categories of elementary particles and how could they be distinguished from one another?**
>
> Answers ⊃ p. 234

The standard model

- The **standard model** of particle physics uses the three quantum force field theories (QED, the electroweak theory and QCD) to describe the interactions between elementary particles.

> **standard model:** the theory that describes the elementary particles of nature and three of the forces that operate between them; the standard model does not include the gravitational force

- The standard model shows that pairs of quarks can combine in quark–antiquark pairs to form particles called **mesons** (e.g. a pion) and in groups of three quarks to form particles called **baryons**. Mesons have quarks with one colour and its anticolour, making them 'white'. Similarly baryons always contain quarks with each of the primary colours, making them 'white'. The colour force and its associated quantum number was introduced to ensure that quarks in hadrons obeyed the Pauli exclusion principle. Figure 16.3 summarises the relationship between hadrons, baryons, mesons and quarks.

> **meson:** a hadron made up of two quarks (a quark–antiquark pair)
>
> **baryon:** a hadron made up of three quarks with different colour charges

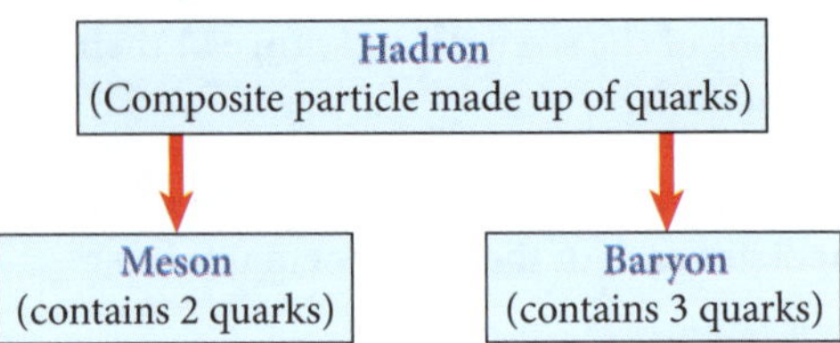

Figure 16.3 Relationship between hadrons, baryons, mesons and quarks

- Protons and neutrons are baryons as they are made up of three quarks, as shown in Figure 16.4. A proton contains two up-quarks and a down-quark, and a neutron contains two down-quarks and an up-quark. The fractional charges on the quarks ensure that the total charge on a neutron is zero and on a proton is +1.

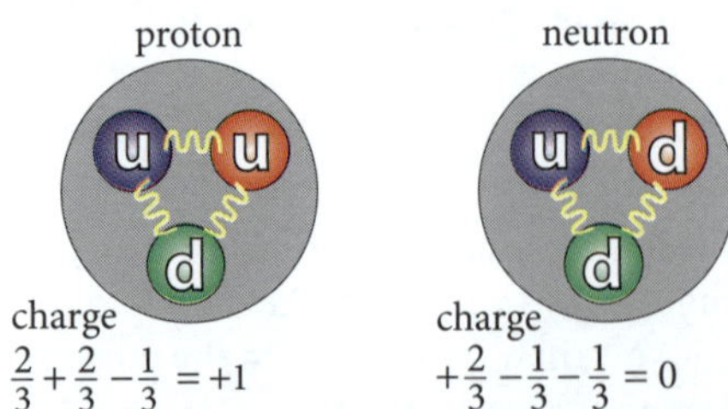

Figure 16.4 Quarks that make up protons and neutrons

- Note the stable particles that make up the atoms and matter in the universe are formed from the first generation of elementary particles; that is, from up- and down-quarks, electrons and electron neutrinos. The second and third generation of elementary particles make up all the short-lived particles that are created in particle accelerators.

- Neutrons and protons also contain gluons that bind the quarks together. When a pair of quarks exchange a gluon they also exchange colours. Note that most of the mass of a neutron or proton is the mass–energy of the gluons and that the mass of the quarks is only a small fraction of the mass of the nucleon.
- When a quark emits a W boson, the quark changes flavour. The W boson then rapidly decays into a beta particle and an antineutrino. This is how beta decay occurs and how a neutron can change into a proton.
- Quarks are so tightly bound together by the colour force that the energy required to separate them creates more quarks. Hence pulling a pair of quarks apart would produce two pairs of quarks, as shown in Figure 16.5. This makes it impossible to isolate single quarks. Physicists call this inability to isolate quarks *quark confinement*. We have seen that the colour force increases with quark separation but interestingly the force between quarks is weak when they are in the nucleon and the quarks appear to drift freely within the nucleon.

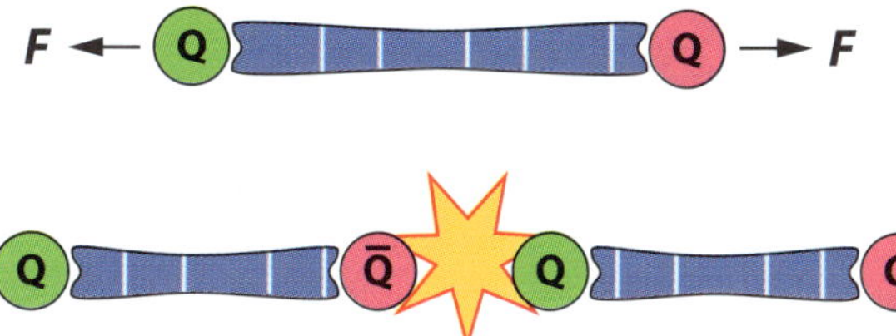

Figure 16.5 Trying to separate the quarks in a hadron produces more hadrons

- The standard model has been incredibly successful in replicating and predicting experimental results. Its predictions of the gluon, the Higgs, and the W and Z bosons, and the six quarks have all now been experimentally confirmed. To date no particles have been detected in particle accelerators that cannot be explained using the standard model. The model has also been used to calculate a range of other parameters that have to be experimentally verified. For example, the standard model gives a value for the anomalous magnetic moment of the electron that has been experimentally verified to one part in a billion.
- However, the standard model is still incomplete because it assumes neutrinos have zero mass, it requires 19 experimentally determined parameters and it does not include the gravitational force, dark matter or dark energy. Testing the limits of the standard model experimentally and theoretically are areas of active current research. Physicists hope to one day develop a grand unified theory (GUT) or a theory of everything (TOE) that will incorporate gravity, dark matter and dark energy.

→ KEY QUESTIONS

8 **What type (flavour) of quarks make up neutrons and protons?**

9 **Why is it impossible to isolate a single quark?**

10 **What is the difference between a hadron and a lepton?**

11 **What are the limitations of the standard model?**

Answers ➲ p. 234

3 Particle accelerators

» Students investigate the operation and role of particle accelerators in obtaining evidence that tests and/or validates aspects of theories, including the standard model of matter.

- The standard model was initially designed to explain the wide range of particles produced in particle accelerators and the forces that act between them. Particle accelerators have also been used to test the predictions of the standard model.
- Particle accelerators use high voltages to accelerate charged particles to give them enough kinetic energy to initiate nuclear reactions.
- An ideal nuclear accelerator has a high beam current and a very high particle energy. We have seen previously that if the electric field is too large between electrodes, current will flow (i.e. cathode rays) between the electrodes. Because this effect limits the potential difference that can be provided to accelerate charged particles, modern accelerators use multiple accelerating stages.
- High energies are required for two reasons. Because of Einstein's mass–energy equivalence equation $E = mc^2$, large energies are required to produce high-mass particles. High particle energies also mean smaller de Broglie wavelengths. As resolving power is proportional to wavelength, increasing the energy of the particle enables smaller structures to be resolved. Thus increasing the energy of particle accelerators has enabled us to investigate the structure of matter on smaller scales and to create more massive particles.
- Experimental, high-energy particle physics today is only carried out in a handful of huge particle accelerators around the world. This is because the cost to build, operate and develop such big machines is huge. The largest particle accelerator in the world, the large hadron collider at CERN, cost about 5 billion dollars to build and costs about 1 billion dollars per year to operate.

 There are many types of particle accelerators that students can investigate; for example, those suggested in the following secondary-sourced investigation. They could also investigate the detectors that have been developed to identify the particles produced in accelerators.

SECONDARY-SOURCED INVESTIGATION

Particle accelorators

This investigation looks at three basic accelerators: the cyclotron, the linear accelerator and the synchrotron.

The cyclotron

The cyclotron device uses two hollow, d-shaped metal half-cylinders called **dees** in a strong magnetic field, as shown in Figure 16.6. A high-voltage AC (radio frequency) supply is connected to the dees and the entire device is evacuated.

dees: hollow metal half-cylinders used in cyclotrons to contain the particles, which are accelerated by the potential difference between the two dees

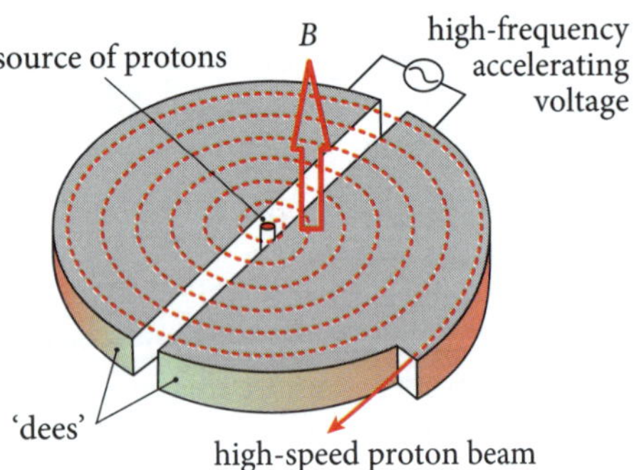

Figure 16.6 Schematic diagram of a cyclotron

A charged particle introduced near the centre of the dees will be accelerated by the voltage across the gap between the dees and will turn in a half-circle when it moves through the magnetic field. Because in this time the polarity of voltage on the dees has reversed, when it returns to the gap between the dees the particle will again be accelerated across the gap. This sequence repeats many times and the particle accelerates to higher and higher velocities. As it does so it turns with a greater radius of curvature in the magnetic field and spirals outwards until it leaves the cyclotron.

The increased path length in the dees exactly compensates for the increased velocity and hence the period to complete one pass through the dees remains constant. We can determine this frequency of the motion as follows:

$F_B = F_C$

$mv^2/r = qvB$ and hence $v = qBr/m$

Now as $v = 2\pi r/T$ we can write $\frac{2\pi r}{T} = qBr/m$,

which simplifies to:

$$T = \frac{2\pi m}{qB} \text{ or } f = \frac{qB}{2\pi m}$$

We call this the **cyclotron frequency**. Note that the cyclotron frequency is a constant that is independent of the radius. If we wish to use the device to accelerate particles beyond a few per cent of the speed of light, we would have to use a changing high-voltage frequency because the frequency will decrease as the relativistic mass of the particle increases. Such devices (synchro-cyclotrons) have been developed but they are much more complex and expensive than devices that operate at constant frequency. Note that the spiralling charges will be accelerating and hence will emit electromagnetic radiation at the cyclotron frequency.

Many large hospitals have a cyclotron to produce short-lived radioisotopes required for the diagnosis and treatment of diseases.

cyclotron frequency: the frequency of particles completing cycles in a cyclotron; for non-relativistic velocities this frequency remains constant

The linear accelerator

As the name implies, linear accelerators (LINACS) accelerate a particle in a straight line. As shown in Figure 16.7, linear accelerators consist of a large number of evacuated, hollow metal cylinders called **drift tubes**. The length of drift tubes increases to ensure the particle will spend the same amount of time in each tube so a constant-frequency AC voltage can be applied to the cylinders.

A high-voltage radio frequency source is connected to each pair of cylinders to accelerate the charged particle as it moves between the cylinders.

drift tubes: hollow metal tubes that contain the particles which are accelerated by the potential difference between the tubes

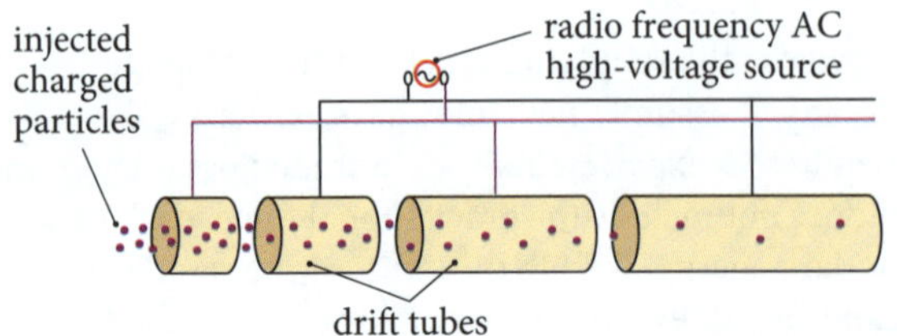

Figure 16.7 Linear accelerator (note that the acceleration occurs only between the drift tubes and that alternate tubes have the same polarity)

Small linear accelerators are used to treat cancer patients using electron beams or the high-energy x-rays produced by colliding high-speed electrons with heavy metal targets. Much larger linear accelerators are used to produce high-speed particles to investigate high-energy physics. For example, the Stanford linear accelerator centre (SLAC) uses a 3.2 km long LINAC that can accelerate electrons and positrons to energies of 50 GeV. The charm quark and the tau lepton were first observed at the SLAC and the accelerator was used in 1990 to identify the quark structure inside protons and neutrons.

The synchrotron

A synchrotron uses a magnetic field to keep the accelerating particles moving in a circle of constant radius. A radio frequency high-voltage source is used to accelerate the particle as it passes through an accelerating stage or stages during each cycle. The device is called a synchrotron because the accelerating voltage must be synched to the period of rotation and the strength of the magnetic field must be synched to the speed of the particle. A schematic diagram of a synchrotron is shown in Figure 16.8.

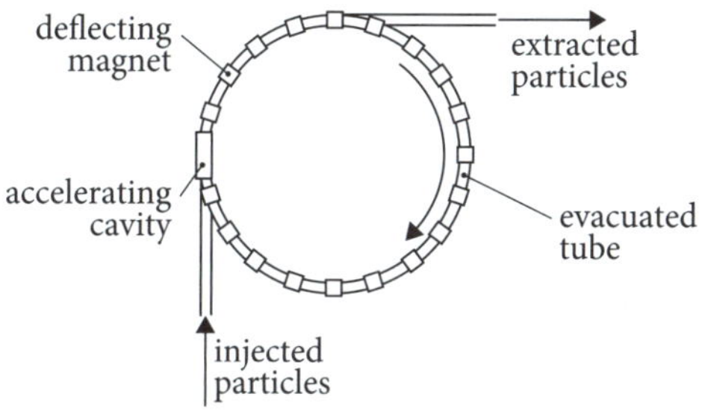

Figure 16.8 Schematic diagram of a synchrotron

Synchrotrons, like the Australian synchrotron in Victoria, are generally used to produce beams of x-rays and other electromagnetic radiation at specific frequencies for scientific research and industrial processes. Very high-energy synchrotrons are used in high-energy particle physics research.

The largest and most powerful synchrotron-type particle accelerator in the world is the large hadron collider (LHC). The LHC is actually a chain of accelerators. Packets of particles are first accelerated in a LINAC to about one-third the speed of light. They then pass into a small synchrotron that accelerates the packet to over 90% of the speed of light. The packets are then injected into a larger synchrotron, which accelerates them to 99.9% of the speed of light. The particles have now reached a transition point where putting more energy into the particles goes into the mass of the particle rather than increasing its velocity. The particles now have 25 times their rest mass and are then injected into an even larger synchrotron with a circumference of 7 km, where they reach energies of 450 GeV. The packets are then sent into the final synchrotron, which has a radius of 27 km and contains two tubes with particles travelling in opposite directions.

After being accelerated to energies of 7 TeV, the particles have effective masses 7000 times their rest mass. Superconducting electromagnets are required to produce the huge magnetic fields needed to keep such energetic particles moving around the synchrotron ring. The counterrotating beams of hadrons are then brought together to collide head on with one another inside massive particle detectors. The energy of the collision is twice the energy of the individual particles as they are travelling in opposite directions. The advantage of colliding two equal mass particles travelling at the same velocity is that the total momentum will be zero and much more energy will be shared between the particles than would be the case if they hit a stationary target.

- When a collision occurs in a particle accelerator it is important to identify all the products and measure their properties. This is achieved with particle detectors that measure quantities such as mass, charge, energy and the momentum of the particles produced.
- Particle accelerators use powerful magnetic fields and layers of subdetectors to measure particle properties. You will recall from our studies of electromagnetism that charged particles travel in curved paths in magnetic fields and measuring the radius of curvature can be used to find the momentum. If the particle moves perpendicularly to the field:

  $F_B = F_C$

  $mv^2/r = qvB$

  $p = mv = qBr$

  Subdetectors track the interaction of charged particles with other substances to determine the charge on the emitted particles.
- Calorimeter subdetectors measure the energy of particles. Absorption calorimeters measure how far particles travel through various materials, while electromagnetic calorimeters measure the energy of electrons and photons by examining how they interact with other charged particles. Hadronic calorimeters measure the interaction of hadrons with other nuclei.
- Particle identification detectors measure radiation emitted from particles. The particle's velocity can be calculated from the Cherenkov radiation that the particle emits when it travels through materials at high speed. A second form of radiation emitted when particles move between different substances can be used to estimate the energy of the particle.
- Figure 16.9 shows the order of the subdetectors in the ATLAS detector used in the large hadron collider. The

ATLAS detector has a diameter of 25 m and a mass of 7000 tonnes. It is one of three detectors placed at different collision points in the main synchrotron ring.

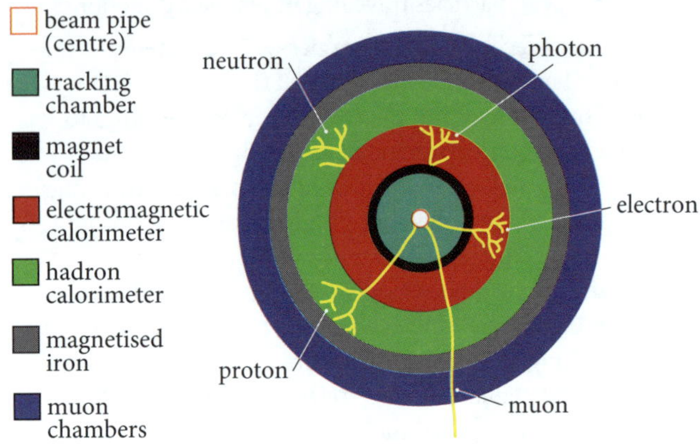

Figure 16.9 Cross-section of the ATLAS particle detector used in the large hadron collider

- Particle physicists put together all these clues about the particle to identify all of the particles that move into the detector after a collision.
- The energies at which collisions occur can also give clues about the particles produced as the mass of the particles produced can never exceed the mass equivalent ($m = E/c^2$) of the energy put into the collision.
- Often the particles identified are decay products of the short-lived particles produced in the collisions; physicists must work backwards to determine the nature of the initial particles produced in the collision.
- Particle physicists analyse the particles produced in a collision to see if any unknown particles are detected and to ensure the results are consistent with the predictions of the standard model.
- For example, the standard model predicts that the Z boson decays rapidly into a lepton–antilepton pair and that there is a total of 24 possible decay paths. It also predicts the mass of the Z boson. When the energies available in the large hadron collider reached the mass equivalent required to produce particles with a mass as great as the Z boson, physicists began to search for the predicted decay products. This is how the Z boson and other particles have been detected in the large hadron collider and other particle accelerators.
- The Higgs boson is another interesting example. It was initially predicted by the standard model in the mid-1960s to give particles mass and was finally detected at the large hadron collider in 2014. The Higgs boson was found to have a mass of 125 GeV/c^2 and was discovered in proton collisions at 7 TeV. One of the unique decay channels for the Higgs boson used to identify it was for it to decay in a pair of Z bosons, which decay into two lepton–antilepton pairs.
- The energies reached in modern particle accelerators are so high that they replicate the conditions that existed a fraction of a second after the Big Bang. Today there is a considerable amount of collaboration between cosmologists and particle physicists. This is because the standard model and the evidence gathered in particle accelerators provide important clues to the creation of matter after the Big Bang. The slight excess of matter over antimatter, dark matter and dark energy are some of the questions that cosmologists and particle physicists are currently seeking to answer collaboratively.
- Theoretical physicists propose theories and models and make predictions that experimental physicists can test. Experimental physicists sometimes make new observations about nature that cannot be explained by existing theories and models. The theoretical physicists then modify existing models or create new theories and models to explain the new observations. We can say that sometimes theory drives experiment and sometimes experiment drives theory. Observations made in particle accelerators drove the development of the standard model and have been instrumental in confirming the model's predictions. Specifically, it was the huge number of hadrons discovered in the 1950s and 1960s that led to quantum chromodynamics and the standard model. Here experiment drove the theory. An example of theory driving experiment would be the theoretical prediction of the existence of W and Z bosons, six quarks and the Higgs boson, all of which were later confirmed by experiments using particle accelerators.

EXAMPLE 3

Consider the positively charged particle with charge (q) and rest mass (m), moving in a synchrotron at velocity (v), shown in Figure 16.10.

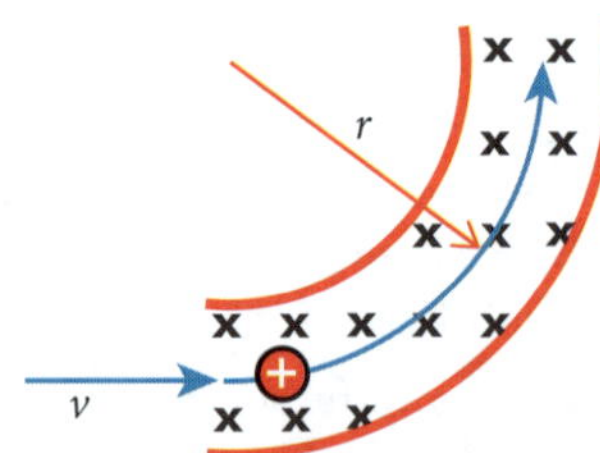

Figure 16.10 Charged particle moving perpendicularly to a magnetic field in a synchrotron

a **Write an expression for the magnetic field strength (B) required to ensure the particle follows a path of radius (r) in the synchrotron. You may assume the particle is moving at non-relativistic speeds.**

b **Write an expression for the momentum of the particle in terms of the charge (q) and the strength of the magnetic field (B). You may assume the particle is travelling at non-relativistic speeds.**

c **If the particle was travelling close to the speed of light, would your answer to part a change? If so, in what way would it change?**

Remember that the magnetic field produces a centripetal force on charges moving perpendicular to the field

Answer:

a The force on the charge due to its motion in the field produces a centripetal force and hence:

$F_c = F_B$

$mv^2/r = qvB$

$B = \dfrac{mv}{qr}$

b **i** Rearranging the expression derived in part a:

$p = mv = qBr$

ii At relativistic velocities the momentum is given by:

$$p_v = \frac{p_0}{\sqrt{1-\frac{v^2}{c^2}}}$$

Substituting this value for mv into the expression derived in part a gives:

$$B = \frac{mv}{qr} = \frac{p_v}{qr} - \frac{mv}{qr\sqrt{1-\frac{v^2}{c^2}}}$$

Therefore the expression for the magnetic field strength must be modified for particles travelling at relativistic speeds. The expression shows that a greater magnetic field strength will be needed to compensate for the increasing effective mass of the particle.

→ KEY QUESTIONS

12 **Why do physicists continue to build more and more powerful particle accelerators?**

13 **How have particle accelerators driven the development of science?**

14 **How have particle accelerators provided evidence to support theories such as the electroweak theory and the standard model?**

15 **Why are cosmologists collaborating with particle physicists at high-power accelerator facilities?**

Answers ➲ p. 234

CHAPTER SYLLABUS CHECKLIST

Are you able to answer these questions from the syllabus for this chapter? Tick each question as you go through the checklist if you are able to answer it. If you cannot answer a question, turn to the relevant page in the study guide to find the answer. For NESA key word meanings, go to www.educationstandards.nsw.edu.au and search 'key words'.

FOR A COMPLETE UNDERSTANDING OF THIS TOPIC:		PAGE NO.	✓
1	Can I explain how investigations of cosmic ray collisions led to the discovery of new particles?	221	
2	Can I explain the advantage of using particle accelerators over cosmic rays to investigate particle collisions?	221	
3	Can I recall that a large number of new subatomic particles were discovered using particle accelerators the 1950s and 1960s?	221	
4	Can I outline the evidence that suggests protons and neutrons are not elementary particles?	222	
5	Can I recall and compare the nature of the four fundamental forces?	222–224	
6	Can I explain how quantum field theories use the exchange of gauge bosons to mediate force?	223	
7	Can I compare the bosons that mediate the four fundamental forces?	224	
8	Can I recall that strong interactions are made up of the colour force between quarks and a residual force (strong nuclear force) between hadrons?	224	
9	Can I outline how the quark model was used to explain the huge number of observed subatomic particles?	225	
10	Can I recall that there are six types (flavours) of quarks and three quark colours?	225	
11	Can I recall the flavours of quarks that make up the proton and the neutron?	225	
12	Can I differentiate between quarks, leptons and gauge bosons?	225–226	
13	Can I recall that hadrons can be made up of three quarks (baryons) or two quarks (mesons)?	226	
14	Can I explain the key elements covered by the standard model?	226	
15	Can I explain why it is impossible to isolate an individual quark (quark confinement)?	227	
16	Can I outline some limitations of the standard model?	227	
17	Can I describe the structure and operation of a particle accelerator?	228–229	
18	Can I explain how particle accelerators made discoveries that led to the development of the standard model?	229–230	
19	Can I outline how particle accelerators have been used to verify predictions of quantum field theories and of the standard model?	230	

HSC EXAM-TYPE QUESTIONS

Objective-response questions

(1 mark each)

1 Which of the following observations implied that nucleons might be composite particles?

A The atom was found to be made from subatomic particles.

B Neutrons were found to have a magnetic moment.

C Protons had charge but neutrons had no charge.

D Nucleons were less massive within the nucleus than outside the nucleus.

2 What are the four fundamental forces of nature?

A mechanical forces, non-mechanical forces, gravity and the strong nuclear force

B the strong nuclear force, the weak nuclear force, gravity and friction

C contact forces, non-contact forces, the strong nuclear force and the weak nuclear force

D the strong nuclear force, the weak nuclear force, gravity and electromagnetic forces

3 Which of the following best explains what happens in beta minus decay?

A An up-quark changes into a down-quark.

B A down-quark changes to an up-quark.

C A lepton changes into a quark.

D A hadron changes into a lepton.

4 What happens when we try to separate the quarks in a hadron?

A The quarks are converted to binding energy and disappear.

B We end up with isolated quarks.

C We end up with more hadrons than we started with.

D It is impossible to pull a hadron apart.

5 Which of the following best describes how particle accelerators are used in physics?

A to initiate nuclear reactions

B to test predictions of the standard model

C to produce new particles

D all of the above

Extended-response questions

6 In the 1960s Gell-Mann introduced new elementary particles called quarks.

a Explain why Gell-Mann introduced the quark model. (2 marks)

b With respect to quarks, what do the terms 'colour' and 'flavour' refer to? (2 marks)

7 There are three types of elementary particles: quarks, leptons and bosons.

a Outline how these particles differ from one another. (3marks)

b List the elementary particles that make up a hydrogen atom. (3 marks)

c Which of the fundamental forces influence quarks? (1mark)

8 A linear accelerator produces protons with a kinetic energy of 15 GeV.

a Find the mass equivalent of 15 GeV. (1 mark)

b Find the velocity of the particle as predicted by classical physics. (1 mark)

c Explain why the actual velocity of the particle is quite different to the value predicted by classical physics. (2 marks)

9 Compare the force that binds nucleons together in the nucleus with the force that holds electrons to the nucleus. (4 marks)

10 The standard model of particle physics has been remarkably successful in predicting the particles found in colliders and particle accelerators.

a List the key elements of the standard model. (4 marks)

b Give an example of a prediction made by the standard model that has been verified experimentally. (1 mark)

c Outline two limitations of the standard model. (2 marks)

ANSWERS

KEY QUESTIONS

Key questions ➲ p. 222

1 Three key observations that indicate protons and neutrons may not be elementary particles are:
   a Neutrons have a magnetic moment even though they have no overall charge.
   b Scattering experiments showed that there were scattering centres inside protons and neutrons.
   c In the 1950s and 1960s over 150 different large particles, such as neutrons and protons, were produced in particle accelerators, implying they were all made up from simpler components.

2 Cosmic rays are high-energy charged particles that bombard the Earth continually from the Sun and other stars. Cosmic rays have enough energy to initiate nuclear reactions and create new particles in collisions. Before high-energy particle accelerators were developed, cosmic rays provided a source of high-energy particles to investigate high-energy collisions of particles at the subatomic scale.

3 Investigations of collisions using cosmic rays and particle accelerators produced many new subatomic particles. The properties of these particles were measured and found to be different to those of neurons, protons and electrons.

Key questions ➲ p. 225

4 The four fundamental forces are: the strong/colour nuclear force, the weak nuclear force, the electromagnetic force and gravity. The strong force acts on particles with colour charge (quarks and gluons). The weak force acts on quarks and leptons. The electromagnetic force acts on particles with electric charge and gravity acts on particles with mass.

5 From strongest force to weakest: strong nuclear, weak nuclear, electromagnetic, gravity. The range of the electromagnetic force and gravity are infinite but the range of the strong and weak nuclear forces is less than the atomic nucleus.

6 The strong force between quarks is mediated by gluons, the weak force by W and Z bosons, the electromagnetic force by photons and gravity by gravitons.

Key question ➲ p. 226

7 Elementary particles can be categorised as hadrons, leptons or bosons. Hadrons experience the strong force and obey the Pauli exclusion principle. Leptons do not interact with the strong force and obey the Pauli exclusion principle. Bosons do not obey the Pauli exclusion principle.

Key questions ➲ p. 227

8 A proton is made up of two up-quarks and one down-quark, and the neutron is made up of two down-quarks and one up-quark.

9 Because the strong force increases with separation (like a rubber band), so much energy is required to separate the quarks that more quarks are produced from the energy. Hence pulling a hadron apart just produces two hadrons.

10 Hadrons interact with the strong force but leptons do not.

11 The standard model does not include gravity, dark matter or dark energy. The model also assumes (incorrectly) that neutrinos have zero mass and it cannot explain the matter–antimatter asymmetry after the Big Bang that led to our matter-dominated universe. In addition, the model requires 19 experimentally determined parameters to be inserted.

Key questions ➲ p. 231

12 Higher particle energies mean more massive particles can be created in collisions. It also enables smaller structures to be examined. This is because increasing the particle energy decreases its de Broglie wavelength, which increases the resolving power in scattering experiments.

13 Particle accelerators have been used to discover many new particles and interactions that theorists have had to explain. The predictions of these new theories can then be tested using accelerators and colliders.

14 The predictions of theories can be tested in accelerators and colliders. For example, the electroweak theory predicted the existence and properties of the W and Z bosons, which were later confirmed experimentally in accelerators. Similarly, the standard model made many predictions about the existence of quarks and other particles that were later confirmed experimentally using accelerators.

15 The energies reached by accelerators today mimic conditions a fraction of a second after the Big Bang. Cosmologists use the standard model to understand how the four fundamental forces and matter were created in the Big Bang. They also use particle accelerators, which recreate conditions shortly after the Big Bang.

HSC EXAM-TYPE QUESTIONS

Objective-response questions

1 **B**. A 'magnetic moment' implies moving charge. The neutron was a neutral particle and hence the magnetic moment was thought to have been caused by moving charges within the neutron. **A** is incorrect as it has nothing to do with the structure of subatomic particles. **C** is incorrect as elementary particles may or may not have charge and hence nucleons may have been elementary particles. **D** is incorrect as it refers to mass defect and has nothing to do with the structure within the nucleons.

2 **D**. These are the four fundamental forces. **A**, **B** and **C** are all incorrect as they contain one or more forces that are composite rather than fundamental forces.

3 **A**. In beta-minus decay, a neutron changes into a proton in the nucleus. This occurs when an up-quark emits a W^- boson (charge −1), which changes the flavour of the quark from up to down. This results in a neutron in the nucleus being changed to a proton. **B** is incorrect because this would change a proton into a neutron and the charge would not be conserved. **C** and **D** are incorrect as neither would change a neutron to a proton in the nucleus.

4 **C**. This is because of quark confinement. When we try to pull quarks apart we make more quarks and hence pulling a pair of quarks (i.e. a hadron) apart simply makes two pairs of quarks (i.e. two hadrons). **A** is incorrect as the mass would increase as energy is put into the hadron. **B** is impossible because quark confinement by the colour force makes it impossible to isolate an individual quark. **D** is incorrect as hadrons are regularly split in particle collisions in accelerators and colliders.

5 **D**. Particle accelerators are used for all of these reasons. **A**, **B** and **C** are all true and therefore **D** provides the best description of how particle accelerators are used in physics.

Extended-response questions

6 EM This question tests students' understanding of the need to find some unifying concept to explain the plethora of new subatomic particles discovered in the 1950s and 1960s. It also tests their understanding of quark colour and flavour.

- a Over 150 new subatomic particles were discovered in particle accelerator collisions in the 1950s and 1960s. ✓ Physicists looked for some way to organise the particles. Gell-Mann suggested a scheme called the eightfold way, which implied that all the particles were made up of a handful of more fundamental particles he called quarks. ✓
- b The strong colour force between quarks requires quarks to have three colour charges. Gell-Mann called these colour charges red, green and blue. ✓ A quark can have any one of these colour charges. There are six distinct quarks that have different masses and charges. Gell-Mann called these types of quarks the quark 'flavours'. The flavours of quark are up, down, top, bottom, strange and charm. ✓

7 EM This question tests students' understanding of elementary particles.

- a Quarks interact with the strong colour force ✓ but leptons do not. ✓ Bosons do not obey the Pauli exclusion principle (as they are spin 1 particles) but leptons and quarks do obey the principle (as they are half-spin particles). ✓
- b A hydrogen atom is made up of an electron, ✓ two up-quarks ✓ and one down-quark. ✓ (Students may also mention photons or gluons.)
- c All four of the fundamental forces influence quarks. ✓

8 EM This question tests students' understanding of relativity in particle accelerators.

- a Converting the energy to joules and applying Einstein's mass–energy equivalence equation:

  $E = mc^2$ and hence

  $$m = \frac{E}{c^2} = \frac{(15\times10^9\times1.602\times10^{-19})}{(3\times10^8)^2}$$

  $$= 2.67\times10^{-26}\ \text{kg}\ ✓$$

- b Classically the kinetic energy is given by:

  $E = mv^2$ and hence

  $$v = \sqrt{\frac{E}{m}} = \sqrt{\frac{15\times10^9\times1.602\times10^{-19})}{1.67\times10^{-27}}} = 1.2\times10^9\ \text{ms}^{-1}\ ✓$$

- c The particle cannot travel faster than the speed of light and the energy put into the particle as it moves nearer the speed of light begins to go into increasing the mass ✓ of the particle rather than into increasing its velocity. ✓ Hence the effective mass of the proton is much greater than its rest mass. Substituting a much larger mass into the equation above would produce a much lower velocity.

9 EM This question tests students' understanding of two of the fundamental forces of nature.

The strong nuclear force that binds nucleons together in the nucleus acts between protons and protons, protons and neutrons, and neutrons and neutrons. ✓ It is the residual of the colour force that binds quarks in the nucleus. The strong force has a very short range (about 3×10^{-15} m) ✓ and increases with separation between nucleons. The colour force in the nucleons is mediated by bosons called gluons, and the strong force between nucleons is mediated by a boson called pions (note that pions are mesons that contain two quarks).

In contrast, the electromagnetic force that attracts the negatively charged electron to the positively charged nucleus is a long-range force ✓ that is much weaker than the strong nuclear force. ✓ The electromagnetic force decreases as the inverse square of the separation between the charged particles and is mediated by a boson called a photon.

10 EM This question tests students' understanding of the standard model of particle physics.

- a The standard model combines the quantum field theories for the electromagnetic, weak nuclear and strong nuclear forces (QED, electroweak theory and QCD) ✓ to explain the interaction between elementary particles. The field theories are based on gauge bosons mediating the forces. ✓ The electromagnetic force is mediated by photons, the weak force is mediated by W and Z bosons, and the colour force is mediated by gluons. ✓ The elementary particles consist of the force mediating bosons plus the Higgs boson, six quarks and six leptons and their antiparticles. ✓
- b The standard model predicted the existence of a boson called the Higgs boson ✓ that gave some particles their mass. This particle was predicted to exist in the 1960s and first seen in a high-energy particle collider (the LHC) in 2014. (Students could also pick any of the flavours of quark or the W and Z bosons.)
- c The standard model does not include the force of gravity, ✓ nor does it explain dark matter. ✓ (Students could talk about the number of experimentally determined factors that must be added into the model; for example, that it does not explain dark energy, that it doesn't explain the matter–antimatter asymmetry in the early universe or that it assumes neutrinos must have zero mass).

SAMPLE HSC EXAMINATION 1

Try to complete these papers as if they are the real thing. These are the instructions you need to follow in the HSC Exam:

General instructions

- Reading time: 5 minutes
- Working time: 3 hours
- Write using black pen.
- Draw diagrams using pencil.
- NESA approved calculators may be used.
- Use the Data Sheet, Formulae Sheet and Periodic Table in this book.

Total marks: 100

Section I: 20 marks

Section II: 80 marks

- Attempt all questions.

Section I: 20 marks

Attempt Questions 1–20.
Allow about 35 minutes for this section.

1 **Sandra and John stand side by side and each throw a cricket ball horizontally from shoulder height. If the balls landed the same distance from Sandra and John, what can we deduce?**

A The balls must have been thrown with the same initial velocity.

B If John was taller than Sandra, he must have thrown the ball with a lower initial velocity.

C If Sandra was taller than John, she must have thrown the ball with a greater initial velocity.

D The balls must have been in the air for the same time.

2 **A ball of mass m on the end of a string turns through a radius r as it rotates around a central rod with velocity v, as shown in Figure E1.1.**

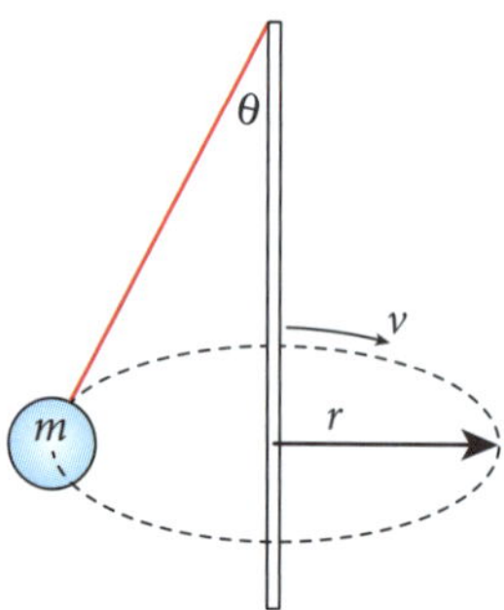

Figure E1.1 Mass on a string moving with uniform circular motion around a vertical rod

What is the angle (θ) between the vertical rod and the string?

A $\sin\theta = v^2/rg$

B $\sin\theta = mv^2/rl$

C $\tan\theta = mv^2/rl$

D $\tan\theta = v^2/rg$

3 **Which of the following statements best describes the energy changes that occur when an Earth satellite moves from a circular orbit of radius r to a circular orbit of radius $2r$?**

A The satellite's kinetic energy halves, its potential energy doubles and the total energy remains constant.

B The satellite's kinetic energy halves, its potential energy halves and the total energy halves.

C The satellite's kinetic energy doubles, its potential energy halves and the total energy halves.

D The satellite's kinetic energy doubles, its potential energy doubles and the total energy remains constant.

4 **An electron travelling with an initial velocity u enters a uniform electric field midway between two charged plates, as described in Figure E1.2. The electron travels a horizontal distance x before colliding with the positively charged plate. (You may ignore gravity in this question.)**

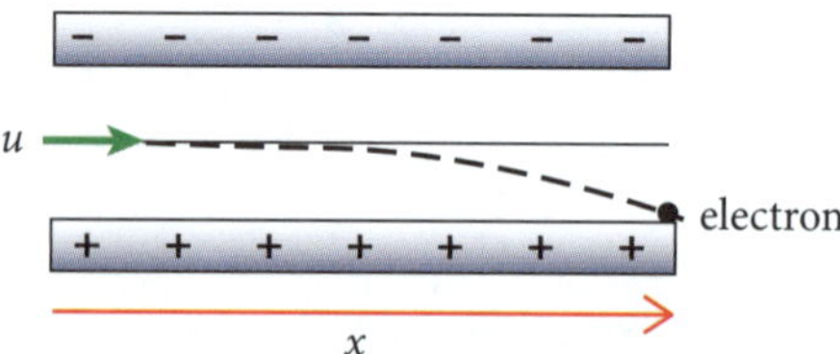

Figure E1.2 An electron fired with velocity u midway between two parallel, charged plates

Where would the electron land if the electric field strength was increased from E to $4E$?

A at $x/2$

B at $x/4$

C at $(\sqrt{2})x$

D at $x/8$

5 **A current I flows in each of the three equally spaced, parallel wires shown in Figure E1.3.**

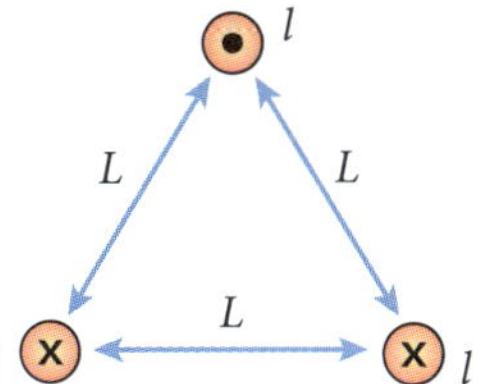

Figure E1.3 Three parallel current-carrying wires

What is the direction of the resultant force on the wire on the lower right?

A
B
C
D

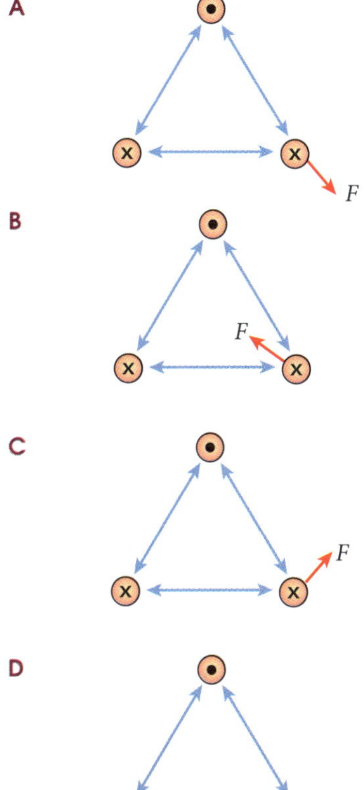

6 **A permanent magnet is dropped through a loop of wire, as illustrated in Figure E1.4.**

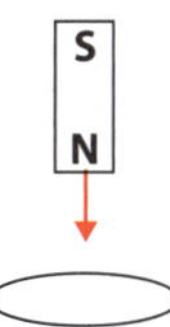

Figure E1.4 Permanent magnet dropped through a loop of wire

When viewed from above the loop, in what direction will the induced current flow in the wire as the magnet falls through the loop?

A momentarily clockwise and then momentarily anticlockwise

B momentarily anticlockwise and then momentarily clockwise

C momentarily clockwise

D momentarily anticlockwise

7 **What would happen to the torque produced by a motor if a load was applied that slowed the rate at which the motor was rotating?**

A The torque produced by the motor would increase.

B The torque produced by the motor would decrease.

C The torque produced by the motor would remain constant.

D Not enough information has been given for the change in torque to be determined.

8 **The details of the absorption spectra of two stars X and Y are set out in Table 1.1.**

Table 1.1 Spectral data for star X and star Y

Star	Absorption spectrum detail	Wavelength shift from line on Earth
X	Contains neutral atomic lines and weak ionic lines	All absorption lines are shifted to shorter wavelength
Y	Contains weak atomic lines but strong ionic lines	No wavelength shift

What can we deduce about the stars from Table 1?

A X is moving away from the Earth and is cooler than Y.

B X is moving towards the Earth and is hotter than Y.

C X is moving away from the Earth and is hotter than Y.

D X is moving towards the Earth and is cooler than Y.

9 **If the red light used to demonstrate interference with a double slit was replaced with a blue light, how would the interference pattern change?**

A The maxima would move further apart if blue light was used.

B The maxima would move closer together with the blue light.

C The intensity of the maxima would decrease if blue light was used.

D The interference pattern would turn blue but otherwise remain unchanged.

10 **A light source with a frequency above the threshold frequency was used in a photoelectric experiment. If the power of the source remained constant but the wavelength was halved, how would the photocurrent produced and maximum kinetic energy of the ejected electrons change?**

A The photocurrent and maximum kinetic energy would both increase.

B The photocurrent and maximum kinetic energy would both decrease.

C The photocurrent would increase and the maximum kinetic energy would decrease.

D The photocurrent would decrease and the maximum kinetic energy would increase.

11 Why do stars move off the main sequence on a Hertzsprung–Russell diagram?

A An iron core forms in the star and fusion no longer produces energy.

B A shell of helium forms in the star, which increases the outwards radiation pressure.

C A core of helium forms in the star, which decreases the outwards radiation pressure.

D Hydrogen fusion begins to occur through the p–p chain reaction.

12 A cathode ray beam was deflected by an electric field, as shown in Figure E1.5. In what direction must a magnetic field be applied between the charged plates to enable the cathode ray to pass through the electric field without being deflected?

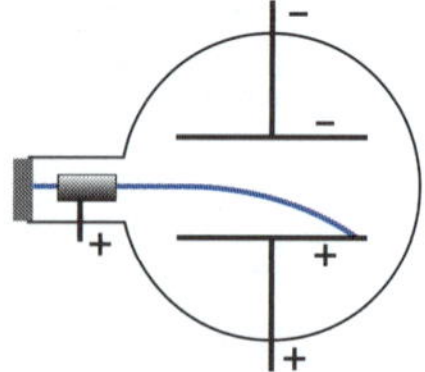

Figure E1.5 Cathode ray deflected by an electric field in an evacuated tube

A into the page | **B** out of the page

C to the right | **D** to the left

13 Bohr's model of the hydrogen atom enabled the ionisation energy of the ground state ($n = 1$) of the hydrogen atom to be related to Rydberg's constant (R), the speed of light (c) and Planck's constant (h). What relationship did Bohr deduce for the ionisation energy of hydrogen?

A $E = Rh/c$ | **B** $E = h/Rc$

C $E = Rc/h$ | **D** $E = Rhc$

14 Two radioactive isotopes A and B both emit alpha particles but B has a half-life 10 times longer than A. What can we conclude about the isotopes?

A B will emit more alpha particles per second than A.

B A will have a bigger radioactive decay constant than B.

C The alpha particles emitted from B will be more energetic than the alpha particles emitted from A.

D After one half-life, 10 times more isotopes in sample A would have decayed than in sample B.

15 Why do uranium fission power reactors produce highly radioactive waste?

A Uranium-238 is very radioactive but not fissionable.

B The fission products have a high proton-to-neutron ratio.

C The fission products have a high neutron-to-proton ratio.

D Uranium-235 is very radioactive.

16 Consider a car of mass m travelling with a constant velocity v around a corner of radius r as shown in Figure E1.6. Which of the following statements best describes the net force on the car when it is at the position shown?

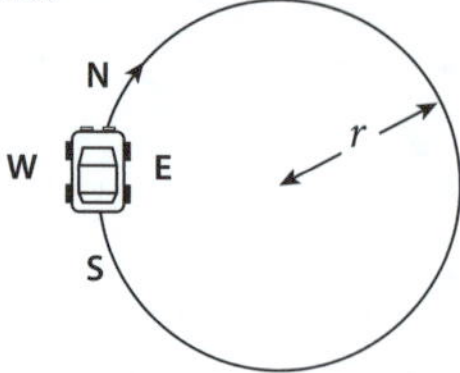

Figure E1.6 Car travelling at constant velocity around a corner of radius r

A There is no net force on the car as it is moving with uniform circular motion.

B The net force depends on the v, m and r and is directed towards the west.

C The net force depends on the v, m and r and is directed towards the east.

D The net force will be the sum of the centripetal force and the gravitational force acting on the car.

17 The radiation emitted from a radioactive isotope is passed into a magnetic field and found to be deflected as shown in Figure E1.7. What type of radioactive isotope is most likely to have emitted this radiation?

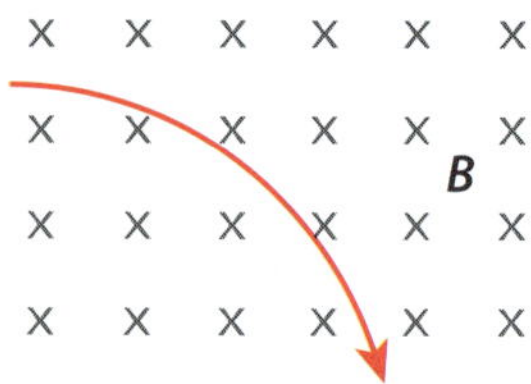

Figure E1.7 Radiation in a magnetic field

A an isotope with a neutron-to-proton ratio that was too high to be stable

B an isotope with a very high binding energy per nucleon

C an isotope with a proton-to-neutron ratio that was too large to be stable

D an isotope with electrons in the nucleus

18 In a Millikan oil drop experiment, an oil droplet of mass 1.635 mg was found to levitate in an electric field of 10^7 NC^{-1}. How many electron changes were on the droplet?

A 10^7 | **B** 10^{10}

C 10^4 | **D** 10^{13}

19 The black-body emission curves from two main sequence stars P and Q were compared. Star Q was found to exhibit a maximum black-body emission at a much shorter wavelength than star P. What does this tell us about the stars?

A Star P is smaller and will leave the main sequence faster than star Q.

B Star P is smaller and will remain on the main sequence longer than Q.

C Star Q is smaller and will the leave the main sequence faster than star P.

D Star Q is smaller and will remain on the main sequence longer than star P.

20 In a high-energy nuclear collision between hadrons in a particle accelerator, dozens of new particles can be produced. These include photons, hadrons and leptons. Why are no free quarks produced?

A because quarks are fundamental particles

B because the energy used to separate quarks in a hadron produces more quark pairs

C because quarks cannot be produced from energy

D because quarks are imaginary rather than real particles

Section II: 80 Marks

Attempt Questions 21–33.

Allow about 2 hours and 25 minutes for this section.

Instructions
- In the HSC Exam you will answer the questions in the spaces provided. These spaces provide guidance for the expected length of response.
- Show all relevant working in questions involving calculations.

---

21 A student turns a 50 g mass on a string in circular motion with a radius of 0.60 m in a vertical plane, as shown in Figure E1.8. The centre of the circle was 1.8 m above ground level and the mass was rotated with a period of 0.20 s.

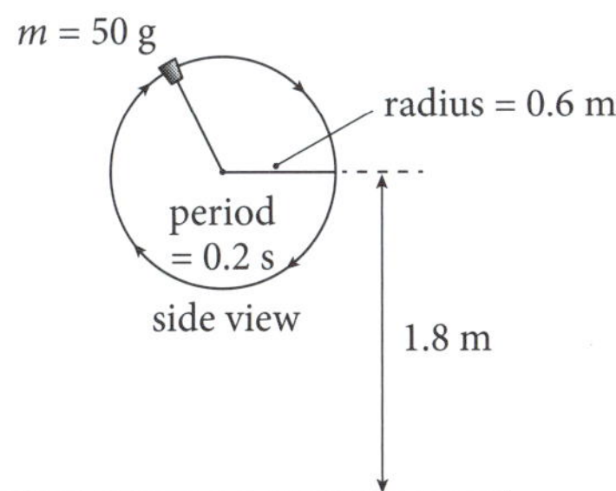

Figure E1.8 Mass on a string turning in uniform circular motion in a vertical plane

a What is the centripetal force required to keep the mass moving in circular motion? (1 mark)

b What is the tension in the string at the bottom of the motion? (1 mark)

c If the string broke when the mass was at the bottom of the motion, how far would it travel horizontally before it hit the ground? (3 marks)

22 A 1200 kg satellite is placed in a circular parking orbit 500 km above the Earth's surface.

a Calculate the velocity and acceleration of the satellite. (2 marks)

b Determine the period of the satellite. (1 mark)

c Explain, qualitatively, how the satellite could be moved efficiently to a circular orbit 1500 km above the surface. (3 marks)

23 The proton synchrotron at CERN is 628 m in circumference and protons circulate around the synchrotron in 2.2 μs as measured from the laboratory frame of reference.

a Find the velocity of the protons. (1 mark)

b Compare the relativistic momentum of the proton to the Newtonian (non-relativistic) momentum at this speed. (2 marks)

c What is the circumference of the synchrotron as measured from the proton's frame of reference? (1 mark)

d What is period as measured from the proton's frame of reference? (1 mark)

e Explain why, even though the speed of the particle is not increasing significantly as energy is imparted to it, the strength of the magnetic field used to keep the proton moving in a circle must be continually increased. (3 marks)

24 Consider the current-carrying wires shown in Figure E1.9.

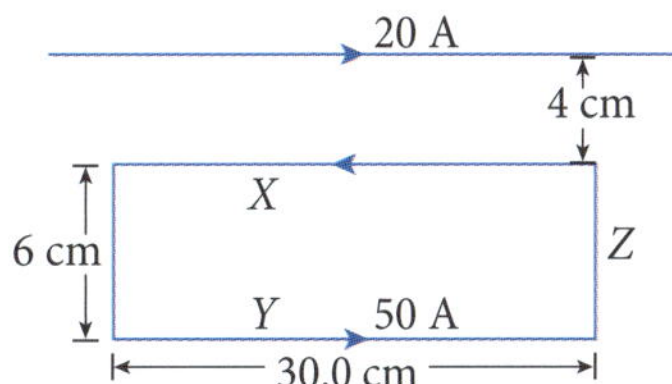

Figure E1.9 Current-carrying wires

a Find the direction of the force on the sides marked X, Y and Z due to the wire above the rectangular loop. (3 marks)

b Find the net force on the rectangular loop of wire due to the wire above it. (3 marks)

25 In an investigation of a DC electric motor a student applies different mechanical loads to the motor and a constant operating voltage, and measures the current that flows through the windings. The results the student obtained are shown in Table 1.2.

Table 1.2 Results of investigation of DC motor operating at 24 V

Load torque (Nm)	Revolutions per minute	Current (mA)
0	200	50
0.5	176	72
1.0	151	143
1.5	124	220
2.0	98	286

a Explain in terms of back EMF why the current increases when the mechanical load on the motor is increased. (3 marks)

b Explain how energy is conserved as the mechanical load on the motor is increased (no calculations required). (2 marks)

26 Figure E1.10 shows a fixed electromagnet and a rotating aluminium disc.

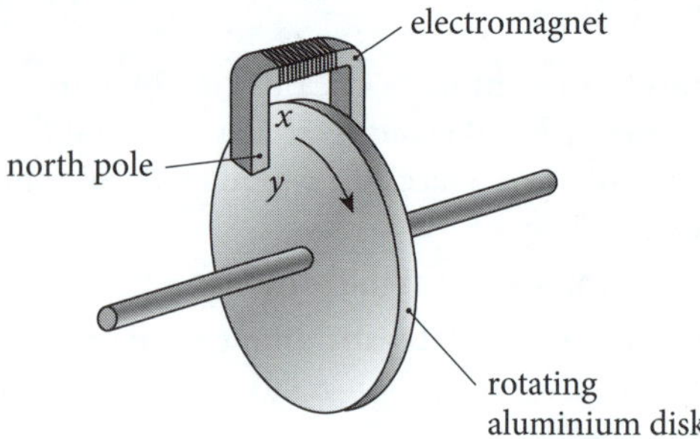

Figure E1.10 Rotating aluminium disc and fixed electromagnet

a Explain why the spinning disc rapidly comes to rest when the electromagnet is switched on. (3 marks)

b Explain how energy is conserved when the disc is brought to rest. (2 marks)

c If the disc was turned when the electromagnet was switched on, would the point marked x be at a higher or lower potential than the point marked y? Explain your answer. (3 marks)

27 This question refers to the Hertzsprung–Russell diagram shown in Figure E1.11.

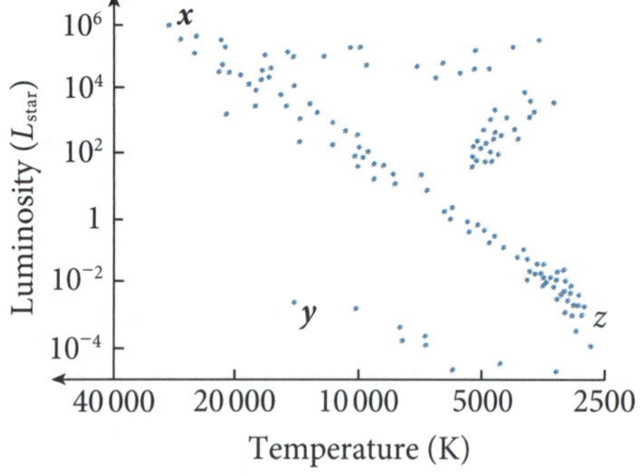

Figure E1.11 A Hertzsprung–Russell diagram

a Name the type of star that would be found at x and y, and compare each star's source of energy. (2 marks)

b Compare the energy source, mass and rate at which mass is converted into energy in stars marked x and z on Figure E1.11. (3 marks)

c Outline the key stages of the lifecycle of a supermassive star. (4 marks)

28 Outline how electricity distribution systems use transformers to efficiently transmit electrical energy. (3 marks)

29 De Broglie made a radical suggestion in his doctoral thesis that had a huge influence on our understanding of atomic physics.

a What was de Broglie's hypothesis? (1 mark)

b Explain how de Broglie's hypothesis provided support for Bohr's stationary state hypothesis. (2 marks)

c Given that the radius of the $n = 4$ orbital of the Bohr hydrogen atom is $r_4 = 8.46 \times 10^{-10}$ m, find the electron wavelength and the momentum of the electron. (2 marks)

30 A diffraction grating is used to separate the visible wavelengths emitted from a discharge tube containing excited hydrogen gas.

a Use Bohr's equation to calculate the wavelength of the two longest wavelength lines emitted from electron transitions that end on the $n = 2$ orbital. (2 marks)

b Calculate the angle of the first order ($m = 1$) maxima of interference for the longest visible wavelength that would be produced if the light from the hydrogen discharge tube was incident on a diffraction grating with 2000 lines per centimetre. (2 marks)

c Outline three limitations of Bohr's model of the atom. (3 marks)

31 Figure E1.12 shows the binding energy per nucleon for the elements in the graph.

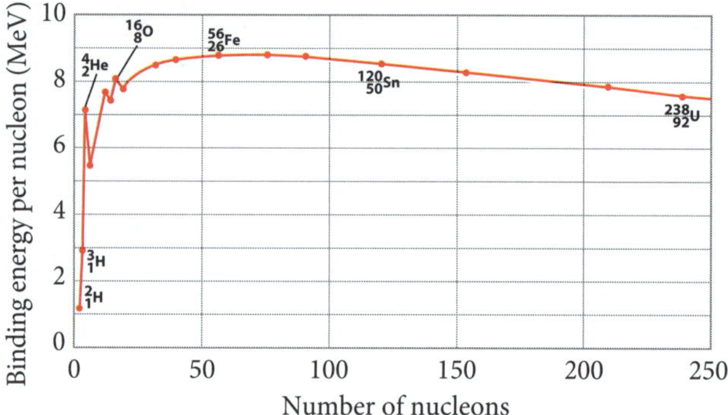

Figure E1.12 Binding energy per nucleon as a function of the number of nucleons

a Use the diagram to determine the binding energy of helium-4 and the mass defect when the nucleus is formed from its constituent nucleons. (3 marks)

b Outline how the diagram can be used to predict which elements will release energy in fusion and fission reactions. (3 marks)

32 a Explain the difference between a polarised electromagnetic wave and an unpolarised wave. (2 marks)

b Explain how a charged particle could be used to produce a polarised electromagnetic wave. (2 marks)

c Unpolarised light of intensity I_0 is passed through three consecutive polaroid filters, each with their axis of polarisation rotated by 30° with respect to the previous filter. Determine the intensity of the light after it passes through each of the filters. (3 marks)

33 Outline the key elements of the standard model of particle physics and explain how particle accelerators influenced the development of the model. (5 marks)

SAMPLE HSC EXAMINATION 2

Section I: 20 Marks

Attempt Questions 1–20.
Allow about 35 minutes for this section.

1 **Projectile *A* is fired at an angle θ above the horizontal with an initial velocity u on a level plane. The projectile is found to have a range of x metres. A second projectile *B* is then fired with the same initial velocity at an angle of $\theta/2$. What can we deduce from this information given?**

A Projectile *A* will have a greater range than projectile *B*.

B Projectile *B* will have a greater range than projectile *A*.

C The time of flight of projectile *A* will be greater than the time of flight of projectile *B*.

D The final velocity of projectile *B* will be greater than the final velocity of projectile *A*.

2 **A mechanic applies a force to a spanner, as shown in Figure E2.1. What is the resulting torque?**

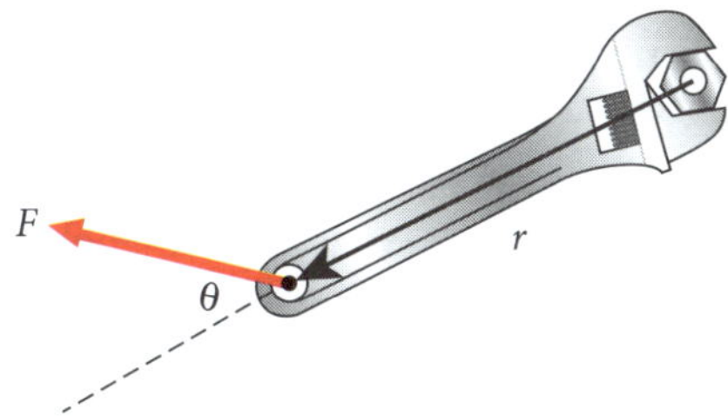

Figure E2.1 Force applied to a spanner

A Torque = Fr

B Torque = $Fr\sin\theta$

C Torque = $Fr/\sin\theta$

D Torque = $Fr\cos\theta$

3 **The gravitational field strength on the surface of planet *X* is g ms^{-2}. What would be the gravitational field strength on the surface of a planet that was half the radius of *X* but four times the mass of planet *X*?**

A 16 g B 8 g

C 4 g D 2 g

4 **A satellite moves in an elliptical orbit, as shown in Figure E2.2. What can we say about the energy of the satellite at the point marked *X* on the diagram?**

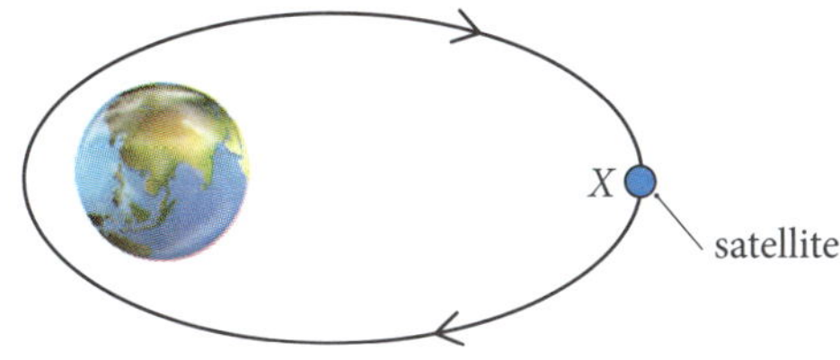

Figure E2.2 Satellite in an elliptical orbit

A The total energy is a maximum.

B The total energy is a minimum.

C The potential energy is a maximum.

D The kinetic energy is a maximum.

5 **A charge enters a magnetic field and collides with a side of the container at point *P*, as shown in Figure E2.3. What changes to the experiment would result in the charge colliding with the wall at point *Q* instead of point *P*?**

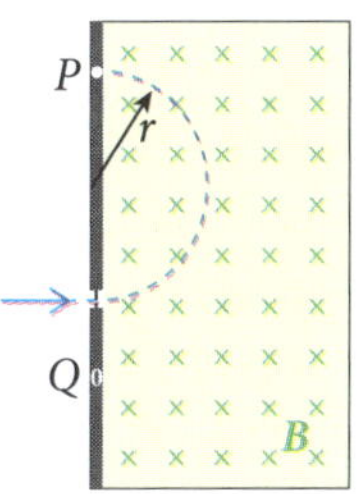

Figure E2.3 A charge moving perpendicularly to a magnetic field

A reversing the magnetic field direction and increasing the particle velocity

B reversing the magnetic field direction and using a negatively charged particle

C using a less massive, negatively charged particle

D using a positively charged particle with a lower initial velocity

6 **A positively charged particle is released at the positive plate of a pair of parallel charged plates, as shown in Figure E2.4.**

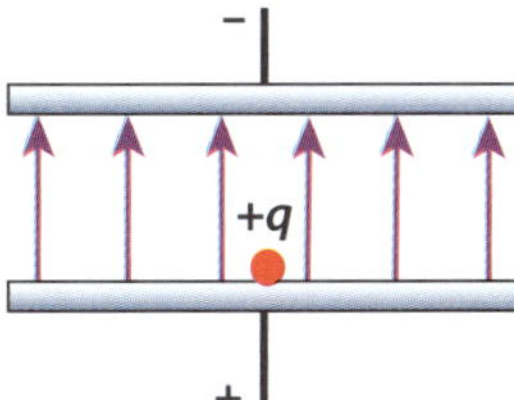

Figure E2.4 A positively charged particle between a pair of charged parallel plates

What quantities do we need to measure to determine the velocity of the particle when it reaches the negatively charged plate?

A The plate separation, the charge on the particle and the electric field strength between the plates.

B The potential difference across the plates and the charge on the particle.

C The mass of the particle, the potential difference between the plates and the charge on the particle.

D The charge on the particle, the electric field strength between the plates and the mass of the particle.

7 A student uses a slip-ring commutator to connect a battery to a coil in a magnetic field, as shown in Figure E2.5. What will happen when he connects the battery to the coil?

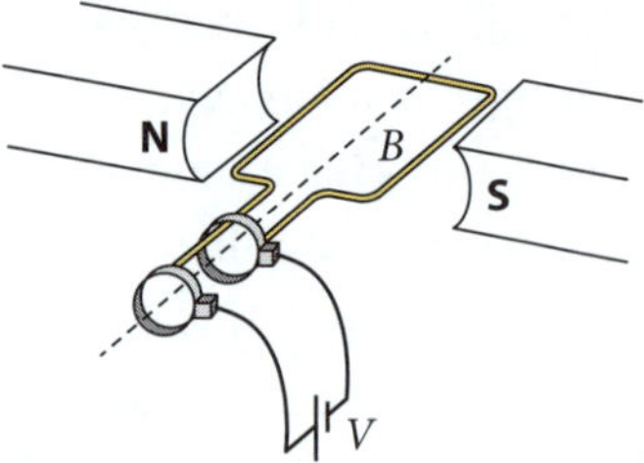

Figure E2.5 Current-carrying coil in a magnetic field

- **A** The coil will rotate continually.
- **B** The side B of the coil will move upwards until the plane of the coil is vertical.
- **C** The coil will not move because the student did not use a split-ring commutator.
- **D** The side B of the coil will move downwards until the plane of the coil is vertical.

8 Two solenoids are placed adjacent to one another, as shown in Figure E2.6. One solenoid is connected to an AC power supply and the other solenoid is connected to a resistor (R).

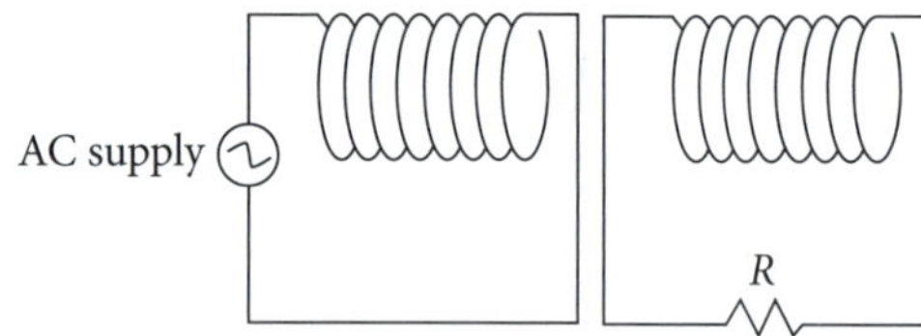

Figure E2.6 Two adjacent solenoids

What would happen if a long, soft iron rod was inserted through the solenoids?

- **A** The iron rod and the resistor would heat up.
- **B** The rod would become magnetised and attract the solenoids together.
- **C** Eddy currents would form in the iron rod, which would decrease the induced current in the resistor.
- **D** The resistor would heat up but the temperature of the iron rod would not change.

9 How was the speed of light first measured?

- **A** by combining the laws of electricity and magnetism
- **B** by observing the orbit of the moons of Jupiter at different times of the year
- **C** by reflecting light from a rotating mirror
- **D** by using radio frequency standing waves

10 How does the rotation of a star affect the absorption lines in the star's spectrum?

- **A** Each absorption line splits into two lines.
- **B** Each absorption line oscillates backwards and forwards around a central position.
- **C** The absorption lines become narrower.
- **D** The absorption lines become broader.

11 In Young's double-slit experiment the light emitted from each slit has the same phase. What would happen to the interference pattern if the light emitted from one slit was one half-cycle (π radians) out of phase with the light emitted from the other slit?

- **A** The interference pattern would not change.
- **B** The distance between the maxima would halve.
- **C** The distance between the interference pattern would double.
- **D** The maxima of interference would become minima and the minima would become maxima.

12 What new assumption did Max Planck make in his derivation of the radiation emitted from black bodies?

- **A** The radiation is made up of photons rather than waves.
- **B** Only radiation with a frequency above the threshold frequency will be emitted from a black body.
- **C** Radiation can only be emitted or absorbed by the black body in whole-number multiples of a specific amount of energy.
- **D** Radiation is produced by oscillating electrons in atoms on the surface of the black body.

13 An observer on the Earth watches a spaceship travel past the Earth at 0.9c. What would the space traveller notice about the observer on the Earth?

- **A** The Earthling's clock would be running slow and the whole planet would appear squeezed up in the direction in which the spacecraft was travelling.
- **B** The Earthling's clock would be running fast and the whole planet would be spread out in the direction in which the spacecraft was travelling.
- **C** The Earthling's clock would be running at the same rate as the clock on the spaceship.
- **D** All light reflected from the Earth would be blueshifted to the ultra violet.

14 What could we conclude if data from a cluster of stars were plotted on a Hertzsprung–Russell diagram and no stars appeared in the upper half of the main sequence?

- **A** The stars in the cluster are very young.
- **B** The stars in the cluster must have been formed from the remains of other stars.
- **C** The cluster was too small for large-mass stars to form.
- **D** The stars in the cluster were very old.

15 Very high energy particle accelerators have been used to produce a huge number of subatomic particles by colliding hadrons together, yet a 'free quark' has never been produced. Why have no free quarks been observed?

- **A** The mass of each of the quarks is so large that the current generation of particle accelerators simply does not have enough energy to produce them.

B Quarks are theoretical constructs used to classify hadrons rather than actual particles.

C Separating a quark pair just produces two pairs of quarks because the strong colour force increases with quark separation.

D Quarks are bound together by leptons, which make them impossible to separate.

16 Young's double slit experiment demonstrates the interference of light. A similar experiment can be conducted with electrons to demonstrate their wave nature. How would the electron interference pattern change if the speed of incident electrons was increased?

A The interference pattern would become more spread out.

B The interference pattern would move closer together.

C The interference pattern would become much less bright.

D The interference pattern would not change.

17 Consider the experiment shown in Figure E2.7. An upwards external magnetic field is reduced to zero and then reversed. When viewed from above, what will be the induced current direction while the magnetic field is changing?

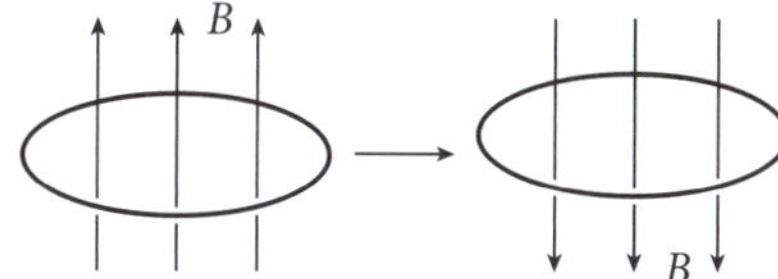

Figure E2.7 A loop of wire and a changing external magnetic field

A The induced current will be clockwise then anticlockwise.

B The induced current will be anticlockwise then clockwise.

C The induced current will be clockwise.

D The induced current will be anticlockwise.

18 The Geiger–Marsden experiment produced a surprising observation. What was the unexpected observation?

A A large number of electrons were diffracted at a specific angle.

B A small number of alpha rays were not scattered at all.

C Some alpha rays were scattered by large angles.

D Alpha particles were scattered by gold nuclei.

19 Oxygen is essential to life on Earth. How were the oxygen atoms on Earth initially produced?

A Oxygen was produced from the condensation of quarks and leptons after the Big Bang.

B Oxygen was produced from fusion in supernova explosions.

C Oxygen is produced by fusion in main sequence stars.

D Oxygen is produced by fusion in supermassive stars.

20 Three electrons *a*, *b* and *c* are projected into perpendicular electric and magnetic fields, as shown in Figure E2.8. How are the velocity of the charges related to one another?

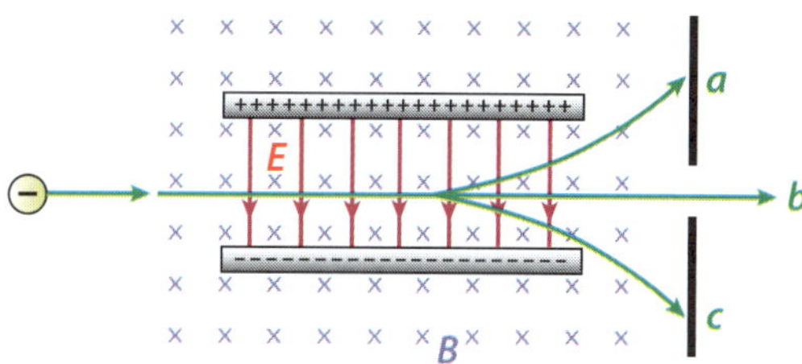

Figure E2.8 Electrons projected into perpendicular electric and magnetic fields

A $v_a = v_b = v_c = \frac{E}{B}$ **B** $v_a \geq v_b \geq v_c$

C $v_a \leq v_b \leq v_c$ **D** $v_b \geq v_c \geq v_a$

Section II: 80 Marks

Attempt Questions 21–32.

Allow about 2 hours and 25 minutes for this section.

Instructions
- In the HSC Exam you will answer the questions in the spaces provided. These spaces provide guidance for the expected length of response.
- Show all relevant working in questions involving calculations.

21 Consider the electrical device shown in Figure E2.9.

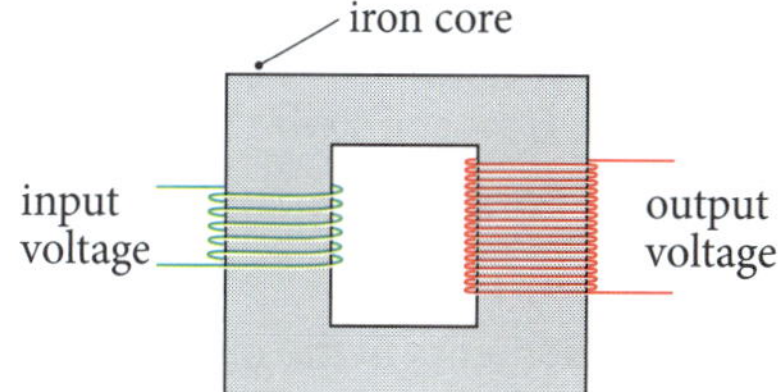

Figure E2.9 Electrical device

a What is the name of this device? (1 mark)

b Explain how the device operates. (4 marks)

c State one source of energy loss in this device and explain how this loss may be minimised. (2 marks)

22 Two football players are standing together. At time $t = 0$, one player starts running north at 5 ms^{-1}, while his team mate remains at rest. The player at rest can kick a ball at 20 m^{-1} at 60° above the horizontal but waits a time t before kicking the ball to ensure the ball will land at the feet of the running player (see Figure E2.10).

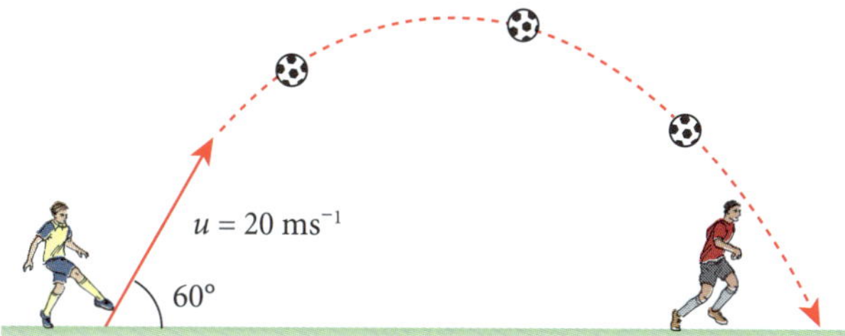

Figure E2.10 Ball kicked at 20 ms^{-1} at 60° above the horizontal

a Find the vertical and horizontal components of the ball's initial velocity. (2 marks)

b Find the time the ball will be in flight when the footballer kicks it. (1 mark)

c Find the range of the ball when it is kicked at 20 ms^{-1} at 60° above the horizontal. (1 mark)

d Determine the time *t* that the player had to wait before he kicked the ball. (2 marks)

23 Two moons are observed to orbit a distant planet. Moon *A* has an orbital period of 16 days and an orbital radius of 40 000 km. Moon *B* has twice the mass of moon *A* and has an orbital radius of 80 000 km.

a Find the orbital period of the second moon. (2 marks)

b Determine the mass of the central planet. (2 marks)

c Qualitatively compare the velocity, centripetal force and total orbital energy of the two moons. (3 marks)

24 To assist braking, some electric trains disconnect their electric motors from the power supply and place a low resistance across the electrical terminals of the motor.

a With reference to conservation of energy, explain how connecting the motor across a low resistance applies a braking force. (3 marks)

b Explain why changing the amount of resistance placed across the motor changes the braking power of the motor. (2 marks)

25 Stellar spectra can be used to investigate many important stellar quantities. How can stellar spectra be used to:

a determine the surface temperature of a star? (3 marks)

b find the velocity of a star? (2 marks)

c identify the elements in the outer atmosphere of a star? (3 marks)

26 The results from an investigation of the maximum kinetic energy of electrons emitted from zinc when it is irradiated with high-frequency light are shown in Figure E2.11.

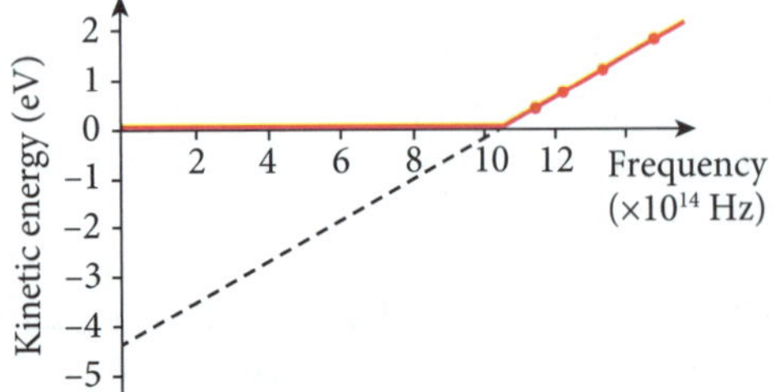

Figure E2.11 The photoelectron kinetic energy as a function of the incident light frequency for zinc

a Explain why no photoelectrons are emitted from the surface when low-frequency light is used. (2 marks)

b Use the graph to determine the threshold frequency and work function for zinc. (2 marks)

c Explain how the gradient of this graph for zinc is related to Planck's constant. (2 marks)

27 Einstein based his special theory of relativity on two postulates.

a State the two postulates. (2 marks)

b Explain how Einstein's postulates account for the null result of the Michelson–Morley experiment. (3 marks)

c Calculate how fast a particle would have to travel for its relativistic momentum to be twice as great as its Newtonian momentum. (2 marks)

28 Consider the following nuclear fission reaction and the masses below.

$$^{1}_{0}n + ^{235}_{92}\text{U} \rightarrow ^{141}_{56}\text{Ba} + ^{92}_{36}\text{Kr} + ?^{1}_{0}n$$

$^{1}_{0}n = 1.008665$ u

$^{235}_{92}\text{U} = 235.043915$ u

$^{141}_{56}\text{Ba} = 140.9139$ u

$^{92}_{36}\text{Kr} = 91.8973$ u

a How many neutrons are released in this reaction? (1 mark)

b Calculate the mass defect and energy released in this reaction. (2 marks)

c Barium-141 is radioactive and emits a beta minus particle followed by an alpha particle. Write nuclear equations for each of these decays. (2 marks)

d The half-life for the first beat decay of barium-141 is 18.5 minutes. What percentage of barium-141 would remain after 1 hour? (2 marks)

29 Bohr developed an atomic model based on Rutherford's nuclear atomic model.

a Outline the two limitations of Rutherford's model. (2 marks)

b Explain how Bohr's model overcame these limitations. (2 marks)

c Use the Rutherford–Bohr model to calculate the minimum energy of a photon that would ionise a hydrogen atom from the $n = 3$ state. (2 marks)

30 In a lecture given in 1920 Rutherford suggested the nucleus might contain uncharged nuclear particles that would have a similar mass to the proton. The following year Rutherford named these particles 'neutrons'. The existence of the neutron was not confirmed experimentally until 1932.

a Why did Rutherford think neutral particles might exist in the nucleus? (1 mark)

b Why was it so difficult to prove neutrons existed? (2 marks)

c How did Chadwick experimentally prove that neutrons existed? (3 marks)

d Explain why the neutron is no longer considered to be an elementary particle. (2 marks)

31 A rotating coil generator is connected to a slip-ring commutator, as shown in Figure E2.12. When the generator is turned at 50 Hz it produces a maximum electromotive force of 2 V.

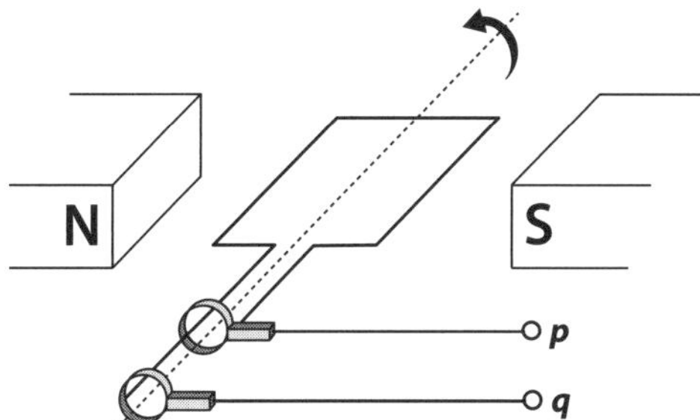

Figure E2.12 Electrical generator with slip-ring commutator

a Explain why an electromotive force is produced by the generator. (2 marks)

b Sketch a graph of the potential difference across the terminals p and q for one complete rotation, starting from the coil position shown in Figure E2.12. (2 marks)

c How would the graph you drew in part b change if the rate at which the generator was turned was halved? (2 marks)

32 Particle accelerators have been extremely important to the development of the standard model of particle physics. While the standard model has been spectacularly successful in explaining interactions between particles, it is still believed to be incomplete.

a With the aid of a diagram, outline how a particle accelerator produces beams of high-energy particles. (4 marks)

b State two shortcomings of the standard model. (2 marks)

c What is the difference between a quark and a hadron? (1 mark)

ANSWERS

SAMPLE HSC EXAMINATION 1

Section I

1 **B**. If John was taller, the ball he threw would be projected horizontally from a greater initial height and hence would stay in the air longer. As the range $s = u_x t$, the only way a ball thrown from a greater height could have the same range as Sandra's ball would be by it having a lower initial velocity (u_x) than Sandra's. **A** is incorrect because it would be true only if the balls were projected from the same height. **C** is incorrect because if a ball had a greater initial velocity and was projected from a greater height it would have a greater range. **D** is incorrect as this would be true only if the balls were projected from the same height.

2 **D**. The tension (T) in the string balances the weight of the ball (mg) and provides the centripetal force (mv^2/r) to move the ball in a circle. Solving the vector addition shown in Figure EA1.1 gives answer **D**. Answers **A**, **B** and **C** are all incorrect as none of the answers is a correct solution for the vector addition shown in Figure EA1.1.

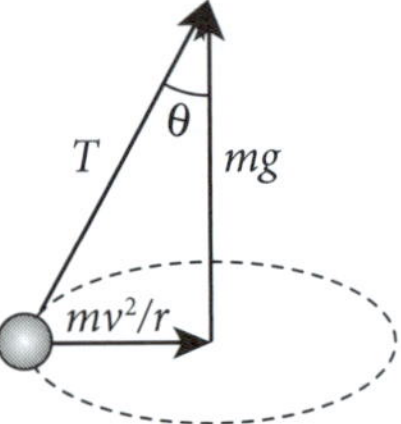

Figure EA1.1 Vector diagram for conical pendulum

3 **B**. Because the kinetic energy is proportional to $\frac{1}{2r}$, the potential energy is proportional to $\frac{-1}{r}$ and the total energy is proportional to $\frac{-1}{2r}$. Doubling r will halve the kinetic energy, potential energy and total energy. Note that because the potential energy and total energy are negative quantities, halving the quantity increases the quantity (making it less negative). **A** and **D** are incorrect because the total energy changes. **C** is incorrect because the kinetic energy decreases with orbital radius.

4 **A**. If the field strength was increased to $4E$, the particle would accelerate downwards four times faster. The time it would take to reach the bottom plate will be given by $s = \frac{1}{2}at^2$ or $t = \sqrt{\frac{2s}{a}}$ and hence, if the acceleration increases by a factor of four, the time to reach the bottom plate will decrease by $\sqrt{\frac{1}{4}} = \frac{1}{2}$. **B**, **C** and **D** are all incorrect numerical answers.

5 **D**. The lower-right wire will be repelled by the top wire (opposite current direction) and attracted by the wire on the lower left. Adding these two vectors gives a force towards the lower left. **A** is incorrect as it would imply both wires were repelling the wire on the lower left. **B** is incorrect as it would imply both wires were attracting the wire on the lower left. **C** is incorrect as it implies the bottom left wire is repelling the bottom right wire.

6 **B**. We solve the problem by applying Lenz's law. The loop acts like a magnet with a magnetic field directed upwards through the loop as the magnet approaches to oppose the change in magnetic flux. Applying the right-hand grip rule gives the induced current as anticlockwise. When the magnet leaves the loop, the loop will act as a magnet with its magnetic field downwards to try to prevent the magnet falling. Applying the right-hand grip rule in this case gives a clockwise induced current flow. **B** is incorrect as this current direction would assist rather than oppose the change. **C** and **D** are incorrect as current must flow in two directions to oppose the change as the bar magnet falls through the loop.

7 **A**. When the motor slows, the back EMF decreases. This increases the current in the coil and hence increases the torque produced by the coil (as the motor torque is proportional to the current in the windings). **B** is incorrect as this would occur only if the current in the motor decreased. **C** is incorrect because changing the speed of operation always changes the back EMF and the current in the motor. **D** is incorrect as there is enough information provided.

8 **D**. The blueshift of the spectrum from star X tells us it is moving towards us and that strong atomic lines and weak ionic lines appear when the outer layer of the star is not hot enough to ionise many of the atoms. **A** and **C** are incorrect because if the star was moving away from the Earth the light would be redshifted. **B** is incorrect because if X was hotter it would have more ionic absorption lines in its spectrum than star Y.

9 **B**. As $n\lambda = d\sin\theta$, decreasing the wavelength will reduce the angle at which interference occurs, which would bring the maxima closer together. **A** and **D** are incorrect as they do not predict a decrease in the separation between the maxima. **C** is incorrect because if the pattern comes closer together, the intensity of the maxima may increase rather than decrease.

10 **D**. Reducing the wavelength increases the photon energy ($E = hc/\lambda$) and hence less photons per second will be in the beam as the power remains constant. This will reduce the photocurrent but give each ejected electron more kinetic energy. **A** and **C** are incorrect as the photocurrent must decrease as less photons per second would hit the metal. **B** is incorrect because the photons have more energy and hence the ejected electrons will have more kinetic energy.

11 **C**. Main sequence stars fuse hydrogen to helium in their cores. When a helium core forms and becomes large enough, hydrogen fusion in the core ceases and gravity collapses the star until the core temperature becomes high enough to fuse helium or, in the case of very low-mass stars, fusion ceases and the star collapses to become a white dwarf. **A** is incorrect as iron cores form only in supermassive stars that are fusing heavy elements in their cores. **B** is incorrect because a helium core decreases the outwards radiation pressure as it restricts the rate at which fusion occurs in the core. **D** is incorrect as most main sequence stars fuse hydrogen to helium using the p–p chain reaction.

12 **B**. An upwards force is required to oppose the force produced by the electric field. Applying the right-hand palm rule to the moving negative charge gives the required magnetic field direction as out of the page. **A**, **C** and **D** specify magnetic field directions that would not produce an upwards force on the negative charge.

13 **D**. When we set $n_i = 1$ and $n_f = \infty$ in Bohr's hydrogen spectra equation we get $1/\lambda = R$. This is the wavelength of the photon that would cause a ground-state hydrogen atom to be ionised. The photon energy is related to the wavelength by $E = hc/\lambda$, and hence $1/\lambda = E/hc$. Combining the two equations for $1/\lambda$ gives $E = Rhc$. **A**, **B** and **C** are incorrect expressions and do not have the units of energy.

14 **B**. As the decay constant is inversely proportional to the half-life $\left(\lambda = \frac{\ln 2}{t_{1/2}}\right)$, a small half-life corresponds to a large decay constant. **A** is incorrect because the rate of decay decreases as the half-life increases. **C** is incorrect as not enough information is given in the question to find the energy of the emitted particles. **D** is incorrect as half of the isotopes in both samples would have decayed in one half-life.

15 **C.** Massive nuclei have a higher neutron-to-proton ratio than lighter nuclei. Hence if a very large nucleus splits in half, the neutron-to-proton ratio will be much higher than the stable value for a medium-sized element. **A** is true but not related to the highly radioactive daughter products produced in fission. **B** is incorrect because the instability is caused by a very low proton-to-neutron ratio. **D** is incorrect because all uranium isotopes are radioactive but much less radioactive than the daughter products of uranium fission.

16 **C.** The net force on the car is the centripetal force ($F_c = mv^2/r$), which is directed to the centre of the circle. **A** is incorrect because the car is changing direction and hence a net force must be applied to the car. **B** is incorrect because the centripetal force is directed towards the centre of the circle, not away from it. **D** is incorrect because the net force on the car is the sum of the gravity and the reaction force from the road on the car.

17 **A.** The right-hand palm rule shows that the radiation is negatively charged and hence is beta radiation. Beta particles are emitted from isotopes with a high neutron-to-proton ratio because the ratio is reduced when a beta is emitted. **B** is incorrect because a high-binding energy per nucleon is characteristic of a stable nucleus. **C** is incorrect because beta-plus decay results from a high proton-to-neutron ratio, not beta minus. **D** is incorrect as nuclei do not contain electrons.

18 **A.** When the electric force balances the gravitational force, $mg = qE$ or $q = mg/E$. Thus the charge on the droplet is $q = \frac{1.635 \times 10^{-6} \times 9.8}{10^7} = 1.6 \times 10^{-12}$ or $\frac{1.6 \times 10^{-12}}{1.6 \times 10^{-19}} = 10^7$ electron charges. **B**, **C** and **D** are all incorrect numerical answers.

19 **B.** A shorter wavelength of maximum emission indicates a hotter star and hence star *Q* has a higher surface temperature than star *P*. As they are main sequence stars, *Q* must be larger (as it is hotter) and will leave the main sequence faster as it has to fuse at a much faster rate to balance the inwards force of gravity on the star. **A** is incorrect because larger, hotter stars use up their hydrogen fuel much faster than smaller, cooler main sequence stars. **C** and **D** are incorrect because *Q* is hotter and hence larger than *P*.

20 **B.** Quark confinement by the strong (colour) force means that attempting to separate quarks simply makes more quark pairs. **A** is incorrect because quarks are fundamental particles but this is not the reason they cannot be separated. **C** is incorrect because quarks can be produced from energy (i.e. $E = mc^2$). **D** is incorrect because quarks are real particles that have been identified in scattering experiments.

Section II

21 EM This question tests students' ability to perform calculations involving circular motion and gravity.

a $F_c = mv^2/r = m\omega^2 r$
$= m(2\pi r/T)^2 r$
$= 4m\pi^2 r/T$
$= 4 \times 0.05 \times \pi^2 \times \frac{0.6}{0.2} = 5.9$ N ✓

b The tension will be the weight plus the centripetal force
$= mg + mv^2/r = 0.05 \times 9.8 + 5.92 = 6.4$ N ✓

c We first calculate the initial velocity:
$v = 2\pi r/T = \frac{2\pi \times 0.6}{0.2} = 18.85$ ms^{-1} horizontally ✓
Now we use the vertical part of the motion to calculate time of flight.
$s = ut + \frac{1}{2}at^2$ and, as the initial vertical velocity is zero:
$t = \sqrt{\frac{2s}{a}} = \sqrt{\frac{2(1.2)}{9.8}} = 0.5$ s ✓
Horizontal distance travelled $= u_x t = 18.85 \times 0.5 = 9.4$ m ✓

22 EM This question tests students' understanding of satellite orbits and their ability to calculate orbital quantities.

a The orbital radius will be $r = 6.371 \times 10^6 + 500 \times 10^3$
$= 6.871 \times 10^6$ m
As gravity provides the centripetal force:
$\frac{GMm}{r^2} = \frac{mv^2}{r}$ and hence
$v = \sqrt{\frac{GM}{r}} = \sqrt{\frac{6.67 \times 10^{-11} \times 6.0 \times 10^{24}}{6.871 \times 10^6}} = 7.632$ kms^{-1} ✓
The centripetal acceleration is given by $\frac{v^2}{r} = \frac{(7632)^2}{6.871 \times 10^6}$
$= 8.47$ ms^{-2} ✓

b Applying the equation from uniform circular motion:
$v = 2\pi r/T$ and hence
$T = 2\pi r/v = \frac{2\pi(6.871 \times 10^6)}{7632}$
$= 5757$ s $= 94$ min 17 s ✓

c Refer to Figure EA1.2.

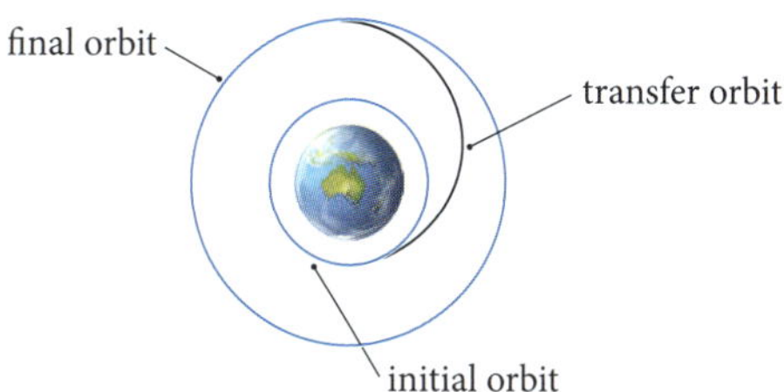

Figure EA1.2 Moving from one circular orbit to another

The satellite first uses a posigrade burn ✓ to increase its velocity and move it into an elliptical transfer orbit. ✓ When the satellite completes half an orbit it uses another posigrade burn to move it from the elliptical transfer orbit to a circular orbit at the higher altitude. ✓

23 EM This question tests students' understanding relativity and their ability to calculate quantities using the equations of special relativity.

a Velocity $= s/t = \frac{628}{2.2 \times 10^{-6}} = 2.85 \times 10^8$ ms^{-1} ✓ $= 0.95c$

b The non-relativistic momentum $p = mv = 1.67 \times 10^{-27} \times 2.85 \times 10^8$
$= 4.77 \times 10^{-19}$ Ns ✓
The relativistic momentum is given by:
$P_v = \frac{m_0 v}{\sqrt{\left(1 - \frac{v^2}{c^2}\right)}} = P_v = \frac{1.67 \times 10^{-27} \times 2.85 \times 10^8}{\sqrt{(1 - 0.95^2)}}$
$= 1.53 \times 10^{-18}$ Ns ✓
At this speed the actual (relativistic) momentum is 3.2 times greater than the Newtonian momentum.

c From the proton's frame of reference it is the synchrotron that is moving and hence the length of the synchrotron will be contracted by:
$l = l_0\sqrt{\left(1 - \frac{v^2}{c^2}\right)} = 628 \times \sqrt{(1 - 0.95^2)} = 196$ m ✓

d The proton is moving at $0.95c$ around a circle 196 m long in their frame and hence the time to go around once will be given by:
$t = s/v = \frac{196}{2.85 \times 10^8} = 0.69$ μs ✓
Or students could apply the time-dilation equation to obtain the same result.

e Because more energy is added to the proton it will not move significantly faster as it is approaching the speed of light, but its effective mass will increase as the energy goes into mass rather than velocity. Now as the momentum is proportional to the mass ($p = mv$), the momentum will continue to increase even though the velocity is not changing significantly. ✓ For a charge moving perpendicularly to a magnetic field we have:

$mv^2/r = qvB$ or $r = mv/qB$ ✓

Thus to ensure the particle continues to move in a circle of radius r, the magnetic field strength must be increased at the same rate as the momentum increases. ✓

24 EM This question tests students' ability to use the right-hand grip rule and right-hand palm rule and to calculate the force between parallel current-carrying wires.

a Applying the right-hand grip rule to show the magnetic field below the top wire is into the page, and then applying the right-hand palm rule to each wire yields the following results:

The direction of the force on X will be down the page. ✓

The direction of the force on wire Y is up the page. ✓

The direction of the force on wire Z is to the left. ✓

b The force on the vertical wires sums to zero. The force on wire X will be given by:

$$F = \frac{\mu_0}{2\pi}\frac{I_1 I_2 l}{r} = 2\times 10^{-7} \times \frac{20\times 50\times 0.3}{0.04}$$

$$= 1.5\times 10^{-3}\text{ N} \checkmark \text{ down the page}$$

For wire Y the force will be given by:

$$F = \frac{\mu_0}{2\pi}\frac{I_1 I_2 l}{r} = 2\times 10^{-7} \times \frac{20\times 50\times 0.3}{0.1}$$

$$= 6.0\times 10^{-4}\text{ N} \checkmark \text{ up the page}$$

The net force on the loop

$= 1.5\times 10^{-3} - 6\times 10^{-4}$

$= 9\times 10^{-4}$ N downwards ✓

25 EM This question tests students' understanding of back EMF in electric motors and its relationship to the conservation of energy.

a When the motor is operating the magnetic flux linking the armature changes, an electromotive force (potential difference) is induced. ✓ By Lenz's law this will oppose the change that created it, which means a back emf will be generated as the motor turns that will oppose the supply emf. ✓ The net effect of the back emf will be to reduce the net emf on the motor and the current that flows in the armature coils. Thus the back emf increases with the speed of rotation and hence the current in the armature decreases with the speed of rotation. Placing a mechanical load on the motor decreases the speed at which it rotates, which reduces the back emf generated and hence increases the current flow in the armature. ✓

b The motor does more mechanical work per unit time when the load is increased. ✓ Because the electrical power supplied to the motor is given by $P = IV$, the increasing current flow (that results in the speed of rotation slowing) increases the electrical power used by the motor. ✓ Thus the electrical power increases as the load increases, ensuring energy is conserved in the system.

26 EM This question tests students' understanding of eddy currents and Lenz's law and their relationship to the conservation of energy.

a The disc is moving rapidly in a magnetic field and the change of magnetic flux experienced by the disc causes eddy currents to be induced in the disc. ✓ By Lenz's law the induced eddy currents will be in a direction that opposes the change that produced them. ✓ That is, the eddy currents will produce magnetic fields that will slow the rate of rotation of the disc. ✓

b Mechanical work must be done to bring the disc to rest. ✓ The induced currents that slow the disc cause resistance heating in the disc and hence the rotational energy of the disc is converted to heat as it is slowed down. ✓

c If we imagine a positive charge moving with the disc through the poles of the magnet and apply the right-hand palm rule ✓ to the moving charge, we see that it will experience a force ($F = qvB$) upwards. ✓ Negative charges moving on the disc would experience a downwards force. Thus charge separation occurs and point x will become more positively charged than point y. As potential difference is defined for positive charge, point x will therefore be a higher potential than point y. ✓

Alternatively, students could justify x being a higher potential by showing that the current must move upwards through the magnetic poles due to Lenz's law.

27 EM This question test students' understanding of the Hertzsprung–Russell diagram and stellar evolution.

a Star y is a star at the end of its lifecycle and is called a white dwarf star. White dwarf stars have ceased fusing light elements and are simply hot embers. The heat in the star was created largely by the gravitational collapse of the star that occurred when fusion ceased. ✓ In contrast, x is a very massive main sequence star, thousands of times larger than a white dwarf star. Main sequence stars like x produce energy by fusing hydrogen to helium in the core of the star. ✓

b Both y and z are main sequence stars that fuse hydrogen to helium in the core of the star. While both fuse hydrogen, z fuses hydrogen through the p–p chain reaction, while x fuses hydrogen through the CNO reaction. ✓ Star x is a massive star and star z is a small, low-mass star. ✓ Because x is much more massive than z, it must fuse hydrogen at a much faster rate to balance the inward pressure from gravity. Thus star x converts mass to energy much faster and has a much shorter lifespan than star z. ✓

c Supermassive stars form when huge clouds of gas and dust condense due to gravity. Eventually the collapse causes the centre of the mass to get hot enough for hydrogen fusion to commence and a protostar appears. When an equilibrium is reached between gravity pushing in and a radiation (fusion) pressure pushing out, the star moves to the main sequence. ✓ Eventually a helium core is formed and hydrogen stops fusing in the core. When this occurs the large star collapses and the temperature of the core increases until helium begins to fuse. A hydrogen shell also begins to fuse and the star expands to become a red giant star. ✓ This pattern repeats as heavier elements fuse in the core until an iron core is formed. When this happens the star collapses and explodes in a supernova. ✓

If the remaining core is large enough it may collapse to become a neutron star or a black hole. ✓

28 EM This question tests students' understanding of how and why transformers are used to transmit electrical power.

The main loss mechanism in electrical distribution systems is line loss due to resistance heating in the lines. Line loss is related to the amount of current that flows and to the resistance of the line ($P_{loss} = I^2R$). ✓ To minimise power loss AC current is transmitted at low currents and high voltages. ✓ Transformers can efficiently increase and decrease AC voltage. By using a transformer to step up the voltage before transmission and then a step-down transformer to reduce the voltage after transmission, line loss is significantly reduced. ✓

29 EM This question tests students' understanding of de Broglie's matter-wave hypothesis and Bohr's model of the atom.

a De Broglie proposed that if waves had particle properties then perhaps particles would have wave properties. De Broglie showed that particles would have an associated wavelength related to the momentum of the particle and Planck's constant $\lambda = h/p$. ✓

b De Broglie showed that the stationary state orbitals of the Bohr atom corresponded to standing electron wave patterns. He showed that the circumference of each orbital was equal to a whole number (n) of electron wavelengths ✓ and wrote:

$2\pi r = nh$ and substituting $\lambda = h/p$ gives the Bohr quantisation condition, $mvr = nh/2\pi$ ✓

Thus Bohr's principal quantum number n corresponded to the number of electron wavelengths in the nth orbital.

c As the circumference of the $n = 4$ orbital contains four electron wavelengths:

$$\lambda = 2\pi r/4 = \frac{2\pi \times 8.46 \times 10^{-10}}{4} = 1.33 \times 10^{-9} \text{ m} ✓$$

Applying de Broglie's equation $p = h/\lambda = \dfrac{6.626 \times 10^{-34}}{1.34 \times 10^{-9}}$

$= 4.99 \times 10^{-25}$ Ns ✓

30 EM

a The longest wavelength will be produced when an electron moves from the $n = 3$ to the $n = 2$ orbital:

$$\frac{1}{\lambda} = R\left[\frac{1}{n_f^2} - \frac{1}{n_i^2}\right] = 1.097 \times 10^7\left[\frac{1}{2^2} - \frac{1}{3^2}\right]$$

$= 1.5236 \times 10^6$ and hence $\lambda = 656$ nm ✓

The next longest wavelength will be emitted when the electron moves from the $n = 4$ to the $n = 2$ orbital:

$$\frac{1}{\lambda} = R\left[\frac{1}{n_f^2} - \frac{1}{n_i^2}\right] = 1.097 \times 10^7\left[\frac{1}{2^2} - \frac{1}{4^2}\right]$$

$= 2.0569 \times 10^6$ and hence $\lambda = 486$ nm ✓

b The slit separation will be given by $d = \dfrac{0.01}{2000} = 5 \times 10^{-6}$ m. ✓

The maxima of interference occur when:

$d\sin\theta = m\lambda$ and, with $m = 1$:

$d\sin\theta = \lambda$ and hence $\theta = \sin^{-1}(\lambda/d)$

$$= \sin^{-1}\frac{656 \times 10^{-9}}{5 \times 10^{-6}} = 7.54° ✓$$

c Bohr's model could not explain: the intensity of the spectral lines, ✓ the spectra of elements with more than one electron ✓ or the Zeeman effect. ✓ (Students could also mention the fine spectra of hydrogen.)

31 EM

a The binding energy per nucleon of helium-4 from the graph is approximately 7.2 MeV per nucleon ✓ and hence the total binding energy (*BE*) is given by:

BE = number of nucleons × binding energy per nucleon

$= 4 \times 7.2 = 28.8$ MeV ✓

The mass defect is therefore:

$$\Delta m = BE \text{ in MeV}/931.5 = \frac{28.8}{931.5} = 0.031 \text{ u} ✓$$

b Energy is produced in nuclear reactions which cause the binding energy per nucleon to increase. ✓ We see from the diagram that the fusion of light elements will produce heavier elements that have a greater binding energy per nucleon. ✓ Providing the daughter product is no heavier then iron, this type of reaction will release energy. Similarly the fission of very large elements will result in daughter products with a higher binding energy per nucleon. ✓ Thus the fission of very heavy elements (e.g. uranium) will result in some mass being converted into energy.

32 EM

a Electromagnetic waves are transverse waves consisting of changing electric and magnetic fields that are perpendicular to one another and to the velocity of the wave. An unpolarised electromagnetic wave has the electric field in all planes perpendicular to the velocity of the wave, ✓ and a polarised wave has the electric field in only one plane. ✓ (Students may also show this with a diagram.)

b If a charged particle was oscillated back and forth in a single direction ✓ in space it would produce polarised electromagnetic waves with their electric field parallel to the direction of oscillation of the charge. ✓ The emitted waves would have a frequency equal to the oscillation frequency of the charged particle.

c Half the intensity of the unpolarised light will be absorbed by the first polarising filter. Hence the intensity after the first filter will be $I_0/2$. ✓ This amount of light will be incident on the second filter and amount of light that passes through will be given by:

$$I = \left(\frac{I_0}{2}\right)\cos^2 30° = \left(\frac{I_0}{2}\right) \times 0.75 = 0.375 I_0 ✓$$

This intensity will be incident on the final filter and hence the amount of light passing through all three filters will be:

$$I = (0.375 I_0)\cos^2 30° = (0.375) \times 0.75 = 0.281 I_0 ✓$$

33 EM

The standard model of particle physics uses the three quantum force field theories (QED, the electroweak theory and QCD) to describe the interactions between elementary particles. ✓ The elementary particles used in the standard model are six quarks, six leptons and the gauge bosons that mediate the three forces (and the Higgs boson, to give particles mass). ✓ The model uses quantum field theories in which force is mediated by the exchange of virtual particles (bosons). ✓

Particle accelerators are machines that accelerate charged subatomic particles to very high energies and collide them into other particles. Studying the particles produced in these reactions by the conversion of energy into mass ($E = mc^2$) has been instrumental in the development and verification of the standard model. For example, a huge number of new subatomic particles were discovered using particle accelerators in the 1950s and 1960s. ✓ These discoveries led physicists towards a theory that was based on elementary particles that were later called quarks. Many of the predictions of the quantum field theories that now form part of the standard model were confirmed in particle accelerators. For example, the W and Z bosons required for the electroweak theory were predicted long before they were proved to exist in accelerators. ✓ Another example is the Higgs boson, which was predicted by the standard model 50 years before particle accelerators had enough energy to produce the particle. Particle accelerators have been instrumental in the creation, development and verification of the standard model.

SAMPLE HSC EXAMINATION 2

Section I

1 **C.** Because the vertical component of the initial velocity is greater for *A*, it will be in flight for a longer time. **A** and **B** are incorrect because the maximum range occurs when $\theta = 45°$. The maximum range will be determined by which of the two projectiles has a launch angle closer to angle θ, which could be either projectile. **D** is incorrect as the final velocity will be the same as the initial velocity and hence will be the same for both projectiles.

2 **B.** The torque is defined as the product of the applied force and the perpendicular distance between the line of action of the force and the pivot. The expressions in **A**, **C** and **D** are all incorrect as they are inconsistent with the definition of torque and would give the wrong value for the torque.

3 **A.** The gravitational field strength is proportional to m/r^2 and hence a planet with four times the mass and half the radius would have a gravitational field strength 16 times greater. **B**, **C** and **D** are all incorrect expressions that are not consistent with the definition of the gravitational field strength on the surface of a planet.

4 **C.** The potential energy increases with altitude. **A** and **B** are incorrect because the total energy remains constant while the satellite is in orbit. **D** is incorrect because the velocity of a satellite decreases as the orbital radius increases.

5 **C.** The radius of curvature is given by $r = mv/qB$ and hence reducing the mass reduces the radius of curvature of the path in the field. In addition the right-hand palm rule shows that the force on a negatively charged particle entering the field would be downwards. **A** is incorrect because increasing the velocity of the charge would cause it to move in a bigger radius. **B** and **D** are incorrect because the right-hand palm rule shows that the force on the charge when it entered the field in both cases would still be upwards.

6 **C.** Because energy is conserved, the kinetic energy gained is equal to the change in electrical potential energy and hence $\frac{1}{2}mv^2 = qV$ and $v = \sqrt{\frac{2qV}{m}} = \sqrt{\frac{2qEd}{m}}$. **A** and **B** are incorrect because the mass of the particle is also required. **D** is incorrect because the plate separation is required to determine the potential difference across the plates.

7 **B.** The motor force on *B* will be upwards by the right-hand palm rule. Because the current direction remains constant, the force on side *B* will also remain constant. **A** is incorrect because this would require a split-ring commutator to reverse the current direction in the coil each half-cycle. **C** is incorrect because a slip-ring commutator will still enable current to flow in the coil and a motor force to appear on the side *B*. **D** is incorrect because the direction of the motor force is incorrect.

8 **A.** Eddy currents would be induced in the rod and would cause resistance heating in the rod. The iron core would increase the voltage induced in the second coil and hence more power would be dissipated in the resistor. **B** is incorrect because when AC is applied to the coil, the magnetic field in the rod would be constantly reversing direction and Lenz's law would cause the coil to be repelled. **C** is incorrect because even with the losses due to the eddy current the second coil will have a much greater flux change due to the iron core intensifying and linking the field to the second coil. This would result in more current in the resistor, not less. **D** is incorrect because eddy currents produced by the changing magnetic field in the rod would cause resistance heating in the rod.

9 **B.** Romer measured the speed of light by observing the change in the time Jupiter's moon went behind the planet when the Earth was different distances from Jupiter. **A**, **C** and **D** were theoretical estimates or experiments that were performed long after Romer's measurement was made.

10 **D.** Some parts of the star will be moving towards the observer (blueshifting the light), some parts will be moving away (redshifting the light) and some parts will be moving at 90° to the line of sight (neither redshifted nor blueshifted). **A** is incorrect as this would happen only if the light was only redshifted and blueshifted. **B** is incorrect because this would occur only if the line was periodically redshifted then blueshifted. **C** is incorrect because a spinning star could emit only a broader range of wavelengths.

11 **D.** The condition for a maximum is that the light from each slit reaches the point on the screen with the same phase. The condition for a minimum is for the light from each slit to reach the screen one half-cycle out of phase. If the light was initially a half-cycle out of phase, the light from each slit would reach the screen one half-cycle out of phase at each of the points where a maximum previously occurred. **A** is incorrect because the phase change will change the interference pattern. **B** is incorrect because this would occur only if the wavelength was halved. **C** is incorrect because this would occur only if the wavelength was doubled.

12 **C.** Planck assumed that radiation can be emitted or absorbed by the black body only in whole number multiples of a specific amount of energy ($E = hf$). **A** is incorrect as this was an assumption that Einstein made many years later. **B** is incorrect because it relates to the photoelectric effect, not black-body radiation. **D** is incorrect because Planck thought the radiation was emitted and absorbed by 'oscillators' on the surface.

13 **A.** From the spaceship's frame of reference, the Earth would appear to be moving past at 0.9*c*. Einstein's theory of special relativity tells us that the moving clock will appear to run slowly and the moving object will squeeze up in its direction of motion. **B** is incorrect because a clock moving with respect to an observer runs slowly. **C** is incorrect because clocks moving with respect to each other run at different rates. **D** is incorrect because the Earth will move towards and then away from the observer in the spacecraft and hence the light from the Earth would be blueshifted and then redshifted.

14 **D.** Large main sequence stars have short lifespans on the main sequence and hence if the cluster was very old it would have no large main sequence stars left. **A** is incorrect because we would expect to see young main sequence stars of all masses. **B** is incorrect because second-generation stars can be both large and small, and hence stars would be found across the main sequence. **C** is incorrect because the size of the cluster has no relationship to the size of stars formed.

15 **C.** Quark confinement prevents free quarks from appearing because pulling a hadron apart simply produces more hadrons. **A** is incorrect because the mass of first-generation quarks is quite low. **B** is incorrect because quarks have been seen in scattering experiments and their reactions have been observed. **D** is incorrect because it is gluons that hold quarks together.

16 **B.** Increasing the velocity decreases the electron wavelength. Now as maxima of interference occur when $d \sin \theta = m\lambda$, decreasing the wavelength (λ) will decrease the separation (θ) between maxima of interference. **A** is incorrect because this would occur only if the velocity was decreased (wavelength increased). **C** is incorrect as this would occur only if the number of electrons per second was decreased. **D** is incorrect as the interference must change if the wavelength (velocity) of the incident electrons is changed.

17 **D.** Lenz's law shows that an anticlockwise current will oppose an upwards field decreasing and a downwards field increasing. **A** is incorrect because an initial clockwise current would enhance rather than oppose a decreasing upwards magnetic field. **B** is incorrect because a clockwise current would enhance rather than oppose an increasing downwards magnetic field. **C** is incorrect because it would enhance rather than oppose both field changes.

18 **C.** This result was not expected from the plum pudding model of the atom that was current at the time. **A** is incorrect as electrons were not used in the experiment. **B** is incorrect as most of the alpha particles passed straight through without being deflected. **D** is incorrect as the nuclear atom was not the current model when the experiment was conducted.

19 **D.** Supermassive stars can produce the core temperatures required to produce energy when smaller nuclei are fused to produce oxygen. **A** is incorrect as only hydrogen and helium and a trace amount of lithium was produced by nuclear synthesis after the Big Bang. **B** is incorrect as nuclei larger than iron are produced in supernova explosions. **C** is incorrect as main sequence stars fuse hydrogen only to helium.

20 **C.** When the forces are balanced $qE = qvB$ or $v = E/B$ and the charge goes straight through. If a charge moves faster than E/B the magnetic force will be greater and it will be deflected downwards. If it moves slower than E/B the electric force will be greater and it will be deflected upwards. **A** is incorrect because the charges would all

move through the crossed fields without being deflected, which is clearly not the case. **B** and **D** are incorrect as the magnetic force is greatest on *c*, and hence *c* must be the fastest moving particle.

Section II

21 **EM** This question tests students' understanding of electromagnetic induction in transformers.

a step-up transformer ✓

b An AC voltage is attached to the primary coil. The changing current in the primary coil produces a changing magnetic field in the iron core. ✓ The iron core intensifies and confines the magnetic field. ✓ Because the iron core also passes through the secondary coil, the secondary coil experiences a changing magnetic flux that induces an AC voltage on the secondary coil. ✓ The ratio of the input and output voltages will be equal to the ratios of primary and secondary turns. ✓

c Because the magnetic field is changing continually in the iron core, eddy currents are induced in the core that cause resistance heating in the core. ✓ To reduce eddy currents the core is laminated. This involves cutting the core into thin slices perpendicular to the plane of the coils and then re-joining the slices with insulating material between the slices. ✓

22 **EM** This question tests students' ability to apply the equations of motion to projectile motion.

a The vertical component will be $u_y = u \sin\theta = 20 \sin 60°$
$= 17.32\ \text{ms}^{-1}$ ✓

The horizontal component will be given by $u_x = u\cos\theta$
$= 20 \sin 60°$
$= 10.0\ \text{ms}^{-1}$ ✓

b Using the vertical component of the velocity and calling upwards positive, we can write that the final vertical velocity will be equal to, but in the opposite direction of, the initial velocity and hence $v_y = -u_y$. Substituting this condition into the appropriate equation of motion:

$v_y = u_y + at$ and hence $-u_y = u_y + gt$ or $t = -2u_y/g$

$t = -2 \times \dfrac{17.32}{-9.8} = 3.53\ \text{s}$ ✓

c Using the horizontal component of the motion:
Range $= u_x t = 10 \times 3.53 = 35.3$ m ✓

d The time for the boy to run 35.3 m is $t = s/v = 35.3/5 = 7.06$ s ✓
But we must account for the time the ball is in flight and hence the ball must be kicked a time $t = 7.06 - 3.53 = 3.53$ s after the boy starts running. ✓

23 **EM** This question tests students' ability to perform calculations using Kepler's law of periods and the satellite equation.

a Applying Kepler's law of periods:

$$\frac{{T_A}^2}{{r_A}^3} = \frac{{T_B}^2}{{r_B}^3}$$ ✓

And hence

$$T_B = \sqrt{\frac{{T_A}^2{r_B}^3}{{r_A}^3}} = \sqrt{\frac{(16^2 \times 80\,000^3)}{40\,000^3}} = 45.25 \text{ days}$$ ✓

b Applying the satellite equation to moon *A* and converting to SI units: ✓

$$\frac{r^3}{T^2} = \frac{GM}{4\pi^2} \text{ and hence}$$

$$M = \frac{4\pi^2 r^3}{GT^2} = \frac{4\pi^2(4 \times 10^7)^3}{6.67 \times 10^{-11}(16 \times 24 \times 60 \times 60)^2}$$
$$= 1.98 \times 10^{22}\ \text{kg}$$ ✓

c The velocity of an orbiting satellite is given by $v = \sqrt{\dfrac{GM}{r}}$ and hence the velocity of moon *B* will be less than the velocity of moon *A*. ✓ Or students could write that as $v = \sqrt{\dfrac{GM}{r}}$ the velocities are related by $v_B = \dfrac{v_A}{\sqrt{2}}$.

The centripetal force is equal to the gravitational attraction between the moon and the central body ($F = \dfrac{GMm}{r^2}$) and, because *B* has twice the mass and twice the orbital radius, the centripetal force will be less on *B* than on *A*. ✓ Or students may write $F_B = \dfrac{F_A}{2}$.

The orbital energy is given by $U + K = -\dfrac{GMm}{2r}$ and, because *B* has twice the mass and twice the orbital radius, the total energy of each moon will be equal. ✓ Or students may write $E_A = E_B$.

24 **EM** This question tests students' understanding of electromagnetic braking and conservation of energy.

a When an electric motor is rotated it acts like a generator, converting the mechanical energy of rotation into electrical energy. ✓ By placing a low resistance across the motor, the electrical energy is converted into heat. ✓ The braking force is produced by the induced current in the motor. By Lenz's law this current is in a direction that opposes the change (the rotating wheels) that created it. ✓

b The emf induced depends on the rate of rotation of the motor but the braking effect and electrical power used depends on the current that flows. ✓ As the power dissipated in the resistor depends on the emf and the current ($P = IV$), using a lower resistance will increase the current, the electrical power dissipated and the braking force produced. ✓

25 **EM** This question tests students' understanding of stellar spectra.

a The radiation emitted from stars is like the radiation emitted by black bodies on Earth. ✓ The temperature of a star can be found by measuring the wavelength of maximum emission intensity (λ) ✓ and then using Wien's black-body relationship, $T = b/\lambda$. ✓ Students may also talk about the relative strength of atomic and ionic lines in the spectrum.

b If a light source moves away or towards an observer, the light is shifted towards the red or blue end of the spectrum, respectively. By comparing the absorption lines from elements on Earth with those in the star, ✓ the amount of redshift or blueshift can be determined and the velocity of the star calculated. ✓

c The absorption lines of each element are unique ✓ and act as a signature for the element. ✓ By comparing the emission/absorption lines from elements on Earth with the absorption lines in the stellar spectrum, the elements in the star's atmosphere can be identified. ✓

26 **EM** This question tests students' understanding of the photelectric effect and their ability to interpret graphical data.

a There is a specific amount of energy (i.e. the work function = ϕ) required to remove an electron from the surface of zinc. Electrons will not be emitted until the photon energy is equal to or greater than the energy required to remove an electron from the surface. ✓ As photon energy is proportional to the frequency ($E = hf$) this means that no electrons will be emitted below a specific threshold frequency. ✓

b The threshold frequency is where the graph crosses the frequency axis. The threshold frequency is therefore $f_0 = 10.5 \times 10^{14}$ Hz. ✓ The energy to remove an electron from the surface corresponds to the point where the graph crosses the energy axis. The work function is therefore $\phi = 4.4$ eV. ✓

c As one photon releases one electron we can use conservation of energy to write $hf = \phi + k$ or the kinetic energy $k = hf - \phi$, ✓ which is the equation that represents the graph. This equation has the form $y = mx + b$ and the gradient is therefore equal to Planck's constant. But because Figure E2.11 has the energy in electron volts (rather than joules), the gradient of the graph will be proportional to Planck's constant. ✓

27 **EM** This question tests students' understanding of Einstein's special relativity postulates and the Michelson–Morley experiment.

a First postulate: All inertial frames of reference are equivalent. ✓ Second postulate: The speed of light in a vacuum is an absolute constant. ✓

b If an ether wind blew past the Earth, light would travel at a different speed parallel to and perpendicular to the direction of the wind. The Michelson–Morley experiment used this assumption in their experiment to measure the speed of the Earth through the ether but they obtained a null result. ✓ The result is explained by either postulate. If the speed of light is an absolute constant that is independent of the speed of the source or the observer, the assumption about light travelling at different speeds upon which the experiment was based is incorrect and the experiment must give a null result. ✓ If all inertial frames are equivalent, then any experiment designed to measure the absolute speed of a frame of reference is doomed to failure. ✓

c The relativist momentum (p_v) is related to the Newtonian momentum (p_0) by:

$$p_v = \frac{p_0}{\sqrt{\left(1-\frac{v^2}{c^2}\right)}}$$

Now as $p_v = 2p_0$, we can write $2 = \frac{1}{\sqrt{\left(1-\frac{v^2}{c^2}\right)}}$ ✓ and rearranging this gives:

$$v = c\sqrt{\frac{3}{4}} = 0.866c = 2.6 \times 10^8 \text{ ms}^{-1} \checkmark$$

28 **EM** This question tests students' understanding of fission, beta decay and half-life.

a To ensure the same number of nucleons is conserved there must be three neutrons released. ✓

b The mass defect (Δm) is given by:

Δm = mass of products − mass of reactants
= (140.9139 + 91.8973 + 3 × 1.008 665)
− (1.008 665 + 235.043 915)
= −0.215 u or 0.215 u of mass is converted to energy ✓

The energy released is 0.215 × 931.5 = 200 MeV ✓

c For the first decay:

$${}^{141}_{56}\text{Ba} \rightarrow {}^{141}_{57}\text{La} + {}^{0}_{-1}\text{e} + {}^{0}_{0}\bar{\upsilon} \checkmark$$

For the next decay:

$${}^{141}_{57}\text{La} \rightarrow {}^{137}_{55}\text{Cs} + {}^{4}_{2}\text{He} \checkmark$$

d The number of half-lives in 1 hour is $n = \frac{60}{18.5} = 3.234$ half-lives. ✓

The fraction remaining will be given by $(1/2)^n = \left(\frac{1}{2}\right)^{3.234}$
$= 0.106 = 10.6\%$ ✓

Or students may work out the decay constant and use $N_t = N_0 e^{-\lambda t}$

29 **EM** This question tests students' understanding of Rutherford's and Bohr's atomic models.

a Rutherford's model was unstable because orbiting electrons should have radiated electromagnetic radiation, lost energy and spiralled into the nucleus. ✓

Rutherford's model also said nothing about atomic emission and absorption spectra. ✓

b Bohr's model was stable because he assumed that when the electron was in specific (stationary state) orbitals it did not radiate electromagnetic radiation. ✓

Bohr's model was able to account for atomic spectra by assuming photons were emitted or absorbed when electrons moved to lower- or higher-energy orbitals, respectively. ✓

c Apply Bohr's equation with $n_i = 3$ and $n_f = \infty$: ✓

$$\frac{1}{\lambda} = R\left[\frac{1}{n_f^2} - \frac{1}{n_i^2}\right] = 1.097 \times 10^7\left[\frac{1}{3^2}\right]$$

$= 1.2188 \times 10^6$ and hence $\lambda = 820$ nm

$$E = hc/\lambda = \frac{6.626 \times 10^{-34} \times 3 \times 10^8}{891 \times 10^{-9}} = 2.42 \times 10^{-19}\text{ J} \checkmark = 1.51\text{ eV}$$

30 **EM** This question tests students' understanding of the discovery and nature of the neutron.

a The mass number was approximately twice the atomic number. ✓ For example, helium was known to have a mass of approximately 4 u but an atomic number of 2.

b Because neutrons have no charge ✓ they are very penetrating and very poor ionisers. Because they interact so weakly, they do not leave a track in a cloud chamber and are not detected by Geiger counters. ✓

c Joliot and Curie had reported a penetrating radiation was emitted when beryllium was irradiated with alpha particles that was so energetic it could knock protons out of paraffin. Chadwick showed that this would require more energy than a gamma ray and measured the recoil velocity of particles hit by the radiation. ✓ By applying the law of conservation of momentum ✓and kinetic energy to the collisions he showed that the radiation consisted of neutral particles with a mass similar to a proton. ✓ That is, he showed the radiation was a stream of neutrons.

d Neutrons are one of a large number of composite particles called hadrons. Neutrons consist of three quarks: ✓ two down-quarks (each with −1/3 charge) and an up-quark (with a charge of +2/3). ✓ The quarks are held within the nucleus by a strong (colour) force mediated by gluons.

31 **EM** This question tests students' understanding of electromagnetic induction.

a As the coil rotates, the magnetic flux linking the coil continually changes. ✓ Faraday's law of electromagnetic induction states that if the magnetic flux in a coil changes, an emf will be induced in the coil ✓ ($\varepsilon = -N\frac{\Delta\Phi}{\Delta t}$).

b Figure EA2.1 shows the induced electromotive force. Note that the induced electromotive force is a maximum at $t = 0$. ✓ (Note this mark can be given for correctly drawing this point on the graph.)

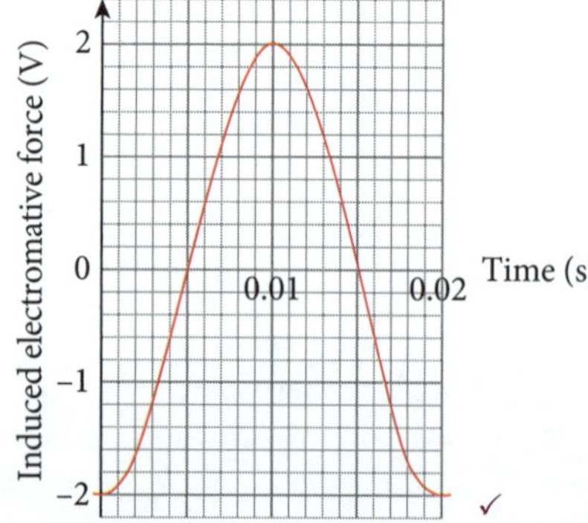

Figure EA2.1 Electromotive force induced in an AC generator

c If the rate of rotation was halved, the maximum induced electromotive force would be reduced to 1 volt ✓ and the period would double to 0.04 seconds. ✓

32 EM This question tests students' understanding of particle accelerators and the influence they have had on the development and verification of the standard model.

a In a linear accelerator (Figure EA2.2) charged particles are accelerated between drift tubes of increasing length. ✓ The length increases to ensure a constant-frequency AC radio frequency voltage can be applied. The entire device is evacuated to prevent collisions. ✓

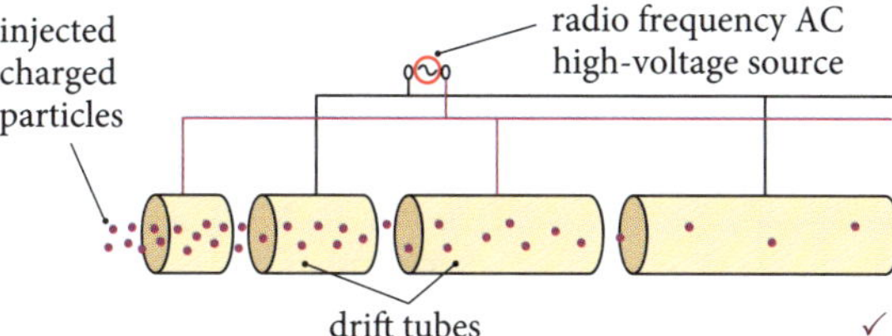

Figure EA2.2 Linear accelerator

Each time a charge particle leaves a drift tube, the polarity of the tube is reversed and the particle is accelerated across a large potential difference before it enters the next drift tube. ✓

b The standard model does not include gravity ✓ and says nothing about dark matter ✓ or dark energy. The model cannot account for the matter–antimatter asymmetry that is critical to the formation of our matter universe in the Big Bang. It requires a large number of experimentally determined parameters to be added before it can be used. (Any two of these answers are suitable for the two marks.)

c Quarks are elementary particles, while hadrons are composite particles composed of two quarks or three quarks. ✓

INDEX

N

O

P

Q

R

FORMULAE SHEET

Motion, forces and gravity

$$s = ut + \frac{1}{2}at^2$$

$$v = u + at$$

$$v^2 = u^2 + 2as$$

$$\vec{F}_{\text{net}} = m\vec{a}$$

$$\Delta U = mg\Delta h$$

$$W = F_{||}s = Fs\cos\theta$$

$$P = \frac{\Delta E}{\Delta t}$$

$$K = \frac{1}{2}mv^2$$

$$P = F_{||}v = Fv\cos\theta$$

$$\sum\frac{1}{2}mv^2_{\text{before}} = \sum\frac{1}{2}mv^2_{\text{after}}$$

$$\sum m\vec{v}_{\text{before}} = \sum m\vec{v}_{\text{after}}$$

$$\Delta\vec{p} = \vec{F}_{\text{net}}\Delta t$$

$$a_{\text{c}} = \frac{v^2}{r}$$

$$\omega = \frac{\Delta\theta}{t}$$

$$F_{\text{c}} = \frac{mv^2}{r}$$

$$\tau = r_{\perp}F = rF\sin\theta$$

$$F = \frac{GMm}{r^2}$$

$$v = \frac{2\pi r}{T}$$

$$\frac{r^3}{T^2} = \frac{GM}{4\pi^2}$$

$$U = -\frac{GMm}{r}$$

Waves and thermodynamics

$$v = f\lambda$$

$$f_{\text{beat}} = \left|f_2 - f_1\right|$$

$$f = \frac{1}{T}$$

$$f' = f\frac{(v_{\text{wave}} + v_{\text{observer}})}{(v_{\text{wave}} - v_{\text{source}})}$$

$$d\sin\theta = m\lambda$$

$$n_1\sin\theta_1 = n_2\sin\theta_2$$

$$n_{\text{x}} = \frac{c}{v_{\text{x}}}$$

$$\sin\theta_{\text{c}} = \frac{n_2}{n_1}$$

$$I = I_{\text{max}}\cos^2\theta$$

$$I_1r_1^2 = I_2r_2^2$$

$$Q = mc\Delta T$$

$$\frac{Q}{t} = \frac{kA\Delta T}{d}$$

Electricity and magnetism

$$E = \frac{V}{d}$$

$$\vec{F} = q\vec{E}$$

$$V = \frac{\Delta U}{q}$$

$$F = \frac{1}{4\pi\varepsilon_0}\frac{q_1 q_2}{r^2}$$

$$W = qV$$

$$I = \frac{q}{t}$$

$$W = qEd$$

$$V = IR$$

$$B = \frac{\mu_0 I}{2\pi r}$$

$$P = VI$$

$$B = \frac{\mu_0 NI}{L}$$

$$F = qv_\perp B = qvB\sin\theta$$

$$F = lI_\perp B = lIB\sin\theta$$

$$\Phi = B_{||}A = BA\cos\theta$$

$$\frac{F}{l} = \frac{\mu_0}{2\pi}\frac{I_1 I_2}{r}$$

$$\varepsilon = -N\frac{\Delta\Phi}{\Delta t}$$

$$\tau = nIA_\perp B = nIAB\sin\theta$$

$$\frac{V_p}{V_s} = \frac{N_p}{N_s}$$

$$V_p I_p = V_s I_s$$

Quantum, special relativity and nuclear

$$\lambda = \frac{h}{mv}$$

$$t = \frac{t_0}{\sqrt{\left(1-\frac{v^2}{c^2}\right)}}$$

$$K_{max} = hf - \phi$$

$$\lambda_{max} = \frac{b}{T}$$

$$l = l_0\sqrt{\left(1-\frac{v^2}{c^2}\right)}$$

$$E = mc^2$$

$$p_v = \frac{m_0 v}{\sqrt{\left(1-\frac{v^2}{c^2}\right)}}$$

$$E = hf$$

$$\frac{1}{\lambda} = R\left(\frac{1}{n_f^2} - \frac{1}{n_i^2}\right)$$

$$N_t = N_0 e^{-\lambda t}$$

$$\lambda = \frac{\ln 2}{t_{\frac{1}{2}}}$$

ISBN 978 1 74125 678 9

Pascal Press
PO Box 250
Glebe NSW 2037
(02) 8585 4050
www.pascalpress.com.au

Publisher: Vivienne Joannou
Project editor: Mark Dixon
Edited by Karen Enkelaar
Answers checked by Adam Sloan
Indexed by Puddingburn Publishing Services
Cover and page design by Sonia Woo
Typeset by Kim Webber
Printed by Vivar Printing/Green Giant Press

Students
All care has been taken in the preparation of this study guide, but please check with your teacher or the NSW Education Standards Authority about the exact requirements of the course you are studying as these can change from year to year.

The validity and appropriateness of the internet addresses (URLs) in this book were checked at the time of publication. Due to the dynamic nature of the internet, the publisher cannot accept responsibility for the continued validity or content of these web addresses.